Permaculture

Permaculture

Edited by **Suede Crawford**

SYRAWOOD
PUBLISHING HOUSE
New York

Published by Syrawood Publishing House,
750 Third Avenue, 9th Floor,
New York, NY 10017, USA
www.syrawoodpublishinghouse.com

Permaculture
Edited by Suede Crawford

International Standard Book Number: 978-1-68286-325-1 (Hardback)

Printed in China.

Contents

Preface

Over the recent decade, advancements and applications have progressed exponentially. This has led to the increased interest in this field and projects are being conducted to enhance knowledge. The main objective of this book is to present some of the critical challenges and provide insights into possible solutions. This book will answer the varied questions that arise in the field and also provide an increased scope for furthering studies.

The focus of permaculture lies within the strategy of utilizing the features of natural ecosystems. As a way of life, it endorses a harmonic coexistence with nature. This book outlines the processes and applications of permaculture in detail. The topics included in the text, such as ecological design principles, natural building, agroforestry, etc., are of utmost significance and are bound to provide incredible insights to readers. Those in search of information to further their knowledge in this field will be greatly assisted by this book. It will also prove to be immensely useful for researchers and students engaged in the study of permaculture.

I hope that this book, with its visionary approach, will be a valuable addition and will promote interest among readers. Each of the authors has provided their extraordinary competence in their specific fields by providing different perspectives as they come from diverse nations and regions. I thank them for their contributions.

Editor

1

Effects of three pre-treatment techniques on dormancy and germination of seeds of *Afzelia africana* (Sm. Ex pers)

Amusa, T. O.

Federal College of Wildlife Management, Forestry Research Institute of Nigeria, P. M. B. 268, New Bussa , Niger State, Nigeria. E-mail: teejayui@yahoo.com.

The study assessed the effects of three pre-treatment techniques on dormancy and germination of seeds of *Afzelia africana* (Sm. ex Pers), an endangered tree species in Savanna ecozone of Nigeria. The three pre-treatment techniques are: soaking in cold water for 1, 12 and 24 h, soaking in hot water (100°C) for 1, 12 and 24 h and soaking in 10, 50 and 98% concentrated sulphuric acid (H_2SO_4) for 5, 10 and 30 min. Experiments were performed based on a completely randomized design with four replicate. The results show that cold water pre-treatment of *A. africana* seeds only gave a fair germination percentage (33.33 to 53.33%) and reduced dormancy period. On the other hand, the effect of hot water pre-treatment on the seeds gave adverse result (26.67 to 36.67%). Although *A. africana* seeds under control treatment gave an impressive germination, which did not differ significantly from the results of other pre-treatment techniques (F-cal = 2.6777; P-level = 0.1180; $P < 0.05$), acid pre-treatments yield a more uniform and regular germination. Thus, while the seeds of *A, africana* could be germinated without pre-treatment, to attain higher percentage germination and reduced dormancy period, the seeds should be pre-treated with sulphuric acid.

Key words: Dormancy, pre-treatment techniques, germination percentage, *Afzelia africana*, *ex situ* conservation.

INTRODUCTION

The rapid rates of forest loss and degradation across the tropics have continued to increase both the fragmentation of many populations and the risk of species' extinction. In Nigeria as elsewhere, conservation of forest genetic resources is achieved through the protection of these resources in their natural habitat (*in situ*) or preservation of samples of the genetic diversity of endangered species away from their field habitats (*ex situ*) in facilities such as botanical gardens, seed gene banks, *in-vitro* gene banks, and field gene banks (Owonubi and Otegbeye, 2004). The success of *ex situ* conservation method is dependent on a number of factors including availability of funds, suitable land, appropriate technology and adequate and viable seeds. Seed is of particular importance and a key element in both *in situ* and *ex situ* plant conservation activities. The propagation of most tropical tree species however, is constrained by recalcitrant seed germination as a result of dormancy (Nwoboshi, 1982). The degree of dormancy makes it difficult for seed to germinate evenly and adequately.

In both the forest and savanna ecosystems of Nigeria *Afzelia africana* (Sm. ex Pers) is an endangered species (Gbile et al., 1978). It belongs to the family Fabaceae (sub family- Ceasalpinoidae) in the order Fabales. It is a large tree with a spreading crown having height that varies from 10 to 20 m and a mean average diameter of 36 cm (Keay, 1989a). *A. africana* is a heliophytic species occurring in the humid and dry forests as well as tree savannahs and forest galleries. It is found in well watered sites with a deep sandy soil and can adapt to lateritic soils.

The tree is an invaluable species. The wood is hard, heavy, durable, termite-proof, light brown to red-brown in colour and an excellent timber. Though difficult to work, it is used in carpentry, canoe, house building and furniture making. It is also used in human medicine: febrifugal, analgaesic, anti-hemorrhageic, laxative, emetic, emmenagogic and aphrodisiac. The foliage is good as cattle forage, particularly before the re-growth of grass in the early rainy season. Wild animals browse the arils, and antelopes eat the young shoots. Pods are rich in ashes used for making soap. The leaves are also rich in nitrogen and are used to enrich the soil. Although locally thought to be inhabited by spirits, it is the hunter's favourite tree (Burkill, 1994). Within the south-eastern zone of Nigeria, seeds of *A. africana* are relished as soup thickener.

Meanwhile, the stock of *A. africana* species has been badly depleted and natural regeneration is constrained by wild animals which rationally feed on it. The trees are not fire resistant and as poacher set fire to the forest to clear the bush for visibility of games, the heat of bush burning destroy the growth of the species. Besides, the nomadic herdsmen that traverse the savanna with their livestock usually lop the species for their livestock to browse. In the light of this, there is an urgent need to consider the *ex situ* conservation option of the species through possible establishment in plantation. There is thus, the need to develop an appropriate pre-treatment technology for breaking dormancy and regularizing germination of *A.africana* seeds, particularly within the savanna ecozone of Nigeria.

MATERIALS AND METHODS

The study area

The study was conducted within the nursery unit of Federal College of Wildlife Management, located on latitude 7°31'- 10°00'N and longitude 4°30'- 4°33'E (Adewetan et al., 1980) in the savanna areas of the Kainji Lake Basin, New Bussa. The climate of the area is tropical with high temperatures, high relative humidity and distinct wet and dry seasons. The average monthly temperature is 34°C with a mean annual relative humidity of 60%. The mean annual rainfall value is 104.45 mm (DRB, 2004). The vegetation of the area has been described as Northern Guinea Savanna (Keay, 1989b).

Experimental procedure and design

Seed collection

A. africana pods were collected under matured trees of the species in field laboratory established by the college within the northern guinea savannah ecozone between December 2007 and February 2008. The collected pods were broken and the seeds harvested decarped and kept under ambient temperature condition for 5 days. The decarped seeds were subjected to viability test through the flotation method; in which case the seeds that floated in water after 24 h of soaking were considered unviable and discarded.

Experiment 1

In this trial, the effects of cold water treatment on A. africana seeds were compared. The treatment involved:

Soaking in cold water (room temperature) for 1 h.
Soaking in cold water (room temperature) for 12 h.
Soaking in cold water (room temperature) for 24 h.

Experiment 2

This trial assessed the effect of hot water treatment on the seeds of *A. africana*. The treatment involved:

Soaking in hot water (100°C) for 1 h.
Soaking in hot water (100°C) for 12 h.
Soaking in hot water (100°C) for 24 h.

Experiment 3

This trial assessed the effect of sulphric acid (H_2SO_4). Three sulphuric acid concentrations (10, 50 and 98%) and three treatment time (5, 10 and 30 min) were compared. The treatment involved:

Soaking in sulphric acid (H_2SO_4) at 10% concentration for 5 min.
Soaking in sulphric acid (H_2SO_4) at 10% concentration for 10 min.
Soaking in sulphric acid (H_2SO_4) at 10% concentration for 30 min.
Soaking in sulphric acid (H_2SO_4) at 50% concentration for 5 min.
Soaking in sulphric acid (H_2SO_4) at 50% concentration for 10 min.
Soaking in sulphric acid (H_2SO_4) at 50% concentration for 30 min.
Soaking in sulphric acid (H_2SO_4) at 98% concentration for 5 min.
Soaking in sulphric acid (H_2SO_4) at 98% concentration for 10 min.
Soaking in sulphric acid (H_2SO_4) at 98% concentration for 30 min.

Control treatment

There was one control treatment, which involved sowing immediately after seed extraction and decarping.

Experimental layout

Since interaction between treatments was not envisaged, the experiment was laid out in a completely randomized design. Thus, each pre-treatment method formed separate experiment with the various levels representing the treatments. A total of 30 seeds was allocated to each treatment and was replicated four times. The seeds were sown in germination pots (of equal dimension) filled with topsoil. Treatments were kept under the weaning shed in order to reduce the rate of evaporation. Watering was done twice daily (morning and evening) throughout the duration of the experiment. Germination count was taken every evening and seedlings were pricked-out after counting to avoid error. Germination was monitored for 42 days, after which the experiment was terminated.

Data analysis

The cumulative germination data were computed and converted to percentage values to determine the germination percentages in each treatment. To ensure conformity with the general assumption of Analysis of Variance (ANOVA) the percentage data were transformed using Arcsine transformation. A one-way ANOVA was carried out to analyze the data and determine the most appropriate pre-treatment method for minimizing dormancy period in *A. africana* seeds within the ecozone.

RESULTS

Effect of cold water treatment

Seed germination under the different cold water treat ment gave a fair germination percentage (33.33 to 53.33%). The longer the cold water treatment, the earlier

Table 1. Germination of *A. africana* seeds in cold water treatment.

Cold water treatment (hour)	Mean number of days taken for first emergence	Mean number of seeds germinated	Germination percentage
1	13	10	33.33
12	12	16	53.33
24	11	12	40.00
Control	14	18	60.00

Table 2. Germination of *A. africana* seeds in hot water treatment.

Hot water treatment (hour)	Mean number of days taken for first emergence	Mean number of seeds germinated	Germination percentage
1	13	11	36.67
12	13	8	26.67
24	13	11	36.67
Control	14	18	60.00

Table 3. Summary of effect of acid concentration and treatment time on days to seedling emergence in *A. africana.*

Treatment time (min)	Acid concentration (%)			Mean number of days taken for first emergence
	10	50	98	
5	12	15	12	**13.00**
10	13	12	10	**11.67**
30	10	10	8	**9.33**
Mean	11.67	12.33	10.00	

the commencement of seed germination, although this was not significant within the treatment. For example, while germination started on the 11th day for seeds soaked for 24 h, it took 12 and 13 days for those soaked in cold water for 12 and 1 h respectively (Table 1). Generally, germination under the control gave higher percentage (60.00%) than those of cold water treatment, though with longer germination commencement period (average of 14 days after sowing).

Effect of hot water treatment

When compared to control and cold water pre-treatment, the effect of hot water on the seeds of *A. africana* could be described as adverse. The highest germination percentage of 36.67% is recorded for both seeds subjected to 1 and 24 h hot water pre-treatment. Seeds subjected to 12 h hot water pre-treatment gave a very low germination percentage of 26.67%. All treatment hours commenced germination on the 13th day (Table 2).

Effect of acid pre-treatment

Under the acid pre-treatment, germination percentage was enhanced by both treatment time and concentration level. Except for the 30 min treatment time, however, commencement of germination when compared to control, cold water and hot water pre-treatments, was not positively impacted by both treatment time and concentration level. The highest germination percentage for the acid pre-treatment was 86.67% at 10% concentration level for 30 min treatment time, while the shortest germination commencement period was 8 days at the 98% concentration level for 30 min treatment time (Tables 3 and 4) (Figures 1 to 4).

From Figure 1, germination of *A. africana* seeds in acid treatment (H_2SO_4) at 10% concentration for 5 min began on the 12th day with total number of 12 seeds germinated (40.00%). Germination at 50% acid concentration began on the 15th day with total number of 7 seeds germinated (23.33%), while at 98% acid concentration germination began on the 12th day with a total of 11 seeds germinated

Table 4. Summary of effect of acid concentration and treatment time on the germination of *A. africana* seeds.

Treatment time (min)	Acid concentration (%)			Mean number of seeds germinated
	10	50	98	
5	40.00	23.33	36.67	33.33
10	66.67	83.33	76.67	75.56
30	86.67	73.33	83.33	81.11
Mean	64.45	60.00	65.56	

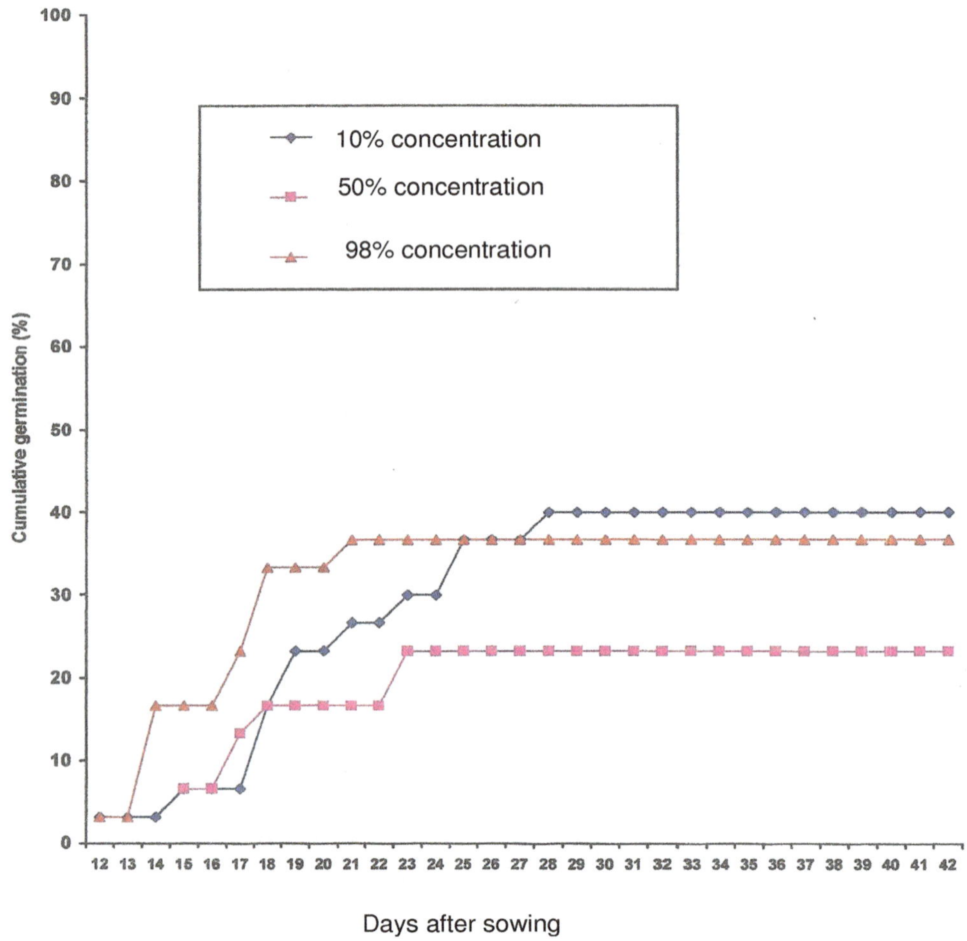

Figure 1. Germination of *A. africana* seeds in acid treatment (H_2SO_4) at 10, 50 and 98% concentration for 5 min.

(36.67%).

From Figure 2, germination of *A. africana* seeds in acid treatment (H_2SO_4) at 10% concentration for 10 min began on the 13th day with total number of 20 seeds germinated (66.67%). Germination at 50% acid concentration began on the 12th day with a total number of 25 seeds germinated (83.33%), while at 98% acid concentration germination began on the 10th day with a total of 23 seeds germinated (76.67%)

From Figure 3, germination of *A. africana* seeds in acid treatment (H_2SO_4) at 10% concentration for 30 min began on the 10th day with total number of 26 seeds germinated (86.67%). Germination at 50% acid concentration began on the 10th day with total number of 22 seeds germinated (73.33%), while at 98% acid concentration germination began on the 8th day with a total of 25 seeds germinated (83.33%).

From Figure 4, germination began on the 14th day with

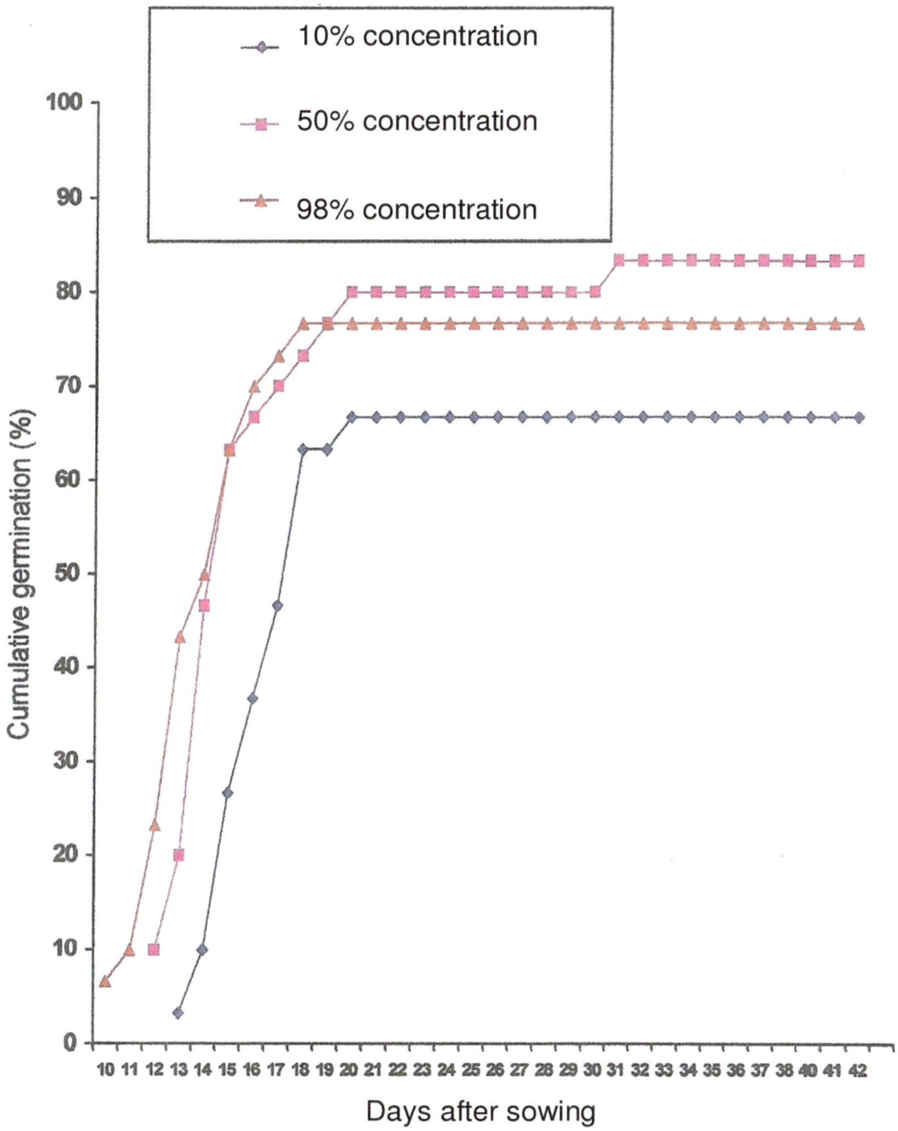

Figure 2. Germination of *Afzelia africana* seeds in acid treatment (H_2SO_4) at 10, 50 and 98% concentration for 10 min.

a total number of 18 seeds germinated (60.00%).

Although acid treatment ranked best among the three pre-treatment techniques assessed in this study, the result of one-way analysis of variance shows that there is no significant different in the overall effect of the three techniques and control treatment (Table 5).

DISCUSSION

Seed dormancy occurs in many tropical tree species to varying degrees (Nwoboshi, 1982). While various pre-treatment methods have been advocated to reduce dormancy and hasten germination, no single pre-treatment technique has been found to be equally effective for all seed species. The preceding findings of this study shows that, although *A. africana* seeds under control treatment gave an impressive germination, which did not differ significantly from the results of other pre-treatment techniques (Table 5), acid pre-treatments yield a more uniform and regular germination. The fact that acid treatment gave highest germination values within shortest time indicates that the more rapidly the seed coat is ruptured, the faster the rate of germination. This is more so, since a very widespread cause of seed dormancy is the presence of hard seed coat which prevents the entrance of water, exchange of gases and /or mechanically constrained the embryo (Mayer and Maber, 1963). Sulphuric acid is thought to disrupt the seed coat and expose the lumens of the macrosclereids

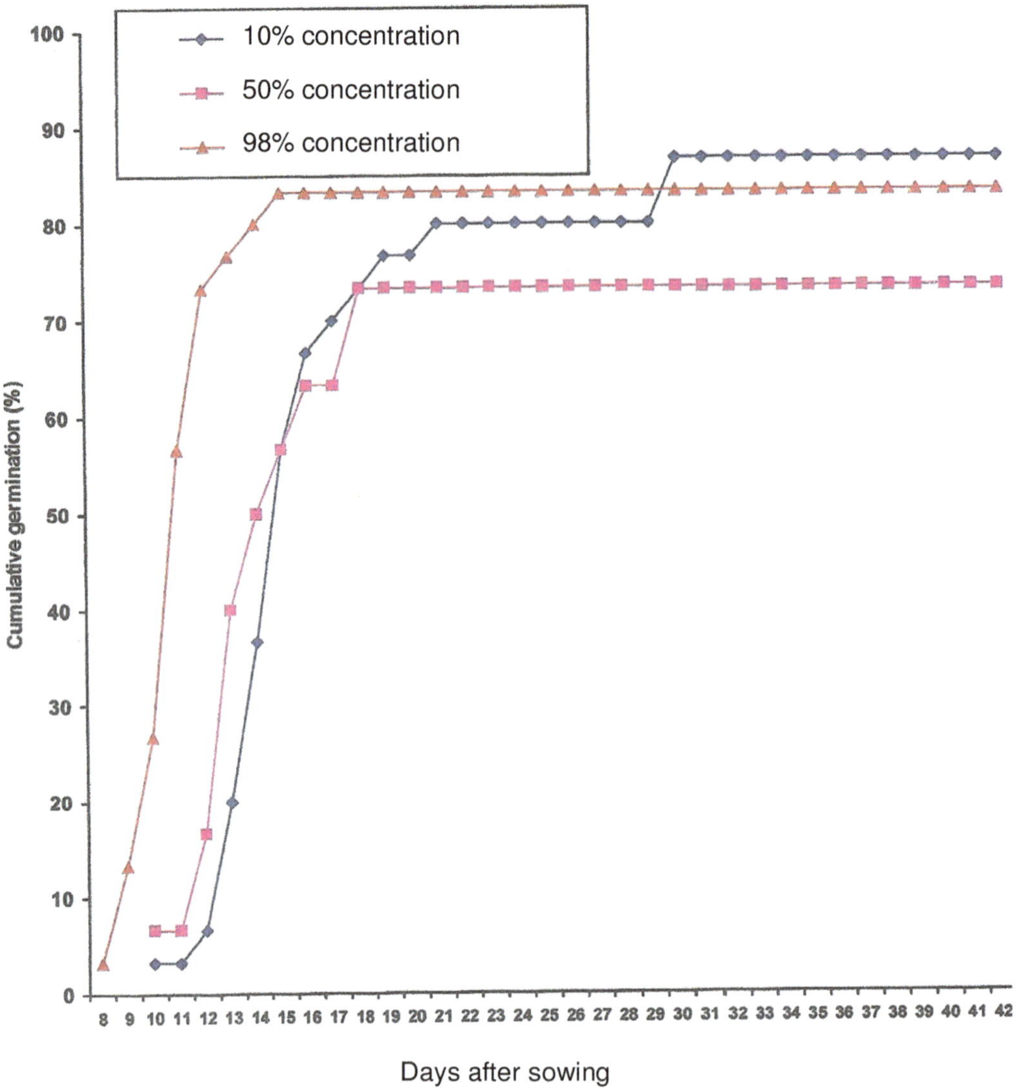

Figure 3. Germination of *Afzelia africana* seeds in acid treatment (H_2SO_4) at 10, 50 and 98% concentration for 30 min.

cells, permitting imbibition of water, which triggers germination (Nikoleave, 1977). Irwin (1982) and Jackson (1994) also found that as soon as the seed coat is softened by scarification, the process of hydrolysis would commence to release simple sugars that could be readily utilized in protein synthesis, thereby encouraging germination.

Cold water pre-treatment of *A. africana* seeds only gave a fair germination percentage and reduced dormancy period (Table 1). The control treatment gave a higher germination percentage. This is in tandem with the observation of Robertson and Small, (1977) that over-soaking seeds in water may reduce germination through oxygen deficiency. However, Owonubi et al. (2005) reported that soaking of *Azadirachta indica* seeds for 1 and 12 h resulted in increasing rate of seed germination corroborating the work of Ibrahim and Otegbeye (2004) on the seeds of *Adansonia digitata.* This implies that different species have varying rates at which their seed coat is permeable to water and gases (Owonubi et al., 2005). Meanwhile, the effect of hot water on the seeds of *A. africana* gave adverse result contrary to what would be obtained in nature (as indicated by the result of the control treatment). This shows that hot water pre-treatment is not an appropriate pre-treatment technology in the seeds of *A. africana.* It perhaps leads to the seed embryo being killed because of prolonged contact with boiled water.

Conclusion

The nature and dimension of hard seed coat dormancy are not fully understood. Pre-germination treatments are

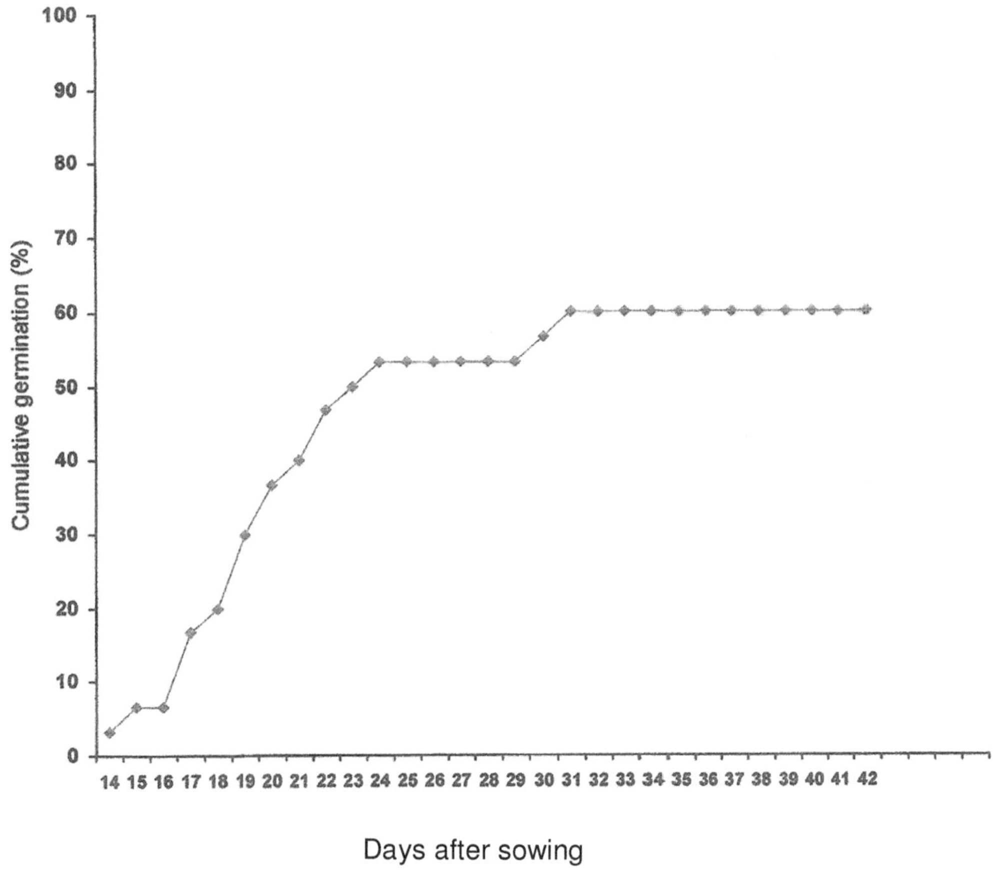

Figure 4. Germination of *A. africana* seeds under control treatment.

Table 5. Analysis of variance for the effect of different pre-treatment techniques on the germination of *Afzelia africana* seeds

Source of variation	Sum of square	Degree of freedom	Mean Square	F-cal	P-level
Treatment	1847	3	615.7	2.677	0.1180
Residual	1840	8	230.0		
Total	3687	11			

F-cal = 2.6777; P-level = 0.1180. Means are not significantly different ($P < 0.05$).

however, needed to break physical dormancy caused by hard seed coat or pericarps. The result of this study has shown that the seeds of *A. africana* could be germinated without pre-treatment, but to attain higher percentage germination and reduced dormancy period, the seeds should be pre-treated with sulphuric acid. The acid treatment, nonetheless, require further research in order to determine the optimum acid concentration level.

RECOMMENDATION

Based on the findings of this study, the following recommendations are made:

1. Wide range of acid (H_2SO_4) concentration level should be used in order to have a close comparison between the various concentration treatments.
2. Wide range of time should also be used in order to know the effect of time on various levels of concentration.

REFERENCES

Adewetan TA, Onyeanusi AE, Osakwe ME (1980). The study of flora and fauna at the Federal College of Wildlife Management, New Bussa. Unpublished research paper, Forestry Research Institute of Nigeria 27p.

Burkill HM (1994). The Useful Plants of West Tropical Africa. Edtion 2. Vol 2. Families E-I. Royal Botanic Gardens, Kew. 635 pp.

Gbile ZO, Ola ABO, Soladoye MO (1978). List of endangered species of

Nigeria flora. Nig. J. For. 8:14-20.

Ibrahim A, Otegbeye GO (2004). Methods of achieving optimum germination in *Adansonia digitata*. Bowen J. of Agriculture. 1 (1) 53-58.

Irwin PT (1982). Plant physiology. Addision Wesley Publishing Co. Inc. USA. pp 501-540.

Jackson MB (1994). Root to shoot communication in flooded plants: involvement of Abscisic acid, ethylene and l-aminocylopropane-l-carboxylic acid. Agron.J. 86(5):775-781.

Keay RWJ (1989a). Trees of Nigeria. Oxford University Press, New York. 476 P.

Keay RWJ (1989b). An outline of Nigerian Vegetation. Government printers, Lagos.

Mayer AM, Maber PA (1963). The Germination of seeds. Pergamon press, Oxford. 236p.

Nikoleave MG (1977). Factors controlling seed dormancy pattern. North Holland Publishing Co., Amsterdam. pp. 51-74.

Nwoboshi LC (1982). Tropical Silviculture: principles and techniques. Ibadan University press, Nigeria. 333p.

Owonubi JJ, Otegbeye GO (2004). Disappearing Forests: A Review of the Challenges for Conservation of Genetic Resources and Environmental Management. J. For. Res. Manage. 1 & 2. December 2004: pp 1-11.

Owonubi JJ, Otegbeye GO, Nwokedi C (2005). Development of Pre-germination technique for Azadirachta indica: preliminary investigation. In: Sustainable Forest Management in Nigeria: Lessons and Prospects (eds..L. Popoola, P. Mfon and P.I. Oni). Proceedings of the 30th Annual Conference of the Forestry Association of Nigeria, held in Kaduna, Kaduna State.7-11th November, 2005. pp 29-38.

Robertson BI, Small JGC (1977). Germination of *Jubaeopsis caffra* seeds. Principles 21: pp114-122

2

Potential of traditional food plants in rural household food security in Botswana

G. M. Legwaila, W. Mojeremane*, M. E. Madisa, R. M. Mmolotsi and M. Rampart

Department of Crop Science and Production, Botswana College of Agriculture, Private Bag 0027, Gaborone, Botswana.

Botswana is endowed with a variety of traditional food plants which grow annually despite erratic rainfall. The most common traditional food plants are leafy vegetables (for example, *Amaranthus, Cleome, Conchorus and Vigna* species) and indigenous fruits (for example, *Azanza garckeana, Strychnos cocculoides, Sclerocarya birrea* etc.). They provide an important source of food for people with either low or middle income and have been consumed for years to supplement diets. They are also a source of income; thereby improve the standard of living conditions for local communities. Traditional food plants contain vital nutrients and essential vitamins important for the proper maintenance of human health, especially for children who are often vulnerable to malnutrition and diseases. Some traditional leafy vegetables are very nutritious with protein contents of up to 36%. Indigenous fruits such as *S. cocculoides* and *A. garckeana* contain more than 30% fat and about 45% crude fibre and total carbohydrates. *S. birrea* fruits and juice contain four times more vitamin C than orange juice. Traditional food plants play an important role in food security because they yield a crop even in poor rainfall years when arable crops fail. The objective of this is synthesis existing information on the status and contribution of indigenous food plants to rural household food security in Botswana.

Key words: Botswana, indigenous fruit trees, leafy vegetables, food security.

INTRODUCTION

Most rural communities in Botswana depend primarily on arable crops as their source of food. Globally, studies have shown that rural communities face food insecurity and are chronically malnourished (Tiisekwa et al., 2004) because of drought stress (Boyer, 1982; Ludlow and Muchow, 1990; Harris and Mohammed, 2003; Babu, 2000), low adoption level of improved crop production technologies (Babu, 2000), poor soils and lack of resources. These have negative effects on arable crop production and result in considerable crop yield reductions (Boyer, 1982; Ludlow and Muchow, 1990). Therefore, most rural households in Botswana depend on various natural resources to sustain their livelihoods. For example, they use natural resources based livelihood strategies as a way of escaping hunger during times of famine. Studies conducted in southern Africa have shown that indigenous food and medicinal plants play a crucial role in the livelihoods of rural communities (Godoy and Bawa, 1993; Arnold and Ruiz, 2001; Ladio and Lozada, 2004; Scherrer et al., 2005; Shackleton and Shackleton, 2004). Their availability is significant, especially during times of famine (Akinnifesi et al., 2008a) as they provide an alternative source of nutrition, medicine (Fleurent, 1979; FAO, 1988; McGregor, 1995; Cunningham, 1997; Muok et al., 2009; Ruiz et al., 1999; Dounias et al., 2000; Shackleton et al., 2002) and cash income (Shackleton et al., 1998; Nesamvuni et al., 2001; Kadu et al., 2006; Akinnifesi et al., 2006).

However, the importance of indigenous food plants in subsistence agriculture as a source of food, a supplement for medicine and a hunger survival strategy by rural communities, especially during times of food shortage in Botswana is often overlooked. Therefore, the objective of this review was to synthesize existing information on the status and contribution of indigenous

*Corresponding author. E-mail: wmojeremane@yahoo.com, wmojerem@bca.bw.

Table 1. Composition of edible portion of selected traditional leafy vegetables compared with some exotic vegetables.

Nutrient content (100 g fresh weight)	Protein (%)	Ca (mg)	Fe (mg)	B-carotene (mg)	Vitamin C (mg)
Traditional vegetables					
Amaranths	4.0	480	10	10.7	135
Cleome	5.1	262	19	8.7	144
Corchorus	4.5	360	7.7	6.4	131
Vigna (Cowpea)	4.7	152	39	5.7	87
Exotic vegetables					
Cabbage	1.4	44	0.8	1.2	33
Spinach	2.3	9.3	32	5.1	28
Kales	2.5	187	32	7.3	93

Sources: FAO (1990) and WHO (2000, 2001).

food plants to rural household food security in Botswana.

The role of indigenous food plants in rural households' food security

Botswana is a semi-arid country of which two third is scrub covered Kalahari sandveld. Rainfall ranges from 250 to 650 mm $year^{-1}$ and occurs mainly in the summer month (November to March) (Brinkhurst, 2010). However, the country is rich in flora despite its arid conditions. An estimated 150 edible wild and semi-wild plant species are reported used by rural communities all the time or during famine (Flyman and Afolayan, 2006). It has been estimated that one year in three consecutive years arable agriculture fail in Botswana due to lack of rainfall (Mojeremane and Tshwenyane, 2004a; Brinkhurst, 2010) and general lack of surface water (Brinkhurst, 2010). It is during these drought years that rural communities in Botswana mostly use indigenous food plants as hunger coping strategy to sustain their livelihoods (Moss, 1988; Twyman, 2001; Ohiokpehai, 2003; Mojeremane and Tshwenyane 2004b).

Botswana is endowed with a wide variety of indigenous vegetables (for example *Cleome*, *Amaranthus*, *Corchorus* and *Vigna* spp), fruit trees (for example *Azanza garckeana*, *Adansonia digitata*, *Sclerocarya birrea*, *Strychnos spinosa, Vangueria infausta, Grewia* spp) and medicinal plants used by both rural and urban communities as diet supplements, medicine and alternative source of cash income. Indigenous food plants are not cultivated but grow naturally in the wild, arable crop land and backyard gardens around homesteads. A greater variety of indigenous vegetables is often found on cultivated arable land. These are collected and consumed at a younger stage of plant development, a stage when their leaves are more palatable and nutritious (Rubaihayo, 1997; Labadarios and Steyn, 2001). Several studies conducted in Botswana (Mojeremane and Tshwenyane, 2004a, b) and elsewhere (Campbell, 1986a, b; Zinyama et al., 1990; Shackleton et al., 2000; Akinnifesi et al., 2004, 2008a; Getachew et al., 2005; Balemie and Kebebew, 2006) have shown that the importance of indigenous food plants is increased during times of food shortage and famine.

For example, Akinnifesi et al. (2004) noted that 60 to 85% of the rural people in southern Africa face food shortages for 3 to 4 months in a year and use indigenous food plants to sustain their livelihoods. Indigenous food plants also provide employment opportunities which are often lacking in rural areas. Mojeremane and Tshwenyane (2004a) noted that some indigenous food plants also provided goods and services such as medicine, building material, firewood, fodder and shade.

The nutritional role and health benefits of indigenous food plants

The nutritional role and health benefits of indigenous food plants are well documented (El and Karakaya, 2004; Ansari et al., 2005; Balemie and Kebebew, 2006; Delang, 2006; Kuhnlein et al., 2006; Shrestha and Dhillion, 2006). They are important dietary supplements and sources of trace elements, proteins, minerals and vitamins for resource poor people (Agte et al., 2000; Parvathi and Kumar, 2002; Gockowski et al., 2003; Shackleton, 2003; Agrahar-Marugkar and Pal, 2004; Redzic, 2006; Singh and Garg, 2006). Most indigenous food plants are comparable or have higher nutrient content than their domesticated or exotic counterparts (Table 1). Most are rich in iron, zinc, vitamin A, C, E (Schmidt, 1971; Chwenya, 1994), folic acid (Tuker, 1986), proteins, carbohydrates and minerals (Agte et al., 2000; Parvathi and Kumar, 2002; Gockowski et al., 2003; Shackleton, 2003; Agrahar-Marugkar and Pal, 2004; Singh and Garg, 2006). Besides indigenous vegetables, rural people in Botswana also use indigenous fruit to supplement their diet. These contain almost all known vitamins and essential minerals than their exotic counterparts (Tables 2, 3 and 4). Most indigenous fruit trees produce more fruits per unit area than some exotic fruit trees and also provide

Table 2. Nutritional values of some exotic fruit trees common in Botswana.

Fruit species	Protein (%)	Carbohydrates (%)	Fat (%)	Fibre (%)	Vitamin B_2 (mg/100 g)	Vitamin C (mg/100 g)	Ca (mg/100 g)	P (mg/100 g)	Fe (mg/100 g)	Energy (Kcal/g)
Orange	0.7	10.9	0.2	0.3	-	30	26	20	0.32	116
Mango	0.6	16.9	0.4	0.7	0.09	16	14	16	1.3	74
Grapes	0.5	16.5	0.3	2.9	-	1	20	30	0.5	71
Banana	1.2	27.2	0.3	0.4	0.08	7	17	36	0.9	116
Papaya	0.6	7.2	0.1	0.8	0.25	57	17	13	0.5	32

Source: Rathore, 2009.

Table 3. Composition (%) of fruits of selected indigenous fruit trees common in Botswana.

Fruit	Composition (%)								
	Dry matter	Ash	Crude protein	ADF	ADL	NDF	pH (25°C)	Acidity	Vitamin C
Adasonia digitata	86.0	4.6	1.3	16.2	11.3	13.1	3.06	7.85	141.3
Sclerocarya birrea	11.6	4.9	3.7	16.3	13.7	16.1	3.98	0.88	128.3
Strychnos spinosa	19.7	4.6	3.3	61.1	4.4	6.2	3.96	0.77	88.0
Vangueria infausta	23.5	3.9	3.0	39.5	35.5	39.4	3.38	1.71	67.7

ADF- Acid detergent fibre; ADL- acid detergent lignin and NDL- neutral detergent fibre. Source: Amarteifio and Mosase (2006).

a lucrative source of cash income. The ripening of indigenous fruits occurs at different seasons; hence provide fresh fruits throughout the year (Akinnifesi et al., 2004, 2006). Campbell et al. (1997) and Akinnifesi et al. (2006) noted that indigenous fruit trees contribute on average about 42% of the food basket for most rural people in southern Africa. According to WHO (2005) daily consumption of indigenous food plants in sufficient quantities can help prevent numerous diseases. Studies have shown that indigenous food plants can improve the nutrition and health of children, the elderly and boost the immune system of HIV/AIDS patients (Barany et al., 2001). Consuming cooked indigenous vegetables with a stiff porridge of maize, sorghum and millet can provide nutrients lacking in these staple food crops.

Harvesting and preserving indigenous food plants

Harvesting and collection of indigenous food plants in most developing countries is done by women and children (Ruiz-Perez et al., 1999; Ohiokpehai, 2003; Styger et al., 1999; Shackleton and Shackleton, 2004). The collection is often combined with other daily activities such as cultivation, weeding arable cropland and collection of firewood. Indigenous food plants in Botswana, particularly vegetables are plentiful in the rainy season but become scarce during the dry season or winter and available only when dried and preserved. Women often process and store indigenous vegetables for use during the dry season when they are the only alternative source of nutrition and cash income in rural areas. The principal method of extending their shelf life is to sun dry fresh or boil in salted water before drying. Various methods in harvesting and collecting indigenous fruits include climbing trees to pick fruits, hand picking from lower branches or short trees, shaking stems or branches and knocking fruits down with sticks as well as picking from the ground following abscission (Kalaba et al., 2009). Some of these fruit harvesting methods can cause fruit damage and losses. For example, Kadzere et al.

Table 4. Mineral contents of fruits of selected indigenous fruit trees common in Botswana.

Species	Ca	K	Mg	Na	P	Fe	Zn
Adasonia digitata	128	1866	121	13.3	50	0.10	0.14
Sclerocarya birrea	94	2183	158	13	69	0.07	0.13
Strychnos spinosa	56	1370	49	21.7	66	0.11	0.22
Vangueria infausta	124	1683	99	13.7	128	0.09	0.02

Source: Amarteifio and Mosase (2006).

(2006a, b) estimated that 25 to 50% of indigenous fruits are lost during and after harvesting. Losses have been attributed to lack of fruit handling and marketing skills (Kabala et al., 2009) and pest damage (Kadzere et al., 2006a, b).

Marketing of indigenous food plants

In Botswana, indigenous food plants are sold by women and children in both formal and informal markets in rural and urban areas. The markets are seasonal and very limited. In urban and peri-urban areas, sales are conducted near shopping centres and from door to door. In rural areas sale is conducted from home, door to door, at bus stops and roadside markets. The sale of indigenous food plants provides cash income (Shackleton et al., 1998; Nesamvuni et al., 2001, Kadu et al., 2006) used to buy food; pay children school fees, uniform and other household needs. However, markets in Botswana (Taylor et al., 1996) and other developing countries favour exotic food plants than their indigenous counterparts because they have been researched and developed. Domesticated or exotic food plants are widely promoted and their uses and management techniques are well known. In contrast, indigenous food plants have received little attention in terms of research, development and promotion in Botswana, despite their large market potential. Taylor et al. (1996) noted that the variety of indigenous food plants sold in formal markets is lower when compared to informal markets. They further indicated that the formalisation of markets is constrained by lack of infrastructure and products of variable quality that are always available in small quantities.

Potential for domestication of indigenous food plants in Botswana

The importance of indigenous food plants in crop production systems has received significant attention in recent years, not only in terms of increasing arable crop production and food security but also in matters related to climate change mitigation through carbon sequestration. Domestication of indigenous food plants could be used as a strategy to improve food security and cash income for people living in rural areas as well as mitigating climate change. Several indigenous fruit tree species have been selected for domestication in Botswana and other countries in southern and eastern Africa (Jaenicke et al., 2000; Akinnifesi et al., 2004, 2006, 2008b; Mateke, 2000, 2002, 2003; Teklehaimanot, 2004; Leakey, 2005; Kabala et al., 2009). It is a common practice for farmers in Botswana to leave indigenous fruit trees when they clear forests/woodlands for arable crop. People have become aware of the importance of indigenous food plants and there is evidence of indigenous fruit tree planting in homesteads and backyard gardens in both the urban and rural areas of Botswana (Taylor and Kwerepe, 1995; Taylor et al., 1996). Apart from providing food, indigenous food plants, particularly fruit trees can be managed in arable land and gardens to capture and store a significant fraction of atmospheric carbon. The planting of indigenous fruit trees may indirectly decrease pressure on indigenous forests and woodlands which are the largest sinks for terrestrial carbon (Albrecht and Kandji, 2003; Montagnini and Nair, 2004). Domestication of indigenous food plants requires gathering them from the wild and cultivation on farmland. Akinnifesi et al. (2006) noted that local planting, product development and marketing are important in the domestication of indigenous food plants.

Participatory research and development of indigenous fruits such as *S. birrea*, *S. cocculoides*, *V. infausta* and *A. garckeana* has been conducted at the Veld Products Research and Development in Botswana (Mateke, 2000, 2003). The domestication research and development involved:

i) Selection of primary indigenous fruit trees based on farmers preferences and market orientation,
ii) Identification of superior or elite trees based on established criteria by users, marketers and market preference,
iii) Development and applying efficient vegetative propagation and nursery management techniques for producing quality propagules of on-farm dissemination,
iv) Integration of improved germplasm into farming systems; and
v) Post-harvesting, processing and marketing research of fresh products from domesticated species.

The domestication research and development initiative has led to the inclusion of some indigenous fruit tree in backyard gardens and to some extent in cropping systems in Botswana. In contrast, indigenous vegetables have not received considerable attention in terms of research and development. The planting of indigenous vegetables in Botswana is probably constrained by lack of knowledge in propagation techniques, seedling production, husbandry skills and their free availability in the wild. To improve the contribution of indigenous vegetables to the rural household livelihoods and food security, participatory domestication and research similar to the one used for indigenous fruit at Veld Products Research and Development is needed.

CONCLUSION

Food security and poverty alleviation in rural communities can be improved by diversifying the existing farming systems. Arable crops alone in dry conditions of Botswana, without incorporating alternative livelihoods such as indigenous food plants would not improve the lives of resource-poor rural households. Experiences in many African farming systems have shown that many rural communities rely on indigenous food plants for food and as source of cash income between cropping seasons. Research has shown that indigenous fruit trees have potential comparable to their domesticated counterparts in providing nutrition, food security and cash income to households. The same can be said about indigenous vegetables, though work is needed on their development, promotion and marketing. Indigenous food plants are adapted to growing in low rainfall, poor soils and have few insect and disease problems than arable crops. The predicted climate change and global warming is likely to affect arable crop production negatively in Botswana, causing food shortages. Therefore incorporating indigenous food plants into existing cropping systems may provide an alternative source of food during years with little no arable crop harvest. Research is therefore needed to screen and develop indigenous vegetables and lesser known indigenous fruits so that they can be incorporated into present cropping systems.

REFERENCES

Agte VV, Tarwadi KV, Mengale S, Chiplonkar SA (2000). Potential of traditionally cooked green leafy vegetables as natural sources for supplementation of eight micronutrients in vegetarian diets. J. Food Comp. Anal. 13(6): 885–892.

Agrahar-Marugkar D, Pal PP (2004). Intake of nutrients and food sources of nutrients among the Khasi tribal woman in India. Nutr. 20(3): 268–273.

Akinnifesi FK, Kwesiga FR, Mhanga J, Mkonda A, Chilanga T, Swai R (2004). Domesticating priority miombo indigenous fruit trees as a promising livelihood option for small farmers in southern Africa. Acta Hortic., 632:15–30.

Akinnifesi FK, Kwesiga F, Mhango J, Chilanga T, Mkonda A, Kadu CAC, Kadzere I, Mithöfer D, Saka JDK, Sileshi G, Ramadhani T, Dhliwayo P (2006). Towards the development of miombo fruit trees as commercial tree crops in southern Africa. For. Trees Livel., 16(1): 103–121.

Akinnifesi FK, Sileshi G, Ajayi OC, Chirwa PW, Harawa R (2008a). Contributions of Agroforestry research and development to livelihood of smallholder farmers in Southern Africa. 2. Fruit, medicinal, fuelwood and fodder tree systems. Agric. J., 3(1): 76–88.

Akinnifesi FK, Mhango J, Sileshi G, Chilanga T (2008b). Early growth and survival of three miombo indigenous fruit tree species under fertilizer, manure and dry-season irrigation in southern Malawi. For. Ecol. Manage., 255(3/4): 546–557.

Albrecht A, Kandji ST (2003). Carbon sequestration in tropical agroforestry systems. A review. Agric. Ecosyst. Environ., 99 (1–3): 15–27.

Amarteifio JO, Mosase MO (2006). The chemical composition of selected indigenous fruits of Botswana. J. Appl. Sci. Environ. Manage., 10(2): 43–47.

Ansari NM, Houlihan L, Hussain B, Pieroni B (2005). Antioxidants activity of five vegetables traditionally consumed by South-Asian migrants in Bradford, Yorkshire, UK. Phytother. Res., 19: 907–911.

Arnold JEM, Ruiz PM (2001). Can non-timber forest products match tropical forests conservation and development objectives? Ecol. Econ., 39: 437–447.

Babu SC (2000). Rural nutrition interventions with indigenous plant foods: a case study of vitamin A deficiency in Malawi. Biotechnol. Agron. Soc. Environ., 4(3): 169–179.

Balemie K, Kebebew F (2006). Ethnobotanical study of wild plants in Derashe and Kucha Districts, South Ethiopia. J. Ethnobiol. Ethnomed., Doi:10.1186/1746-4269-2-53.

Barany M, Hammett AL, Sene A, Amichev B (2001). Non-timber forest benefits and HIV/AIDS in sub-Saharan Africa. J. For., 99(12): 36–41.

Boyer JS (1982). Plant productivity and environment. Science, 218: 443–448.

Brinkhurst M (2010). Fruit of sand: Complexities of Botswana's veld resources. Studies by undergraduate researchers at Guelph, Department of Geography, College of Social and Applied Human Sciences 3(2). (http://www.criticalimprov.com/index.php/surg/viewarticle/1092/1664)

Campbell BM (1986a). The importance of wild fruits for peasant households in Zimbabwe. Food Nutr., 12(1): 38–44.

Campbell A (1986b). The use of wild food plants and drought in Botswana. J. Arid Environ., 11: 81–91.

Campbell B, Luckert M, Scoones I (1997). Local evaluation of savannah resources. Econ. Bot., 51: 57–77.

Chwenya J (1994). Potential for agronomic improvement of indigenous plant Germplasm in African Agriculture: A case study of indigenous vegetables in Kenya. In Putter A. (eds): Proceedings of a CTA/IPGRI/UNEP seminar, 59 October 1992, Nairobi, Kenya. CTA, The Netherlands/IPGRI, Rome, Italy.

Cunningham AB (1997). Review of the ethnobotanical literature from eastern and southern Africa. Afr. Ethnobot. Netw. Bull., 1: 23–88.

Delang CO (2006). Indigenous systems of forest classification: Understanding land use patterns and the role of NNTP's in shifting cultivators' subsistence economies. Environ. Manage., 37: 470–486.

Dounias E, Rodrigues W, Petit C (2000). Review of the ethnobotanical literature for Central and West Africa. Afr. Ethnobot. Netw. Bull., 2: 5–117.

El SN, Karakaya S (2004). Radical scavenging and iron-chelating activities of some greens used as traditional dishes in Mediterranean diet. Int. J. Food Sci. Nutr., 55(1): 67–74.

Fleurent A (1979). The role of wild foliage plants in the diet: A case study from Lushoto, Tanzania. Ecol. Food Nutr., 8: 87–93.

Flyman MV, Afolayan AJ (2006). A survey of plants used as wild vegetables in four districts of Botswana. Ecol. Food Nutr., 45(6): 405–415.

FAO (1988). Traditional food plants. FAO Food and Nutrition Paper FAO, Rome, No. 42.

FAO (1990). Utilization of Tropical Foods: Fruits and Leaves. FAO Nutr. Pap. FAO, Rome, 47/7.

Getachew A, Urga K, Dikasso D (2005). Ethnobotanical study of edible wild plants in some selected districts of Ethiopia. Hum. Ecol., 33(1): 83-118.

Gockowski J, Mbazo'o J, Mbah G, Moulende TF (2003). African traditional leafy vegetables and the urban and peri-urban poor. Food

Pol., 28(3): 221–235.
Godoy RA, Bawa KS (1993). The economic value and sustainable harvest of plants from the tropical forest: Assumptions, hypothesis and methods. Econ. Bot., 47(3): 215–219.
Harris MA, Mohammed S (2003). Relying on nature: wild foods in Northern Nigeria. Ambio, 32(1): 24–29.
Jaenicke H, Simons AJ, Maghembe JA, Weber JC (2000). Domesticating indigenous fruit trees for agroforestry. Acta Hortic. 523: 45–52.
Kabala FK, Chirwa PW, Prozesky H (2009). The contribution of indigenous fruit trees in sustaining rural livelihoods and conservation of natural resources. J. Hortic. For., 1(1): 1–6.
Kadu CAC, Imbuga M, Jamnadass R, Dawson IK (2006). Genetic management of indigenous fruit trees in southern Africa. A case study of *Sclerocarya birrea* based on nuclear and chloroplast variation. S. A. J. Bot. 72: 421–427.
Kadzere I, Watkins CB, Merwin IA, Akinnifesi FK, Hikwa D, Hove L, Mhango J, Saka JDK (2006a). Harvesting and post-harvest handling practices of *Uapaca kirkiana* (Muell. Arg.) fruits: A survey of roadside markets in Malawi. Agro. Syst., 6(2): 133–142.
Kadzere I, Watkins CB, Merwin IA, Akinnifesi FK, Saka JDK (2006b). Post-harvest damage and darkening in fresh fruit of *Uapaca kirkiana* (Muell. Arg.). Postharv. Biol. Technol., 39: 199–203.
Kuhnlein H, Erasmus B, Creed-Kanashiro H, Englberger L, Okeke C, Turner N, Allen L, Bhattacharjee L (2006). Indigenous peoples' food systems for health: Finding interventions that work. Public Health Nutr., 9(8): 1013–1019.
Labadarios D, Steyn NP (2001). South African food-based dietary guidelines: guidelines for whom? S. Afr. J. Clin. Nutr., 14(1): 5–6.
Ladio AH, Lozada M (2004). Patterns of use and knowledge of wild edible plants in distinct ecological environment: A case study of a Mapuche community in Patagonia. Biodiv. Conserv. 13(6): 1153–1173.
Leakey RRB (2005) Domestication potential of Marula (*Sclerocarya birrea* subsp. *caffra*) in South Africa and Namibia: 3. Multiple trait selection. Agrofor. Syst., 64(1): 51–59.
Ludlow MM, Muchow RC (1990). A critical evaluation of the traits for improving crop yield in water limited environments. Adv. Agron. 43: 107–153.
Mateke S (2000). Tree improvement in the domestication programme of indigenous fruit tree native to southern Africa. The Botswana experience. In: Shumba EM, Lusepani E, Hangula R (eds). The domestication and Commercialization of Indigenous Fruit Trees in the SADC Region. SADC Tree Seed Centre Network, Gaborone, Botswana, pp. 29–35.
Mateke SM (2002). Results of eight years studies on propagation, growth and production levels of four indigenous fruit trees in the wild and under cultivation in Botswana. Regional Agroforestry Conference "Agroforestry Impacts on Livelihoods in Southern Africa: Putting research into practice". May 2002. Warmbaths, South Africa.
Mateke S (2003). Cultivation of native fruit trees of Kalahari sandveld: Studies on commercial potential, interactions between soil and biota in Kalahari sands of southern Africa. Veld Products, Botswana, pp. 49.
McGregor J (1995). Gathered produce in Zimbabwe's communal areas changing resource availability and use. Ecol. Food Nutr., 33(3): 163–193.
Mojeremane W, Tshwenyane SO (2004a). The resource role of morula (*Sclerocarya birrea*): A multipurpose indigenous fruit tree of Botswana. J. Biol. Sci., 4(6): 771–775.
Mojeremane W, Tshwenyane SO (2004b). *Azanza garckeana*: A valuable edible indigenous fruit tree of Botswana. Pak. J. Nutr., 3(5): 264–267.
Montagnini F, Nair PKR (2004). Carbon sequestration: An underexploited environmental benefit of Agroforestry systems. Agrofor. Syst., 61/63(1–3): 281–295.
Moss H (1988). Under-exploited food plants in Botswana. Unpublished report for Veld Products Research, Gaborone, Botswana.
Muok BO, Matsumura A, Ishii T, Odee, DW (2009). The effect of intercropping *Sclerocarya birrea* (A. Rich.) Hochst, millet and corn in the presence of arbuscular mycorrhizal fungi. Afr. J. Biotechnol., 8(5): 807–812.
Nesamvuni C, Steyn NP, Potgieter MJ (2001). Nutritional value of wild, leafy plants consumed by the Vhavenda. S. Afr. J. Sci., 97(1/2): 51–54.
Ohiokpehai O (2003). Promoting the Nutritional Goodness of Traditional Food Products. Pak. J. Nutr. 2(4):267–270.
Parvathi S, Kumar VSF (2002). Studies on chemical composition and utilization of the wild edible vegetable Athalakkai (*Momordica tuberose*). Plant Foods Hum. Nutr., 57: 215–222.
Rathore M (2009). Nutrient content of important fruit trees from arid zone of Rajasthan. J. Hortic. For., 1(7): 103–108.
Redzic SJ (2006). Wild Edible plants their traditional use in the human nutrition in Bosnia-Herzegovina. Ecol. Food Nutr., 45(3): 189–232.
Rubaihayo EB (1997). Conservation and use of traditional vegetables in Uganda. In: Guarino L (eds). Traditional African Vegetables, Proceedings of the IPGRI International Workshop on Genetic Resources of Traditional Vegetables in Africa. Conservation and Use. International Plant Genetic Resources Institute, Nairobi, pp. 104–116.
Ruiz Perez M, Ndoye O, Eyebe A (1999). Marketing of non-timber forest products in the humid forest zone of Cameroon. Unasylva, 50: 12–19.
Scherrer AM, Motti R, Wecherle CS (2005). Traditional plant use in the areas of Monte Vesole and Ascea National Park (Campania, Southern Italy). J. Enthopharmacol., 97(1): 129–143.
Schmidt DR (1971). Comparative yields and composition of eight tropical leafy vegetables grown at two fertility levels. Agron. J., 63: 546–550.
Shackleton SE, Shackleton CM, Dzerefos CM, Mathabela FR (1998). Use and trading of wild edible herbs in central the Lowveld savanna region, South Africa. Econ. Bot., 52(3): 251–259.
Shackleton SE, Shackleton CM, Cousin B (2000). Re-valuing the communal lands of South Africa: A new understanding of rural livelihoods. ODI Nat. Res. Perspect., 62: 1–4.
Shackleton CM, Shackleton SE (2004). The importance of non-timber products in rural livelihood security and as safety-nets: A review of evidence from South Africa. S. Afr. J. Sci., 100: 658–664.
Shackleton SE, Shackleton CM, Netshiluvhi TR, Geach BS, Ballance A, Fairbanks DHK (2002). Use patterns and value of savannah resources in three rural villages in South Africa. Econ. Bot. 56(2): 130–146.
Shackleton CM (2003). The prevalence of use and value of wild edible herbs in South Africa. S. Afr. J. Sci., 99(1/2): 23–25.
Shrestha PM, Dhillion SS (2006). Diversity of traditional knowledge concerning wild food species in a local managed forest in Nepal. Agrofor. Syst., 66(1): 55–63.
Singh V, Garg AN (2006). Availability of essential trace elements in Indian cereals, vegetables and species using INAA and the contribution of species to daily dietary intake. Food Chem., 94: 81–89.
Styger E, Rakotoarimanana JEM, Rabevohitra R, Fernandes ECM (1999). Indigenous fruit trees of Madagascar: Potential components of agroforestry systems to improve human nutrition and restore biological diversity. Agrofor. Syst., 46(3): 289–310.
Teklehaimanot Z (2004). Exploiting the potential of agroforestry tree: *Parkia biglobosa* and *Vitellaria parodoxa* in Sub-Saharan Africa. Agrofor. Syst., 61(1): 207–220.
Taylor F, Kwerepe B (1995). Towards domestication of some indigenous fruit trees in Botswana: In Maghembe JA., Ntupanyama Y, Chirwa PW (eds). Improvement of indigenous trees of miombo woodlands of southern Africa. ICRAF, Nairobi, Kenya.
Taylor F, Mateke SM, Butterworth KJ (1996). A holistic approach to the Domestication and commercialization of non-timber products. In Leakey RRB, Temu AB, Melnyk M, Vantomme P (eds). Domestication and commercialization of non-timber forest products in Agroforestry systems. Non-Wood Prod., FAO, Rome, 9.
Tiisekwa BPM, Ndabikunze BK, Samson G, Juma M (2004). Suitability of some indigenous tree fruits for manufacturing juices and jams in Tanzania. In: Rao MR, Kwesiga FR (eds). Agroforestry impacts on livelihoods in southern Africa: Putting research into practice. Proceedings of the Regional Agroforestry Conference held in Warmbaths, South Africa 20–24 May, 2002. World Agroforestry Centre (ICRAF), Nairobi, Kenya, pp. 331–335.
Tucker, JB (1986). Amaranth - The one and future crop. BioScience, 36(1): 9–13.

Twyman C (2001). Natural resource use and livelihoods in Botswana's wildlife management areas. Appl. Geogr., 21(1): 45–68.
WHO (2000). A global agenda for combating malnutrition: Progress Report, WHO, Geneva, Switzerland.
WHO (2001). Childhood nutrition and progress in implementing the International Code of Marketing of Breast-milk Substitutes. WHO, Geneva, Switzerland.
WHO (2005). Fruit and Vegetables for Health. Report of a Joint FAO/WHO Workshop, 1–3 September, 2004, Kobe, Japan.
Zinyama LM, Matiza T, Campbell DJ (1990). The use of wild foods during periods of food shortage in Zimbabwe. Ecol. Food Nutr., 24(4): 251–265.

3

Effect of polyethylene mulching on the spatial variability of soil physical properties and growth parameters of taro (*Colocasia esculantum*)

Nsalambi V. Nkongolo[1]*, Mitate Yamada[2] and Izumi Nakagawa

[1]Department of Agriculture and Environmental Science, College of Agricultural and Natural Sciences, Lincoln University, Jefferson City, MO 65102-0029, U.S.A.
[2]Crop Production and Postharvest Technology Division, Japan International Research Center for Agricultural Sciences (JIRCAS), Tsukuba Science City, Japan 305-8686.
[3]Vegetable Cultivation Technology System Laboratory, Department of Farmland Utilization, National Agriculture Research Center, Tsukuba, Ibaraki 305-8666, Japan.

The objective of this study was to investigate the spatial variability of soil physical properties and growth of taro (*Colocasia esculantum*) in an unfertilized 42 m ridge covered with black plastic mulch. The study was conducted at the National Agriculture Research Centre Tsukuba (Japan). The experimental field was made of 3 adjacent ridges of 42 m long and 0.90 m width each. Uniform taro corms were planted in the middle of each ridge at 0.30 m spacing. At the end of growing season, soil samples were taken for measurements of air content (f_a), bulk density (ρ_b), gravimetric (θ_g) and volumetric water contents (θ_v), total pore space (TPS), pore tortuosity (τ), gas diffusion coefficient (D_SD_O) and water-filled pore space (WFPS). Taro plants were harvested, separated into leaves (LFW), petioles (PFW), corms (CFW), roots (RFW) and tubers (TFW). All autocorrelation functions (r(*h*)) for soil properties and taro growth dropped sharply after one lag. Except for θ_g, (r(*h*)) for θ_a, ρ_b, Ds/Do and WFPS dropped below zero at about 16 lags, which revealed that these soil properties could be sampled at 4.8 m for independent observations. For taro growth parameters, only PFW, TFW and RFW could be sampled together at 14 lags or 4.2 m for uncorrelated observations. Except for RFW, all taro growth parameters significantly correlated among themselves and with θ_g, f_a, ρ_b, Ds/Do, WFPS and τ with correlation coefficients ranging from 0.16 to 0.99. Plastic mulch improved soil properties which in turn improved the relationship between these properties and taro growth parameters.

Key words: Soil physical properties, taro growth parameters, plastic mulch, volcanic ash soil.

INTRODUCTION

Taro (*Colocasia esculantum*) is an herbaceous, perennial, tropical root crop composed of a main plant and suckers. It is one of the most important staple crops in the Pacific Islands (Ctahr and Taro, 2009) but also widely grown throughout Africa, Asia, West Indies and South America. It is harvested primarily for its corm, which is a starchy, underground stem. Taro is adapted to moist environments and can be grown under rain-fed or irrigated upland (that is, non flooded) as well as flooded conditions (Miyasaka et al., 2003). Taro requires deep planting for optimum corm set, because shallow soils provide a restrictive environment for corm and root development. In addition, Taro has important water requirements, which are not always met by soil moisture. In upland taro cultivation, the ability of a soil to store water becomes an important issue due to the high variability and unpredictability of rainfall and the cost of irrigation. Therefore, cultural practices, like mulching with black polyethylene are necessary to mitigate the impact of limited moisture. The use of plastic materials for mulching is a very common practice for vegetable crops and black polyethylene

*Corresponding author. E-mail: nkongolo@lincolnu.edu.

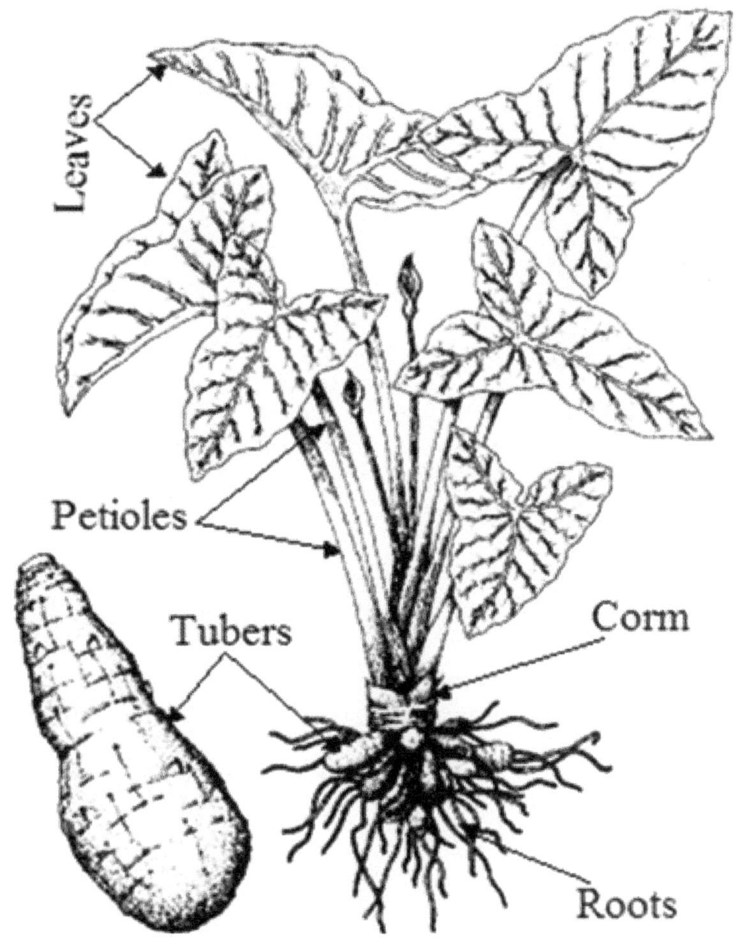

Figure 1. Taro plant morphology.

is widely used due to its excellent properties and low cost (Moreno and Moreno, 2008). In fact, plastic mulches have many positive advantages for the user, such as increased yields, earlier maturing of crops, crops of higher quality, enhanced insect/ pest management, and weed control. Furthermore, black polyethylene cover-sheets increases soil moisture retention, soil temperature, reduces evapo-transpiration from the soil and enhance crop yield. The effect of black polyethylene on soil properties and crop yield has been investigated by severalworkers (Miyasaka et al., 2001; Moreno and Moreno, 2008; Oñate and Peco, 2005; van der Meulen et al., 2006; Hulugalle et al., 1990; Anikwe et al., 2007; Ossom and Matsenjwa, 2007; Chaudhry et al., 2008; Singh et al., 2008; Subrahmaniyan et al., 2008; Ndubuisi, 2009; Takashi and Hajime, 2000; Subrahmaniyan et al., 2006; Islam et al., 2007; Isalam et al., 2006). However, it is still unclear how the use of black polyethylene as a mulching technique affects the spatial variability of soil properties and in-turn, how this variability in soil properties relate to plant growth parameters. The objective of this study was to investigate the spatial variability of soil physical properties and the corresponding variability in taro growth parameters on a volcanic ash covered with black polyethylene.

MATERIALS AND METHODS

Soil preparation and taro management

This study was conducted at the National Agriculture Research Center, Tsukuba, Japan. The soil of the experimental site is a light-colored Kuroboku soil (Low-humic Andosols) originating from volcanic ash. The experiment was composed of three adjacent ridges of 42 m long and 0.90 m width. Soil preparation includes mowing and subsoiling to a depth of 0.20 - 0.30 m with a regular size Shibuya D-23 tractor attached with a Rotary Cx 1402 cultivator. No plowing was done. Three parallel ridges of 42 m long and 0.90 m large crossing all the field were made. Each of the 42 m ridge was planted with taro at 0.30 m spacing and covered with a black plastic mulch and One hundred thirty nine samples of each soil and taro plant were collected along one of the three 42 m ridges at the above mentioned spacing. Soil samples were taken with a 0.05 m diameter and 0.05 m height cylinder while the entire taro plant at each soil sampling location was harvested, cleaned, separated into leaves, petioles, corms, roots, tubers (Figure 1) and fresh weight was recorded. Fresh wet weight of soil was also recorded and transferred into a three phase meter (Daiki) for soil and air volume determination. They were oven dried at 105°C for 24 h and the

Table 1. Summary of simple statistics for soil physical properties

Simple statistics	f_a(m^3 m^{-3})	ρ_b(kg.m^{-3})	θ_g(kg.kg^{-1})	θ_v(m^3 m^{-3})	TPS(m^3 m^{-3})	τ(m m^{-1})	D_SD_O(m^2 s^{-1}/ m^{-2} s)	WFPS(m^3 m^{-3})
Mean	0.48	0.59	0.50	0.30	0.78	2.08	0.23	38.09
SD	0.03	0.03	0.02	0.02	0.01	0.12	0.02	2.67
C.V.	5.27	4.30	4.58	6.07	1.31	5.61	10.48	7.02
Minimum	0.38	0.54	0.43	0.26	0.74	1.89	0.14	32.91
Median	0.49	0.59	0.50	0.29	0.78	2.04	0.24	37.66
Maximum	0.53	0.68	0.57	0.37	0.80	2.63	0.28	49.33
Skew	-0.93	0.95	0.68	1.01	-0.98	1.45	-0.71	1.07
Kurtosis	1.66	1.15	0.74	1.65	1.38	3.70	0.94	2.00

f_a = air content (air content), ρ_b = bulk density, θ_g = gravimetric water content, θ_v = volumetric water content, TPS = total pore space, τ = pore tortuosity, D_SD_O = relative gas diffusion coefficient, WFPS = water-filled pore space.

volume of dry soil was recorded by three phase meter. Soil bulk density, total pore space, gravimetric water content, volumetric water content and air filled porosity were calculated as described by Nkongolo et al. (2007a)

RESULTS AND DISCUSSION

Soil physical properties

During the time of sampling, 62% of the pore space of volcanic ash soil was occupied by air as revealed by an air-filled porosity (0.48 m^{3}m^{-3}) while 38% of the pore space was filled with water (Table 1). Values for soil air and total porosity in volcanic ash soil were higher as compared to regular mineral soil. These results were in agreement with other authors who have suggested that the volcanic ash soils exhibit unique soil properties such as high water retention, large total porosity due to noncrystalline materials like allophane and good drainage, all of which are favorable for plant root growth (Shoji et al., 1993; Moldrup et al., 2003). The bulk density was very low (0.59 kg m^{-3}) as expected for volcanic ash soils. In fact, as suggested earlier for soil air content and total pore space, the low bulk density of volcanic ash soils is also due to high porosity caused by well-developed aggregate structures made of non crystalline minerals (Nanzyo et al., 2001; Shoji et al., 1993). The coefficient of variation (CV) varied between 1.31 to 10.48% and was confined within the range of normally reported for soil physical properties. The highest CV value was recorded for the relative gas diffusion coefficient (10.34%), followed by water filled pore space (7.02%). The pore tortuosity factor was also low as the result of well-developed aggregate structures for this type of soil (Nanzyo et al., 2001; Hamamoto et al., 2008). All means for soil physical properties were equal or closer to their corresponding medians, suggesting that the data was normally distributed. This is also confirmed by low CV, kurtosis and skew values.

Taro growth

The total weight of taro plant varied from 508 to 15816 g on a fresh weight basis. Corm fresh weight ranged from 30 to 1020 g with an average of 347.84 g/plant (Table 2). Assuming a plant spacing of 30 × 30 cm in a regular field for taro, each plant would have occupied an area of roughly 900 cm^2. In addition, considering the fact that taro corm contains 63 to 85% of water (Huang et al., 2007; Onwueme, 1999, 1994), taro's assumed corm yield was therefore ranged between 5.86 and 14.30 ton/ha (on dry matter basis) with an average yield of 9.66 ton/ha. These value fall within the range of taro yield reported in Japan, but short below the yields of taro in China, Salmon Islands and Fiji (Onwueme, 1999). However, this assumed yield was obtained without fertilizer input, only as the results of soil physical properties and other growth parameters improvement by the use of black polyethylene mulch. Our results agree with those reported by nikwe et al. (2007) who studied tillage and plastic mulch effects on soil properties and growth and yield of cocoyam (*Colocasia esculenta*) on an ultisol in Southeastern Nigeria. They reported that the highest corm yield in tilled black plastic mulched plots, and concluded that the plastic mulched plots had provided a better soil environment for cocoyam than un-mulched plots and in tilled mulched plots, especially tilled black plastic mulched plots provide superior edaphic environment for cocoyam when compared to other

Table 2. Summary of simple statistics for Taro's growth parameters.

Simple statistics	LFW ($g.plant^{-1}$)	PFW ($g.plant^{-1}$)	CFW ($g.plant^{-1}$)	RFW ($g.plant^{-1}$)	TFW ($g.plant^{-1}$)	AGB ($g.plant^{-1}$)	BGB ($g.plant^{-1}$)	TTB ($g.plant^{-1}$)
Mean	234.68	791.37	347.84	288.20	888.78	1800.30	1521.60	3321.90
SD	21.44	690.98	214.57	57.94	575.55	1410.20	785.67	2038.20
C.V.	9.14	87.32	61.69	20.11	64.76	78.33	51.64	61.36
Minimum	203.00	10.00	30.00	198.00	20.00	204.00	282.00	508.00
Median	231.00	710.00	310.00	280.00	800.00	1639.00	1360.00	3066.00
Maximum	302.00	4760.00	1020.00	634.00	4340.00	9822.00	5994.00	15816.00
Skew	0.88	3.26	0.86	2.85	2.94	3.17	2.08	2.54
Kurtosis	0.61	15.16	0.58	14.84	15.44	14.55	9.23	11.61

LFW = leaves fresh weight, PFW = petioles fresh weight, CFW = corms fresh weight, RFW = roots fresh weight, TFW = tubers fresh weight, AGB = above ground biomass (leaves + petioles), BGB = below ground biomass, (corms + tubers + roots) and TTB = taro total biomass (AGB + BGB).

treatments used. Studying the influence of mulch on agronomic characteristics, soil properties, disease and insect pest infestation of dry bean (*Phaseolus vulgaris* L.) in Swaziland, Ossom and Matsenjwa (2007) reported that un-mulched plants and plant under black polyethylene mulch had significantly higher pod yield as compared to plants under grass, newspaper and clear plastic mulches. Better soil properties and higher plant yield in black polyethylene mulch treatment was reported by various workers (Singh et al., 2008; Subrahmaniyan et al., 2008; Ndubuisi, 2009). Chaudhry et al. (2008) studied mulching impact on moisture conservation, soil properties and plant growth. They observed maximum saving/ conservation of water (45%) and high plant growth under polyethylene sheet and concluded that polythene sheet is more efficient to reduce water losses, but it is a costly option.

Spatial variability of soil physical properties

The highest fluctuations in soil physical properties recorded between 36 and 41 m for air filled porosity, gas diffusion coefficient, gravimetric water content and water filled pore space, as showed by time series plots (Figure 2). Air filled porosity and gas diffusion coefficient decreased sharply as gravimetric water content and water filled pore space increased. However, soil bulk density increased sharply from 0.53 to about 0.67 kg/ m^3 between 18 and 24 m along the 42 m transect. Autocorrelation plots for air-filled porosity, gas diffusion coefficient, gravimetric water content and water filled pore space recorded similar patterns. The autocorrelation function (r(*h*) dropped quickly to less than unity after only one lag and continued to decrease without leveling off asthe lag and dropped below zero at about 16 lags for air-filled porosity, gas diffusion coefficient and water filled pore space. However, r(*h*) for gravimetric water content continued to decrease without dropping to zero. Physical systems which may exhibit such type of autocorrelogram include sediment deposits for periodic flooding or compacted non compacted patterns due to wheel traffic in a row crop (Warrick et al., 1986). In this study, the presence of compacted-non compacted patterns were due to use of tractor during field preparation might be one of the reasons. In addition, posing the plastic film mulch on the ridge might also have brought some compaction The autocorrelation function (r(*h*)) for soil bulk density recorded cyclic patterns It dropped to 0.22 after one lag, continued to drop, then leveled off to 0.22 after 11 lags or 3.3 m, and continued its up-low patterns. Given that when r(*h*) drops to 0 the samples are no longer were related to one another, air-filled porosity, bulk density, gas diffusion coefficient and water filled pore space could therefore be sampled together at 4.80-m or 16 lags for independent observations.

Spatial variability of taro's growth parameters

All taro growth parameters recorded two peak zones, first between 10 and 12 m and the second zone between 35 and 37 m of the 42 m transect (Figure 3). Between these two zones, the fresh weight of leaves, petioles, tubers, corms and roots fluctuated with a series of high and low values, but did not exceed the peak zones. Autocorrelation function (r(*h*)) for taro petioles, tubers and roots recorded similar cyclic patterns: they dropped after one lag and continued with high and low

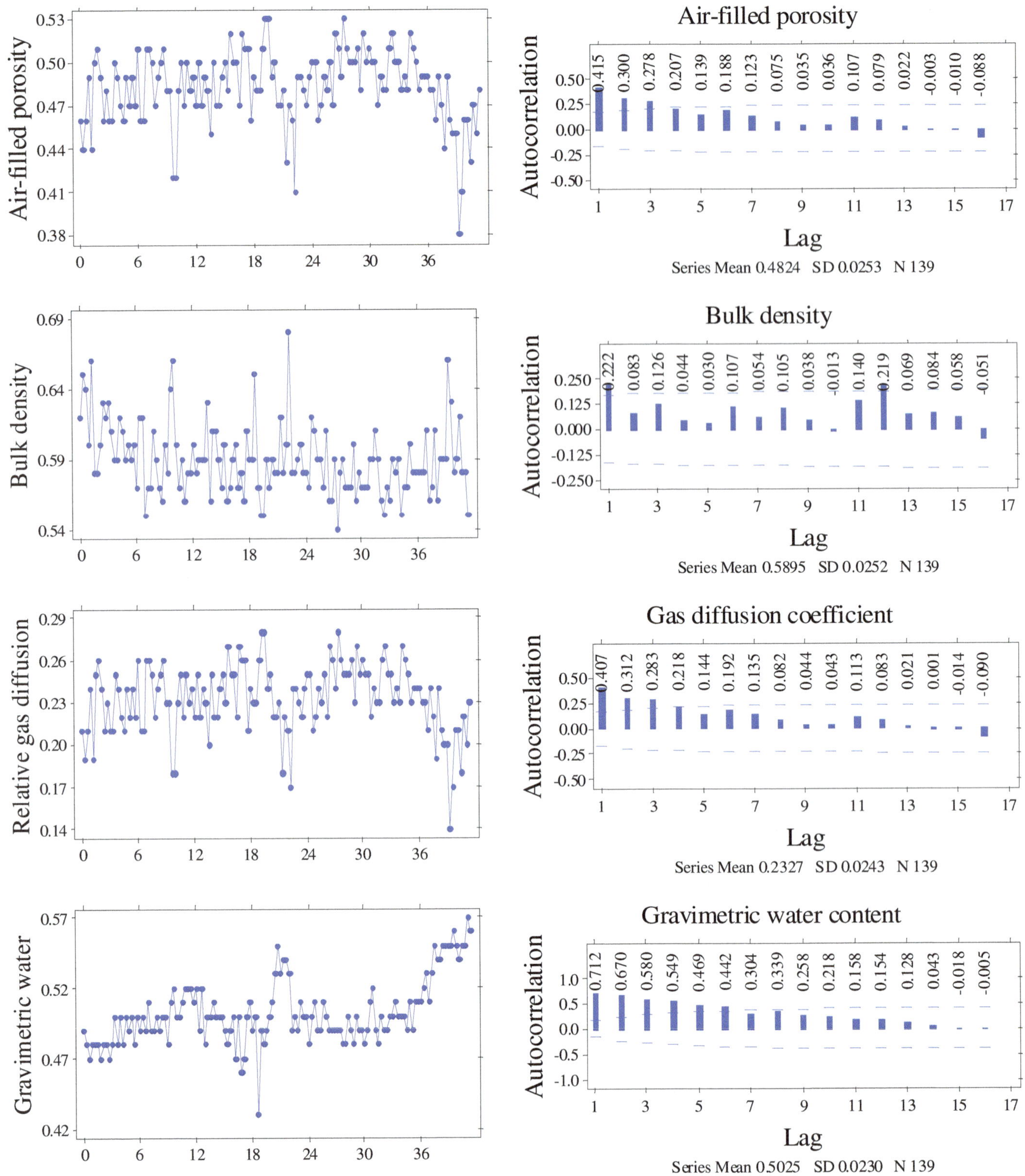

Figure 2. Time series and autocorrelation plots of soil physical properties.

values to finally drop below zero after 14 lags. Since the distance at which r(h) drops to 0 indicates the limiting distance in which samples are related to one another, taro petioles, tubers and roots could therefore be

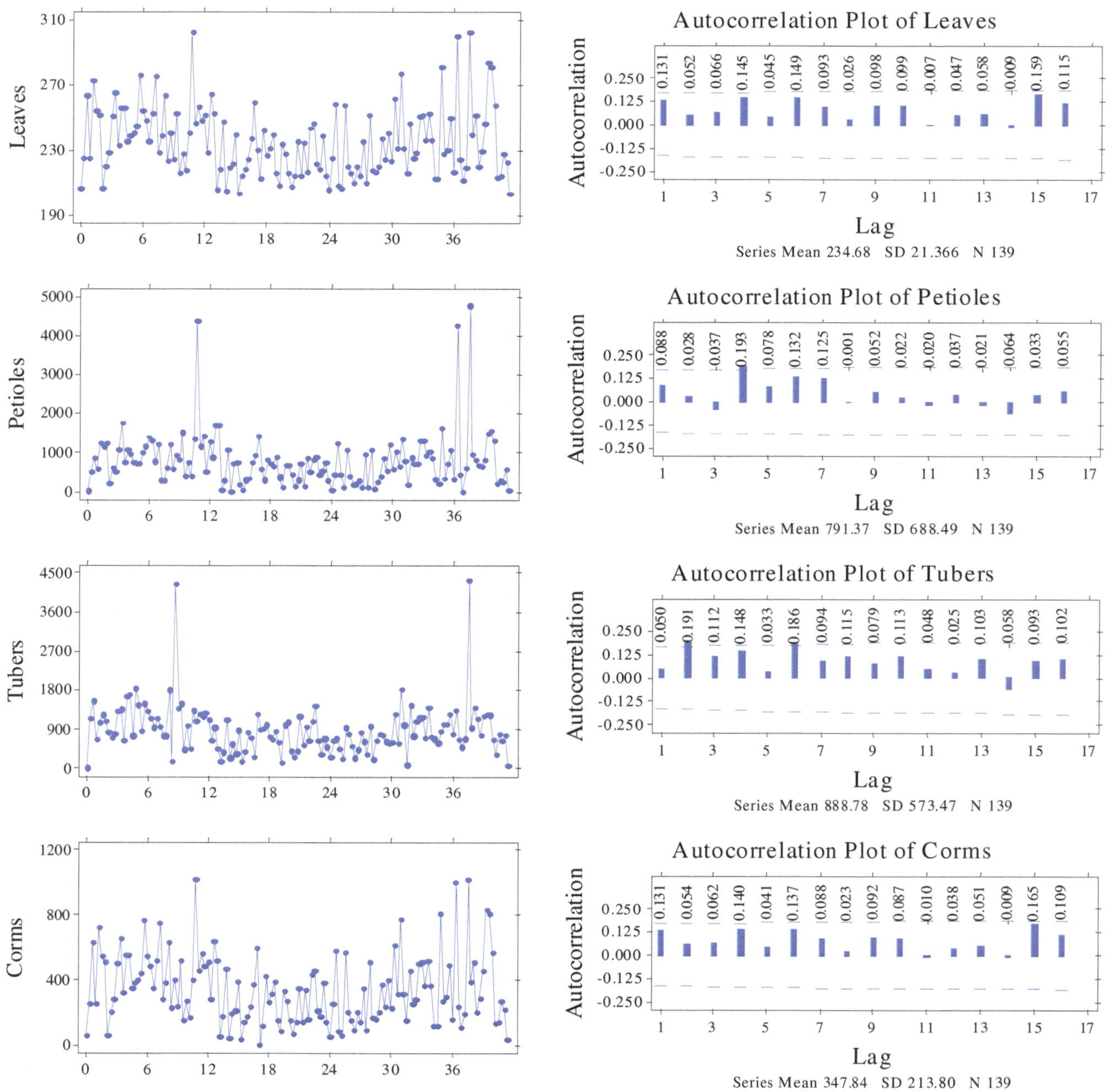

Figure 3. Times series and autocorrelation plots of taro growth parameters.

sampled together at 14 lags or 4.2 m for independent observations. Autocorrelation function (r(*h*)) for taro leaves and corms also showed cyclic patterns, but they shifted between high and low values without dropping to zero. As discussed above, physical systems which may exhibit such type of autocorrelogram include sediment deposits for periodic flooding or compacted non compacted patterns due to wheel traffic in a row crop (Warrick et al., 1986). Perhaps, compaction patterns caused by tractor during field preparation might have a direct impact on plant growth and therefore influenced the autocorrelogram.

Correlation between soil physical properties and taro growth parameters

Soil air filled porosity (f_a) was significantly correlated with leaf fresh weight (LFW) (p = 0.0429) and corms fresh weight (CFW) (Table 3). Conflicting results have been reported for the relationships between soil air content and plant growth. Ouimet et al.(1990), Caron et al. (2001), Nkongolo et al. (2007b) have also not reported any relationship between f_a and plant growth. However, De Rouin et al. (1988) found a minimal relationship between f_a and plant growth. Wall and Heiskanen (2003) examined

Table 3. Correlation between soil physical properties and taro growth parameters.

	LFW	PFW	ABG	CFW	RFW	f_a	θ_g	D_sD_o
PFW	0.822							
p-value	*0.0001*							
AGB	0.8385	0.9922						
p-value	*0.0001*	*0.0001*						
CFW	0.9982	0.8198	0.8371					
p-value	*0.0001*	*0.0001*	*0.0001*					
RFW	0.6217	0.5707	0.5803	0.6134				
p-value	*0.0001*	*0.0001*	*0.0001*	*0.0001*				
f_a	-0.1720	-0.1234	-0.1144	-0.1664	-0.0603			
p-value	*0.0429*	0.1479	0.1800	*0.0503*	0.4806			
θ_g	0.1239	0.1734	0.1701	0.1166	0.1132	-0.4659		
p-value	0.1463	*0.0412*	*0.0453*	0.1715	0.1846	0.0001		
D_sD_o	-0.1686	-0.1243	-0.1146	-0.1629	-0.0600	0.998	-0.4713	
p-value	*0.0472*	0.1450	0.1790	*0.0554*	0.4829	0.0001	0.0001	
τ	0.1783	0.1201	0.1121	0.1731	0.0607	-0.995	0.4634	-0.9884
p-value	*0.0357*	0.1591	0.1889	*0.0416*	0.4781	0.0001	0.0001	0.0001

the effect of air-filled porosity and organic matter concentration of soil on growth of *Picea abies* seedlings after transplanting. They found that the shoot height and mass growth as well as root mass were significantly higher in 20, 30 and 40% than in 5 and 10% air-filled porosity. The air content in this study ranged from 0.38 to 0.48 m^3 m^{-3}. Soil gravimetric water content was significantly correlated with plant fresh weight (PFW) (p = 0.0412) and above ground biomass (AGB) (p = 0.0543). Our results conflict with those reported by Bilderback and Lorscheider (1995) who studied the physical properties of double-processed pine bark on rooting of *Photinia × 'fraseri'* cuttings and found that regression and correlation analyses indicated no relationship between percent volume of air space (air-filled porosity) or container capacity (water content) with rooting responses measured for any of the three species propagated. The relative gas diffusion coefficient (D_sD_o) significantly correlated with taro leaf fresh weight (LFW) (p = 0.0472) while the pore tortuosity factor (τ) significantly correlated with both LFW (p = 0.0357) and CFW p = 0.0416). These results agree with those reported in other studies. Caron et al. (2001) found that root growth of *Euphorbia pulcherima* was significantly and positively correlated to gas relative diffusivity. Similarly, the existence of correlations between plant growth and the exchange properties of media (pore tortuosity factor and gas diffusion coefficient) has been reported by Nkongolo et al. (2007c); Nkongolo and Caron (2006). Taro growth parameters were also significantly correlated among themselves and coefficients of correlation ranged between 0.58 to 0.99. Finally, soil physical properties also correlated among themselves with correlation coefficients ranging from 0.47 to 0.99.

REFERENCES

Anikwe MAN, Mbah CN, Ezeaku PI, Onyia VN (2007). Tillage and plastic mulch effects on soil properties and growth and yield of cocoyam (*Colocasia esculenta*) on an ultisol in southeastern Nigeria. Soil Tillage Res., 93(2): 264-272.

Bilderback TE, Lorscheider MR (1995). Physical properties of double-processed pine bark: Effects on rooting. Acta Hortic., 401:77-83.

Caron J, More PH, Rivière LM (2001). Aeration in growing media containing large particle size. Acta Hortic., 548: 229-234.

Chaudhry M R, Malik AA, Sidhu M (2008). Mulching Impact on Moisture Conservation, Soil Properties and Plant Growth. Pakistan J. Water Resourc., 8 (1): 1-8.

Ctahr, Taro (2009). Taro Research by the College of Tropical Agriculture and Human Resources. Background paper. University of Hawai at Manoa, HI. 40p.

De Rouin N, Caron J, Parent LE (1988). Influence of some artificial substrates on productivity and DRIS diagnosis in greenhouse tomatoes (lycopersicon esculentum L. Mill., cv 'Vedettos'). Acta Hortic., 221: 45-52.

Hamamoto S, Tokida T, Miyazaki T, Mizoguchi M (2008). Dense gas flow in volcanic ash soil: Effect of pore structure on density-driven flow. Soil Sci. Soc. Am. J., 72(2): 480-486.

Huang CC, Chen WC, Wang CR (2007). Comparison of Taiwan paddy-

and upland-cultivated taro (*Colocasia esculenta* L.) cultivars for nutritive values. Food Chem., 102(1): 250-256.
Hulugalle NR, Lal R, Gichuru M (1990). Effect of Five Years of No-tillage and Mulch on Soil Properties and Tuber Yield of Cassava on an Acid Ultisol in South-eastern Nigeria. Exp. Agric., 26(2): 235-240.
Islam MJ, Hossain AKMM, Khanam F, Majumder UK, Rahman MM, Rahman MS (2007). Effect of mulching and fertilization on growth and yield of garlic at dinajpur in Bangladesh. Asian J. Plant Sci., 6(1): 98-101.
Islam MS, Karim AJMS, Hossain MS, Masud MM (2006). Tillage and mulch effects on some soil physical properties in shallow red brown terrace soils of Bangladesh. Bull. Inst. Trop. Agric. Kyushu Univ., 29(1): 69-82.
Miyasaka SC, Ogoshi RM, Tsuji GY, Kodani LS (2003). Site and Planting Date Effects on Taro Growth Comparison with Aroid Model Predictions. Agron. J., 95(3): 545-557.
Miyasaka SC, Hollyer JR, Kodani LS (2001). Mulch and compost effects on yield and corm rots of taro. Field Crops Res., 71(2): 101-112.
Moldrup P, Yoshikawa S, Olesen T, Komatsu T, Rolston DE (2003). Air permeability in undisturbed volcanic ash soils: Predictive model test and soil structure fingerprint. Soil Sci. Soc. Am. J., 67(1): 32-40.
Moreno MM, Moreno A (2008). Effect of different biodegradable and polyethylene mulches on soil properties and production in a tomato crop. Sci. Hortic., 116(3): 256-263.
Nanzyo MH, Tsuzuki H, Otsuda H, Yamasaki S (2001). Origin of clay-size vermiculite in sandy volcanic ash soils derived from modern
Pinatubo Lahar deposits in central Luzon, Philippines. Clay Sci., 11(4): 381-390.
Ndubuisi MC (2009). Physical Properties of an Ultisol Under Plastic Film and No-Mulches and their Effect on the Yield of Maize. J. Am. Sci., 5(5): 25-30.
Nkongolo NV, Yamada M, Nakagawa I (2007a). Spatial variability selected soil physical properties and root growth of sweet corn. J. Environ. Mon. Rest. 3(1): 276-285.
Nkongolo NV, Yamada M, Yamasaki A (2007b). Effect of animal manure on the physical properties of a volcanic ash soil and cabbage growth response. J. Environ. Mon. Rest, 3(1): 304-313.
Nkongolo NV, Gauthier F, Caron J (2007c). Effects of fresh and composted organic wastes on media physical properties and growth of three ornamental species. Int. J. Soil Sci., 2(4) :235-246.
Nkongolo NV, Caron J (2006). Pore space organization and plant response in peat Substrates. I. Prinus x Cistena. Sci. Res. Essays, 1(3): 77-89
Onwueme IC (1999). Taro Cultivation in Asia and the Pacific. FAO Regional Office for Asia and the Pacific, RAP Publication No. 1999/16, Bangkok, Thailand.
Onwueme IC (1994). Tropical root and tuber crops - Production, perspectives and future prospects. FAO Plant Production & Protection Paper 126, FAO, Rome, 228p.
Oñate JJ, Peco B (2005). Policy impact on desertification: stakeholders' perceptions in southeast Spain. Land Use Pol., 22(2):103-114.
Ouimet R, Charbonneau J, Parent LE, Blain J, Gosselin A (1990). Effect of peat substrate composition and growing bags volume on the productivity of greenhouse tomato. Can J. Plant Sci., 70(2):585-590.
Ossom EM, Matsenjwa VN (2007). Influence of Mulch on Agronomic Characteristics, Soil Properties, Disease and Insect Pest Infestation on Dry Bean (*Phaseolus vulgaris*) in Swaziland. World J. Agric. Sci., 3(6): 696-703.
Singh AK, Singh S, Appa Rao VV, Meshram DT (2008). Effect of mulching on soil properties, growth and yield of 'NA7' aonla (Emblica officinalis) in semi-arid ecosystem. Indian J. Agric. Sci., 78(2): 193-197.
Shoji S, Nanzyo M, Dahlgren R (1993). Volcanic ash soils: Genesis, properties and utilization. Elsevier, Amsterdam.Dev. Soil Sci. 21.
Subrahmaniyan K, Kalaiselvan P, Balasubramanian TN, Zhou W (2008). Soil Properties and Yield of Groundnut associated with Herbicides, Plant Geometry, and Plastic Mulch. Comm. Soil Sci. Plant Anal., 39 (7/8): 1206-1234.
Subrahmaniyan K, Kalaiselvan P, Balasubramanian T, Weijun Z (2006). Crop productivity and soil properties as affected by polyethylene film mulch and land configurations in groundnut (*Arachis hypogaea* L.) Arch. Agron. Soil Sci., 52(1): 79-103.
Takashi F, Hajime A (2000). Soil Properties and Tomato Growth and Yield in No-till and Hairy Vetch Mulch Field with Sandy and Volcanic Ash Soils. Bull. Fac. Agric., Niigata Univ., 52(2):157-168.
van der Meulen ES, Nol L, Cammeraat LH (2006). Effects of Irrigation and Plastic Mulch on Soil Properties on Semiarid Abandoned Fields. Soil Sci. Soc. Am. J., 70(3): 930-939.
Wall A, Heiskanen J (2003). Effect of air-filled porosity and organic matter concentration of soil on growth of *Picea abies* seedlings after transplanting. Scand. J. For. Res., 18(4): 344-350
Warrick AW, Myers DE, Nielsen DR (1986). Geostatistical Methods Applied to Soil Science. SSSA, Agronomy Monograph No. 9.

4

Seasonal availability and consumption of wild edible plants in semiarid Ethiopia: Implications to food security and climate change adaptation

Debela Hunde Feyssa[1]*, Jesse .T. Njoka[2], Zemede Asfaw[3] and M. M. Nyangito[2]

[1]Department of Land Resources and Agricultural Technology, College of Agriculture and Veterinary Sciences, University of Nairobi, Kenya, Home Institute, Jimma University, Ethiopia.
[2]Department of Land Resources and Agricultural Technology, College of Agriculture and Veterinary Sciences, University of Nairobi, Kenya.
[3]Department of Biology, College of Natural Sciences, Addis Ababa University, Ethiopia.

Quantitative ethnoecological analysis of seasonal availability and implication to food security of wild edible plants (WEPs) was conducted in Boosat and Fantalle districts of semiarid east Shewa, Ethiopia from October, 2009 to September, 2010. Semistructured interview, focus group discussions, key informants discussions, seasonal record of fruits abundance were used to collected data on gathering and consumption of WEPs to cope with food shortage and adapt to climate change. Collected data was summarised into frequency tables, graph and qualitatively described under each subtopic. Thirty seven WEPs were identified for use as human food, and livestock feed and other multipurpose uses. About 24.3 % of WEPs were locally marketed, 75.7% were not marketed. All wild fruits were not included in official production system in the study area. It has indicated the underutilized existing potential of WEPs. Wild edible plants were preferred by local people of the study area not only for their food value, but also for their availability during dry seasons and shortage of food, potential for dryland agrobiodiversity and multipurpose to human wellbeing, livestock and environmental services they provide. Pairwise ranking by key informants was in agreement with direct matrices ranking for multiple uses of WEPs. The pairwise ranking, market survey and participant observations, community preference has confirmed the real potential of top seven priority WEPs species for dryland agrobiodiversity and agroforestry. Hence, these WEPs can be potential for dryland agrobiodiversity and agroforestry, to enhance people's livelihoods in semiarid areas. This result can shed light on further research and promotion work on WEPs utilization and management.

Key words: Wild, edible plants, seasonal, availability, food security, use, management.

INTRODUCTION

Wild edible plants (WEPs) are available as food source for humans and animals. Wild edible plants availability varies by seasons (Asfaw, 2006; Asfaw, 2009). Indigenous people are endowed with knowledge of seasonal availability of WEPs and implications to climate change adaptations. The available information on the seasonal availability and their contribution to climate change adaptation was poorly documented in Ethiopia.

Ethnobotany of Ethiopia's WEPs is also patchy that consists of listing of plant names without details on their use and management (Balaemie and Kebebew, 2006; Teketay and Eshete, 2004). Increased acculturation, mobility and displacement of indigenous communities, diminishing biodiversity, due to population pressure and climate change also led to a decline in use of WEPs and conveyance of associated knowledge (Asfaw, 1997; Fantahun et al., 2009). The crucial role these plants play in transhumants and settled farmers communities of east Shewa, Ethiopia, paucity of knowledge among scientists and policy makers about traditional knowledge were

*Corresponding author. E-mail: feyssahunde@yahoo.com.

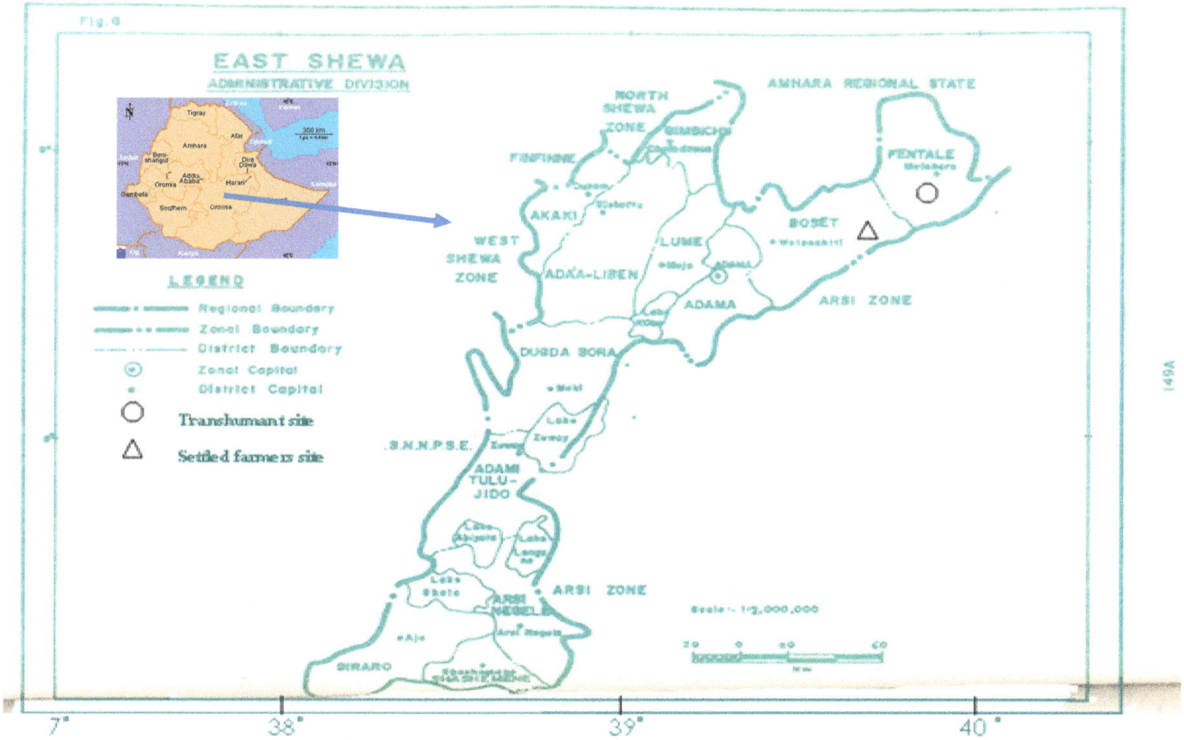

Figure 1. Map of the study area showing Fantalle and Boosat districts of east Shewa, adapted from BPED (2000).

potential risks, the plants were disappearing before their role is recognized.

Given climate change, level of poverty and environmental degradation there is high risk of biodiversity loss at large scale. Under such circumstances, knowledge and skills of uses and nutritious climatically adapted WEPs will be irreversibly lost. Given the impeding climate change, WEPs are expected to provide options for screening species that are likely to be best adapted to climate change and means of sustenance. Thus, it needed research attention to add new knowledge to science and hence, increase awareness on use and management WEPs by transhumants and settled farmers in semiarid areas of east Shewa, Ethiopia. The specific objectives were to: 1) identify seasonal availability of WEPs and preferences in semiarid east Shewa, 2) analyse their contribution to food security of transhumants and settled farmers and 3) determine prioritized list of WEPs for inclusion in dryland agrobiodiversity and agroforestry

MATERIALS AND METHODS

Study area

The study was conducted in the semiarid part of east Shewa in Fantalle and Boosat (whereas- administrative level equivalent to district in Ethiopia) located between 7°12' to 9°14'N latitudes and 38°57' to 39°32'E longitudes in the northern part of Great East African Rifty Valley in Ethiopia (Figure 1) from October, 2009 and September, 2010. The area lies in the Somalia-Maasi Center of plant endemism (White, 1983) described as the Acacia-Commiphora woodland vegetation type by Demissew and Friis (2009). The climate of the area is harsh with erratic and variable rainfall. The highest mean annual rainfall of semiarid east Shewa was 171.05 mm and lowest is 23.04 mm. The highest mean monthly rainfall 243.11 (July and August) and the lowest mean monthly rainfall was, 5.78 mm November. The main rainy season is from June to September known as GANNA (KIREMT) in Oromo and Amaharic languages and low rainfall from February to May known as AFRASA (BELG). The highest mean maximum temperature is 36.73°C in June and the lowest (31.24°C) in December. Economic activities of the area can be described as the agropastoral type with Boosat district marked with mixed agriculture of livestock and crop production (BPED, 2000; BDFED, 2009). Fantalle wereda is livestock production with transhumant and rudimentary crop agriculture practiced in favourable years. The vegetation of the area includes many WEPs. However, the vegetation is declining due to the effects of anthropogenic factors and climate change. This has affected both natural resources and the livelihoods of people in the study area.

Methodology

Six study sites were selected based on topographic variations and watershed alignment. Accordingly, Diglau Tiyo, Tri Bireti and Xadacha form Boosat, Galcha, Qobo and Dheebiti from Fantalle districts (Cook and Stubbendieck, 1986). Informants were selected randomly by the use of village registrar based on difference in lifestyles of people. Key informants were selected by the help of development workers and community leaders (Martin, 1995; Cotton,

Table 1. Seasonal availability of WEPs in Boosat and Fantalle districts.

Scientific name	Local name	Famliy	Flowering and fruiting months											
			Ja	F	Ma	A	M	Ju	J	Ap	S	O	N	D
Acacia senegal (L.) Willd.	Saphansa adi	Fabaceae	+	+	+	+	+	+	0	0	+	+	+	+
Acacaia seyal Del.**	Wacu	Fabaceae	+	+	+	+	+	0	0	o	+	+	+	+
Acacia prasinata Schweinf**	Dodoti	Fabaceae	+	+	+	+	+	0	0	0	0	0	+	+
*Acacia tortolis(*Forssk.*)***	Xadacha	Fabaceae	+	+	+	+	+	0	0	0	+	+	+	+
B. aegyptiaca(L.)Del.	Baddanno	Balanatiaceae	+	+	+	+	+	0	0	0	+	+	+	+
Bercemia discolor(klotzsch)Hemsl.	Jajjaba	Rhamnaceae	0	0	0	0	+	+	+	+	+	+	+	+
Carissa spinarum L.	Agamsa	Apocynaceae	+	+	+	+	+	0	0	+	+	+	+	+
Celitis toka (Forssk.)Hepper and Wood**	Matoqoma	Ulmaceae	+	+	+	0	0	0	0	0	0	+	+	+
Combretum molle R.Br. ex G. Don*	Birecha	Combretaceae	+	+	+	0	0	+	+	+	+	0	0	0
*Commmiphora africana(*Arich.)Engl.	Hamessa	Buraseraceae	0	0	0	0	0	0	0	0	+	+	+	+
Commmiphora schimperi(Berg.)Engl.**	Challnaka	Buraseraceae	+	+	+	+	+	+	+	+	+	+	+	+
Cordia africana Lam.**	Wdecha	Boranginaceae	+	+	+	+	+	+	0	0	0	+	+	+
Cordia monica Roxb./*Cordia ovalis* R.ex DC.	Mendhero	Boraginaceae	0	+	+	+	0	0	+	+	+	+	+	0
Cordia sinensiss Lam.**	Leedii	Boraginaceae	0	+	+	+	0	0	0	0	0	+	+	+
*Dobera glabra(*Forssk.)poir.**	Ade	Salvadoraceae	0	0	0	0	+	+	+	+	0	0	0	0
Eulea racemosa Murr.susp.schimperi(A.Dc.)White	Mi'essa	Ebenaceae	+	+	+	+	0	0	+	+	+	+	+	+
*Flueggea virosa(*Willd)Voigl.	Kecacule	Rhamnaceae	0	+	+	+	0	+	+	+	+	0	0	0
Ficus vasta Forssk.**	Qilxu	Moraceae	+	+	+	+	+	+	+	+	+	+	+	+
Ficus sycomorus L.	Oda	Moraceae	+	+	+	+	+	0	0	0	0	+	+	+
Ficus thonningi Blume.*	Dambi	Moraceae	+	+	+	+	+	0	0	0	0	+	+	+
G.flavescens A. Juss.	Amurjii	Tiliaceae	+	+	+	+	+	0	0	0	+	+	+	+
Grewia bicolor A.Juss.	Harorecha	Tiliaceae	0	+	+	+	0	0	+	+	+	+	+	+
Grewia schweinfurthii Burret	Mudhegurre	Tiliaceae	+	+	+	+	+	0	0	0	+	+	+	+
Grewia tenax (Forssk.)Vahl.	Eka fila	Tiliaceae	+	+	+	+	+	0	0	0	+	+	+	+
Grewia villosa Willd.	Ogomdi	Tiliaceae	+	+	+	+	+	0	0	0	+	+	+	+
Lantana camara L.	Midhaan dubara	Verbenaceae	+	+	+	+	+	+	+	+	+	+	+	+
Meriandra benegalensis Benth.	Hadha toke	Lamiaceae	0	+	+	+	0	0	+	+	+	+	+	+
Myrica salicifolia Hochst ex.A.Rich	Biiqa	Loganiaceae	0	+	+	+	0	0	0	0	+	+	+	0
Opuntia ficus- indica	Hadami	Cactaceae	+	+	+	+	0	0	0	0	+	+	+	+
Permna resinosa (schimperi)(Hochest)Schauer.	Urgecha	Lamiaceae	0	+	+	+	+	+	+	+	+	+	+	0
Rhus natalensis Krauss *	Dabobecha	Anacardiaceae		+	+	+	0	0	+	+	+	+	0	0
Rumex nervosus Vhal.	Dhangago	Polygonaceae	+	+	+	+	+	+	+	+	+	+	+	+
Salvadora persica	Riga Ilkani	Salvadoraceae Wall.	0	0	0	0	0	+	+	+	+	+	+	+
Sterculia africana(Lour.)Fiori	Qararu	Sterculiaceae	+	+	+	+	+	+	+	0	+	+	+	+
*Tamarindus indica*L	Rooqa	Fabaceae	+	+	+	+	0	0	0	0	+	+	+	+

Table 1. Contd.

X. americana L.	Hudha	Olacaceae	0	0	+	+	+	+	+	0	+	+	+	0
Z.spina-christi(L.)Desf.	Qurqura Mi'o	Rhamnaceae	+	+	+	+	+	0	0	0	+	+	+	+

**=in Fantalle district only; *=in Boosat district only ; J, F,Ma,Ap,M,Ju,J,A,S,O,N,D= Months from January to December respectively, Or= Oromo language (local name of all plants is given by this same language).

1996). Guided transect walks along six study sites was conducted to identify availability of WEPs by ethnoecological methods (Martin, 1995; Cotton, 1996). Seasonal inventory was conducted to record abundance of WEPs in the study districts. In the beginning list of WEPs in two districts were obtained by semistructured interview of 120 randomly selected informants and guided field walks with key informants. After obtaining informant consensus on the list of WEPs, once every month for 12 months, the researchers visited the study sites for a week to observe and record multipurpose WEPs identified with key informants field walk. Record was taken as present (+) when wild plant was observed with flowers and fruits in the specified month (0) for absence in the data sheet against each month to get data on seasonal availability of WEPs. Finally, the record was re-displayed to key informants from each district and consensus was obtained for the recorded data. Following *in situ* identification of WEPs, voucher specimens were collected to verify the identification at National Herbarium, Addis Ababa University, Ethiopia. Key informants pair wise ranking of the most available WEPs was done, following procedures of Martin (1995), Cotton (1996) and IIRR (1996).

Data analysis

Data from interviews and seasonal records of WEPs availability were quantitatively analysed by SPSS version 16 and summarized into frequency tables in percentages. Graph was plotted for seasonal distribution of WEPs. Chi-square tests were performed for the mean difference of IK between two districts, plant species related information were summarized in tables, percentages and ranked. Jaccard's coefficient of similarity was calculated for species use similarity between transhumant and settled farmers. JCS= c/c+b+a, where, a= number of species found only in habitat A (settled farmers), b= is number of species found only in habitat B (transhumant), c= number of species in habitat A and B. Finally, JCS was multiplied by 100, in order to obtain the percentage species composition similarity between the transhumant and settled farmers area (Ladio et al., 2006) comparison between cultures. Voucher specimens were collected, properly identified and deposited at National Herbarium, Addis Ababa University, Ethiopia.

RESULTS

Wild edible plants used as food

A total of 37 WEPs were identified for food and other multipurpose uses in semiarid, east Shewa, Ethiopia. These WEPs were adapted to specific seasons of the year (Table 1). According to informants, this was common before 3 decades. Informants explained that, in early times there was clearly defined seasonal availability of WEPs with fruiting and ripening time. They have indicated that currently, there is decreasing tendency of observing clearly defined seasonality of some WEPs, such as *Dobera glabra*. Key informants ascribed it to shortage of rain water; they called it jijirama qilensaa, which means climate change in Oromo language. However, they still know the fruiting and ripening time of some WEPs (Table 1).This has enabled them to use the plants for household food, livestock feed, income generation and medicine all over the year.

Balanities aegyptiaca, *Ziziphus spina-christi*, *Ximenia americana* and *Grewia* spp were identified as all season WEPs, producing fruits during dry and short rain seasons and abundantly growing in dry seasons. Respondents informed that bonas gannas lalisa means evergreen and productive in winter and summer seasons in Oromo language. *Grewia tenax* produce fruits during April to June known as badheesa in Oromo language and is named famine plant called beela baalesa in Oromo language. *Grewia* spp gives fruit in July to September main rainy season called hagaya in Oromo language. When there is shortage of food from farm managed crops *Grewia schweinfurthii* produce fruits and edible by animals and from children to adults. They are wet season WEPs. The WEPs species identified by present study, demonstrated a large body of knowledge passed from one generation to another, reflecting the presence of a rich lore in semiarid east Shewa. Local people had good knowledge of wild fruit species, including their time of fruiting and ripening. There is considerable overlap in ripening among the different species, both within and between localities, resulting in year-round fruit availability.

Species use composition in transhumant and settled farmers land use systems

Twenty-five (67.57%) of WEPs were commonly identified by both transhumants and settled farmers. The species composition similarity coefficient was 0.7. Eight (8.11%) and 9 (24.32%) of WEPs were used only by transhumants and settled farmers respectively (Table 2).

Table 2. Species use similarity between transhumant and settled farmers for WEPs and Jaccard's Coefficient of Similarity (JCS).

Use categories of plants	Total no. of spp. identified	Total no. of WEPs reported by:			JCS	Percentage similarity in species used by A and B
		Transhumants A	Settled farmers B	Both A and B		
WEPs	37	34	28	25	0.7	70

Seasonal availability of wild edible plants in the study area

Two of 70 and 75 year old key informant from Qobo study site explained that, *D. glabra* germinate and mature when there is rain only. They also indicated that, this plant did not give good fruit since the time known as the year of nabi known as bara nabyi in Oromo language which was 48 years ago when there was sever hunger and people has survived partly eating its fruits and other plants fruits. The year was also known as amata ade in Afaan Oromo language the year of *D. glabra* and it was famine year bara beela in Oromo language. *X. americana* ripen twice (September to November and March to July). Key informants explained that, it needs small rain to flower and fruit.

Key informants explained that, the time/ season and frequency of harvesting vary from plant to plant depending on the availability of edible plants and parts. It varied from place to place due to ecological and climatic conditions. For example *X.* and best collected for consumption within 1 to 2 months time. On the other hand, some weedy vegetables such as *Amaranthus caudatus* and chemerda (local name) were available only on seasonal basis. They were available during main rain and short rainy seasons July and September, or short rain February to April respectively. But harvesting of WEPs depends on availability of food in local people's stock. Sometimes these species were also available in irrigated fields even during dry season. Hence, production of WEPs need to know their seasonal availability. Cross tabulation analysis of interview administered to 120 informants using SPSS Version16 indicated that, WEPs were available throughout all seasons of the year for people to use. *Z. spina*-christi, *B. aegyptiaca*, and *Grewia* spp. were most available in all seasons (Table 3).

The most commonly used and preferred WEPs by the community for their dry season availability were *Z. spina*-christ *B. discolor*, *G. flavescnse*, *B.* aegyptiaca, *X. Americana, Grewia* spp (Table 4).This indicated that people use these WEPs in all the seasons. It also implies their drought resistance characteristics and their contribution to community food. Hence, there exists biological security to livelihood of people in semiarid east Shewa. Field inspections in 4 seasons of the year and focus group discussions with key informants were conducted to verify the interview data. Discussants unanimously (100%) agreed with the list of plants and seasons (Table 5) and identified additional WEPs. The WEPs were verified by researchers field inspection. *X. americana* and *Grewia* spp were listed less. From informants response and field inspection this was ascribed to the rarity of *X. americana* in the vegetation of the study area. *Grewia* spp. was good in abundance in the area. *Grewia villossa*, *G. tenax* and other WEPs were used but it become difficult to harvest the fruits due to browsing effect specially camels and goats

Seasonal availability of wild edible plants from researcher's 12 month field record

Seasonal record of the availability of WEPs indicated that their availability varied in months and seasons. They were most abundant from February to April in short rainy season afrasa (belg) and dry season boana (bega) in Oromo and Amharic languages respectively (November to January) and less abundant in main rainy season, ganna (kiremt) (June to August). Number of species recorded ranged from 30 to 31; 24 to 31 and 13 to 16 species respectively for the seasons (Figure 2). Why main rainy season were less abundant may pose a challenge to researchers and readers. This may be because, plants accumulate energy, water for the dry season growth and development and become ready for flowering and fruiting after main and in short rainy seasons.

Periods of consumption of WEPs by different age groups

An interview response of 120 transhumants and settled farmers has shown that people consume WEPs at normal time to supplement their diets and to cope with times of food shortage. However, there is significant variation between the time of consumptions (P= 0.000). In settled farmers area of Boosat districts consumption at normal time is higher than transhumant of Fantalle. In*americana* produce edible parts between twice transhumant pastoralists consumption of WEPs in both normal time and times of food shortage were relatively higher than settled farmers (Table 6). Chi-square (χ^2) values for 3 categories of consumption

Table 3. Responses of informants for seasonal availability of WEPs (N=120).

Districts of respondents		Seasons of availability/abundance				Total
		Autumn	Spring	Summer	Winter	
Boosat	Frequency	20	14	18	8	60
	% of total	16.7	11.7	15.0	6.7	50.0
Fantalle	Frequency	7	29	23	1	60
	% of total	5.8	24.2	19.2	0.8	50.0
Total	Frequency	27	43	41	9	120
	% of total	22.5	35.8	34.2	7.5	100.0

Autumn = September, October, November: Spring = March, April, May; Summer = June, July, August; Winter = December, January, February.

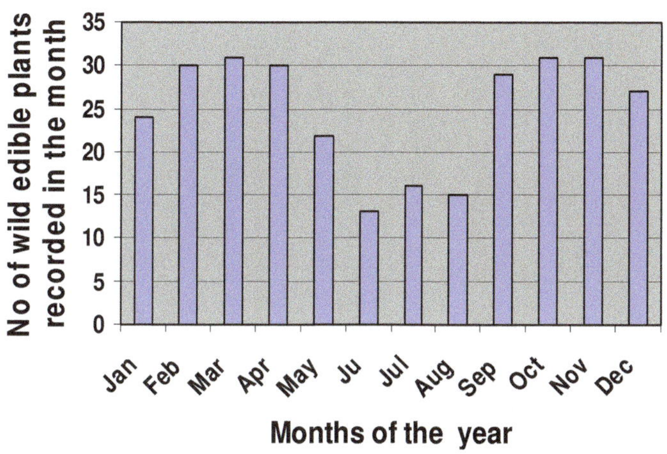

Figure 2. Seasonal availability of WEPs.

periods has significant variation (P<0.05). This indicated that WEPs consumption become part of the normal diet and time of food shortages. The year round availability and an increased consumption habit signifies the contribution of WEPs to households' food security.

Marketability and price of wild edible plants and socioeconomic implications

Marketability

Three rounds market survey, focus group discussions and participatory observation in local markets at Metahara, Addis Ketema in Fantalle district; Welenchiti and Bole in Boosat district in dry and wet seasons survey to assess marketability of WEPs fruits has revealed that majority of the WEPs (75.7%) in east Shewa were not marketed. It has been found that 24.3% were widely marketed from the wild harvest. This demands the integration of these products in dryland agrobiodiversity or agroforestry to properly utilize the potential. One of the reasons for less marketability of WEPs mentioned by key informants was little production available from unmanaged wild pool due to livestock pressure and agricultural expansion and increased selective consumption of WEPs by urban people. These are indications for a need for awareness creation on use and management of WEPs.

Market chain and price of wild edible fruits and seeds

Two routs of market chains of WEPs in east Shewa were identified. Transhumants and settled farmers collect wild fruits from the wild. The two chains identified were: 1) transhumants and settled farmers sell fruits directly to retailers in the formal markets for cheap; 2) retailers-town people get profit by selling to consumers in towns in better price. From market assessment it was observed that people consume fruits of *B. discolor*, *Z. spina*-christi, *X. americana* and others (Table 7). Fruits of these plants were observed in the market and in actual collecting practice by local people and by the researcher during this study.

Focus group discussions made with transhumant and settled farmers have revealed that, the abundance of wild edible fruits is decreasing but the price is increasing than earlier years. From market survey and interaction made with business residents in the study area at local markets showed that, the price of fruits of WEPs such as *Z. spina*-christi, *X. americana* has increased from time to time, that is from 0.10 cents 15 years, to 0.25 cents ten years back. 2 years before data collection and during the research, the average price of WEP fruits rose to 0.50 Ethiopian cents per 10 g.

Gender dimension and division of labour of use and management of WEPs

According to key informants' response, children and

Table 4. Largely available WEPs in the different seasons (N=120) by districts.

District of the respondent		List largely available WEPs in the above seasons							Total
		B. aegyptiaca	*Z.spina*-christi	*X.americana*	*B. discolor*	*Permena recinosa*	*G. flavescnse*	*Grewia* spp	
Boosat	Frequency	8	14	1	4	3	23	7	60
	% of total	6.7	11.7	0.8	3.3	2.5	19.2	5.8	50.0
Fantalle	Frequency	4	30	5	6	4	9	2	60
	% of total	3.3	25.0	4.2	5.0	3.3	7.5	1.7	50.0
Total	Frequency	12	44	6	10	7	32	9	120
	% of total	10.0	36.7	5.0	8.3	5.8	26.7	7.5	100.0

*= *G.villossa* and G. tenax, χ =19.264, df=6, P=0.004 at 5% probability.

women were believed to carry the responsibilities of collecting, preparing and serving the family food in the past. The present study has revealed that, all household members participate in collecting, preparing for home consumption, managing and selling WEPs to maximize household income and food. This was confirmed by 50 and 34.17% informants' from Boosat and Fantalle respectively (Table 8). However, there were still prevailing attitudes in the community in favour of the still prevailing more responsibility on women and children. All WEPs reported were edible by inhabitants in the area. In normal circumstances they were mostly eaten by children. *G. tenax* and *G. villosa* are mostly eaten by children even today. But all respondents know their edibility. Gender category in eating WEPs was not common. But *Lantana camara* was reported to be eaten by girls and named as midhaan dubara, which means females crop in Oromo language. The edibility of *L. camara* was relatively less in pastoral land uses system. Some people also showed their negative perception for its edibility. Hence, study on chemical composition including toxicity of the edible fruit of *L. camara* for future use and management is left for research. Currently, there was an observed trend by households to maximize the management and utilization of WEPs both directly as human and livestock food, income generation and environmental services.

DISCUSSION

Potential contribution to food security

Seasonal food shortages, when household stocks were empty and the new crop were still in the field were common times to dowel on collecting, selling and consuming WEPs in the study area. Addis (2005) and Fantahun et al. (2009) indicated that, it last from July to September in the Northern part of Ethiopia. Many wild fruits that ripen at this time (Figure 2), are therefore important back ups of household food. Generally, the year-round availability of fruits of different species within and across the study areas provides supplementary food and nutrition and presents an opportunity for trade if properly supported by extension services. Balmeie and Kebebew (2006) gave an account of seasonality in general but did not deal with species seasonal abundance. These could be basic motives for local people to conserve wild fruit species and encourage their domestication.

Implication of seasonal availability of WEPs to human well being and environmental integrity

Thirty six WEPs identified were used as food by semi arid people of east Shewa. They increase diversity of income, food, and healthcare system. Hence, increase biological insurance for food and local healthcare system being year round available. They also satisfy the need of local material, spiritual and cultural needs. They can increase resilience of ecological niches contribution to environmental integrity. Contribute to the continued existence of dryland agroforestry. These in turn maintain multiple uses of WEPs. It also sustains existence of biocultural heritage which include indigenous knowledge, indigenous people's confidence on natural resources as their own property. In general priority, WEPs have

Table 5. Seasonal availability of wild edible plants.

Species abundantly available		Seasons of availability of WEPs				Total
		Autumn	Spring	Summer	Winter	
B. aegyptiaca	Frequency	2	2	5	3	12
	% of total	1.7	1.7	4.2	2.5	10.0
Z. spina-christi	Frequency	12	18	13	1	44
	% of total	10.0	15.0	10.8	0.8	36.7
X. americana	Frequency	0	3	2	1	6
	% of total	0.0	2.5	1.7	0.8	5.0
B. discolor	Frequency	2	6	2	0	10
	% of total	1.7	5.0	1.7	0.0	8.3
Permina recinosa	Frequency	2	2	3	0	7
	% of total	1.7	1.7	2.5	0.0	5.8
G. flavescense	Frequency	5	11	13	3	32
	% of total	4.2	9.2	10.8	2.5	26.7
Grewia spp.	Frequency	4	1	3	1	9
	% of total	3.3	0.8	2.5	0.8	7.5
Total		27(22.5%)	43(35.8%)	41(34.2%)	9(7.5%)	120(100.0%)

X^2=20.253, df =18, P=0.319.

potential for value addition and industrial processing for. better use (FAO, 1999; Gamado et al., 2005). They can also be used for rehabilitation of degraded lands. They can serve as shade, mulch to the soil thereby enhancing soil microbial activities and biodiversity. In turn this will enhance productivity of the area.

Fruiting calendar and innate resilience of WEPs to climate change adaptation

All WEPs identified by the present research were available in dry seasons, either for human and livestock feed. This was proven by field exploration and key informants explanations. It has indicated the innate resilience of wild species to water stress environments, which often missed in exotic species, makes them important for transhumant and settled framers. Key informants explained that, seasonal variation of rain has affected the fruiting time and fruit harvest. But since most of the plants were drought resistant, they were found producing some fruits, though not like during sufficient rain. If some management inputs including the provision of little water, during dry seasons can be done, it can bean advantage over some regular crop which need relatively higher amount of water and management inputs to be productive. Hence, integrating both productions can enhance production and food security.

People are also living in fragile environment (rangelands). As population increases, resources degradation increases and decrease in carbon fixation to vegetation and increase in surface CO_2, with decrease in land cover; climate is also changing from time to time. Therefore, since people cannot continue migrating to other places because of shortage of land. Instead of inhabiting new areas, they cope with climate variability and change by using vegetation of the area for multiple uses. One of this can be consumption of WEPs which were climatically adapted. Not only by using them as they are found, but also enhancing their productivity availability by putting appropriate managerial practices, such as dryland agroforestry, home garden, live fences. Promoting riverine vegetation could be an available opportunity with irrigation.

Effect of consumption patterns and processing by local people

Focus group discussions and interaction made with households has shown that, WEPs can contribute to people's livelihood and food security, but cannot significantly replace conventional agricultural products.

Table 6. Times of consumption of WEPs by different age groups.

Times of consumption	Adult				Children				Adult and children			
	Bos	Fen	Tot	%	Bos	Fen	Tot	%	Bos	Fen	Tot	%
Famine/Drought	3	3	6	5	6	8	14	11.7	4	2	6	5
Normal time to supplement	27	3	30	25	33	5	38	31.7	32	5	37	30.8
Normal and drought	30	54	84	70	21	47	68	56.7	24	53	77	64.2
Total	60	60	120	100	60	60	120	100	60	60	120	100
Mean+SEM				2.65±0.052				2.45±0.064				2.59±0.054
X				26.057				30.858				31.291
P				0.000				0.000				0.000
df				2				2				2

Local people were observed using fresh fruits. The current consumption pattern of WEPs enabled getting high nutritional value from the plants without altering the chemical nature of fruits. The consumption of WEPs for biological security of food sources was attested by Wube (2009). But unprocessed fruits also have impact on their palatability. It has been indicated that processing these foods will add value to their palatability and add charisma to the consumers. This was due to technological gaps to enable people process and add value to these wild products and generate adequate income, to support their livelihoods than the way they use now. Hence, extension service needs to integrate to the general development plan. The WEPs were managed by Boosat and Fantalle people due to multipurpose uses (food, forage to livestock, medicinal, wood and other environmental services). These uses should be promoted and supported by appropriate extension services to obtain sufficient economic and environmental benefits from WEPs. Promotion of sustainable utilization of WEPs could be an alternative to reduce food shortage in semiarid Africa (Jens et al., 2002; Asfaw and Taddesse, 2001). Hence, earlier studies support the present focused findings as the existence of high potential of WEPs to food security.

The danger of focusing on high yielding verities as a sole solution to food security

Ethiopia is an important center of diversity for many domesticated crops (Harlan, 1969). The country is also a reservoir to ancient farming systems, farmers' varieties of many crops and rich plant lore. Harlan (1969) estimated that, Ethiopia is a center of diversity to 38 crops, which were later reduced to 11 (Zohary, 1970; cited in Asfaw and Tadesse, 2001). However, the crop diversity is in constant threat of being irreversibly lost, due to replacement of the farmers varieties by improved cultivars. Under such precarious situation, the indigenous knowledge associated with the plants can be lost irreversibly. The situation is compounded by acculturation, modernization and lack of full understanding on the benefit of the indigenous wild fruits and associated indigenous knowledge to the present and future generations. Indigenous knowledge faces another threat of disregard, through rapidly shifting dietary habit of the rural community from traditional dishes that constitute indigenous food sources. These situations exacerbate food shortage and increase the rate of malnutrition. To avert the problem, there is a need to focus on the available and ecologically beneficial food sources including the underutilized WEPs. Narrowing down of food sources has a potential threat to sustainable crop production in view of the impeding climate change and emergence of new virulent pathogens. Diversifying food sources through the use of ecologically adapted crops including WEPs, contribute to the fight against food insecurity and malnutrition and help to adapt to climate change

Comparison of farm wild fruits and farm managed crops

Average market price of marketed WEPs was less than that of farm crops (Table 7). This does not indicate that farm crops were cheaper than WEPs. It can be the reverse, if inputs of production and time required managing the farm crops are

Table 7. Price of WEPs compared with equivalent farm crops in local markets.

No	WEPs in local market	Average prices in Ethiopian Birr per Kg				Remark
		Wild fruits	*Sorghum bicolor*	*Zea mayse*	*Eragrostis teff*	
1	*B.aegyptiaca*	2.00	5.30	3.20	9.00	Less marketed
2	*Z.spina*-christi	2.50	5.30	3.20	9.00	Most preferred fruit
3	*X.americana*	2.25	5.30	3.20	9.00	Fruits were rare
4	*B.discolor*	2.00	5.30	3.20	9.00	Fruits were rare
5	*G.flavescense*	1.75	5.30	3.20	9.00	Mostly for hh use
6	*G. tenax*	2.50	5.30	3.20	9.00	Fruits were rare
7	*G. vilossa*	2.50	5.30	.20	9.00	Fruits were rare
8	*Grewia schweinfurthii*	2.25	5.30	.20	9.00	Fruits were rare
9	*Opuntia ficus indica*	2.00	5.30	3.20	9.00	Popular

hh=household.

Table 8. Division of labour and gender dimension in collecting, managing and utilization of WEPs.

Management type	Responses	Husband		Total	Wife		Total	Boys		Total	Girls		Total
		District			District			District			Districts		
		Boos	Fen		Boos	Fen		Boos	Fen		Boos	Fen	
Collect	Yes	25	27	52	39	20	59	53	43	96	43	28	71
	No	35	33	68	21	40	61	7	16	23	17	32	49
Collect prepare for consumption and manage	Ye	8	22	30	41	21	62	32	30	62	39	19	58
	No	52	38	90	19	39	58	28	30	58	21	41	62
Preserve and store	Yes	6	12	18	30	26	56	12	15	27	6	15	21
	No	54	48	102	30	34	64	48	45	93	54	45	99
Selling	Yes	3	5	8	52	29	81	45	6	51	42	31	73
	No	57	55	112	8	31	39	15	54	69	18	29	47
Decision making on issues of WEPs	Yes	11	41	52	33	16	49	23	5	28	23	13	36
	No	49	19	68	27	44	71	37	55	92	37	47	84
Mange money	Yes	25	2	27	37	22	59	8	10	18	4	0	4
	No	35	58	93	23	38	61	52	50	102	56	60	116

calculated. The WEPs do not require input and are cheaper, accessible and affordable. In spite of an increase in the utilization of wild fruits, processing and value addition were not improved, being an outstanding area that needs emphasis in development plans. This also needs further policy and local level awareness creation, to really value WEPs to improve their uses and enhance their conservation for sustainable utilization. Otherwise when the price remain little, a large amount of harvest from the wild will be required to get more money and this can threaten the wild population and become a challenge to conservation.

CONCLUSIONS

The year round availability of wild edible fruits in the study area provided supplementary food and nutrition. This has also provided an opportunity for local trade to generate income to households. It could be strong motive for local people to conserve wild fruit species and encourage their domestication as dryland agrobiodiversity and agroforestry; live fence and in area closure of pasture (kalo) areas. This in turn improves their contribution to food security by direct consumption as food and enhancing livestock production.

RECOMMENDATIONS

The preferred WEPs at community level should be given development attention in light of dryland agrobiodiversity and agroforestry by complementing local knowledge and modern practices. If properly mobilized, use of traditional technologies and farming practices wild edible plants can enormously contribute to the ecological balance and economic resilience of people. It is therefore, necessary to use the documented WEPs, and properly use the indigenous knowledge surrounding before irreversibly lost. Policy enhancement is recommended to properly utilize and manage the existing potential of WEPs.

ACKNOWLEDGEMENTS

We like to thank RUFORUM-IDRC, RUFORUM-Rockefeller foundation and Association of African Universities for funding the research work. We also thank local people and district management of east Shewa, Ethiopia for sharing valuable information and facilitation of the field work. We thank University of Nairobi for unreserved administrative support and guidance. Finally, Jimma University, Ethiopia for availing field vehicle, ICT facilities, hosting principal researcher during field work.

REFERENCES

Addis G, Urga K, Dikasso D (2005). Ethnobotanical Study of Edible Wild Plants in Some Selected Districts of Ethiopia. J. Hum. Ecol., Springer Netherlands.http://www.springerlink.com, (accessed June 2 2009), 33(1): 83-118.

Asfaw Z (1997). Indigenous African Food Crops and Useful Plants: Survey of Indigenous Food Crops, Their Preparations and Home Gardens. The United Nations University Institute for Natural Resources in Africa. ICIPE Science Press, Nairobi, Kenya, p. 65.

Asfaw Z, Taddesse M (2001). Prospects for Sustianble Use and Management of Wild Food Plants in Ethiopia. Econ. Bot., 55(1): 47-62.

Asfaw Z (2006).Towards a dynamic indigenous knowledge practice: Optimization of the use and management of plant resources in drylands of Ethiopia. In: Drylands Ecosystems: Challenges and opportunities for Sustainable Natural Resources Management. Proceeding of the Regional Workshop held at Hotel Impala, ArushaTanzania, (Nikundiwe, A. M. and Kabigumila J. D. I.,eds). June 7-9, pp. 64-71.

Asfaw Z (2009). The Future of Wild Food Plants in Southern Ethiopia Ecosystem Conservation Coupled With Management of the Roles of Key Social Groups. In: International Symposium on Underutilized Plants for Food Security, Nutrition, Income and Sustainable Development International Society for Horticultural Science-ISHS, January 2009, pp. 86-87.

Balemie K, Kebebew F (2006).Ethnobotanical Study of Wild Edible Plants in Derashe and Kucha Districts, South Ethiopia. J. Ethnobot. Ethnomed., 2(53): 1-9.

BDFED (Boosat District Finance and Economic Development Office) (2009). Socioeconomic Profile of Boost District. East Shewa, Boosat, pp. 1-61.

BPED (Bureau of Planning and Economic Development) (2000). Physical Socioeconomic Profiles of 180 Districts of Oromiya Region. Finfinne, Ethiopia, pp. 1-300.

Cook CW, Stubbendieck J (1986). Range Research: Basic Problem and Techniques. Society for Range Mangement, Denver Colorado 80204, p. 317.

Cotton CM. (1996). Ethnobotany: Principles and applications. John Wiley and Sons Ltd., Chichester, New York.

Demessew S, Friis I (2009). Natural Vegetation of the Flora area. IN: Flora of Ethiopia and Eritrea. General Part and Index, In: Hedberg I, Friis I, Persson E (eds). National Herbarium, Biology Department, Science Faculty, Addis Ababa, University, Addis Ababa and Department of Systematic Botany, Uppsala University, Uppsala, Sweden, 1-7&8: 27-32.

FAO (Food Agricultural Organization) (1999).Use and potential of wild plants in farm household. FAO Information Division 00100 Rome, Italy. http://www.fao.org/documents/ (Accessed. 1994. State of the world's forests.

Fentahun MT, Hager H (2009). Exploiting locally available resources for food and nutritional security enhancement: Wild fruits diversity, potential and state of exploitation in the Amhara region of Ethiopia. J. Food, 1(2): 207-219.

Gemedo-Dalle T, Maass BL, Isselstein J (2005). Plant Biodiversity and Ethnobotany of Borena Pastoralists in Southern Oromia, Ethiopia. Econ. Bot., NewYork Bot. Garden, 59(1): 43-65.

Harlan JR (1969). Ethiopia: A centre of diversity. Econ. Bot., 23: 309-314.

Jens G, El-Siddig K, Eberta G (2002). The Potential of Under-utilized Fruit Trees in Central Sudan. Conference on International Agricultural Research for Development. Deutscher Tropentag 2002 Witzenhausen, October 9-11, 2002.

IIRR (1996). Recording and Using Indigenous Knowledge: A Manual. International Institute of Rural Reconstruction, Silang, Cavite, Philippines, p. 211.

Martin GJ (1995). Ethnobotany: A methods manual. London, UK: Chapman and Hall.

Teketay D, Eshete A (2004). Status of indigenous fruits in Ethiopia. In: Chikamai B, Eyog-Matig O, Mbogga M (eds.) Review and Appraisal on the Status of Indigenous Fruits in Eastern Africa: A Report Prepared for IPGRI-SAFORGEN in the Framework of AFRENA/FORENESSA, Kenya Forestry Research Institute, Nairobi, Kenya, pp. 2-35.

White F (1983). The Vegetation of Africa: A Descriptive Memoir to Accompany the UNESCO/AETFAT/UNSO Vegetation of Africa,

UNESCO, Botanic Institute, BIBL, UPPSALA, pp. 44-117.
Wube M (2009). Promoting Underutilized Food Plants Species for Food Security in Ethiopia: ISHS Acta Horticulture 806: International Symposium on Underutilized Plants for Food Security, Nutrition. http://www.actahort.org, 806_5, (Accessed 3 August, 2009).

Growth and yield responses of ginger (*Zingiber officinale*) to three sources of organic manures in a typical rainforest zone, Nigeria

Egbuchua, C. N. and Enujeke, E. C.

Department of Agronomy, Delta State University, Asaba Campus, Asaba, Delta State, Nigeria.

A field study was carried out in the 2011/2012 cropping seasons at the Teaching and Research, Farms of the Delta State University, Asaba Campus, Asaba, a typical rainforest zone in Nigeria. The aim was to evaluate the responses of ginger to three sources of organic manures namely; cow-dung manure, poultry manure and pig manure at the rate of 20 t/ha. The experiment was a randomized complete block design with three replicates. Growth and yield parameters of ginger were taken at the 4, 6, 8, 10, 12 and 14 weeks after planting (WAP), while fresh weight was taken at 16th week after planting (WAP). The growth parameters observed were plant height, number of leaves, leaf area and number of tillers. Data collected were subjected to analysis of variance (ANOVA) and significantly different means in the F-test were separated using the Least Significant Difference (LSD) at 5% level of probability. Results showed significant increase in plant height, number of leaves and leaf area at the 6th to 12th WAP compared to the control. Growth parameters were not significantly different ($P > 0.05$) at the 4th WAP. Plots treated with poultry manure produced the highest plant height of 12.67 cm, highest number of leaves of 14.87 and leaf area of 231.8. This was followed by pig manure with values of 12.12, 14.25 and 222.5 cm. The least mean values of 7.14, 12.58 and 210.5 cm were obtained from the control. The general results indicated that organic manures in the forms of cow dung, poultry and pig manures have great tendency to increase growth characters and yield of ginger in a rainforest zone, Nigeria.

Key words: Growth and yield responses, ginger, organic manures rainforest zone, Nigeria.

INTRODUCTION

Ginger is the rhizome of the plant *Zingiber officinale* that is well consumed as a delicacy, medicine and spices. It is a perennial reed-like plant with annual leafy stems that is about a meter (3 to 4 feet) tall. The plant is usually propagated vegetatively by planting rhizome pieces which produce clusters of white and pink flower buds that bloom into yellow flowers. The mature roots of ginger are fibrous and the juice from old ginger roots is extremely potent and often used as spices and a quintessential ingredient of Chinese, Korea, Japanese and many South Asian cuisines for flavouring dishes (Jakes, 2007). It is also used largely as recipes such as ginger bread, cookies, crackers, cakes, ginger-ale and ginger beer (Asumugha et al., 2006; Jakes, 2007). The medicinal values of these great ancient spices are widely recognized across the continents to contain a number of unique organic phytochemical ingredients that can take care of some human ailments. Recent studies on health related effects of ginger which have also stimulated farmers concern on the growth of the plant have shown the efficacy of the plant in some life challenging ailments such as entero toxin induced diarrhea, diabetic nephropathy, nausea, plasma antioxidant, vomiting, high cholesterol, high blood pressure and inflammation (Chen et al., 2007; Ernest and Pittler, 2008; Kim et al., 2008).

Like any other plant, ginger requires the right kind of nutrients to sustain its growth and maximum yield especially in the humid environment where rainfall is high and nutrient reserves are low due to leaching, and erosion effects. Plant nutrients usually supplied by the soil in most Sub-Saharan environment are often inadequate and sometimes in plant unavailable form hence, they need to be augmented with other sources that are cheap and environmentally friendly. The use of organic manures is one technology that have been exploited overtime and across ages because of its ability to restore soil fertility, supply major plant nutrients, such as N. P. K., Ca, Mg and also stabilizer soil pH (Sanchez and Miller, 1986). Increase in soil chemical properties which are quite essential in crop growth and yield have also been associated with organic manures (Adetunji, 1990). Organic manures however are without their limitations. These include inadequate availability, transportation an handling problems, slow nutrient release, high C:N ratio and sometimes heavy metal pollution (Ayeni et al., 2010). Now that emphasizes are gradually shifting to organic agriculture to maintain soil productivity and limit the use of synthetic fertilizers some of which have contributed to the changing climate, the objective of this study therefore was to evaluate the responses of three sources of organic manures to the growth and yield of ginger in a typical rain forest zone where the crop is aptly grown.

MATERIALS AND METHODS

Description of the study area

The study was conducted at the Teaching and Research Farms of the Faculty of Agriculture, Delta State University, Asaba during the 2011/12 cropping seasons. The experimental area is located within latitude 06° 14'N and longitude 06° 49'E of the equator, and lies significantly in the tropical rainforest zone with over 1,650 mm of rainfall per annum. The rainfall is bimodal in character with peaks in July and September. The mean temperature is about 37.3°C; and a relative humidity of about 73.2%. The land area is relatively flat and influenced seriously by seasonal flooding due to 'its proximity to River Niger that naturally demarcated Delta and Anambra States by the Eastern axis. By nature of its geomorphological settings, the study area overlies the ancient metamorphic crystalline basement complex formation which are considerably more acid than base. They are generally gneisses and pegmatites that gave rise to coarse-textured soils that are deficient in dark ferro-magnesian minerals (Egbuchua, 2012). Land use in the study area is virtually based on rain-fed agriculture and the natural vegetation is typically of rainforest but have been reduced to derived savanna due to repeated clearing and cultivation over the years.

Field work

An experimental plot measuring 13 × 6 m was selected for the study. The plot was manually cleared, ploughed and made into 16 beds of 2 × 2 m dimensions. From each bed, composite soil samples were collected for pre-planting soil analysis. The treatments which consisted of cow dung manure, poultry manure and pig manure were applied two weeks before planting using incorporation method. The rate of application across board was 20 t/ha^{-1} Seed pieces of ginger rhizomes were cut to contain at least two buds and weighing about 10 g. Each was planted on the beds at a spacing of 20 × 20 cm at a depth of 5 cm. The planted rhizomes were carefully covered with sand and stubble mulch materials. Watering and weeding were done at intervals. Three weeks after planting, KARATE 2.5 EC was sprayed at the rate of 500 ml/ha to control insect pests. The experimental layout was a randomized complete block design (RCBD) with 3 replicates.

Pre-planting soil analysis

The pre-planting soil analysis was carried out in the Faculty of Agriculture Standard Laboratory Anwai Campus. The most important physico-chemical properties of the soil samples evaluated were: The particle size distribution using hydrometer method as described by Gee and Bauder (1986). The pH was determined in a 1:1 soil/water suspension using digital pH metre. Organic carbon was determined by Walkley-Black dichromate wet-oxidation method (Nelson and Sommers, 1982). Total nitrogen was by micro-kjeldahl distillation technique as described by Bremmer and Mulvaney (1982). Available phosphorus was determined by Bray No 1 method as described by IITA (1982). Cation Exchange Capacity (CEC) was determined by I N NH_4AC method (Rhoades, 1982). The chemical contents of the organic manures used in the study were also routinely analysed using appropriate methods.

Data collection

The growth parameters were determined at 4, 6, 8, 10 12 and 14 weeks after planting (WAP). The parameters essentially determined were; plant height which was measured from the base of the plant to the apex using measuring tape. Number of leaves per 10 randomly selected beds was measured by counting manually. Leaf area was determined by taking the leaf area per plant, and substituting the number of leaves of each plant by the length and width of the middle leaf. Number of tillers was taken from four plants tagged at the middle of each bed by counting at 14 weeks after planting (WAP). Fresh weight of harvested ginger was done at 16 weeks after planting (WAP) using sensitive balance.

Data analysis

Data collected were subjected to analysis of variance (ANOVA). The significantly different means in the F-test were separated using the Least Significant difference at 5% level of probability.

RESULTS

Initial pre-planting soil analysis

The results of the initial soil analysis is shown in Table 1. The textural class of the soil showed that it was

Table 1. Pre planting soil analysis of the study area.

Soil properties	Values
Particle size distribution (%)	-
Sand	83.70
Silt	12.00
Clay	4.30
pH (H_20)	5.5
Organic carbon (gkg^{-1})	15.6
Total nitrogen (gkg^{-1})	1.16
Available phosphorus ($mgkg^{-1}$)	4.35
Cation exchange capacity ($cmolkg^{-1}$)	6.75
Textural class	Loamy sand

Table 2. Chemical properties of the organic manures used for the study.

Characteristics	Organic sources/values obtained		
	CDM	PM	PgM
N (%)	1.06	2.57	1.67
P (%)	0.52	3.08	2.36
K (%)	0.97	2.47	0.75
Ca (%)	1.07	12.68	3.83
Mg (%)	0.88	0.93	0.54
Fe ($mgkg^{-1}$)	572	1756	1691
Mn ($mgkg^{-1}$)	344	573	505
Zn ($mgkg^{-1}$)	123	722	623
Cu ($mgkg^{-1}$)	22	82	510

CDM = Cowdung manure; PM = poultry manure; PgM = pig manure.

sandyloam in texture. The soil pH was strongly acidic with a pH of 5.5. The organic carbon (15.6 gkg^{-1}), total nitrogen (1.16 gkg^{-1}); available phosphorus (4.35 $mgkg^{-1}$) and cation exchange capacity (6.75 $cmolkg^{-1}$) were all seemingly low depicting the low fertility status of the study area.

Chemical content of the organic manures

The analytical values of the chemical content of the organic sources used as treatments are shown in Table 2. The results showed that the three organic sources were high in major macro and micronutrients that can support growth and increase yield of crops.

Morphological growth characters

Plant height

There was a significant effect in the shoot height of ginger as affected by different sources of organic manures. Plant height was found to be progressively increased with the treatment compared to the control (Table 3). The increase in plant heights were most obvious from the 8th week after plant where a significant effect was observed. Plant height got to its highest peak of 12.67 cm with poultry manure treatment. The least value of 7.14 was obtained with the control.

Number of leaves

The effects of treatments on the number of leaves (Table 4) showed that treatment also produced significant highest number of leaves than the control at 14 weeks after planting (WAP). The highest number leaves (14.87) were obtained with the application of poultry manures. This was followed by pig manure (14.25) and cow dung manure (13.22). The control experiment recorded the least value of 12.58.

Leaf area

Leaf area of plants treated with organic manures was not

Table 3. Effects of three sources of organic manures on plant height of ginger (cm).

Treatment (20 t/ha^{-1})	Weeks after planting (WAP)					
	4	6	8	10	12	14
Control (O)	4.55	5.12	6.03	6.32	7.05	7.14
Cow dung manure	4.58	6.65	8.20	9.74	10.23	11.75
Poultry manure	4.59	6.93	8.75	10.78	11.13	12.67
Pig manure	4.59	6.73	8.55	10.17	10.87	12.12
F - LSD (0.05)	NS	NS	0.75	0.91	1.08	0.58

Table 4. Effects of three sources of organic manures on number of leaves of ginger.

Treatment (20 t/ha^{-1})	Weeks after planting (WAP)					
	4	6	8	10	12	14
Control (O)	3.0	4.41	6.02	8.21	10.34	12.58
Cow dung manure	3.0	4.45	6.04	9.33	11.23	13.22
Poultry manure	3.0	4.48	7.13	10.13	13.76	14.87
Pig manure	3.0	4.47	6.82	9.85	12.38	14.25
F - LSD (0.05)	NS	NS	6.02	0.27	0.01	0.05

Table 5. Effects of three sources of organic manures on leaf area of ginger.

Treatment (20 t/ha^{-1})	Weeks after planting (WAP)					
	4	6	8	10	12	14
Control (O)	153.2	155.2	197.5	202.7	210.3	210.5
Cow dung manure	153.4	155.3	210.3	215.8	218.4	218.7
Poultry manure	153.7	155.7	215.8	222.7	231.7	231.8
Pig manure	153.5	155.5	214.4	219.5	222.4	222.5
F - LSD (0.05)	NS	NS	32.75	29.48	32.72	32.74

statistically different (P < 0.05) at the 4^{h} and 6 WAP (Table 5). However, progressive increase in total leaf area was observed as from the 8 week up to the 14th week after planting. The highest value of 231.8 was obtained with the application of poultry manure, followed by pig manure (222.5) and the least of value 210.5 was in the control experiment.

Number of tillers

Irrespective of treatment applied, the number of tillers were not significantly different (P < 0.05) at 14^{t} WAP (Table 6). The best tillers of 2.8 was obtained with the application of poultry manure while the least (1.3) was in the control experiment.

Fresh weight

The fresh weight of harvested ginger (Table 7) at 16 WAP produced significantly highest yield of 114.7 kg/ha with the application of poultry manure. The control treatment had the least yield of 10.7 kg/ha.

DISCUSSION

The efficacy of organic manures in promoting crop growth and yield components have been variously studied and reported. Hsieh and Hsieh (1990) has reported high nutrient contents of cowdung, pig and poultry manures which are capable of improving soil quality and increase yield of cultivated crops. Adetunji (2004) has reported reasonable high content of nitrogen, potassium and organic carbon in cow dung manure; high content of copper micro nutrient and lower content of fibrous material in pig manure, and very high content of N.P.K, Ca and micro nutrients in poultry-based manure. The high contents of these macro and micro nutrients have the capacity to improve morphological characters and yield of cultivated crops. Organic

Table 6. Effects of three sources of organic manures on the number of tillers at 14 WAP.

Treatment (20 t/ha^{-1})	Weeks after planting (WAP)
Control (O)	1.3
Cow dung manure	2.3
Poultry manure	2.8
Pig manure	2.5
F - LSD (0.05)	0.06

Table 7. Effects of three sources of organic manures on fresh weight of ginger at 16 WAP.

Treatment (20 t/ha^{-1})	Weeks after planting (WAP)
Control (O)	10.7
Cow dung manure	11.2
Poultry manure	14.7
Pig manure	12.3
F - LSD (0.05)	0.6

manures also have strong tendency to neutralize soil acidity, raise soil buffering capacity and provide micro nutrients such as Zn, B, Cu and Fe that can influence crop production positively.

In this study, most ginger morphological characters such as plant heights, number of leaves, leaf area were significantly ($P \leq 0.05$) increased with the application of the treatments compared to the control experiment where no treatment was applied. The delay in observing significant effects especially at the 4th week after planting (WAP) could be attributed to low rate of initial decay which is controlled by the C: N ratio and lignin contents of the organic sources. The improvement of ginger growth over the control experiment could be vividly explained by the various nutrient contents of the organic manures. For instance, the chemical quality of poultry manure used as treatment showed high contents of organic carbon (176%), total nitrogen (19.2%), phosphorus (28.7%), potassium (2.46%) calcium (21.0%) and magnesium 3.52% coupled with adequate levels of micro nutrients. Hence, poultry manure gave the best performance in terms of growth parameters and yield indices. Similar studies by Hsieh and Hsieh (1990) and Ojeniyi (2011) showed the potency of poultry manure in improving crop quality, quantity and yield when incorporated into cultivated soil. Ayeni et al. (2010) have equally reported that organic manures when properly used have proven to be very efficient in increasing soil nutrient contents, ensuring positive residual effects and enhancing soil's physico-chemical properties.

Although, the quality of any organic manure is very difficult to quantify due to differences in the quality of the sources. The results of the study showed that poultry manure was the most impressive on the growth and yield parameters evaluated. This was followed by pig manure and cow dung manure. All the organic manure sources proved better than the control experiment in all the parameters evaluated. This could be attributed to the various nutrient contents of the organic manures.

Conclusion

The cultivation of ginger in recent years especially in the rainforest zone of Nigeria is at the increase. This is because of high demand of the rhizomes not only as dietary spices but in tackling some ailments of great worries to humanity. These include high blood pressure, high cholesterol level, insomania and various nausea conditions. The general belief in this part of the World is that ginger has the potency of either reducing or eliminating these ailments and as such its consumption is on the increase. Results of the study showed the efficiency of organic manures in ginger production in all aspects of growth and yield parameters evaluated. Although, the use of organic manures is associated with such problems as slow in nutrient release, high C:N ratio and pollution problem. However, its uses will somehow minimize total reliance on mineral fertilizers which are not only too costly for poor resource farmers to acquire, but are associated with problems relating to soil acidity, nutrient inbalance, inadequate supply of macro and micro nutrients and ineffectiveness due to the blanket method of application.

REFERENCES

Adetunji MT (1990) Soil test and fertilizer recommendations report. A comprehensive soil fertility study for a s45 ha farm at Balogun village, Abeokuta.

Adetunji MT (2004). Integrated soil nutrient management options for Nigerian Agriculture. Proceedings of the 29ffi Annual Conference of the Soil Science Society of Nigeria. 6-10th December, 2004. pp. 27-34.

Asumugha GN, Ayaegbunam HN, Ezulike TO, Nwosu KL (2006). Guide to ginger production and marketing in Nigeria. National Root Crops Research Institute, Umudike Nigeria. Ext. Guide 7:1-7.

Ayeni LS, Omole T, Adeleyej O, Ojeniyi SO (2010) Integrated application of poultry manure and NPK fertilizer on performance of tomato in derived savanna transition zone of South-West Nigeria. Sci. Nat. 8(2):50-54.

Bremmer JM, Mulvaney CS (1982). Total nitrogen, In: Page A. L; Miller,. R. H. and Keeney, D. R(ed): Methods of soil analysis Part 2. Agron. 9. Madison, WI, USA. pp. 149-157.

Chen JC, Huang L, Shiblu W, Hsiang C (2007). Ginger and its bioactive component in inhibiting entero toxigenic escherichia, Coli-heat-labile. J. Agric. Food Chtm. 55(21):8390-8397.

Egbuchua CN (2012). Pedological characterization of some alluvial complex soils in Delta State, Nigeria. Nig. J. Soil Environ. Res. 9:65-70.

Ernest E, Pittler MH (2008). Efficacy of Ginger for nausea and vomiting: A systematic review of randomized clinical trial. Br. J. Anesth. 84(3):367-371.

Gee GW. Bauder JW (1986). Particle size analysis. In: Klute, A. (ed.). Methods of soil analysis Part 2. Chemical and microbiological properties. 2"d edition Agron. Monograph 9. Madison, W.I. USA.

Hsieh SC, Hsieh CF (1990). The use of organic manures in crop production. FFTC Ext. Bull. 360:20.

IITA-Intemational Institute of Tropical Agriculture (1982). Automated and semi-automated of andplant analysis. Manual series No. 7. IITA, Ibadan, Nigeria.

Jakes JS (2007). Ginger as "Beverage of Champion". J. Plant Nutr. 38(6):45-56.

Kim JS, Park S, Hyewon Y, Jaetteon S (2008). Cytotoxic components from the dried rhizomes of *Zihgiber officinale* Roscoe. Archiv. Pharmaceut. Res. 31(4):415-418.

Nelson DW, Sommers LE (1982). Organic carbon In: Page A. L., Miller, and Keeney, D. R (eds.) Methods of soil analysis Part 2. Agron. 9 Madison, W. L. 538-580.

Ojeniyi SO (2011). Advances in integrated nutrient management for crop production in Nigeria. Proceedings of the 35`i' Annual Conference of the Soil Science Society of Nigeria 7th -11th March, 2011, Minna, Nigeria.

Rhoades JD (1982). Cation exchange capacity. In: Page A. L., Miller, R. H. and Keeney, D.R. (ed), Methods of soil analysis. Part 2. Agron. 9 Madison. W. I. pp. 149-157.

Sanchez PA, Miller RH (1986). Organic matter and soil fertility management in acid soils of the tropical soils. XVIII Congress of Int. Soil Sci. Soc. P. 10.

Assessment and sustainable management of non-timber forest products used as food and medicine among urban dwellers in Oyo State, Nigeria

Famuyide O. O., Adebayo O., Bolaji-Olutunji K. A. , Awe F., Owoeye A. Y., Awodele D. O. and Adeyemo, A.

Forestry Research Institute of Nigeria (FRIN), P. M. B.5054, Jericho hills, Ibadan, Oyo State, Nigeria.

This study assessed the Non Timber Forest Products (NTFPs) used as food and medicine by the urban dwellers of Oyo State, Nigeria. It identified the NTFPs, sources and perception about the effectiveness of utilization as medicine. Data were sourced on the basis of senatorial districts. The three senatorial districts are Oyo South, Oyo Central and Oyo North with a population size of 1,764,217, 1,906,814 and 1,909,863 respectively. Both random and proportionate sampling techniques were adopted with 217 questionnaires processed. Descriptive analysis such as frequency, percentage and histograms with Chi square were used to describe the socioeconomic variations and to assess the various utilizations of NTFPs. The Chi-square analysis showed that occupation (p = 0.001), educational status (p=0.002), income (p=0.013) and age (p=0.011) were statistically significant in determining how the respondents feel about the effectiveness of the use of NTFPs as medicine, but variables like gender (p=0.071) and religion (p=0.121) were not significant. Efforts should therefore be made by government and relevant research institutes to train the people on the domestication of these NTFPs to achieve sustainability, pharmaceutical involvement to make more refine and more awareness about NTFPs. Some of the reasons given by the respondents for using the NTFPs include cheapness, availability and accessibility and some of the sources where they obtain the NTFPs were given as market, hawkers, forest and friends.

Key words: Non-timber forest products, utilization, availability, accessibility, domestication.

INTRODUCTION

Over the years, forestry and its products have contributed immensely to the economic development of Nigeria; the importance of the products cannot be over-emphasized (Fonta et al., 2010). Forests products can be classified into two: Timber, which constitutes the bulk of forests based materials used for economic purposes, and the Non-Timber Forest Products (NTFPs). During the 1960s and 1970s, forest products earned large amounts of foreign exchange and the sector was ranked highest in employment generation. The forest sector earned annual foreign exchange of between 308 million to 412 million naira or about 4.2% of GDP (World Bank, 1988). The situation, however, turned around between 1970 and 1985, due to the discovery of oil. NTFPs have been studied by researchers from many different academic fields and each field used a slightly different definition of NTFPs. They are any product or service other than timber that is produced in a forest (CIFOR, 2004). They include fruits, nuts, vegetables, fish medicinal plants, resins, essences, a range of barks and fibers, bamboo, rattans, honey, insects, animals, fodder, fertilizers, medicinal extracts, construction materials, cosmetic and cultural products, natural dyes, tannin, gums, latex and other exudates, essential oils, spices, edible oils, decorative articles, horns, tusks, bones, pelts, plumes, hides and skins, non-wood ligno-cellulosic products, phytochemicals and aroma chemicals. NTFPs are indispensable part of the livelihood strategy of communities living in and near forests and constitute an important source of livelihood for millions of people across the globe.

The role and contribution of NTFPs have been crucial in subsistence as livelihood support, in rural economics and biodiversity conservation since times immemorial due to their richness of variety. About 80% of the population of the developing world depends on NTFP for their primary health and nutritional needs (FAO, 1995). Osemeobo (1991) noted that rural women were found to be making between N115 and N500 in fruit gathering and sale. It is therefore paradoxical that in spite of their real and potential value, most NTFPs remain grouped as minor products of forests. However, in the recent time, there has been increasing recognition of the fact that this approach to forest management is neither conducive to sustainable management of the forests, particularly of the tropical moist forest nor is it in the best economic interest of the predominantly rural societies in the tropics. Due to the relative scarcity of most of the NTFPs now as a result of deforestation, as noted by Nwoboshi (1986) and the present awareness of their importance, more value is being added which has made them highly marketable. Research at a global scale has identified that rural households draw from a diversity of income sources, adopt a range of livelihood strategies in order to achieve and maintain a sustainable livelihood. These include the use of NTFPs both for household consumption and for sale. The contribution of forest to poor people's livelihoods goes largely unrecorded in national statistics. This is due mainly to the role of forest products in subsistence economies and the informal sector, from which reliable and quantifiable data are inherently difficult to obtain.

Various uses of NTFPs have been revealed and described in literature; prominent among them are the uses as food and medicine. However, the utilization varies from region to region. In other words, some used as medicine in a particular region may serve as food in other region and the species used for a particular purpose may differ with regions. It is the realization of these that this study assessed the utilization of NTFPs used as food and medicine in Oyo State. The empirical study was on the assessment of NTFPs among the urban dwellers, who are the so called elites. There is, however, great variation in the extent to which forest products are used from area to area and even between households within a community. Indeed, this variation reflects the extent to which NTFPs are an integral part of rural livelihoods.

Forest foods

NTFPs species are used as food in the form of wild fruits, vegetables, and nuts, edible roots, as bush meat, snails, edible insects and honey. Others are used as food additives in form of spices, flavorings, and food colorants and as fermentation agents, various animal foods such as folder for livestock, straw, baits to catch animals and bee plants. Similar reports on the use of NTFPs as food and food condiments have been made by Andel (2006). Jimoh and Haruna (2007); Tee and Amonum (2008). Other edible food materials found in the forest include insects, rodents, wild game and fish. These have been found to have superior nutritional quality, when compared with domesticated varieties. Besides, processed and stored forest food products help to insure a year round food supply (Jimoh and Adebisi, 2005). Furthermore Jimoh and Adebisi (2005) maintained that NTFPs include a vast number of edible and now edible products gathered from the forest. Some of the products are eaten raw while others are processed in various forms through boiling, drying and other methods.

Medicinal plants

Various NTFPs species have medicinal value for the treatment of various ailments including the treatment of stomach aches, cut/wounds, diarrhea, ulcer, infertility, Malaria, fever, blood purification and others. The roots, seeds, bark, resin, leaves are used for these purposes. Others are used for fishing and to control insects. Abere and Lameed (2008) reported that the African giant land snails (*Achatina achatina* and *Archachatina marginata*) are used to cure whooping cough, anaemia, ulcer, asthma, aphrodisiac and hypertension. Abere and Lameed (2008) further maintained that the fluid of the snails is used to treat headache, treatment of dysentery, eye problems, and small pox. The meat cures bone fracture, infertislity in women while the shell is used to prepare talismans for protection and used culturally to appease the gods as well as to ward off evil spirits. Snails have also been successfully used to curtail aggression, malformation of bone structure and promotion of easy child birth, nourishment of lactating women, suppression of convulsion, healing of amputated fingers and circumcision of male children (Abere and Lameed, 2008). Medicinal and Aromatic Plants (MAPs) include plants used to produce pharmaceuticals, dietary supplement products and natural health products, beauty aids, cosmetics, and personal care products, as well as some products marketed in the culinary/food sector. MAPs have been an important resource for human healthcare from prehistoric times to the present day. According to the World Health Organization, the majority of the world's human population, especially in developing countries, depends on traditional medicine based on MAPs (WHO, 2002). About 50,000 and 70,000 plant species are known to be used in traditional and modern medicinal systems throughout the world (Schippmann et al., 2006). The importance of NTFPs was raised for the past few decades as a result of many factors such as the dependence of rural communities of NTFPs, new market preferences for natural products, increasing concerns about forests and biodiversity conservation, and occurrence of many NTFPs among the biological richness and ecological complexities of natural forests (Grimes et al., 1994).

Sustainable management of NTFPs

Sustainability issues

The increasing demand for natural products in the sectors of food and medicinal ingredients poses major ecological and social challenges. High pressures on wild resources are threatening the survival of populations and species, while also endangering local ecosystems. Overharvesting of selected plants for commercialization, premature collection along with habitat destruction, open grazing, forest fire and soil erosion are major threats to the sustainability of NTFP conservation. Sustainable management of NTFPs is important because of their value as a perennial source of subsistence income and to conserve biodiversity. For the past few decades, these resources have been highly exploited for trade. Harvesting usually takes place before the plants mature. Moreover, bark is harvested by cutting plants of all size classes within the available area. These practices not only hamper the regeneration of the concerned species, but also pose threats to their long-term survival (Edwards, 1996; Olsen and Larsen, 2003). Most high-valued NTFPs are collected from the wild without paying attention to the quantity and quality of harvested material. Being the least benefitting groups, collectors often tend to harvest more than the harvestable quantity to get more money. Similarly, competition among collectors compels them to collect NTFPs prematurely, resulting in their gradual disappearance from the wild. Soil erosion and forest fires are issues that occur mostly by anthropogenic causes.

Specific objectives

1. To identify non timber forest products (NTFPs) in the study area,
2. To identify the sources of NTFPs to urban dwellers,
3. To determine the relationship between the socioeconomic variables of the respondents and their perception about the effectiveness of the utilization of NTFPs as medicine,
4. To identify the constraints faced by the respondents (urban dwellers) in getting the NTFPs.

METHODOLOGY

The research was carried out in Oyo State, which is an inland state in south-western Nigeria, with a population of about 5,580,894 (NPC, 2006) and its capital at Ibadan. It is bounded in the south by Ogun State and in the north by Kwara State, Nigeria. In the west it is bounded partly by Ogun State and partly by the Republic of Benin while in the east it is bounded by Osun State. The State is located on Latitude 8° and Longitude 4° east and covers a total of 27,249 km^2 of landmass. The vegetation pattern of the state is that of rainforest in the South and guinea savannah to the North. Thick forest in the South gives way to grassland interspersed with trees in the North.

Oyo state is divided into three senatorial districts: Oyo South, Oyo North and Oyo Central, Oyo North has 13 Local Governments viz: Saki West, Saki East, Atisbo, Irepo, Olorunsogo, Kajola, Iwajowa, Itesiwaju, Ogbomoso North, Ogbomoso South, Orire, Oorelope and Iseyin. Oyo Central comprises 11 Local Governments of Afijio, Akinyele, Egbeda, Ogo-Oluwa, Surulere, Lagelu, Oluyole, Ona-Ara, Oyo East, Oyo west and Atiba. While Oyo South consists of 9 Local Governments. They are Ibadan North, Ibadan North East, Ibadan North-west, Ibadan South-East, Ibadan South-West, Ibarapa Central, Ibarapa North, Ibarapa East and Ido.

Method of data collection

Data were collected on the basis of Senatorial District and the three senatorial districts are Oyo South, Oyo Central and Oyo North. Both random and proportionate sampling techniques were adopted. Random sampling technique was adopted in selecting the Local Government Areas from each Senatorial District. The proportionate sampling technique was used to determine the exact number of questionnaire that was administered to respective communities in the chosen Local Governments in relation to the population of the Senatorial District they belong. A total of two hundred and thirty (230) copies of questionnaire were administered at household level out of which two hundred and seventeen (217) copies were found analyzable.

Method of data analysis

Descriptive statistics such as frequency, histogram and percentage were used to describe the socioeconomic characteristics of the respondents. Chi-.square was used to test the relationship between socioeconomic variables of respondents and their utilization of non timber forest products as medicine.

Chi Square $(X^2) = \sum (O-E/E)^2$

Where O = Observed Frequency of users of NTFPs who feel they are more effective than orthodox medicine.

Hypothesis

The null hypothesis (Ho) states that there is no significant relationship between the selected socioeconomic variables (occupation, educational status, income, age, gender and religion)

Table 1. Socioeconomic characteristics of respondents.

Variable	Frequency	Percentage
Sex		
Male	130	62.8
Female	77	37.2
Total	207	100.0
Age (years)		
≤ 30	25	12.1
31-40	63	30.4
41-50	71	34.3
Above 50	48	23.2
Total	207	100.0
Marital status		
Single	28	13.5
Married	174	84.1
Divorced/Widow/Separated	5	2.4
Total	207	100.0
Religion		
Christianity	109	52.7
Islam	98	47.3
Total	207	100.0
Occupation		
Civil Servant	37	17.9
Trading	93	44.9
Artisanship	48	23.2
Farming	13	6.3
Others	16	7.7
Total	207	100.0

Source: Field Survey, 2012.

Table 2. Socioeconomic characteristic continued.

Variable	Frequency	Percentage
Educational background		
No formal	30	14.5
Primary	60	29.0
Secondary	68	32.9
Tertiary	49	23.6
Total	207	100.0
Household size		
1-5	54	26.1
6-10	85	41.1
Above 10	68	32.8
Total	207	100.0
Household monthly income		
<N10,000	26	12.6
N10,000-N30,000	128	61.8
N31,000-N50,000	34	16.4
>N50,000	19	9.2
Total	207	100.0

Source: Field Survey, 2012.

of the respondents and their perception about the effectiveness of the utilization of NTFPs as medicine while the alternative hypothesis (H_1) states that there is a significant relationship between the socioeconomic variables and respondents perception about the effectiveness of the utilization of the NTFPs as medicine and food.

RESULTS AND DISCUSSION

The socioeconomic characteristics of the respondents are summarized in Tables 1 and 2. From the table, it is shown that majority (62.8%) of the respondents or users of NTFPs found in the urban centers were males and 37.2% were females. This may not be unconnected to the fact that the male folk are mainly the household head and the major controller of household resources, as attested to by Edeh and Mbam (2010) in their research on constraints limiting efficient utilization of improved cassava technologies in Ebonyi State. It was discovered from the field that males uses NTFPs as medicine more as they move in their daily activities. For instance the hawkers of the traditional medicine are seen more on the street, in front of offices and Motor Park in the urban area. About sixty five percent (64.7%) of the respondents were between age range of 31 to 50 years; an indication that the respondents were within the active workforce, with majority (84.1%) married. The respondents that were Christians accounted for 52.7% while 47.3% were Muslims. This implies that both Christians and Muslims use Non timber Forest Products (NTFPs) both as food and medicine in the study area. Larger proportion (about 45%) of the respondents was traders 22% were artisan 37% were civil servant and remaining were farmers. From Table 2, about 62% of them either had primary or secondary school education the remaining had post secondary education. This is an indication that education does not really constitute a barrier in using NTFPs. The World Health Organisation (WHO, 2002) confirmed that 80% of the world's population presently uses herbal medicine and tend towards organic value for some aspects of primary care. It has become indispensable practice. Adekunle and Samwobo (2004) documented 103 plants species of different life forms traditionally used to treat different sickness and diseases. About 74% of the respondents had household size of six or more people, with monthly income of N30, 000 and below accounting for about 74% of the respondents. This implies that majority of the respondents are within the low income class.

Table 3 depicts the various species of NTFPs used as medicine by the respondents. Different reasons were given by the respondents for their use of these species

Table 3. NTFPs used as medicine.

Local name	Scientific name	Part used	Uses
Bamboo	*Bambusa vulgaris*	Leaves	Syphilis/gonorrhea
Honey	*Apis mellifera*	Fluid	wound dressing/cough
Bitter kola	*Garcinia cola*	Fruit	Healthy liver/gall bladder
Kola nut	*Cola nitida*	Nut	Asthma/stimulant
Walnut	*Juglans nigra*	Nut	Detoxification
Snail	*A. achatina*	Whole part/fluid	Ulcer/Hypertension
Alligator pepper	*Aframomum melegueta*	Leaves	Malaria fever
Shea butter	*Vitellaria paradoxa*	Fruit/seed	Cosmetics
Locust bean	*Parkia biglobosa*	Roots	Infertility/poison
Cashew	*Anacardium*	Leaves/bark	Diabetes/hypertension
Irvingia	*Irvingia gabonensis*	Leaf/fruit/bark	Diarrhoea
Lemon grass	*Cymbopogon Spp.*	Leaves	Perfume
Neem	*Azadirachta indica*	Leaf,/root/ bark	Malaria/gonorhoea
Pawpaw	*Carica papaya*	Fruit, leaves	Poison/nausea/typhoid
Ugwu	*Telfairea occidentalis*	Leaves	Anaemia
Cassia	*Cassia alata*	Leaves	Skin infections
African pear	*Dacryodes edulis*	Stem bark	Cough/body ache
Scent leaf	*Occimum gratissimum*	Leaves, tender stem	Diarrhoea/typhoid
Mango	*Mangifera indica*	Leaves/bark/fruit	Fever/hypertension
Garden egg	*Solanum melongena*	Fruit	Heart disease/glaucoma
Pineapple	*Ananas comosus*	Fruit	Healthy bone
Tea Leaf	*Camellia sinensis*	Leaves	Heart disease/cancer
Bitter leaf	*V. amygdalina*	Leaves/root	Diarrhoea
Guava	*Psidium guajava*	Leaves	Diarrhoea
Baobab	*Adansonia digitata*	Bark/Seed	Low sperm count
Jatropha	*Jatropha curcas*	Leaves	Stomach ache

Source: Field Survey, 2012.

as medicine. These include cheapness, addiction, availability, accessibility and others. This corroborates the works of Amusa et al. (2010) and Jimoh et al. (2012) in which factors accounting for the dependency of communities on traditional medicine were given as culture, efficacy, cost, availability, accessibility and poverty. However, when asked if they observed any side effects from the use of NTFPs as medicine, less than 10% of the respondents claimed they had observed one form of side effect or the other. Some of the side effects include excessive sleeping, dizziness, running stool, weakness of the body, vomiting, and stomach ache. This may not be unconnected to lack of specification on the required dosage or quantity of the medicinal plants to take as many of them claimed they took as much as they can while others claimed they took 2 to 3 cups (with no specific unit of measurement) twice or thrice a day. This also confirms the earlier work by Famuyide et al. (2011) in which the negative effects from the medicinal use of NTFPs were stated as vomiting, nausea, stomach ache and body itching.

Table 4 shows the species that are used as food by the respondents. Some of these species are eaten raw as snacks while others are cooked or transformed into edible form. Some of the species that are eaten raw as snacks include *Carica papaya* (pawpaw), *Psidium guajava (*Guava), *Anacardium occidentale* (Cashew), *Irvingia gabonensis* (Bush mango), *Musa accuminata (*Bannana), while those cooked or transformed into edible form include *Thryonomys swinderianus* (Grasscutter), *Agaricus bosporium (*Mushroom), *Juglans nigra* (Walnut), *Parkia biglobosa* (Locust bean). Several reasons were given by the respondents for using non timber forest products as food, as shown in Figure 1. Some (26.2%) of the respondents said that they eat NTFPs because of the taste and flavor while about 27% claimed they eat the NTFPs because of their nutritional value, as evident in Figure 1. This corroborates the study by Awe et al. (2009a) in which majority (76%) of the respondents claimed they consumed grasscutter meat (a non timber forest product) because of its taste and quality. Figure 2 shows the different sources where the respondents obtained their NTFPs. About 43% of them claimed they got theirs from the market, while those that got theirs directly from the forest accounted for about 33%. Different reasons were given by the respondents as

Table 4. NTFPs used as food.

Local name	Scientific name	Part used	Uses
Pawpaw	*Carica papaya*	Fruit	Snacks
Guava	*Psidium guajava*	Fruit	Snacks
Mango	*Mangifera indica*	Fruit	Snacks
Grasscutter	*Thryonomys swinderianus*	Whole part	Meat
Cashew	*Anacardium occidentale*	Fruit	Snacks
Mushroom	*Agaricus bosporium*	Strip/pileus	Condiment
Honey	*Apis mellifera*	Fluid	Sweetener
Walnut	*Juglans nigra*	Fruit	Snacks
Snail	*Archachatina marginata*	Whole part	Cuisine
Locust Bean	*Parkia biglobosa*	Fruit pump, seeds	Condiment
Bush mango	*Irvingia gabonensis*	Fruit	Snacks
Orange	*Citrus sinensis*	Fruit	Snacks
Banana	*Musa accuminata*	Fruit	Snacks
Ugwu	*Telfairea occidentalis*	Leaves	Cuisine
Garden Egg	*Solanum melongena*	Fruit	Snacks/condiment
Pineapple	*Ananas comosus*	Fruit	Snacks
Coconut	*Cocos nucifera*	Fruit/nut	Snacks
Baobab	*Adansonia digitata*	Fruit	Snacks
Bitter leaf	*Vernonia amygdalina*	Leaves	Cuisine
Sheabutter	*Vitellaria paradoxa*	Fruit/seed	Cooking oil

Source: Field Survey, 2012.

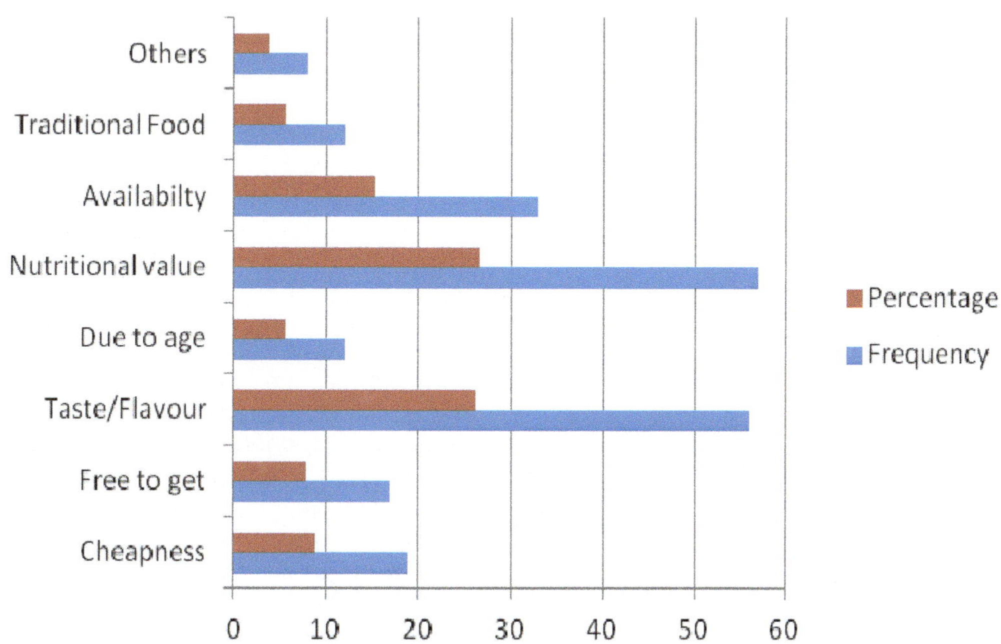

Figure 1. Reasons for using NTFPs as food.

problems being encountered in getting the NTFPs when they needed them. These include seasonality (63.1%), perishability (22.9%), transport cost (8.9%), with others such as proximity, affordability, e.tc, accounting for 5.1%.

Table 5 shows Chi-square analysis testing the relationship between socioeconomic variables of respondents and their perception about the effectiveness of the utilization of NTFPs as medicine. In other words, the test was to ascertain whether there was any significant relationship between the selected

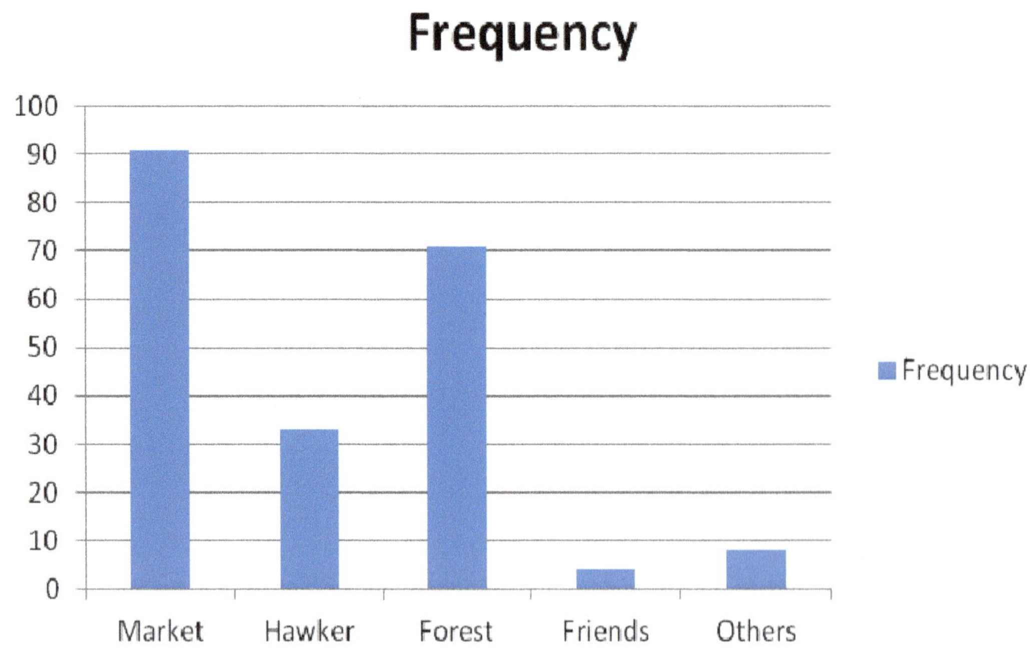

Figure 2. Sources of NTFPs in the area.

Table 5. Chi-square result of relationship between socioeconomic variables of respondents and their perception about the effectiveness of the utilization of NTFPS as medicine.

Variable	X^2	DF	P-value	Decision
Occupation	12.217	4	0.001	Significant
Educational status	16.014	3	0.002	Significant
Income	18.337	3	0.013	Significant
Age	19.194	3	0.011	Significant
Gender	3.102	1	0.071	Not significant
Religion	3.006	1	0.121	Not significant

Source: Chi-square analysis.

socioeconomic variables of the respondents and what they felt about the effectiveness of the medicinal use of NTFPs. It was found that occupation, educational status, income and age were statistically significant while religion and gender were not significant. This implies that religion and gender have nothing to do with what the respondents feel about the effectiveness of medicinal use of non timber forest products in the study area. This conforms with the work of Odebode et al. (2011) in which religion was found not to be significant in determining the perception of consumers about the consumption of grasscutter meat within Ibadan Metropolis.Hence the null hypothesis that there is no significant relationship between the socioeconomic variables of the respondents and their perception about the effectiveness of the utilization of NTFPs as medicine was rejected for occupation, educational status, income and age, while it was upheld for religion and gender.

CONCLUSION AND POLICY RECOMMENDATION

This study assessed the Non Timber Forest Products (NTFPs) used as food and medicine by urban dwellers of Oyo State, Nigeria. The study revealed that majority (62.8%) of the respondents was male while 37.2% were female. It was also discovered that the utilization of NTFPs has no religious barrier as both Christians (52.7%) and Muslims (47.3%) used them. Some reasons given by the respondents for using the NTFPs include cheapness, availability and accessibility and some of the sources where they obtain the NTFPs were given as market, hawkers, forest and friends. Some of the constraints encountered by the respondents in obtaining the NTFPs included seasonality (63.1%), perish ability (22.9%), transportation (8.9%), with others such as proximity, affordability, e.tc, accounting for 5.1%. In view of this, efforts should therefore be made by governments at all levels and relevant research institutes to train the people on the domestication of these NTFPs so as to achieve sustainability. Pharmaceutical companies should work on processing of NTFPs medicinal plants to make it more available in a refined form, less chemical and still natural. More awareness should be made about utilization of some NTFPs as food for their nutritional value and medicinal value especially when organic food is advocated for globally.

REFERENCES

Abere SA, Lameed GA (2008). The medicinal utilization of snails in some selected states in Nigeria. In: Onyekwelu JC Adekunle VAJ, Oke DO (eds.). Proceeding of the first National conference of the Forests and Forest Products Society (FFPs) held in Akure, Ondo State between 11 th and 18 th of April, 2008. pp. 233-237.

Adekunle MF, Sam-Wobo SO (2004). Ethno botany and Prioritization of Forest Plant used in Traditional Treatment of Guinea worm Infection in Ogun State Nigeria. J. For. 34(2):71-80.

Amusa TO, Jimoh SO, Aridanzi P, Haruna M (2010). Ethno botany and Conservation of Plant Resources of Kainji Lake National Park, Nigeria. Ethno Bot. Res. Appl. 8:181-194.

Andel TV (2006). NonTimber the of wild plants Agromica Publication and C.T.A, the Netherlands.

Awe F, Idumah FO, Imoagene E, Eniola TS (2009a). Socio-economic characteristics and Rural Households' Perception about the Consumption of Grasscutter (*Thryonomys swinderianus*) Meat in Selected LGAs of Kogi State. J. Agric. For. Soc. Sci. 7(1):124-133

CIFOR (2004). Forest Products Livelihood and Conservation, CIFOR Annual Report. P. 12.

Edeh HO, Mbam BN (2010). Constraints Limiting Efficient Utilization of Improved Cassava Technologies in Abakaliki Local Government Area of Ebonyi State: A Factor Analysis Approach. J. Agric. For. Soc. Sci. 8(1):36-44.

Edwards DM (1996). Non –Timber Forest Products from Nepal: Aspects of the trade in medicinal and aromatic plants. FORESC occasional paper 2/93. Ministry of Forest and Soil conservation. Forest Research and Survey center, Kathmandu, Nepal.

Famuyide OO, Adebayo O, Ajayi CA, Azeez FA (2011). Impact Assessment of Selected Non-Timber Forest Products in Ibadan Metropolis. Cont. J. Sustain. Dev. 2(1):27-33.

FAO (1995). Nonwood Forest Products for Rural Income and Sustainable Forestry. Food Agric. Organ. United Nations Rome 7:4-9.

Fonta MW, Ogujiuba KK, Amakom U (2010). 'Modeling and understanding the interrelationship between natural resource extraction, poverty and inequality: The case of forestry in sub Sahara Nigeria, forthcoming (November, 2010) in environmental modeling for Sustainable Regional Development: System Approaches and Advanced Methods, (eds.,) Vladimir Olej, Ilona Obršálová and Jiří Křupka, IGI Global Publishers, www.igi-global.com.

Grimes A, Loomis S, Jahnige P, Burnham M, Onthank K, Alarcon R, Cunenca WP, Martinez CC, Nell D, Balick M, Bennett B, Mendelsohn R (1994). Valuing the Rainforest: the Economic Value of Non-timber Forest Products in Equador. Ambio 23(7):405-410.

Jimoh SO, Amusa TO, Azeez IO (2012). Prevalence, Utilization and Conservation Strategies for Non-Timber Forest Products in South western Zone of Nigeria. Resour. Environ. 2(1):46-54.

Jimoh SO, Adebisi SR (2005). Nontimber Forest Products and Sustainable Forest Management in Nigeria. InPopoola, L., Mfon, P. and Oni, P. I. (eds). Proceedings of the 30th Annula conference of Forestry Association of Nigeria, held in Kaduna, Kaduna State, Nigeria. pp. 266-271.

JImoh SO, Haruna EA (2007). Contribution of Nontimber forest Products to Household Food Security and Income around Onigambari Forest Reserve, Oyo State, Nigeria. J. Environ. Exten. 6:28-33.

National Population Commission of Nigeria (NPC) (2006). Aworkshop report on the review of 2006 population census www.informationng.com/tag/national-population-commission.

Nwoboshi (1986). Meeting the Challenges of Deforestation in Nigeria through Effective Reforestation. In Oguntala AB (ed) 1986. Annual Conference of the Forestry Association of Nigeria, Minna. pp. 134-139.

Odebode AV, Awe F, Famuyide OO, Adebayo O, Ojo OB, Daniel G (2011). Households' Consumption Patterns of Grasscutter (*Thryonomys swinderianus*) meat within Ibadan Metropolis, Oyo State, Nig. Contin. J. Food Sci. Technol. 5(2):49-57.

Olsen CS, Larsen HO (2003). Alpine medicinal plant trade and Himalayan mountain livelihood strategies. Geograp. J. 169:243-254.

Osemeobo GJ (1991). Effects of Common Property Resources Utilization on wildlife conservation in Nigeria. Geojournal 23:241-148.

Schippmann U, Leaman D, Cunningham AB (2006). Cultivation and wild collection of medicinal and aromatic plants under sustainability aspects. In: Bogers RJ, Craker LE, Lange D (eds). Medicinal and Aromatic Plants. Springer, Dordrecht. Wageningen UR Frontis Series no. 17. (http://library.wur.nl/frontis/medicinal_aromatic_plants/05_schipmann.pdf).

World Bank (1988). World Development Report Washington D.C

World Health Organisation (WHO) (2002). WHO for promoting the use of traditional medicine. Geneva: p. 2 (fact sheet P. 134).

7

Assessment of plants grown and maintained in home gardens in Odeda area Southwestern Nigeria

Aworinde D. O.[1], Erinoso S. M.[2], Ogundairo B. O.[3] and Olanloye A. O.[3]

[1]Department of Biological Sciences, Ondo State University of Science and Technology, Okitipupa, Ondo State, Nigeria.
[2]Department of Botany, University of Ibadan, Ibadan, Oyo State, Nigeria.
[3]Department of Biological Sciences, Federal University of Agriculture, Abeokuta, Ogun State, Nigeria.

A survey of thirty (30) home gardens in Odeda Local Government Area (LGA) of Ogun State, Southwestern Nigeria was conducted to assess plants grown and maintained by household members and the diversity of vegetal species and their uses. Semi-structured questionnaire and structured interviews were used to collect useful information. A total of 120 plant species belonging to 50 botanical families were documented. From the data, Euphorbiaceae, Solanaceae, Rutaceae, Malvaceae, Caesalpiniaceae, Poaceae and Apocynaceae (in order of decreasing number of species) were the most frequent families. Taxa such as *Musa* species, *Vernonia amygdalina*, *Citrus* species, *Psidium guajava* and *Terminalia catapa* were found to be the common food/medicinal plants as evidenced by their densities in the study sites. The household members cited most of the plants as food; others as medicinal and ornamentals. Miscellaneous uses include cosmetics, ceremonial and scouring. Species composition, shapes/structures as well as plant uses are discussed. Home garden products serve alimentary purposes and represent promising base materials for poverty alleviation and may also help to augment "fresh" food nutrient intake. The wealth of the home gardens may be related to the rural origin and to the culture of the owners who still keep a cultivation tradition. Some plant species are frequent. This may suggest uniformity in plant use and could contribute to the conservation of local species. Home gardens can also be regarded as 'humble' germplasm bases; hence, the need to re-energize the productive capacity of home gardens.

Key words: Agro-forest systems, home garden, ethnobotany, diversity, Odeda, Southwestern Nigeria.

INTRODUCTION

Traditional agro-systems are constantly undergoing reforms. Of all the agro-systems, home garden is regarded to be the least understood and commands the least attention. Home gardens represent land-use systems involving deliberate management of multipurpose trees and shrubs in intimate association with annual and perennial agricultural crops and invariably livestock within the compounds of individual houses (Fernandes and Nair, 1986) and are the central points of biodiversity conservation. Home gardens form a dominant and promising land use system in many parts of the tropics (Michon et al., 1983; Singh, 1987) and are viewed as an alternative to the ever increasing demand for food (Soemarwoto et al., 1985) particularly in West Africa. Hoogerbrugge and Fresco (1993) defined home garden as "a small scale, supplementary food production system by and for household members that mimics the natural multilayered ecosystem". An addition to the definition to

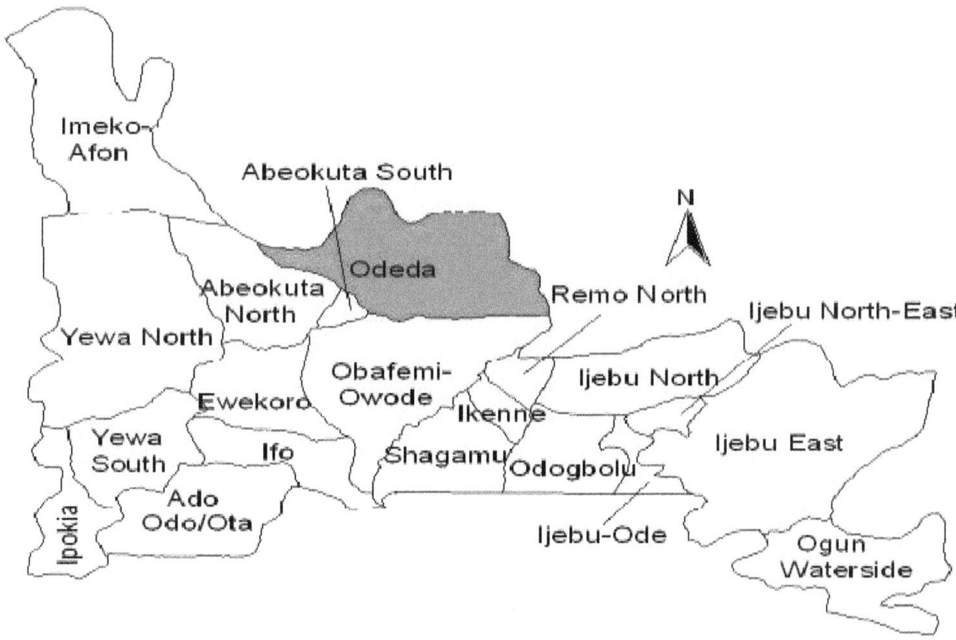

Figure 1. Map of Ogun State, Nigeria.

include the role of medicinal plants by Agelet et al. (2000) is worth considering.

Investigations on agro-forestry in tropical areas have focused on home gardens; reports from these have been described as prototype for sustainable ecological systems (Padoch and De Jong, 1991; Lamont et al., 1999; Albuquerque et al., 2005). Species diversity in tropical home gardens is reported to be very high due to species having different life forms, height and canopy structure (Babu et al., 1982; Soemarwoto and Conway, 1991). Research involving biological diversity in rural communities has intensified in recent years and its scope now extends to the semi-urban centres but mainly to locales that maintain richness of vegetation and animals, such as parks, arboretum and flower gardens. The species diversity sheltered by these locales provides resources for food, decoration, construction and medicinal uses as well as contributing to flood control and protection against soil erosion (Soemarwoto, 1987). With the development of the environmental sciences, scientists have now understood the significance of the relationship between humans and nature in the rural and/or urban environment.

Recently, just as in agro-forestry, ethnobotany (as a discipline) has dedicated increasing attention to the theme of conservation of home garden biodiversity. The diversity found in home gardens is a source of genetic variability that has been accumulated by the local populations, constituting a valuable patrimony for food security and even a source of genetic material for the improvement of species for commercial purposes (Soemarwoto et al., 1985). Structure and floristic composition vary in homegarden establishment. This study, part of a wider project on ethnobotany on Nigeria biome, is designed to contribute to existing knowledge about the structure, usage and floristic composition of homegardens in the Southwestern Nigeria and the relationship between homegarden sizes and species composition.

METHODS

Study areas

This study was carried out in Odeda Local Government Area (LGA) of Ogun State (Figure 1) located in the moderately hot, humid tropical climatic zone of Southwestern Nigeria. There are two distinct seasons: rainy season (March/April to October/November) and dry season (October/November to March/April). There are two main vegetation types in the study sites: tropical rainforest and guinea savanna. The LGA has a land area of 1547.29 km^2 with a population of 109449 (NBS, 2006). The headquarters of the LGA is at Odeda on the A5 highway 7°13′00″N 3°31′00″E. The climate favours the cultivation of a wide range of food crops such as rice, maize, cassava, yam, cocoyam, oil palm, vegetables and fruit trees. The economy of the areas studied is subsistence. However, surpluses are taken to markets or neighbouring towns.

Informed consent

The purpose of the study was explained to household members in the LGA. Informed consent was obtained from each of the respondents.

Ethnofloristic study

Field work was conducted between March, 2012 and June, 2012 in Odeda, Southwestern Nigeria. Locations selected for the survey

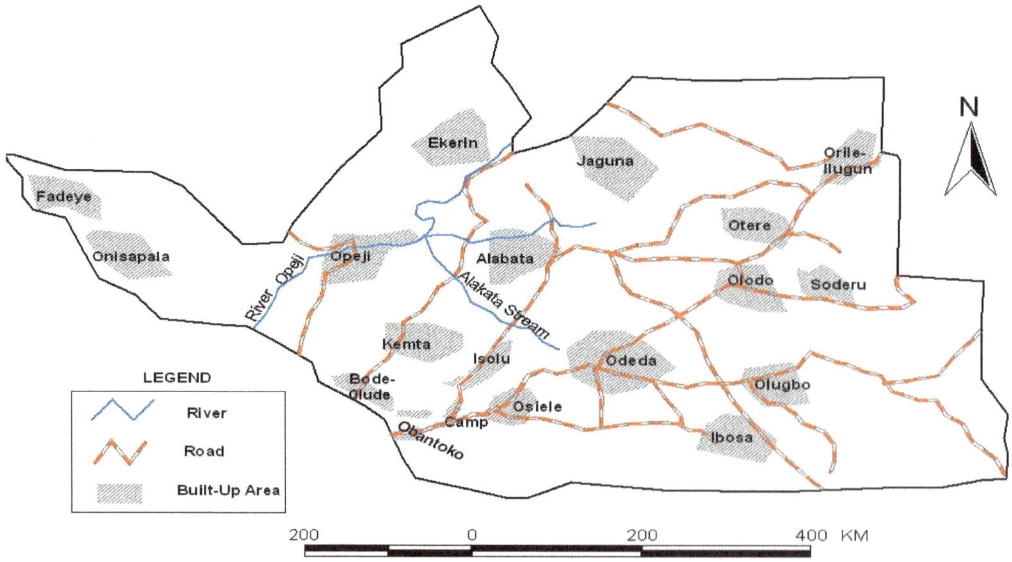

Figure 2. Geographical localization of the study sites. Source: Taiwo et al. (2011).

Table 1. Demographic structure of respondents.

Parameter	Specification	N (%)
Age	1 - 20	0 (0)
	21 - 40	8 (26.7)
	41 - 60	12 (40)
	>60	10 (33.3)
Gender	Male	12 (40)
	Female	18 (60)
Religion	Christianity	15 (50)
	Islam	8 (26.7)
	Traditional	7 (23.3)
Level of education	PSC	16 (53.4)
	JSSC	1 (3.3)
	SSC	7 (23.3)
	NCE	2 (6.7)
	FD	4 (13.3)
Nationality	Nigerian	30 (100)
	Non-Nigerian	0 (0)
Ethnic Identity	Yoruba	27 (90)
	Igbo and Nupe	3 (10)
Occupation	Teaching	7 (23.3)
	Transporting	7 (23.3)
	Farming	11 (36.7)
	Trading	9 (30)
	Technician	2 (6.7)

N = Number of respondents; % = Percentage of respondents.

were Obantoko, Osiele, Odeda, Olodo, Ageri, Adewusi, Kila and Orile-Ilugun (Figure 2). A total of 30 home gardens were surveyed and household members interviewed using semi-structured questionnaire to gather information on the uses of the species present in the gardens. Interviews were conducted in English Language (since all the respondents are literate). The questionnaire was in two sections. Section 1 dealt with basic demographic information. Section 2 consists of plant species present in the different home gardens, their local names, plant forms, parts used and uses. Plant specimens were collected and identified in the field. Plant specimens that could not be immediately identified in the field were taken to Forestry Herbarium, Ibadan (FHI) for identification. Voucher specimens of all species mentioned were deposited in the FHI herbarium.

Data analysis

Data obtained from the questionnaires were entered into the computer and analyzed using Epi6-info version 6.04 (CDC, Atlanta, GA, USA) (Dean et al., 1994).

RESULTS

Findings from the study indicated that majority of the respondents are females (Table 1). The level of education of household members interviewed ranged from Primary School Certificate (PSC), Junior Secondary School Certificate (JSSC), Senior School Certificate (SSC), Nigeria Certificate in Education (NCE) to First Degree (FD). The ethnic groups identified are the Yoruba and Igbo. The latter incorporates the Nupe (from Benue State). All the interviewed members are literate but a good percentage is self-employed. Home gardens studied proved to be a basic agro-forestry system mainly for food, medicinal and ornamental plants and occupied an area between 53 and 583 m^2 with a mean of 158.5 m^2

and standard deviation of 162.42. The shape, size and number of species varied for each home garden but were most commonly rectangular (63%), square (13.3%) (not perfect), triangular (3.3%). Irregular shapes (20%) were also identified. Structurally, vertical dimension took the largest percentage (53.3%), while horizontal (26.7%) and random settings (20%) were also encountered. All the species sighted were reported to be useful for several purposes. The respondents cited most as alimentary (40%). Others as medicinal (28%), ornamental (25%), shade (4.7%) and miscellaneous uses (2.3%). Some plants considered as food are also used in culinary, while some others serve both as fruit plants and shade such as *Terminalia catapa, Mangifera indica,* and *Cocos nucifera.* From the 30 home gardens surveyed, 120 different species of plants were found and these are distributed among 50 botanical families (Tables 2 and 3). The most represented families are Euphorbiaceae, Solanaceae, Rutaceae, Malvaceae, Caesalpiniaceae, Poaceae and Apocynaceae in order of decreasing number of species.

Two groups of plants were found in the studied sites. Group 1: plants grown and maintained. Group 2: naturally established plants but maintained. In the latter case, plants cited were *Ficus exasperata, Nicotiana tabaccum, Newbouldia laevis, Gmelina arborea, Azadirachta indica*, and *Glyricidia sepium*. Others include *Talinum triangulare* and *Phyllanthus amarus* (only used, not maintained). Some of the plants are grown because of their seasonal importance. Such plants include *Zea mays, Saccharum officinarium, Sorghum bicolor and Dioscorea* species. Variation in species composition was not large throughout the 30 home gardens studied. Taxa such as *Musa* spp., *Vernonia amygdalina*, *Citrus* spp., *Psidium guajava* and *T. catapa* were found to be the common food/medicinal plants across home gardens surveyed and had the highest density. All plant life forms were duly represented (Table 4) with herb taking the largest share (31.6%) and culm (0.8%) the lowest. Table 5 shows the plant parts used and their frequencies. The leaves form the major part of plant used followed by whole plant - as ornamentals. Ornamental species such as *Delonix regia, Dieffenbachia amoena, Acalypha ornata, A. ciliata*, and *Aloe vera* are often arranged in courtyards or at the frontage of the building.

DISCUSSION

Since the beginning of agriculture, people have cultivated many varieties of useful plants. Botany is a subject developed out of the study of agriculture and medicine (Dutta, 2008). The first medical uses of plants are lost in antiquity but evidence exists that plants were widely used for such purposes long before history (Hufford, 1978). Ethnobotany makes clear the connection between human cultural practices and plant utilization. In the study of Ethnobotany today, home gardens and their roles as biodiversity conservation centres cannot be over-emphasized.

Plant domestication represents a major achievement. The study of indigenous cultivation, food production, local medicinal knowledge and varied use of vegetal species has implications for food nutrient augmentation as well as discovering new medicines. The diverse uses of plants indicated in this study point to the fact that home gardens have been used to cultivate not only vegetables, but also medicinal and other miscellaneous plants. Cosmetics, scouring and ceremonial are classified in this work as miscellaneous or rare uses. All the alimentary species from the family Lamiaceae were used as condiments, indicating their important role in diversifying the flavour of food. The gardens studied are subsistence in nature [based on their mean size and according to the classification of Fernandes and Nair (1986). Home gardens have the function of guaranteeing subsistence as well as complementing the household income, for instance, taxa such as *Amaranthus* and *Celosia* species are cultivated on small commercial scale. In the same way, as the family income of the respondents did not change too much, the richness of species also did not change significantly.

The most representative families in this study have also been cited in previous studies that were focused on old urban home gardens by Eichemberg et al. (2009) and Albuquerque et al. (2005). The most widely used medicinal plants (*V. amygdalina, Jatropha* spp., *Cympobogon citratus*) have again been implicated in this study. The leaves form the most used plant part. This is in line with similar ethnobotanical studies by Adekunle (2008), Ayodele (2005) and Erinoso and Aworinde (2012).

The physical appearance of home gardens is species composition dependent. The horizontal and vertical arrangements of species form the two structural components of home gardens (Millat et al., 1996). A systematic distribution pattern was not recognized in this study as against the pattern reported by Caballero (1992) as some gardens have plant species haphazardly distributed; typical of random pattern. The lack of unified methodology in the study of home garden (Albuquerque et al., 2005) makes home garden structure to vary in different regions (Soemarwoto et al., 1985).

Women play a significant role in home garden maintenance, although the idea of establishment is solely that of men. This is evidenced from the number of females interviewed. They tend to maintain the aesthetic feature of their environment. Most of the ornamentals especially the herbaceous species are potted. Not all the medicinal plants sighted are grown. Some established naturally but were later maintained when the use value was discovered. These plant species may have been introduced by agents of dispersal such as birds, wind and explosive mechanism, especially those ones that are propagated by seeds. This study has once again proved

Table 2. List of plant species found in home gardens in Odeda Area, Southwestern Nigeria.

Species	Family	Habit	Part(s) used	Use(s)
Citrus sinensis	Rutaceae	Tree	Fruit	f
Lawsonia inermis	Lythraceae	Tree	Leaf	cs
Calotropis procera	Asclepiadaceae	Shrub	Milky latex	fp
Jatropha gossipifolia	Euphorbiaceae	Shrub	Leaf	m
Jatropha curcas	Euphorbiaceae	Shrub	Leaf	m
Acalypha ornata	Euphorbiaceae	Herb	Whole plant	or
Acalypha ciliata	Euphorbiaceae	Herb	Whole plant	or
Aloe vera	Asphodelaceae	Herb	Leaf	or, m
Aloe barbadensis	Asphodelaceae	Herb	Leaf	or. m
Psidium guajava	Myrtaceae	Tree	Fruit/Whole plant	f, sh
Ocimum gratissimum	Lamiaceae	Shrub	Leaf	c, m
Musa paradisiaca	Musaceae	Herb	Fruit	f
Musa sapientum	Musaceae	Herb	Fruit	f
Corchorus olitorius	Tiliaceae	Herb	Leaf	f
Vernonia amygdalina	Asteraceae	Shrub	Leaf	m, c
Amaranthus hybridus	Amaranthaceae	Herb	Leaf	f
Celosia argentea	Amaranthaceae	Herb	Leaf	f
Capsicum chinensis	Solanaceae	Shrub	Fruit	c
Capsicum fruitescens	Solanaceae	Shrub	Fruit	c
Ananas comosus	Bromeliaceae	Herb	Fruit	f, m
Carica papaya	Caricaceae	Tree	Fruit/Leaf	f, m
Abelmoschus esculenta	Malvaceae	Herb	Fruit	f, c
Nicotiana tabacum	Solanaceae	Herb	Leaf	sn, m
Cajanus cajan	Papilionaceae	Tree	Leaf	m
Zea mays	Poaceae	Grass	Seed	f
Dieffenbachia amoena	Araceae	Herb (Cane)	Whole plant	or
Theobroma cacao	Sterculiaceae	Tree	Seed	b
Citrus medica	Rutaceae	Tree	Fruit	m
Morinda lucida	Rubiaceae	Tree	Leaf	m
Lycopersicon esculentum	Solanaceae	Shrub	Fruit	c. f
Cocos nucifera	Arecaceae	Tree	Fruit (Endosperm)	f. m
Cymbopogon citratus	Poaceae	Grass	Leaf	m
Ficus exasperata	Moraceae	Tree	Leaf	m, sc
Mangifera indica	Anacardiaceae	Tree	Fruit	f
Bougainvillea speciosa	Nyctaginaceae	Climber	Whole plant	or
Tabebuia rosea	Bignoniaceae	Tree	Whole plant	or
Bryophillum pinnatum	Crassulaceae	Herb	Leaf/Whole plant	m, or
Duranta repens	Verbenaceae	Shrub	Whole plant	or
Croton zambesicus	Euphorbiaceae	Tree	Whole plant/Leaf	or, m
Annona muricata	Annonaceae	Tree	Fruit	f
Eugenia uniflora	Myrtaceae	Shrub	Whole plant	or
Eucalyptus toreliana	Myrtaceae	Tree	Whole plant	or
Hibiscus rosa-sinensis	Malvaceae	Shrub	Whole plant	or
Caesalpinia pulcherima	Caesalpiniaceae	Shrub	Whole plant	or
Delonix regia	Caesalpiniaceae	Tree	Whole plant	or
Spondias mombim	Anacardiaceae	Tree	Fruit	f
Gossypium hirsutum	Malvaceae	Shrub	Fruit	m
Zingiber officinale	Zingiberaceae	Herb	Rhizome	m, c
Anacardium occidentalis	Anacardiaceae	Tree	Fruit	f
Telfaria occidentalis	Cucurbitaceae	Climber	Leaf	m, c
Tectona grandis	Verbenaceae	Tree	Whole plant	sh

Table 2. Contd.

Annona squamosa	Annonaceae	Shrub	Fruit	f
Artocarpus communis	Moraceae	Tree	Fruit	f
Kalanchoe tomentosa	Crassulaceae	Herb	Whole plant	or
Glyphaea brevis	Tiliaceae	Shrub	Stem	ce
Citrus aurantifolia	Rutaceae	Tree	Fruit	m
Abrus precatorius	Papilionaceae	Climber	Leaf	m
Newbouldia laevis	Bignoniaceae	Tree	Leaf	ce
Ricinus communis	Euphorbiaceae	Shrub	Seed (oil)	m
Cycas circinalis	Cycadaceae	Palm	Whole plant	or
Cycas revoluta	Cycadaceae	Palm	Whole plant	or
Colocasia esculenta	Araceae	Herb	Corm	f
Saccharum officinarium	Poaceae	Culm	Stem	f
Citrus reticulata	Rutaceae	Tree	Fruit	f
Terminalia catapa	Combretaceae	Tree	Fruit/Whole plant	f, sh
Basella alba	Chenopodiaceae	Climber	Leaf	c
Citrulus lunatus	Cucurbitaceae	Creeper	Fruit	f
Sorghum bicolor	Poaceae	Grass	Seed	f
Ipomoea batatas	Convolvulaceae	Creeper	Tuber	f
Euphorbia milli	Euphorbiaceae	Herb	Whole plant	or
Acalypha wilkesiana	Euphorbiaceae	Herb	Whole plant	or
Sanseveria laurentii	Liliaceae	Shrub	Whole plant	or
Hibiscus sabdarifa	Malvaceae	Shrub	Leaf	m, be
Talinum triangulare	Portulacaceae	Herb	Leaf	c
Moringa oleifera	Moringaceae	Tree	Leaf	m
Ficus benjamina	Moraceae	Tree	Whole plant	sh
Gmelina arborea	Verbenaceae	Tree	Whole plant	sh
Ixora coccinea	Rubiaceae	Shrub	Whole plant	or
Ixora ludiflora	Rubiaceae	Shrub	Whole plant	or
Elaeis guinensis	Arecaceae	Tree	Seed (Oil)	c
Murraya paniculata	Rutaceae	Shrub	Whole plant	or
Codiaeum variegatum	Euphorbiaceae	Herb	Whole plant	or
Thuja occidentalis	Cupressaceae	Tree	Whole plant	or
Senna alata	Caesalpiniaceae	Shrub	Leaf	m
Azadiractha indica	Meliaceae	Tree	Leaf	m
Solanum macrocarpon	Solanaceae	Herb	Fruit	f
Dioscorea alata	Dioscoreaceae	Climber	Tuber	f
Heliotropicum indicum	Boraginaceae	Herb	Leaf	m
Thymus bifida	Lamiaceae	Herb	Leaf	c
Leucena leucocephala	Mimosaceae	Tree	Leaf	m
Basella rubra	Chenopodiaceae	Climber	Leaf	c
Parquetina nigrescens	Asclepiadaceae	Climber	Leaf	m
Lactuca taraxacifolia	Asteraceae	Herb	Leaf	m
Glyricedium sepium	Papilionaceae	Tree	Leaf	m, sh
Solanum dasyphylum	Solanaceae	Shrub	Fruit	m
Talinum paniculata	Portulacaceae	Herb	Leaf	c
Hura crepitans	Euphorbiaceae	Tree	Whole plant	or
Senna siamea	Caesalpiniaceae	Tree	Whole plant	sh
Agave americana	Amaryllidaceae	Herb	Whole plant	or
Indian forma	Lauraceae	Shrub	Whole plant	or
Manihot esculentum	Euphorbiaceae	Shrub	Underground stem	f
Croton variegatum	Euphorbiaceae	Shrub	Whole plant	or
Crassocephalum crepidiodes	Asteraceae	Herb	Leaf	c

Table 2. Contd.

Canna indica	Cannaceae	Herb	Whole plant	or
Euphorbia laterifolia	Euphorbiaceae	Herb	Whole plant	or
Mirabilis jalapa	Nyctaginaceae	Herb	Whole plant	or
Cordyline terminalis	Liliaceae	Shrub	Whole plant	or
Phyllanthus amarus	Euphorbiaceae	Herb	Leaf	m
Thevetia peruviana	Apocynaceae	Shrub	Whole plant	or
Polyalthia longifolia	Annonaceae	Tree	Whole plant	or
Sesamum indicum	Pedaliaceae	Herb	Seed (Oil)	m
Alstonia boonei	Apocynaceae	Tree	Leaf/Whole plant	m, sh
Cnidoscolus acontifolius	Euphorbiaceae	Shrub	Leaf	c
Rauvolfia vomitoria	Apocynaceae	Shrub	Leaf	m
Plumera alba	Apocynaceae	Herb	Whole plant	or
Cactus spp.	Cactaceae	Herb	Whole plant	or
Dracaena manni	Liliaceae	Herb	Whole plant	or
Hyptis bracteolate	Lamiaceae	Herb	Leaf	m
Thaumatococcus danielli	Maranthaceae	Herb	Leaf	m
Crecentia cujete	Bignoniaceae	Tree	Fruit	m

m, Medicinal; c, culinary/condimental; or, ornamental; sh, shade; sc, scouring; sn, snuff; be, beverage; ce, ceremonial; f,food; co, cotton; fp, food processing.

Table 3. Plant distribution according to families.

Family	No. of species	Family	No. of species
Euphorbiaceae	15	Nyctaginaceae	2
Solanaceae	6	Crassulaceae	2
Rutaceae	5	Cucurbitaceae	2
Apocynaceae	4	Cycadaceae	2
Malvaceae	4	Portulacaceae	2
Caesalpiniaceae	4	Chenopodiaceae	2
Poaceae	4	Lythraceae	1
Myrtaceae	3	Bromeliaceae	1
Lamiaceae	3	Caricaceae	1
Asteraceae	3	Sterculiaceae	1
Papilionaceae	3	Zingiberaceae	1
Rubiaceae	3	Combretaceae	1
Moraceae	3	Convolvulaceae	1
Anacardiaceae	3	Moringaceae	1
Bignoniaceae	3	Cupressaceae	1
Verbenaceae	3	Meliaceae	1
Annonaceae	3	Dioscoreaceae	1
Liliaceae	3	Boraginaceae	1
Asclepiadaceae	2	Mimosaceae	1
Amaranthaceae	2	Amaryllidaceae	1
Asphodelaceae	2	Lauraceae	1
Musaceae	2	Cannaceae	1
Tiliaceae	2	Pedaliaceae	1
Araceae	2	Cactaceae	1
Arecaceae	2	Maranthaceae	1

Table 4. Plant life form and their frequency.

Habit	Frequency
Herb	38
Tree	37
Shrub	30
Climber	7
Grass	3
Creeper	2
Palm	2
Culm	1

Table 5. Frequency of plant parts used.

Plant part	Frequency
Leaves	42
Whole plant	39
Fruits	25
Seeds	6
Stem	2
Tuber	2
Milky latex	1
Corm	1
Rhizome	1
Underground stem	1

that home gardens are useful sources of plant germplasm. But together with the germplasm conserved in the home gardens, it is extremely important to maintain the knowledge regarding these species and varieties. And as such new generations should be encouraged to recover and re-energize the productive capacity of homegardens.

ACKNOWLEDGEMENT

The authors would like to thank all the people interviewed in the rural/semi-urban community of Odeda for the opportunity to use their home gardens as study sites; for information given as well as hospitality and accommodation.

REFERENCES

Adekunle MF (2008). Indigenous uses of plants leaves to treat malaria fever at Omo Forest Reserve (OFR), Ogun State, Nigeria. Ethiopia. J. Environ. Stud. Manage. 1(1):31-35.

Agelet A, Bonet MA, Valles J (2000). Homegardens and their role as a main source of medicinal plants in mountain regions of Catalonia (Iberian Peninsula). Econ. Bot. 53(4):295-309.

Albuquerque UP, Andrade LHC, Caballero J (2005). Structure and floristics of homegardens in Northeastern Brazil. J. Arid. Environ. 62(3):491-506.

Ayodele AE (2005). The medicinally important leafy vegetables of South-Western Nigeria. (http//www.siu.edu/ebl/leaflets/ayodele/html).

Babu KS, Jose D, Gokulapalan C (1982). Species diversity in a Kerala home garden. Agroforestry Today 4:15-20

Caballero J (1992). Maya homegardens: past, present and future. Etnoecológica 1:35-54.

Dean AG, Dean AJ, Coloumbier D (1994). Epi info: A word processing database and statistics program for epidemiology on microcomputers. Centre for disease control and prevention, Atlanta, GA, USA.

Dutta AC (2008). Botany for Degree Students. 6th Edition. Oxford University Press, New Delhi, India. p. 15.

Eichemberg MT, Amorozo MC, Moura LC (2009). Species composition and plant use in old urban homegardens in Rio Claro, Southeast of Brazil. Acta Bot. Bras. 23(4):1057-1075.

Erinoso SM, Aworinde DO (2012). Ethnobotanical survey of some medicinal plants used in traditional health care in Abeokuta areas of Ogun State, Nigeria. Afr. J. Pharm. Pharmacol. 6(18):1352-1362.

Fernandes ECM, Nair PKR (1986). An evaluation of the structure and function of tropical homegardens. Agrofores. Syst. 21:279–310.

Hoogerbrugge ID, Fresco LO (1993). Homegarden Systems: Agricultural Characteristics and Challenges. International Institute for Environment and Development (Gatekeeper series, 39), London.

Hufford TL (1978). Botany: Basic Concepts in Plant Biology. Harper and Row Publishers, USA. p. 19.

Lamont SR, Eshbaugh WH, Greenber G (1999). Species composition, diversity and use of homegardens among three amazonian villages. Econ. Bot. 53:312–326.

Michon GJ, Bompart P, Hecketsweiler A, Ducatillion C (1983). Tropical forest architectural analysis as applied to agroforests in humid tropics: The example of traditional village-agroforests in West Java. Agrofores. Syst. 11:17-129.

Millat E, Mustafa MD, Hall JB, Teklehaimanot Z (1996). Structure and floristics of Bangladesh homegardens. Agrofores. Syst. 33(3):263-280.

NBS (2006). National Bureau of Statistics, National Population Commission, Abuja, Nigeria.

Padoch C, De Jong W (1991). The house gardens of Santa Rosa: diversity and variability in an amazonian Agric. Syst. Econ. Bot. 45:166–175.

Singh GB (1987). Agroforestry in the Indian sub-continent: past, present and future. pp. 117-138. In: Steppler HA, Nair PKR (eds.) Agroforestry: A Decade of Development. ICRAF, Nairobi, Kenya.

Soemarwoto O (1987). Home gardens: a traditional agroforestry system with a promising future. In: Steppler HA, Nair PKR (eds.) Agrofrestry: A Decade of Development. ICRAF, Nairobi, Kenya.

Soemarwoto O, Conway GR (1991). The javanese homegarden. J. Farm. Sys. Res. Ext. 2(3):95-117.

Soemarwoto O, Soemarwoto I, Soekartadiredja KEM, Ramlan A (1985). The javanese home garden as an integrated agro-ecosystem. Food Nutr. Bull. 7(3):44-47.

Taiwo AM, Adeogun AO, Olatunde KA, Adegbite KI (2011). Analysis of Groundwater Quality of Hand-Dug Wells in Peri-Urban Area of Obantoko, Abeokuta, Nigeria for Selected Physico-Chemical Parameters. Pacific J. Sci. 12(1):527-534.

8

Soil seed banks of a rangeland area White Nile State, Sudan

A. K. Elsafori[1]*, A. N. Guma'a[2] and M. A. El Nour[3]

[1]College of Forestry and Range Science, Sudan University of Science and Technology, P. O. Box 6146, Post Code 11113, Khartoum, Sudan.
[2]Faculty of Education, University of Khartoum, Sudan.
[3]Faculty of Forestry, University of Khartoum, Sudan.

Three soil depths (0 to 5, 5 to 10 and 10 to 15 cm) were chosen at the study area and the soil seed bank was analyzed for the number of live and dead seeds for each of them. The analysis revealed the following: The seed bank density was higher in the upper soil depths (0 to 5 and 5 to 10 cm) as compared to the lower ones (10 to 15 cm). It was found that the seed density had decreased with increasing depth. The live seed density ranged from 1015 to 5371 seeds/m^2, whereas that of the dead seeds ranged from 3215 to 6957 seeds/m^2. The dominant plant species to which the live seeds belonged were grasses including *Schoenefeldia gracilis, Brachiaria* spp., *Dactyloctenium aegyptium* and *Aristida spp.* The dominant species to which the dead seeds belonged were mixed life- forms including *Panicum turgidum, Euphorbia aegyptiaca* and *Cyperus rotundus.*

Key words: Seed bank, density, species composition, semi-arid environments, rainfall variability, dispersal, viable seeds and dead seeds.

INTRODUCTION

Sudan is the largest country in Africa and the nineth largest in the world with an area of about 2.5 million square kilometers. It exhibits a wide range of variation in its topography, climate, soil, and hydrology.

The study area (Um Rimmitta) lies between latitudes 14°36/ and 14°49/N and longitude 32°05/ and 32°11/ E. It is bordered from the east by the White Nile and from the west by North kordofan State and from the north by Guetaina province about seventy kilometers north of Ed Dueim town.

LITERATURE REVIEW

Soil seed banks are important components of vegetation dynamics affecting both ecosystem resistance and resilience. In arid ecosystems, seeds are characterized by high spatial and temporal variability and are particularly affected by spatial patterns of vegetation. Rainfall unpredictability is the underlying factor causing the huge soil seed banks found in arid environments.

According to Roberts (1981) the term soil seed bank has been used to designate the live seed reservoir present in a soil. For Baker (1989), this reservoir corresponds to seeds not germinated but, potentially capable of replacing the annual adult plants, which had disappeared by natural death or otherwise and perennial plants that are susceptible to plant diseases, disturbance and human or animal consumption. Simpson et al. (1989) defined the soil seed bank as all the viable seeds present in the soil or mixed with soil debris.

The persistence of seeds in the soil is a major component of plant succession and plays a substantial role in the evolution of plant communities (Robert, 1981). Seeds that remain incorporated in the soil form a reserve that can be depleted for potential regeneration over a period of time that may extend to a century.

The seasonal pattern of rainfall, as it influences the size and composition of the soil seed bank is the major factor affecting recruitment of species (Orr, 1991). Kropac (1966) divided the techniques for estimating the population of the soil seed bank into two groups. The first

*Corresponding author. E-mail: Elsafori2@yahoo.com.

group is the physical extraction which includes sieving, flotation and subsequent viability determination. The second group is the germination techniques that rely on direct assessment of seedlings arising from soil samples. Both have merits as well as disadvantages.

The physical extraction of seeds from the soil tended to overestimate the number of germinating seeds, since extraction counts may include dormant and dead seeds. The seed numbers are normally expressed as number of seeds/m^2 related to certain depth. The determination of the proportion of dead seeds present in the soil seed bank is important in studying population dynamics and consequently methods of determining viabilities have been developed.

The germination technique, although undoubtfully useful, may be less efficient than the extraction method (Jensen, 1969). The time of sampling in relation to the vegetation cycle must also be taken into account, either by occasional samples at different sites or sampling at the same time in different years (Barralis, 1972).

To determine the soil weed seed bank, Fay and Olson (1978) described a simple technique for rapid determination of weed seed, rhizome, and corm or bulb population in the soil. Soil samples were placed in nylon mesh bags and washed by machine, leaving only small residues including the desired propagules. These were hand-separated and the soil population of a given plant species was then calculated. The total manipulation time per sample was approximately 10 min.

The size of soil samples is an important factor in soil seed bank determinations. Different soil sample sizes were used by different workers, depending on the condition under investigation. Malone (1976) worked on chemical extraction of seeds using aqueous sodium hexa- meta-phosphate. He adopted individual soil samples of 500 gm, but he mentioned that 100 gm are more conveniently manipulated. In his study, individual soil samples of 250 gm were used for extraction by calcium chloride ($CaCl_2$).

The size of weed seed bank varies to a great extent. Jensen (1969) found that the average seed bank size within the upper 20 cm was 50258 living seeds/m^2 in 57 Danish fields. In a British investigation (Roberts and Stokes, 1966), values between 1600 and 86000 live seeds/m^2 were recorded in 58 fields within the upper 15 cm. Von Hofsten (1947) reported an estimate of 1777 seeds/m^2 for *Sinapis arvensis* in Sweden. Another way for estimating changes of the size of the seed bank is to register the number of emerged seedlings in the field. According to Ebregt et al. (1988) weed surveys give a poor estimation of the seed bank due to large annual fluctuations in emergence and growth conditions. Thus soil sampling is a more reliable method than field surveys to estimate the seed bank.

The soil depth is another determination factor in soil seed bank studies. Demel and Granstrom (1995) reported that in dry Afromontane forests of Ethiopia at least 167 plants species were identified in the 0 to 9 cm soil layer with total densities ranging between 12300 and 24000 seeds/m^2. In Sudan, Mustafa (1997) found about 20 seeds/m^2 under the canopy of *Acacia seyal*. In Kenya, Schimidit in Tybrik et al. (1994) found 1510 seeds/m^2 whereas Kaarakka (1996) found a maximum of 6 seeds/m^2 under the canopy of *Acacia zanzibrica*. Ibrahim (2005) recorded seed bank densities of about 19700 dead seeds/m^2 and 7238 live seeds/m^2 in the low rainfall savannah.

The weed species have survived through time, because of their ability to resist several adverse climatic conditions including tolerance to high or low temperatures, dry or humid environments and variations in oxygen supply (Hafliges, 1990).

The composition of seed bank is variable and is classified as temporary or persistent with respect to the regeneration of the vegetation during different times of the year temporary banks are composed of seeds of short life, which do not have dormancy (Garwood, 1989). Persistent seed banks are composed of seeds that live for more than one year and can buried remain in the soil for many years. Generally, seed bank composition depends on seed longevity and viability.

Seed longevity in the soil varies with plant species, burial depth, soil moisture in addition to internal and external seed factors. Freitas (1990) reported a seed longevity of 40 years for *Portulaca oleracea*. Seeds in the soil last longer in the deeper layers than on the surface (Taylorson, 1970; Qi et al., 1996) and those with impermeable coats are mostly likely to last longer in the soil (Owen, 1950). Bekker et al. (1997) found that change in soil moisture contents could have a large effect on seed longevity.

Several internal and external factors prevent seed germination. Among the internal seed factors, the most important are the presence of the seed coat, a biochemical inhibitor in the seed, and immature embryo. Among the external factors the most common are soil water content and temperature (Fernandez et al., 1991). The dormancy represents a main mechanism of species preservation in the seed bank, thus distributing germination through the year (Carmona, 1992).

Different factors affect the viability of seed bank in the soil. Grazing and cutting intensity affect the seed bank, through effects on the seeds return. O'Connor et al. (1992) studied the seed bank of *Aristida bipartia* and other spp. in savanna grassland and reported that the seed bank was dominated by less palatable species in areas subjected to heavy grazing.

The dynamics of a seed bank involves a series of events including time and land preparation that influences the seed dispersion in the soil profile (Simpson et al. 1989). Clements and Bentoit (1996) studied the influence of land preparation types on the seed bank. They found that more than 70% of the seeds were present in the layers of 0 to 5 cm in plots where no mechanical method was used and 30% for plots mechanically managed. Other events include management of soil depth and fair or uniform distribution of seeds in the soil profile (Dessaint et al., 1991).

MATERIALS AND METHODS

Preparation of soil samples

Six transects were made and soil samples were taken from 42 sample plots in the study area from depths (0 to 5, 5 to 10 and 10 to 15 cm) along each transect. The samples were mixed thoroughly and sub-samples of 250 gm were prepared for seed extraction.

The soil sub- samples were placed in a set of sieves with pores of 1, 0.5, 0.25, and 0.1 mm and washed for 10 to 15 min under continuous flow of water. The soil was washed away and the seeds there in were transferred to a 500 ml beaker containing water. The dead seeds were observed to float and the water containing the floating dead seeds was immediately filtered in a Bunchner funnel. The residue (dead seeds) was air-dried.

The live seeds at the bottom of the beaker were extracted as follows: A weight of 1.5 g of $CaCl_2$ was accurately weighed and dissolved in 250 ml of distilled water. The solution was added to the live seeds in the beaker and left for 40 min.

The live seeds were observed to float in the $CaCl_2$ solution. These were then filtered in a Buchner funnel and air-dried.

Determination of soil seed density

The density of seeds was determined by the following formula:

$$\text{Density of seeds} = \frac{\text{Number of seeds/soil depth x 2 x 10000}}{\text{Quadrat area x number of quadrats/soil depth}}$$

Identification of soil seeds

The extracted seeds were identified by comparison with a reference seed collection from known plants growing in the study area. Magnifying lenses and an Mbc-10 dissection microscope were used for seed identification.

RESULTS

Species composition at Site A

At Site A, 21 species were identified from the live seeds whereas 20 species were identified from the dead seeds. The most dominant species identified from the live seeds included *Indigofera arrecta*, *Cyperus rotundus*, *Dactyloctenium aegyptium*, *Tephrosia uniflora*, *Brachiaria mutica Indigofera hochstetteri*, *Aristida adescensionis*, *Panicum turgidum*, *Scheonefeldia gracilis* and *Euphorbia aegyptiaca*. The most dominant species identified from dead seeds included *Brachiaria sp.*, *Euphorbia aegyptiaca*, *Panicum turgidum*, *Aristida adescensionis*, *Schoenefeldia gracilis*, *Cleome escaposa*, *Mullugo nudicaulis*, and *Corchorus spp* (Table 1).

Soil seed density for Site A

The number of seeds/m^2 was calculated for each soil depth by the following equation:

$$\text{No. of seeds/m}^2 = \frac{\text{Number of seeds/depth x 2 x 100 x 100}}{\text{Quadrat area (m}^2\text{) x Number of quadrats/depth}}$$

The following results represent the number of live and dead seeds/m^2 for each of the three soil depths at Site A:

Depth 0 to 5 cm:
Number of live seeds/m^2 = 163 x 2 x 10000/ 1400 = 2329
Number of dead seeds/m^2 = 283 x 2 x 10000/1400= 4043
Depth 5 to 10 cm:
Number of live seeds /m^2 = 17 x 2 x 10000/ 1400 = 243
Number of dead seeds/m^2 = 75 x 2 x 10000/ 1400 = 1071
Depth 10 to 15 cm:
Number of live seeds/m^2 = 10 x 2 x 10000/ 1400 = 143
Number of dead seeds/m^2 = 70 x 2 10000/ 1400 = 1000

Species composition at Site B

At Site B, 26 species were identified from the live seeds whereas 28 species were identified from the dead seeds. The most dominant species identified from live seeds included *Indigofera arrecta, Dactyloctenium aegyptium, Tephrosia uniflora, Brachiaria mutica*, (*Indigofera hochstetteri, Aristida adescensionis, Cenchrus biflorus, Panicum turgidum.* The most dominant species identified from dead seeds included *Cyperus rotundus, Dactyloctenium aegyptium, Brachiaria eruciformis, Cenchrus biflorus, Aristida adescensionis, Panicum turgidum* and *Indigofera hochstetteri* (Table 2).

Soil seed density for Site B

The following is a summary of the number of live and dead seeds/m^2 for each of the three soil depths at site B:

Depth 0 to 5 cm:
Number of live seeds/m^2 = 201 x 2 x 10000/ 1400 = 2871
Number of dead seeds/m^2 = 199 x 2 x 10000/1400= 2843
Depth 5 – 10 cm:
Number of live seeds /m^2 = 126 x 2 10000/ 1400 = 1800
Number of dead seeds/m^2 = 205 x 2 x1000/ 1400 = 2971
Depth 10 to 15 cm:
Number of live seeds/m^2 = 49 x 2 x 10000/ 1400 = 700
Number of dead seeds/m^2 = 80 x 2 x 10000/ 1400 = 1143

Species composition at Site C

At Site C, 16 species were identified from live seeds whereas 18 species were identified from dead seeds. The most dominant species identified from live seeds included *Aristida adescensionis, Schoenefeldia gracilis, Mullugo nudicaulis.* The most dominant species identified

Table 1. Soil seed bank at Site A: Depths (cm).

Species	0 - 5		5 - 10		10 - 15		Total	
	Live	Dead	Live	Dead	Live	Dead	Live	Dead
Acacia oerfota	-	2	-	-	1	-	1	2
A. tortilis	1	-	-	-	-	-	1	-
Abutilon figarianum	-	-	-	-	-	2	-	2
Aristida adscensions	9	39	-	1	-	7	9	47
Brachiaria mutica	5	2	-	-	-	-	5	2
B. reptans	12	75	3	26	-	8	15	99
Chloris virgata	-	2	1	-	-	-	1	2
Cleome escaposa	2	10	-	1	-	6	2	17
Corchorus sp	4	8	-	3	2	-	6	11
Cyperus rotundus	20	3	2	3	-	6	22	12
Cyperus spp.	36	2	-	1	-	-	36	3
Dactylectonium aegyptium	15	3	3	-	-	1	18	4
Eragrostis sp.	1	3	-	-	-	1	1	4
Euphorbia aegyptiaca	8	60	-	6	4	26	12	92
Fimbristilis biss-umbellta	7	-	-	-	-	-	7	-
Indigofera hochstetteri	3	2	4	-	1	-	8	2
Mollugo nudicaulis	3	3	1	9	-	3	4	15
Ocimum sp	5	-	1	1	-	-	6	1
Panicum turgidum	11	41	2	20	2	10	15	71
Schoenefeldia gracilis	9	21	-	1	-	-	9	22
Senna alexandrina	6	4	-	3	-	-	6	7
Solanum dubium	5	-	-	-	-	-	5	-
Tribulus terrsitris	-	2	-	-	-	-	-	2
Zornia sp	1	1	-	-	-	-	1	1
Total	163	283	17	75	10	70	190	418

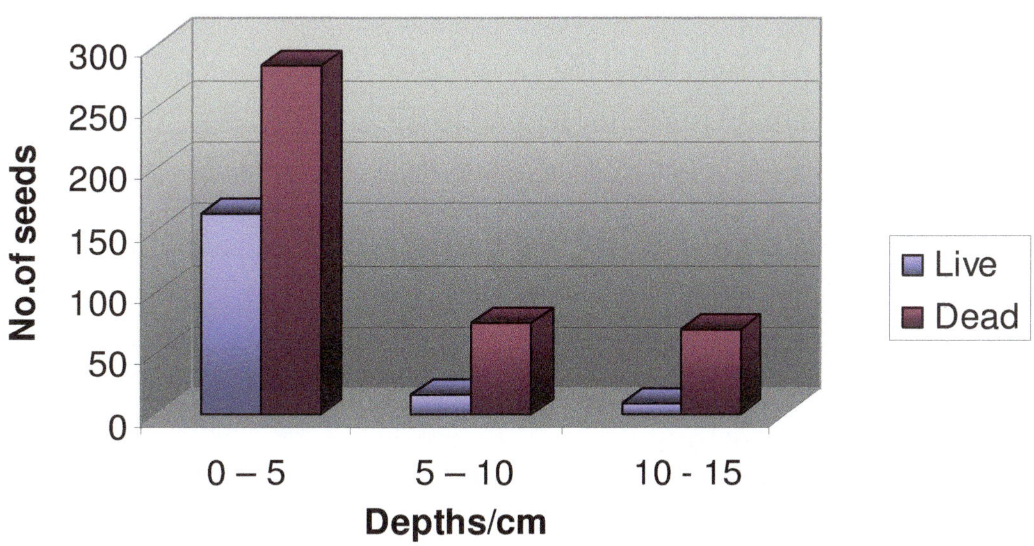

Figure 1. Histogram showing the number of live and dead seeds at each soil depth for Site A.

from dead seeds included *Aristida adescensionis, Cyperus rotundus, Schoenefeldia gracilis* and *Mullugo nudicaulis* (Table 3).

Soil seed density at Site C

Here is a brief summery of the numbers of dead and

Table 2. Soil seed bank at Site B: Depths (cm).

Species	0 - 5		5 - 10		10 - 15		Total	
	Live	Dead	Live	Dead	Live	Dead	Live	Dead
Abutilon figarianum	-	-	-	-	4	-	4	-
Achyranthes aspera	-	-	-	1	-	-	-	1
Aristida adscensionis	13	18	6	14	8	7	27	39
Brachiaria mutica	3	-	1	2	-	-	4	2
B. reptans	18	19	20	20	3	2	41	41
Cenchrus biflorus	8	19	11	19	3	12	22	50
Chloris virgata	-	-	3	2	-	-	3	2
Cleome escaposa	1	-	-	8	-	-	1	8
Corchorus spp.	4	4	5	4	5	-	14	8
Cyperus rotundus	7	42	6	13	4	13	17	68
Cyperus spp.	5	10	4	15	-	7	9	32
Dactyloctenium aegyptium	28	27	43	34	8	18	79	79
Eragrostis sp.	3	5	-	2	2	-	5	7
Euphorbia aegyptiaca	1	2	-	-	-	2	1	4
Fagonia cretica	1	2	-	-	-	-	1	2
Fimbristylis biss-umbellata	1	5	-	1	2	-	3	6
Hibiscus vitifollus	-	1	-	-	-	-	-	1
Indigofera arrecta	42	5	1	11	-	-	43	16
I. hochstetteri	17	12	13	17	2	1	32	30
Leucas urticifolia	1	-	-	-	-	-	1	-
Mollugo nudicaulis	6	-	2	1	-	2	8	3
Ocimmum spp.	-	1	3	-	-	-	3	1
Panicum turgidum	10	18	8	34	5	14	23	66
Schoenefeldia gracilis	9	7	2	4	1	1	12	12
Senna alexandrina	1	1	-	1	-	-	1	2
Sesamum alatum	-	1	-	-	-	-	-	1
Solanum dubium	-	-	-	-	1	-	1	-
Tephrosia uniflora	19	-	-	-	-	-	19	-
Tribulus terrestris	-	-	-	-	1	-	1	-
Zornia spp.	3	-	1	2	-	1	4	3
Total	201	199	129	205	49	80	379	484

liveseeds for each of the three ssiol depths at Site C:

Depth 0 to 5 cm:
Number of live seeds/m^2 = 44 x 2 x 10000/ 1400 = 629
Number of dead seeds/m^2 = 128 x 2 x 10000/ 1400
Depth 5 to 10 cm:
Number of live seeds/m^2 = 17 x 2 x 10000/ 1400
Number of dead seeds/m^2 = 46 x 2 x 10000/ 1400 = 657
Depth 10 to 15 cm:
Number of live seeds/m^2 = 10 x 2 x 10000/ 1400 = 143
Number of dead seeds/m^2 = 51 x 2 x 1000 / 1400 = 729

The results of soil seed bank presented in Tables 1, 2 and 3 were statistically analyzed using the variance ratio (F) test. The results of the statistical analysis for live and dead seeds have been presented in Tables 4 and 5, respectively.

Summary of soil seed bank in the study area

From the results aforementioned, it can be seen that the seed bank density was higher in the upper depths as compared to the lower ones. The live seed density in the study area (semi-arid) ranged between 1015 and 5371 seeds/m^2 while that for the dead seed density was ranged between 3215 to 6957 seeds/m^2. The soil seed bank density for both live and dead seeds in the study area was in the range 3034 to 8462 seeds/m^2.

DISCUSSION

The soil seed bank of the study area was quantitatively determined at each of Sites A, B and C. The herbs and grasses recorded the highest number of seeds, as

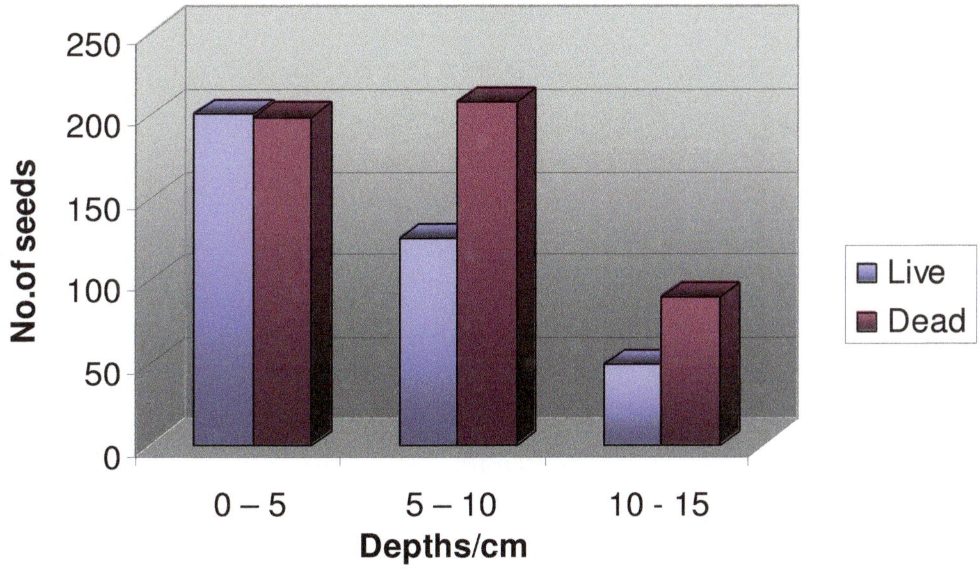

Figure 2. Histogram showing the number of live and dead seeds at each soil depth for Site B.

Table 3. Soil seed bank at Site C: Depths (cm).

Species	0 - 5		5 - 10		10 - 15		Total	
	Live	Dead	Live	Dead	Live	Dead	Live	Dead
Aristida adscensionis	14	65	5	8	1	6	20	79
Brachiaria mutica	2	-	1	1	-	3	3	4
B. reptans	1	-	-	1	-	-	1	1
Cenchrus biflorus	-	2	1	-	-	4	1	6
Cenchrus ciliaris	-	-	1	2	-	-	1	2
Cleome escaposa	-	-	-	1	-	-	-	1
Corchorus sp.								
Cyperus rotundus	4	23	4	9	-	17	8	49
Cyperus spp.	5	-	-	1	-	2	5	3
Dactyloctenium aegyptium	-	6	1	-	2	-	3	6
Echinochloa spp.	-	-	-	1	-	1	-	2
Eragrostis spp.	-	-	1	-	1	-	2	-
Indigofera arrecta	-	-	-	-	-	2	-	2
I. hochstettri	2	-	-	-	-	1	2	1
Mollugo nudicaulis	2	10	1	11	3	5	6	26
Ocimmum spp.	1	-	-	-	-	2	1	2
Panicum turgidum	-	-	-	-	-	3	-	3
Schoenefeldia gracilis	13	21	-	8	-	5	13	34
Tribulus terrestris	-	1	-	-	-	-	-	1
Total	44	128	17	46	10	51	71	225

compared to the other life forms across the three sites (Tables 1, 2 and 3 respectively). The woody species recorded a low number of seeds, which agrees with Mustafa (1997), Tybrik et al. (1994) and Kaarakka (1996). This may be attributed to land over-use through a number of practices including the use of fruits as forage, marketing of seeds, excessive felling of trees, predation, pathogenicity, seed suffocation in water-logged areas and wild fires used in land clearance for shifting cultivation.

In depth A (0 to 5 cm), the density of live seeds was 5829 seeds/m^2 whereas the dead seeds recorded a density of 8715 seeds/m^2. In depth B (5 to 10 cm), the density of live seeds was 2286 seeds/m^2 as compared to

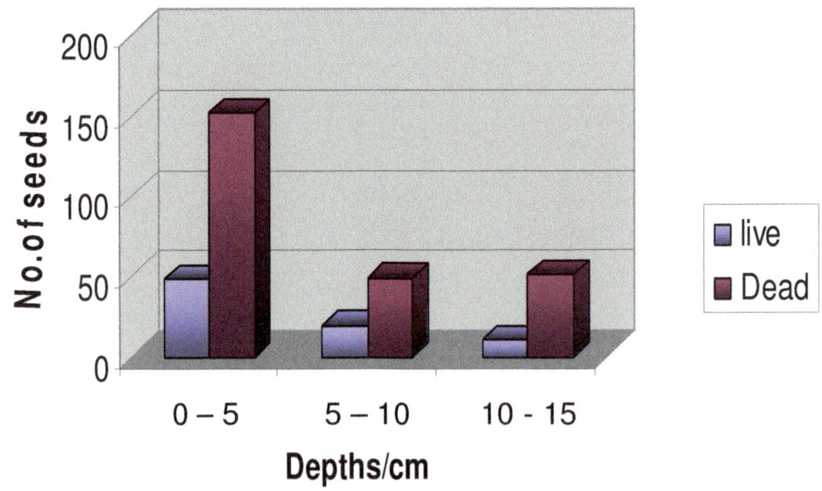

Figure 3. Histogram showing the number of live and dead seeds at each soil depth for Site C.

Table 4. ANOVA table for the live soil seed bank data in the study area.

Site	Source	df	Mean S	F. value	P.
A	Species	23	23.940 34.000	0.7041	0.8178
B	Species	29	136.454 44.422	3.0717	0.0001
C	Species	18	8.957 6.456	1.3874	0.1938

Table 5. ANOVA table for the dead soil seed bank data in the study area.

Site	Source	df	Mean S	F. value	P.
A	Species	23	309.295 117.750	2.6267	0.0024
B	Species	29	180.523 25.933	6.9610	< 0.0001
C	Species	18	147.121 67.649	2.1748	0.0218

a density of 4699 seeds/m^2 for the dead seeds. In depth C (10 to 15 cm) the density of live seeds was 986 seeds/m^2 while that of the dead seeds was 2872 seeds/m^2. From these results, it can be seen that the vertical distribution and soil seed density were higher in the upper soil layers and that both vertical distribution and soil seed density decreased with increasing depth. This is in line with what was found by Clements and Bentiot (1996) and Dessaint et al. (1991). The total soil seed bank density in the study area was low when compared with Demel and Granstrom (1995). That is explained by the fact that they determined the density under forested areas, whereas the present study was conducted at non-forested areas.

The density of live seeds was low in the study area as compared to Roberts and Stocks (1966). This may probably be due to the differences in site conditions and environmental factors.

The study also revealed that the percentage of dead seeds was higher as compared to that of live seeds. The low viability of seeds may be attributed to suffocation resulting from water- logging, predation, shedding of seeds before maturity and high temperatures.

The study has also revealed that the live seeds in depth A (0 to 5 cm) belonged to 21 species, while the dead seeds belonged to 20 species. In the depth B (5 to

10 cm) the live seeds belonged to 8 species and the dead seeds belonged to 12 species. However, in depth C (10 to 15 cm) the live seeds belonged to 5 species and the dead seeds belonged to 10 species. The number of species to which the live and dead seeds belong confirm the previous findings with respect to vertical distribution and seed density for the three soil depths. That is the number of species to which the seeds belong decrease with increasing depth.

The dominant soil seed flora were *Indigofera arrecta, I. hochstetteri, Dactyloctenium aegyptium, Schoenefeldia gracilis, Aristida adscensionis, Brachiaria spp., Panicum turgidum, Euphorbia aegyptiaca, Cenchrus biflorus,* and *Cyperus* spp. The dominance of the aforementioned species in the soil seed bank may be attributed to their prolific seeding and good growth and establishment. Moreover, these species are: tolerant to decay, protected from predators and resistant to several adverse climatic conditions.

There are differences among the dominant species in the percentages of the total live seeds. For instance, the frequencies of the total live seeds for *Dactyloctenium aegyptium, Cyperus spp., Indigofera spp.,* and *Brachiaria spp.* were 15, 14.9, 13.1 and 10.2% respectively. From field observations and soil seed bank identification at the study area, it was found that some grasses and herbs were dominant and this agreed with Hafliges (1990). This indicates that some of the dominant species have different seed production capacities at the different sites and can survive for a long time because of their ability to resist several adverse climatic conditions. However, some species have different percentages of the total number of dead seeds as follows: *Aristida spp.* and *Panicum turgidum*, have frequencies 14.4 and 12% respectively. This may be due to the intolerance of these species to adverse climatic conditions, predation, suffocation, decay besides the shedding of seeds before maturity.

The results of statistical analysis for soil seed bank data (Tables 1and 2) showed a very highly significant difference for the live seeds in site B and non- significant differences for sites A and C. The very highly significant difference of live seeds at site B significant difference was probably due to site conditions, diversity of species, prolific seeding and resistance to decay. The non-significant differences in sites A and C may be attributed to different seed production capacities of the species under different sites.

As for the dead seeds at the three sites, a very highly significant difference was found at site B, a highly significant difference was found at site A and a significant difference was found at site C. This may probably be due to the site conditions and floristic composition. The soil seed analysis had also shown that the less palatable species were dominant.

This agrees with O'Conner and Pickett (1992), who reported that heavy grazing results in the dominance of the less palatable species.

REFERENCES

Baker HG (1989). Some Aspects of the Natural History of Seed Banks. In, Leck, M. A., Parcker VT, Simpson RL (ed) (1989). Ecol. of Soil Seed Banks. Academic Press, London.

Barrali G (1972). Evlition comparative de la flore adventice avecou Sans de's herbage chimique weeds Res., (12): 115- 127.

Bekker RM, Verweij GL, Smith REN, Reine R, Bakker RP, Schineider S (1997). Soil Seed Banks in European Grasslands: Does Land Use Affect Regeneration Perspectives. J. Appl. Ecol., 34(5): 1293-1307.

Carmona R (1992). Problematica manejo de bancos de sementes de invasoras em solos agricolas. Planta Daninha, V. 10(1/2): 5-16.

Clements DR, Bentoit DL (1996). Tillage Effects on Weed Seed Bank Composition. Weed Sci., 5(44): 314-322.

Demel T, Granstrom A (1995). Soil Seed Banks in Dry Afromontane Forests of Ethiopia. J. Veg. Sci., (6): 777-786. Opulus Press Uppsala, Sweden.

Dessaint F, Cgadocuf R, Barralis G (1991). Spatial Pattern Analysis of Weed Seeds in the Cultivated Soil Seed Bank J. Appl. Ecol., (28): 271-370.

Ebregt EM, Zanstra GH, Van ZJC (1988). Weed Seed Banks from Neighbouring Farms with and without Long Term Use of Herbicides. Centre for Agrobiological Research. CABO Report 96, 23 pp.

Fay RK, Olson WA (1978). Techniques for Separating Weed Seed from Soil: Weed Sci., (26): 530-533.

Fernandez QC, Saavedra MS, Malas H (1991). In Christoffoleti, P. J. and R. S. X. Caetano, (1998). Soil Seed Banks. Sci. Agric., (55). Special Issue Piracicaba (1998).

Freitas RR (1990). Dinamica do banco de sementes em uma comounidade plantas daninhas com aspectos da germinacaoe dormenica de sementes de caplm mar. melada (Brachiaria plantaginea(Link) Hitc. Lavras, 118 p. Dissertaco (Mestrado)- Escola Superiorode de Agriculture de lavaras.

Gardwood NC (1989). Tropical Soil Seed Bank: A Review. In: Leck, M.A.; pae, V.T. and R. L. Simpson, (Ed). Ecology of Soil Seed Banks. London Academic Press: pp. 149-209.

Halfiges Scholz, cited by Freitas (1990). Soil Seed Banks. Scientia Agricola. Vol. (55). Special Issue Piracicaba (1998). Brazil.

Jensen HA (1969). Content of Buried Seed in Arable Soil in Denmark and Its Relation to Weed Population. Danskbot Art, 27(2): 1- 56.

Kaarakka V (1996). Management of BushLand Vegetation using rain Water Harvesting in Eastern Kenya. Acta Forestalia Fennica 353-93.

Kropac Z (1966). Management of Bush Land Vegetation Using Rain Water Harvesting in Eastern Kenya. Acta Forestalia Fennica: pp. 353-93.

Malone CR (1976). A rapid Method for Estimation of Viable Seeds in the Soil. Weeds Res., (15): 381-382.

Mustafa AF (1997). Regeneration of *Acacia seyal* Forests on the Dryland of the Sudan Clay Plain. University of Helsinki Tropical Forest. Rep., 103 pp.

O'Connor TC, Pickett GA (1992). The influence of Grazing on Seed Production and Seed Bank of Africa Savanna Grasses. J. Appl. Ecol., (29): 247-260.

Orr DM (1991). Trends in the recruitment of Astrebla spp. In relation to seasonal rainfall. Rangeland J. Australia, 13(2): 107-117.

Owen FV (1950). Viability of Sugar Beet Seed held in Cold Storage for 22 years. Proc. Amer. Soc. Sugar Beet Techno. 127-9, Bibl. 1.

Qi M, Upadhyaya MK, Turkington R (1996). Dynamics of Seed Bank and Survivorship of Meadow Salsify Populations. Weed Sci., 44(1): 100-108.

Roberts HA (1981). Seed Bank in the Soil. Advances in Applied Biology. Weed Res., 78: 253-256.

Roberts HA, Stokes FG (1966). Studies on the Weeds of Vegetable Crops. VI. Seed Population of Soil Under Commercial Cropping. J. Appl. Ecol., (3):181-190.

Simpson RL, Leck MA, Parcker VT, Simpson RL (ed.) (1989). Ecology of Soil Seed Banks. London Academic Press. London.

Taylorson RB (1970). Changes in Dormancy and Viability of Weed Seeds in soils. Weed Sci., (18): 265-69.

Tybrik K, Schimidit LH, Kauser T (1994). Seed Bank of African Acacias. Afr. J. Ecol. (32): 327-330.

9

Effects of anthropogenic activities on genetic diversity and population structure of American ginseng (*Panax quinquefolius* L.) growing in West Virginia

Samuel G. Obae[1]* and Todd P. West[2]

[1]Department of Plant Science and Landscape Architecture, University of Connecticut, 1376 Storrs Road, Unit 4067, Storrs, CT 06269, USA.

[2]Department of Plant Sciences, North Dakota State University, Dept # 7670, PO Box 6050, Fargo, ND 58108, USA.

Using random amplified polymorphic DNA (RAPD) markers, genetic variation and population structure of wild and cultivated American ginseng growing in West Virginia (WV) was assessed. Also, the effects of cultivation intensity and harvest pressure on genetic diversity of ginseng populations were evaluated. Eight primers were used to amplify DNA samples from 468 plants, generating a total of 98 band fragments of which 84 were polymorphic. Overall mean genetic diversity measures were lower in West Virginia populations compared to populations from Pennsylvania and Wisconsin. Among West Virginia populations, mean genetic diversity indices were higher in cultivated than in wild populations but differences were not significant. Means of genetic diversity indices were higher in wild populations from low harvest pressure region [percent polymorphism (P) = 33.33%, Nei's (1973) gene diversity (H) = 0.1172, and Shannon's index (I) = 0.1743] compared to those from high harvest pressure region (P = 28.27%, H = 0.1019, I = 0.1513), however, these indices were not significantly different (P > 0.05, Mann-Whitney test). Analysis of molecular variance (AMOVA) however showed significant genetic differentiation (P < 0.001) within and among wild populations (48.37 and 54.10% respectively). Cultivated ginseng populations from a region with low cultivation intensity had lower levels of genetic diversity indices (P = 29.93%, H = 0.0948, I = 0.1609) compared to populations from a region with high cultivation intensity (P = 60.60%, H = 0.2593, I = 0.3243), these indices were significantly different (P < 0.05, Mann-Whitney test). AMOVA further revealed that in cultivated populations, 53.68 and 42.11% of the total genetic variation was attributed to within and among population differentiation in regions respectively and 4.20% of genetic variation existed between regions. Based on these data, it is evident that genetic diversity of American ginseng is substantially increased with increase in its cultivation intensity. This can be ascribed to higher levels of gene flow associated with sourcing of seeds from various sources to meet cultivation needs of large scale growers. Even though genetic variability among populations experiencing different harvest pressure was not statistically significant in this study, the lower genetic diversity indices exhibited in wild populations from high harvest pressure regions underscores the potential impacts of harvesting pressure on genetic diversity of this medicinal plant. Thus, the data presented in this study will be useful in guiding conservation strategies for this economically important but threatened medicinal species.

Key words: Cultivation intensity, harvest pressure, analysis of molecular variance (AMOVA), random amplified polymorphic DNA (RAPD).

INTRODUCTION

The immediate threats to medicinal plants are anthropogenic activities resulting in habitat loss, habitat degradation and over-harvesting exacerbated by increasing demand for herbs and herbal products both regionally and internationally (Cunningham, 1993; Hamilton, 1997; Kuipers, 1997; Ahmad, 1998; Lange,

*Corresponding author. E-mail: samuel.obae@uconn.edu.

1998). About 75% of medicinal plants used worldwide are collected from the wild (Canter, 2005) and the other 25% are from cultivated sources.

Cultivation of medicinal plants has been on the rise lately, mainly as a response to declining wild populations and the potential economic gains anticipated from the growing demand of herbal products globally. Whereas increasing cultivation of medicinal plants minimizes the pressure on their wild populations, more often, the cultivation process results in change of genetic diversity and structure of plant populations (Doebley, 1989). Such changes in population gene pool associated with the cultivation process are often a concern to plant conservationists as there is always a fear that this will lead to erosion of 'locally' adapted genotypes of a given species, especially if 'non-local' gene pools are used as seed sources for cultivation. For instance, collection of seeds from a narrow source for use in establishing a cultivated population can cause genetic drifts or bottlenecks that can result in populations of relatively low genetic diversity and that have diverged significantly from their wild progenitor gene pools (Zohary, 2004). On the other hand, if cultivated populations are located in close proximity to their wild counterparts, the possibility of gene flow back and forth between cultivated and wild populations is more likely and that will cause reduction in genetic isolation therefore resulting in non-distinct cultivated and wild populations (Otero-Arnaiz et al., 2005).

American ginseng (*Panax quinquefolius* L.), Araliaceae, is a perennial herb native to the eastern deciduous woodlands of North America (Catling et al., 1994). It is among the top 12 commonly used herbs in the United States (O'Hara et al., 1998) and one of the most widely used medicinal herb in the world (Christensen et al., 2006). Growing popularity of ginseng among consumers worldwide has led to its extensive collection from the wild and its increasing cultivation within and outside of its native range (Cruse-Sanders and Hamrick, 2004; Christensen et al., 2006). West Virginia (WV) is at the core of American ginseng's native range, and collection of American ginseng from the wild has been a part of the folklore of West Virginians for hundreds of years (Anon, 2008). In the US, WV ranks second in wild ginseng production (Anon, 2008; WVDOF, 2008). Even though wild American ginseng occurs statewide and ginseng could potentially be cultivated throughout the state, it's harvesting and cultivation is regionalized. According to data records of ginseng production in WV over a 30 year period (1978 to 2006), it is evident that more than half of the total harvested wild ginseng originates from eight counties in the state's southwestern corner, in contrast, 88% of all cultivated ginseng production during this same period came from one county on the northern part of the state (Anon, 2008; WVDOF, 2008). From this data, it can be inferred that ginseng populations in WV are subjected to different levels of harvest pressure and cultivation intensity. Under these varied anthropogenic influences, one may speculate that genetic diversity and population structure will be different or at least are bound to change especially since the species has a long reproductive cycle and its harvesting entails destruction of the entire plant.

American ginseng is listed as a threatened species under Convention of International Trade of Endangered Species (CITES) and therefore, its trade is closely monitored at federal and state levels (Robbins, 1998). Documenting the range of its genetic diversity and population structure, and evaluating how they are influenced by human activities is integral in decision making process to protect this economically important medicinal species. Genetic diversity is a valuable resource from which populations derive short-term adaptation to environmental stochasticity and long-term evolutionary changes (Ellstrand and Elam, 1993), and is a good reservoir from which breeding varieties can be derived.

Genetic diversity and population structure under varied levels of harvest pressure is not extensively explored, and very little is known about how the cultivation process has impacted genetic variability of American ginseng populations growing in WV. The objectives of this study were to assess genetic diversity and population structure of American ginseng populations growing in WV, and evaluate how genetic diversity persist under varied levels of anthropogenic activities (cultivation intensity and harvest pressure). The results presented in this study illuminate the extent of genetic diversity of American ginseng regionally and provide valuable data that would be useful in guiding conservation strategies for this economically important but threatened medicinal species.

MATERIALS AND METHODS

Study species

American ginseng (*P. quinquefloius* L.) is a long-lived perennial herb with palmate compound leaves and a fleshy tap root. The species is self-fertilized (Schlessman, 1985); however, outcrossing by generalist insects has been reported (Carpenter and Cottam, 1982; Lewis and Zenger, 1983). Plants typically do not flower until they are old enough to produce two leaf whorls, usually in the fourth or fifth year (Carpenter and Cottam, 1982; Lewis and Zenger, 1983; Charron and Gagnon, 1991; Anderson et al., 1993). Seeds usually undergo a dormancy period of 18 to 20 months before germinating in early spring of the second growing season (Lewis and Zenger, 1982; Proctor and Bailey, 1987; Pritts, 1995). The species has no known asexual reproduction using root or stem cuttings (Lewis, 1988).

Sampling

Populations in specifically demarcated regions reflecting different levels of cultivation intensity and harvest pressure were sampled. Regions were demarcated based on ginseng harvest records obtained from West Virginia Department of Forestry (WVDOF, 2008) for the period between 1978 and 2006. Regarding cultivation intensity, two regions were categorized; region 1 (WVCR1), low cultivation intensity, mainly comprised of small scale ginseng farms typically less than 1 acr of land under ginseng cultivation with reported ginseng production averaging less than 10 pounds per

Figure 1. United States map showing sampled populations of *P. quinquefolius*. A map of West Virginia State is enlarged to show population sites and regions where samples were collected. WVCR1 = low cultivation intensity region; WVCR2 = high cultivation intensity region; WVWR1 = low harvest pressure region; and WVWR2 = high harvest pressure region. Sample collection areas are not drawn to scale.

year; and region 2 (WVCR2), high cultivation intensity, characterized with large scale ginseng farms reportedly producing on average 300 pounds or more per year over a 30 year period (Figure 1). With regard to harvest pressure on wild ginseng populations, two regions were categorized; region 1 (WVWR1), a low harvest pressure region, where ginseng collection averages less than 500 pounds per year; and region 2 (WVWR2), a high harvest pressure region, where ginseng collection averages over 500 pounds per year over a 30 year period (Figure 1). Two wild and two cultivated populations from Pennsylvania (PA) and Wisconsin (WI) respectively were also sampled and included in the study for comparison (controls). Wild plants from PA originated from populations that have had a long-time protection from harvest, whereas cultivated samples from WI originated from large scale farm extensively cultivating American ginseng.

Plant materials

Leaf samples from 468 plants representing a total of 26 populations of wild (14) and cultivated (12) American ginseng were collected in 2007 and 2008 from WV, PA, and WI (Table 1). Leaf samples from cultivated populations were donated by growers, whereas samples of wild populations were collected from the wild with the help of ginseng diggers in various regions. At least 9 randomly selected plants within a 50 m radius of a contiguous patch of ginseng plants were sampled. Upon collection, individual leaves were wrapped separately in moist paper towels then placed in Ziploc bags and transported back to the laboratory within 24 h. Once in the laboratory, leaves were rinsed with deionized water then immediately frozen in liquid nitrogen and stored in Ziploc bags at -80°C until DNA extraction.

DNA isolation

Approximately 20 mg of leaf material was placed in 1.5 ml micro centrifuge tube and ground with a sterile disposable plastic pestle (Fisher Scientific, Pittsburgh, PA). DNA extraction was done using a GenElute Plant Genomic DNA Mini Prep Kit (Sigma-Aldrich, St. Louis, MO) following the manufacturer's instructions. Quality of isolated genomic DNA was assessed by measuring its 260/280 nm absorbance ratio using a spectrophotometer (GeneQuat, Pharmacia, LKB Biochrom, England) and by running DNA samples through 1.5% agarose gel in 0.5× TBE (tris borate ethylenediaminetetraacetic acid) buffer. Gels were stained with ethidium bromide (EtBr), then visualized on electronic UV transilluminator (ULTRA-LŪM, Inc., Claremont, CA) and digitally photographed with Canon Powershot G6 camera (Canon USA, Inc.). Only DNA samples with absorbance ratio of > 1.5 and an intact high molecular weight band on the gel were used in RAPD analysis. All DNA samples were diluted to 20 ng/µl with milliQ waterand stored at -20°C until RAPD

Table 1. Population, ecotype, sample size, and population code of populations used in this study.

Population	Ecotype	Sample size	Population code
WV populations			
POP_1	Cultivated	16	WV-C_1
POP_2	"	16	WC-C_2
POP_3		20	WC-C_3
POP_4	"	10	WC-C_4
POP_5	"	10	WC-C_5
POP_6	"	15	WC-C_6
POP_7	"	25	WC-C_7
POP_8	"	20	WC-C_8
POP_9	"	25	WC-C_9
POP_10	"	40	WC-C_10
POP_11	Wild	14	WV-W_11
POP_12	"	18	WV-W_12
POP_13	"	10	WV-W_13
POP_14	"	16	WV-W_14
POP_15	"	15	WV-W_15
POP_16	"	18	WV-W_16
POP_17	"	20	WV-W_17
POP_18	"	12	WV-W_18
POP_19	"	9	WV-W_19
POP_20	"	10	WV-W_20
POP_21	"	25	WV-W_21
POP_22	"	25	WV-W_22
Non-WV populations			
Pennsylvania	Wild	16	PA-W_1
	Wild	20	PA-W_2
Wisconsin	cultivated	21	WI-C_1
	cultivated	22	NC-C_2

analysis.

Primer selection

Initially, 29 decamer primers were selected from a list of those that had been reported successful in RAPD analysis of *P. quinquefolius* in previously published studies. These primers included; UBC-98, UBC-203, UBC-297 (Bai et al., 1997); OPD-03, OPD-05, OPH-04, OPH-05, OPO-15, OPU-10, OPU-15 (Boehm et al., 1999); UBC-06, UBC-18, UBC-81, UBC-164, UBC-177, UBC-210, UBC-227, UBC-262, UBC-326, UBC-398, UBC-419, UBC-464, UBC-497 (Schluter and Punja, 2002), and OPAD-01, OPAD-02, OPAD-11, OPAD-15, OPN-02 , OPN-19 (Lim, 2004). All primers were prescreened for polymorphism and reproducibility using a representative sample from all 26 populations. Eight of the 29 primers exhibited high polymorphism and reproducibility and were therefore selected for further use in RAPD analysis of all samples (Table 2). All primers were synthesized by Operon Biotechnologies (Huntsville, AL).

PCR amplification and gel electrophoresis

Amplification conditions were initially optimized by varying the amount of DNA template and primer concentrations. Upon establishing optimal PCR conditions that yielded discernible and reproducible bands, all samples were amplified twice with duplicate amplifications run on separate days. A HotStarTaq® Master Mix Kit (Qiagen®, Germantown, MD) was used for PCR with each 25 µl mixture containing: 12.5 µl Hotstart mix, 11 µl of RNase free H_2O, 0.5 µl of primer (0.2 µM), and 1 µl of DNA template (~20 ng). Amplifications were done on a single thermo cycler (GeneAmp® PCR System 9700, Applied Biosystems, Carlsbad, CA) with PCR thermocycling based on Schluter and Punja (2002) protocol with some modifications. At the initial cycle, the reaction mixture was heated at 94°C for 15 min to activate the HotStarTaq DNA polymerase, as recommended by the manufacturer, then 10 min at 36°C, and 2 min at 72°C. Subsequent 46 cycles included denaturation at 94°C for 30 s, annealing at 36°C for 1 min, and elongation at 72°C for 1 min. In the final cycle, the reaction was extended for 10 min at 72°C. Ten micro liters of amplified products were loaded into 1.5% agarose gels alongside 1 kb DNA ladder (Promega, Madison, WI) and electrophoresed in 0.5× TBE buffer at 105 V for 40 min. Gels were stained with EtBr, visualized with electronic UV transilluminator and digitally photographed.

Data analysis

Amplification products were scored manually and each band in

Table 2. Primer code, nucleotide sequence, total number of bands, number of polymorphic bands, and size ranges of polymorphic bands generated by each primer used for RAPD analysis of *P. quinquefolius* populations.

Primer	Sequence (5'→ 3')	No. of bands	No. of polymorphic bands	Size ranges(bp) of polymorphic bands
OPD-05	TGAGCGGACA	21	15	300-2000
OPH-04	GGAAGTCGCC	10	9	250-2500
OPH-05	AGTCGTCCCC	8	8	250-2500
OPO-15	TGGCGTCCTT	14	12	300-2000
OPAD-11	CAATCGGGTC	13	12	300-2500
UBC-81	GAGCACGGGG	14	12	250-3000
UBC-98	ATCCTGCCAG	10	8	250-2500
UBC-164	CCAAGATGCT	8	8	250-2000

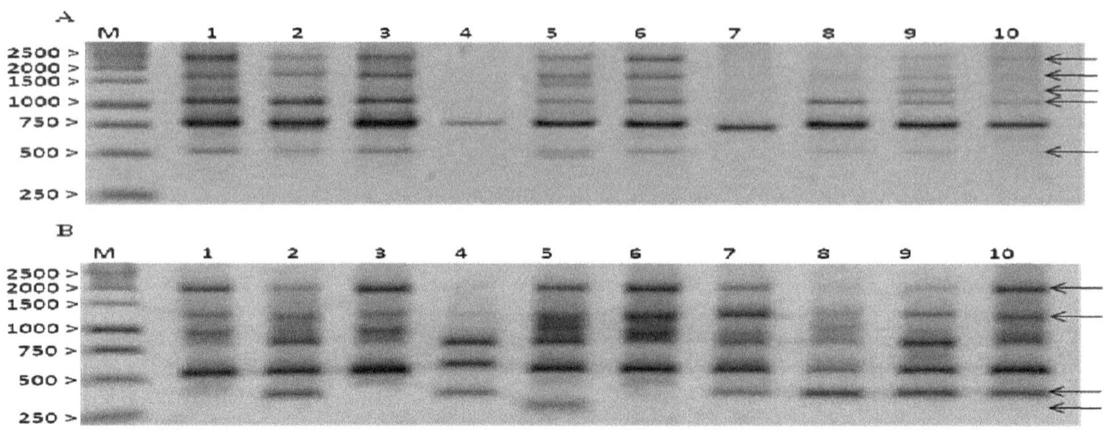

Figure 2. DNA fragments generated by primers OPAD-11(A), and OPO-15 (B) in samples 1-10 of POP_12. M = 1kb DNA ladder, arrows indicate polymorphic bands.

RAPD profile was treated as an independent character (locus) with two states (alleles) scored as either presence (1) or absence (0) of a band at a given size location on the gel picture for each primer. Monomorphic bands across all samples and bands below 250 bp or above 3000 bp were omitted from subsequent statistical analyses. Bands outside of this range (250 to 3000 bp) are considered unreliable for RAPD analysis (Stewart and Porter, 1995). To estimate genetic diversity and population structure, various parameters were estimated using different software programs for population genetics analysis. POPGENE Ver. 1.31 software (Yeh et al., 1997) was used to calculate, percent polymorphic loci (*P*), Nei's (1973) gene diversity (*H*), and Shannon's diversity index defined as $I = -\sum p_i \log_2 p_i$, where p_i is the frequency of a given RAPD band (Lewontin, 1972). The advantage of Shannon's index is that it does not assume that populations are in Hardy-Weinberg equilibrium. Statistical significance of genetic diversity measures among different levels of comparisons was determined using Mann-Whitney test.

Hierarchical population genetic structure and fixation index (Φ_{ST}) values were determined by analysis of molecular variance (AMOVA) using ARLEQUIN Ver. 3.11 software (Excoffier et al., 2005) with 1000 permutations. AMOVA uses phenotypic distances to describe how the RADP variance is partitioned among and within populations, and tests for the significance against the null hypothesis of no population structure (Excoffier et al., 1992; Stewart and Excoffier, 1996). AMOVA was based on Euclidean squared distance matrix which was constructed using AMOVA-PREP software (Miller, 1997).

Genetic relationships of different population groups was determined based on Nei's 1978 genetic distance measure using TFPGA software (Miller, 1998). A phenogram was generated using unweighted pair group method with arithmetic averages (UPGMA) and 1000 bootstraps to measure robustness of different clusters in in the phenogram.

RESULTS

RADP polymorphism

The eight primers used in RAPD analysis of 26 populations (468 samples) generated a total of 98 distinct loci (average 12.25 loci per primer) of which 84 were polymorphic (85.71%), an average of 10.5 polymorphic loci per primer (Table 2). The 84 polymorphic bands were clear, unambiguously scorable, and were between the range of 250 and 3000 bp. Examples of polymorphic bands generated by different primers are shown in Figure 2.

Genetic diversity and structure in all populations

For the 26 populations, mean *P, H,* and *I* values were

Table 3. Summary of within-population genetic diversity.

Group, region, population	n	*P*	*H*	*I*
WV-group				
WVCR1				
WV-C_1	16	36.90	0.1589	0.2275
WC-C_2	16	27.38	0.0829	0.1286
WC-C_3	20	28.57	0.0998	0.1504
WC-C_4	10	36.90	0.1287	0.1925
WC-C_5	10	28.57	0.1066	0.1568
WC-C_6	15	16.67	0.0470	0.0748
WC-C_7	25	34.52	0.1296	0.1916
WVCR2				
WC-C_8	20	64.29	0.2941	0.3691
WC-C_9	25	63.81	0.2508	0.3077
WC-C_10	40	53.71	0.2329	0.2960
WVWR1				
WV-W_11	14	34.52	0.1240	0.1839
WV-W_12	18	39.29	0.1270	0.1914
WV-W_13	10	32.14	0.1177	0.1742
WV-W_14	16	20.24	0.0687	0.1032
WV-W_15	15	35.71	0.1312	0.1932
WV-W_16	18	30.95	0.0923	0.1417
WV-W_17	20	48.81	0.1675	0.2506
WV-W_18	12	25.00	0.1090	0.1560
WVWR2				
WV-W_19	9	27.38	0.1184	0.1697
WV-W_20	10	32.14	0.1130	0.1679
WV-W_21	25	15.48	0.0570	0.0655
WV-W_22	25	38.10	0.1193	0.1820
Non-WV group				
WI-C_1	21	52.58	0.1700	0.2577
WI-C_2	22	52.38	0.1675	0.2525
PA-W_1	16	70.24	0.2028	0.3135
PA-W_2	20	76.19	0.2060	0.3226
Cultivated (12 populations)	20	41.36	0.1482	0.2171
Wild (14 populations)	16	37.59	0.1253	0.1883
All (26 populations)	18	39.33	0.1359	0.2016

n = sample size; *P* = percent polymorphic loci; *H* = gene diversity (Nei, 1973); *I* = Shannon's index (Lewontin, 1972). Group = WV or non-WV; regions: WVCR1 = low cultivation intensity, WVCR2 = high cultivation intensity, WVWR1 = low harvest pressure, WVWR2 = high harvest pressure.

39.33%, 0.1359, and 0.2016 respectively. Among wild populations, mean *P, H,* and *I* values were 37.59%, 0.1253, and 0.1883 respectively and lowest genetic variability was in population WV-W_21 (*P* = 15.48%, *H* = 0.0570, *I* = 0.0655) whereas population PA-W_2 had the highest genetic variability (*P* = 76.19%, *H* = 0.2060, *I* = 0.3226). Among cultivated populations, mean *P, H,* and *I* values were 41.36%, 0.1482, and 0.2171 respectively and lowest genetic variability was in population WV-C_6 (*P* = 16.67%, *H* = 0.0470, *I* = 0.0748) whereas population WV-C_8 had the highest genetic variability (*P* = 64.29%, *H* = 0.2941, *I* = 0.3691) (Table 3).

Table 4. Summary of mean genetic diversity of populations in groups, ecotypes and regions.

Group, ecotype, region	n	*P*	*H*	*I*
WV (22 populations)	389	35.05	0.1267	0.1861
WV cultivated (10 populations)	197	39.13	0.1441	0.2095
WVCR1 (7 populations)	112	29.93	0.0948	0.1603
WVCR2 (3 populations)	85	60.60	0.2593	0.3243
WV wild (12 populations)	192	31.65	0.1121	0.1666
WVWR1 (8 populations)	123	33.33	0.1172	0.1743
WVWR2 (4 populations)	69	28.27	0.1019	0.1513
Non-WV (4 populations)	79	62.85	0.1866	0.2870
Cultivated (2 population)	43	52.48	0.1688	0.2551
wild (2 populations)	36	73.21	0.2044	0.3181

n = sample size; *P* = percent polymorphic loci; *H* = gene diversity (Nei, 1973); *I* = Shannon's index (Lewontin, 1972). Group = WV or non-WV; ecotype = wild or cultivated; regions: WVCR1 = low cultivation intensity region, WVCR2 = high cultivation intensity region, WVWR1 = low harvest pressure region, WVWR2 = high harvest pressure region.

Overall, cultivated populations had higher genetic diversity than wild populations, however, there were no significant differences ($P > 0.05$, Mann-Whitney test) in all genetic diversity indices. The fixation index ΦST analogous to FST (Wright, 1951) obtained by AMOVA using pairwise distances among populations was equal to 0.473 (range 0.082 to 0.678), meaning that 47.3% of total variation was attributed to among population differentiation and the rest (52.7%) of variation to individuals within populations and all values were significantly different ($P < 0.001$) from zero (no differentiation).

Comparison of genetic variations between WV and non-WV populations (groups)

Overall, non-WV populations had significantly higher ($P < 0.05$, Mann-Whitney test) mean genetic variations (P = 62.85%, H = 0.1866, I = 0.2870) than WV populations (P = 35.05%, H = 0.1267, I = 0.1861) (Tables 4 and 5, analysis 1). There was significant differentiation among and within populations in the two groups ($P < 0.001$, AMOVA) (Table 6, analyses 1 and 2). For WV group, among population differentiation accounted for 48.48% of the total variation and the rest (51.52%) was attributed to individual differentiations within populations (Table 6, analysis 1). For non-WV group, 25.33% of the total variation was attributed to among population diffrentiation and 74.67% was due to individual differentiations within populations (Table 6, analysis 2). When total population differentiation was partitioned to a three level hierarchy to account for among groups (WV and non-WV) differentiation, highest genetic differentiation was attributed to individuals within populations (46.50%), among population differentiation accounted for 36.84%, and variation between groups accounted for 16.66% of the total variation (Table 6, analysis 3). UPGMA phenogram based on Nei's (1978) unbiased genetic distances between WV and non-WV groups revealed two well defined clusters separating WV and non-WV populations. Non-WV group formed two distinct sub clusters, one having Pennsylvania (PA) populations and the other having Wisconsin (WI) populations together. These clusters were supported by high bootstrap values suggesting a strong genetic differentiation between groups (Figure 3).

Genetic diversity within and among wild and cultivated populations (ecotypes) in WV

In cultivated ecotype, lowest mean genetic variations was in population WV-C_6 (P = 16.67%, H = 0.0470, I = 0.0748) whereas population WV-C_8 had the highest genetic variability (P = 64.29%, H = 0.2941, I = 0.3691). For wild ecotype, lowest genetic variability was in population WV-W_21 (P = 15.48%, H = 0.0570, I = 0.0655) and population WV-W_17 had the highest genetic variability (P = 48.81%, H = 0.1675, I = 0.2506) (Table 3). Overall, cultivated ecotype exhibited higher mean genetic diversity values (P = 39.13%, H = 0.1441, I = 0.2095) than wild ecotypes (P = 31.65%, H = 0.1121, I = 0.1666) (Table 4), however, differences in these diversity values were not statistically significant (Table 5, analysis 2).

AMOVA revealed significant ($P < 0.001$) molecular variations within ecotypes, with cultivated populations having lower among population differentiation than wild populations (Table 6, analyses 4 and 5). However, when variation was partitioned to account for variation between ecotype, a significant ($P < 0.001$) proportion of variation was attributable to among population differentiation (49.10%) and to individual differentiation within populations (51.72%) but not between ecotypes (P = 0.070). Variation between ecotypes was very small (close to zero), so a negative value was obtained (Table 6, Analysis 6).

Table 5. Mann-Whitney test for differences in mean genetic diversity indices among populations of *P. quinquefolius* in various groups, ecotypes, and regions comparisons.

Analysis	n	*P*	*H*	*I*
Analysis 1: WV vs. non-WV populations (groups)				
WV (22 populations)	18	35.05	0.1267	0.1861
non WV (4 populations)	20	62.85	0.1866	0.2866
Significance		*	*	*
Analysis 2: WV cultivated vs. WV wild populations (ecotypes)				
Cult. (10 populations)	20	39.13	0.1441	0.2095
Wild (12 populations)	16	31.65	0.1121	0.1666
Significance		NS	NS	NS
Analysis 3: WV cultivated populations from low vs. high cultivation intensity regions				
WVCR1 (7 populations)	16	29.93	0.0948	0.1609
WVCR2 (3 populations)	28	60.60	0.2593	0.3243
Significance		*	*	*
Analysis 4: WV wild populations from low vs. high harvest regions				
WVWR1 (8 populations)	15	33.33	0.1172	0.1743
WVWR2 (4 populations)	17	28.27	0.1019	0.1513
Significance		NS	NS	NS

n = average sample size; *P* = percent polymorphic loci; *H* = gene diversity (Nei, 1973); *I* = Shannon's index (Lewontin, 1972). * Significant; NS = not significant at $P < 0.05$.

Genetic diversity of populations in low (WVCR1) versus high (WVCR2) cultivation intensity regions in WV

For populations in WVCR1, *P* values ranged from 16.67 to 36.90%, *H* ranged from 0.0470 to 0.1589, and *I* ranged from 0.0748 to 0.2275 for WV-C_6 and WV-C_1 populations respectively (Table 3). Whereas, for populations in WVCR2, *P* ranged from 53.71 to 64.29%, *H* ranged from 0.2329 to 0.2941, and *I* ranged from 0.2960 to 0.3691 for WV-C_10 and WV-C_8 populations respectively (Table 3). Mean estimates of genetic diversity of populations in WVCR2 were higher (P = 60.60%, H = 0.2593, I = 0.3243) than those of populations in WVCR1 (P = 29.93%, H = 0.0948, I = 0.1603) (Table 4) and were significantly different ($P < 0.05$, Mann-Whitney test) (Table 5, analysis 3).

When genetic differentiation was partitioned to account for between regions variation using AMOVA, it was revealed that a higher proportion of differentiation (53.68%) was attributed to individuals in populations within regions, followed by among populations within regions (42.11%) and differentiation between regions was the lowest (4.20%) but significant (Table 6, analysis 7).

Genetic diversity of populations in low (WVWR1) versus high (WVWR2) harvest pressure regions in WV

For populations in WVWR1, *P* ranged from 20.24 to 48.81%, *H* ranged from 0.0687 to 0.1675, and *I* ranged from 0.1032 to 0.2506 for WV-W_14 and WV-W_17 populations respectively (Table 3). Whereas for populations in WVWR2, *P* ranged from 15.48 to 38.10%, *H* ranged from 0.0570 to 0.1193, and *I* ranged from 0.0655 to 0.1820 for WV-W_21 and WV-W_22 populations, respectively (Table 3). Mean estimates of genetic variation within populations in WVWR1 region were higher (P = 33.33%, H = 0.1172, and I = 0.1743) compared to populations in WVWR2 region (P = 28.27%, H = 0.1019, and I = 0.1513) (Table 4) but were not significantly different ($P > 0.05$, Mann-Whitney test) (Table 5, analysis 4).

Furthermore, AMOVA revealed significant ($P < 0.001$) genetic differentiation among populations within regions and individuals in populations within regions, however, variation between regions was not significant ($P = 0.791$) and resulted in a negative value (Table 6, analysis 8).

DISCUSSION

Genetic diversity and structure of *P. quinquefolius*

Using RAPD markers, this study evaluated genetic variation and structure among 26 populations of *P. quinquefolius*. Mean values of genetic diversity indices showed that cultivated populations had higher levels of diversity compared to wild populations. This observation concurs with a report by Grubbs and Case (2004) for 44

Table 6. Summary of analysis of molecular variance (AMOVA) for groups, ecotypes and regions of *P. quinquefolius* using RAPD markers. Significance tests based on 1000 random permutations.

Source of variation	df	SS	variance components	% of total variance	*P-value*
Analysis 1: WV group					
Among population	21	1938.86	4.96	48.48	< 0.001
Within populations	367	1935.28	5.27	51.52	< 0.001
Analysis 2: Non-WV group					
Among population	3	209.85	3.09	25.33	< 0.001
Within populations	75	684.03	9.12	74.67	< 0.001
Analysis 3: WV vs. non-WV group					
Between groups	1	379.07	2.12	16.66	< 0.001
Among population	24	2148.70	4.69	36.84	< 0.001
Within populations	442	619.31	5.93	46.50	< 0.001
Analysis 4: WV cult. populations					
Among population	9	862.60	4.67	45.30	< 0.001
Within populations	187	1055.33	5.64	54.70	< 0.001
Analysis 5: WV wild populations					
Among population	11	984.85	5.34	52.22	< 0.001
Within populations	180	879.95	4.89	47.78	< 0.001
Analysis 6: WV population ecotypes					
Between ecotypes	1	91.41	-0.08	-0.82	0.070
Among population	20	1847.45	5.01	49.10	< 0.001
Within populations	367	1935.28	5.27	51.72	< 0.001
Analysis 7: WV cult. regions					
Between regions	1	159.56	0.44	4.20	< 0.001
Among population	8	703.04	4.43	42.11	< 0.001
Within populations	187	1055.33	5.64	53.68	< 0.001
Analysis 8: WV wild harvest regions					
Between regions	1	86.93	-0.25	-2.47	0.791
Among population	10	897.92	5.47	54.10	< 0.001
Within populations	180	879.95	4.89	48.37	< 0.001

df = degrees of freedom; SS = sum of squares.

populations of *P. quinquefolius* sampled from different regions across ginseng's native range. On average, genetic diversity values obtained in our study were higher than those reported for 21 populations (Cruse-Sanders and Hamrick, 2004) and 44 populations (Grubbs and Case, 2004) of *P. quinquefolius* using allozymes. This can be attributed to the vast numbers of loci that can be examined with RAPD markers compared to allozymes, therefore, they are able to reveal high degree of DNA level variation. Allozymes are only limited to protein coding regions and may not be representative of genome wide diversity (Stewart and Excoffier, 1996). However it is prudent to note that differences in these values could also be a result of sampling from different populations, even though this study and the other two studies sampled plants from the same geographic area in WV. Genetic differentiations among and within populations were highly significant. According to Wright (1978), F_{ST} values above 0.25 indicate substantial genetic differentiation. In this study mean fixation index Φ_{ST} value was 0.473, implying that the populations we evaluated were highly differentiated genetically. This observation is consistent to that reported previously (F_{ST} = 0.547) for 18 wild populations of *P. quinquefolius* (Mooney, 2007).

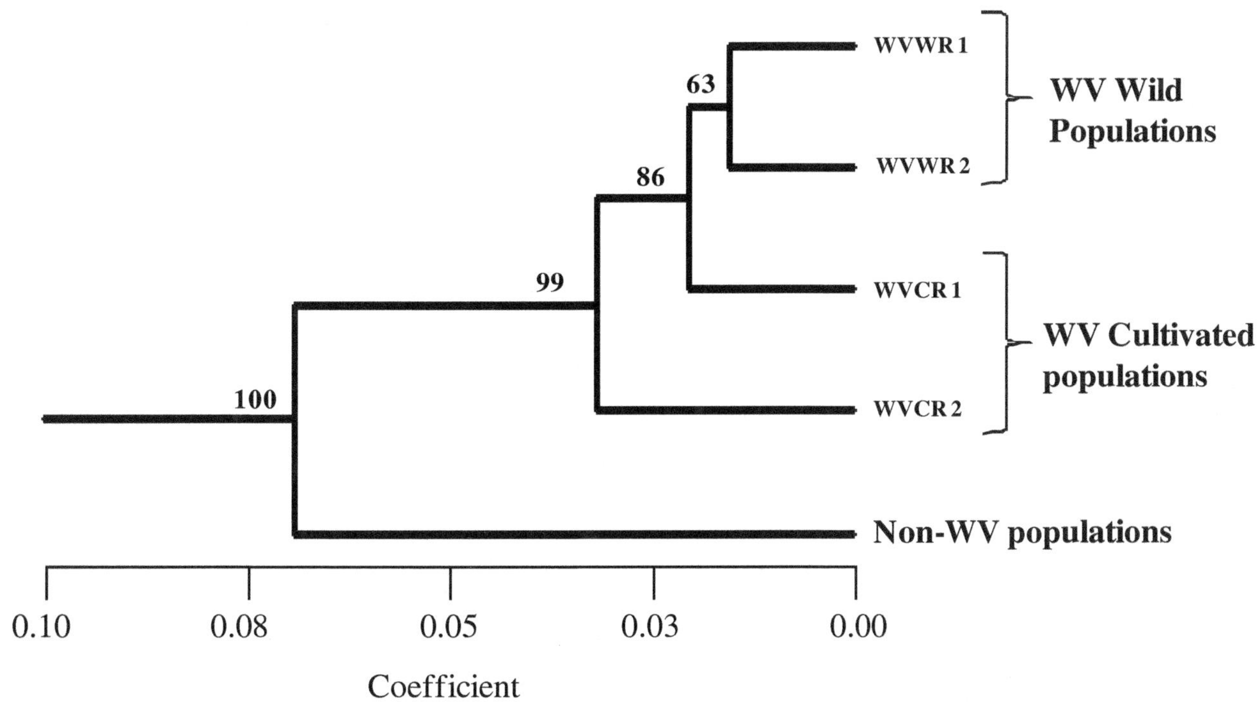

Figure 3. Relationship tree of population groups of *P. quinquefolius* revealed by UPGMA cluster analysis using Nei's 1978 genetic distance measure based on RAPD data. Only bootstraps values greater than 60% are indicated. WVCR1 = low cultivation intensity region; WVCR2 = high cultivation intensity region; WVWR1 = low harvest pressure region; and WVWR2 = high harvest pressure region.

Comparison of genetic diversity between WV and non-WV populations

Overall, we observed lower mean genetic diversity indices for WV populations (P = 35.05, H = 0.1267, I = 0.1861) compared to non-WV populations (P = 62.85, H = 0.1866, I = 0.2866). This could be attributed to harvest pressure associated with the long history of wild ginseng harvesting in WV. American ginseng populations facing harvest pressure have been reported to have reduced genetic diversity when compared to populations in protected areas (low harvest pressure) (Cruse-Sanders and Hamrick, 2004). Indeed, when wild populations in WV were compared directly to wild populations from PA which have not been harvested for several years (protected), measures of genetic diversity in WV wild populations were approximately half that of PA wild populations. The partitioning of genetic structure of WV populations was almost equal for among and within population differentiation (Table 6, analysis 1); however, for non-WV populations, within population differentiation was almost three times that of among population differentiation (Table 6, analysis 2). An almost equal genetic differentiation among and within populations from WV could be explained by lack of significant genetic differences observed among wild and cultivated populations particularly from low cultivation intensity region.

Genetic diversity and structure of population ecotypes in WV

With regard to ecotypes (wild versus cultivated), there were no significant differences in all diversity indices between ecotypes (Table 5, analysis 2). We attributed this lack of significant differences to a high level of genetic similarity observed between wild populations and cultivated populations in low cultivation intensity region (WVCR1). This high level of genetic similarity could be attributed to growers in this region recruiting into their farms seeds collected from wild populations. Although it is highly recommended that ginseng diggers plant seeds at locations where they collect wild roots in order to sustain future ginseng populations, it is not uncommon for diggers to take along with them some wild seeds to start ginseng crops in their own property. After all, these seeds have no associated costs (as opposed to expensive commercial seeds) and establishing ginseng populations on their personal property ensures easy accessibility and most importantly security of their crop from other diggers.

From AMOVA, it was revealed that most of the genetic variation in cultivated ecotype was attributed to individuals within populations, whereas in wild ecotype greater differentiation was partitioned to differentiation among populations. This observation concurs with other reports on *P. quinquefolius* (Schluter and Punja, 2002; Mooney, 2007). Cultivated populations typically have a high degree of gene flow among population mainly due to movement of seeds between growers as opposed to wild populations which have mostly become isolated because of fragmented habitats therefore have limited gene flow among populations.

Effects of cultivation intensity and harvest pressure on genetic diversity

Significantly higher genetic diversity indices were observed in populations from high cultivation intensity region than in populations from low cultivation intensity regions (Table 5, analysis 3). This could be explained by the higher gene flow in high cultivation intensity region which is a consequence of growers recruiting seeds from a broader source in order to meet their cultivation scale. In contrast, lower gene flow in lower cultivation intensity region could be attributable to small scale growers obtaining seeds from a narrow source or even collecting from a single source in the wild to start their ginseng crop. Surprisingly, harvest pressure did not have a significant effect on genetic diversity of WV wild populations evaluated in this study (Table 5, analysis 4). However, mean genetic diversity measures were slightly higher in populations from low harvest pressure region (WVWR1) than populations from a high harvest pressure region (WVWR2). This trend is in agreement with both field and simulation studies of *P. quinquefolius* (Cruse-Sanders and Hamrick, 2004; Cruse-Sanders et al., 2005). Lack of significant effect of harvest pressure on genetic diversity of wild populations observed in this study could be a consequence of small sample size especially from high harvest pressure region (only 4 populations, 69 samples were evaluated).

Conservation implications

Genetic variation is a valuable resource from which populations derive short-term adaptation to environmental stochasticities and long-term evolutionary changes (Ellstrand and Elam, 1993). This study indicates that there is a considerable amount of genetic diversity at population level for *P. quinquefolius* growing in WV. Despite not finding significant effects of harvest pressure on genetic diversity in populations we evaluated, it would be unwise to rule it out as driving force behind reduced genetic diversity of wild populations in WV. Further research involving extensive field sampling is needed to assess the effect of harvest pressure on wild *P. quinquefolius* populations growing in WV. For now, it is recommended that good stewardship practices for wild populations continue to be strictly adhered to by ginseng diggers in order to prevent further reduction of genetic diversity in wild populations as their populations decline due to harvesting. Such good stewardship practices include harvesting only mature plants and replanting seeds upon harvesting of roots.

There were no significant differences in genetic diversity between wild and cultivated ginseng from small farms in low cultivation intensity regions. This implies that most of the seeds used by these small farms to cultivate ginseng are likely collected from wild populations within this region. Whereas this shows that some growers are using 'local' seeds to cultivate ginseng thus conserving 'local' gene pool, the potential downside to this is that it will eventually lead to non distinguishable wild and cultivated population genotypes in WV. Data from this study also showed that increase in cultivation intensity resulted in increase in the genetic diversity of American ginseng populations. While this is beneficial as it expands the gene pool, the potential downside to this is that the wide sourcing of seeds beyond the 'local' gene pool to meet cultivation need for large scale growers may end up introgressing 'non-local' genotypes into 'local' gene pool thereby eroding regional specific genotypes. To minimize this, we recommend growers to obtain seeds locally but from different sources within a region to help conserve regional genotypes.

ACKNOWLEDGEMENTS

The authors are thankful to ginseng growers and diggers for generously helping us with samples. Financial support for this project was provided by HATCH funds through the Division of Plant and Soil Sciences, West Virginia University.

REFERENCES

Ahmad B (1998). Plant exploration and documentation in view of land clearing in Sabah. In: Nair MNB and Ganapathi N (Eds). Medicinal Plants. Cure for the 21st Century. Biodiversity Conservation and Utilization of Medicinal Plants. Serdang, Malaysia, pp. 161-162.

Anderson RC, Armstrong JE, Benjamin PK, Fralish JS (1993). The ecology and biology of *Panax quinquefolium* L. (Araliaceae) in Illinois. Am. Mid. Natural, 129: 357-372.

Anon (2008). American ginseng and the idea of the Commons. In: Tending the Commons: folklife and landscape in Southern West Virginia. The Library of the Congress. http://memory.loc.gov/ammem/collections/tending/essay1a.html.

Bai D, Brandle J, Reeleder R (1997). Genetic diversity in North American ginseng (*Panax quinquefolius* L.) grown in Ontario detected by RAPD analysis. Genome, 40(1): 111-115.

Boehm CL, Harrison HC, Jung G, Nienhuis J (1999). Organization of American and Asian ginseng germplasm using randomly amplified polymorphic DNA (RAPD) markers. J. Am. Soc. Hortic. Sci., 124: 252-256.

Canter PH (2005). Bringing medicinal plants into cultivation. Focus Altern. Compl. Ther., 10: 167-168.

Carpenter SG, Cottam G (1982). Growth and reproduction of American ginseng (*Panax quinquefolius*) in Wisconsin. Can. J. Bot., 60: 2692-2696.

Catling PM, Haber E, Small E (1994). Poorly known economic plants of Canada: ginseng, *Panax quinquefolius* L. Can. Bot. Assoc. Bull., 27: 70-72.

Charron D, Gagnon D (1991). The demography of northern populations of *Panax quinquefolium* (American ginseng). J. Ecol., 79: 431-445.

Christensen LP, Jensen M, Kidmose U (2006). Simultaneous determination of ginsenosides and polyacetylenes in American ginseng root (*Panax quinquefolium* L.) by high performance liquid chromatography. J. Agric. Food. Chem., 54: 8995-9003.

Cruse-Sanders JM, Hamrick JL (2004). Genetic diversity in harvested and protected populations of wild American ginseng, *Panax quinquefolius* L. (Araliaceae). Am. J. Bot., 91 (4): 540-548.

Cruse-Sanders JM, Hamrick JL, Ahumada JA (2005). Consequences of harvesting for genetic diversity in American ginseng (Panax quinquefolius L.): a simulation study. Biodivers. Conserv., 14: 493-504.

Cunningham AB (1993). African medicinal plants. Setting priorities at the interface between conservation and primary healthcare. People and Plants Working Paper 1. UNESCO, Paris, France.

Doebley J (1989). Isozymic evidence and evolution of crop plants. In: Soltis ED and Soltis PM (Eds), Isozymes in Plant Biology. Dioscorides, Portland, OR, pp. 165-191.

Ellstrand NC, Elam DR (1993). Population genetic consequences of small population size: implications for plant conservation. Annual Rev. Ecol. Syst., 24: 271-242.

Excoffier L, Smouse PE, Quatro JM (1992). Analysis of molecular variance inferred from metric distances among DNA haplotypes: application to human mitochondria DNA restriction data. Genetics, 131(2): 479-491.

Excoffier L, Laval G, Schneider S (2005). ARLEQUIN ver. 3.11: An integrated software package for population genetics data analysis. Evol. Bioinfo. Online, 1: 47-50.

Grubbs JH, Case MA (2004). Allozyme variation in American ginseng (*Panax quinquefolius* L.): Variation, breeding system, and implications for current conservation practice. Conserv. Genet., 5: 13-23.

Hamilton AC (1997). Threats to plants: An analysis of centers of plant diversity. In: Conservation into the 21st Century. Proceedings of 4th International Botanic Gardens Conservation Congress (Eds. Touchell DH and Dixon KW), Kings Park and Botanic Garden, Perth, Australia, pp. 309-322.

Kuipers SE (1997). Trade in medicinal plants. In: Medicinal plants for forest conservation and healthcare, Non-Wood Forest Products (Eds. Bodeker G, Bhat KKS, Burley J and Vantomme P),. FAO, Rome, Italy, 11: 45-59

Lange D (1998). Europe's medicinal and aromatic plants: their use, trade and conservation: an overview. TRAFFIC International, Cambridge, United Kingdom.

Lewis WH (1988). Re-growth of a Decimated Population of (*Panax quinquefolium*) in a Missouri Climax Forest. Rhodora, 90: 1-5.

Lewis W, Zenger V (1982). Population dynamics of American ginseng (*Panax quinquefolium*) Araliaceae. Am. J. Bot., 69(9): 1483-1490.

Lewis W, Zenger V (1983). Breeding systems and fecundity in the American ginseng, *Panax quinquefolium* (Araliaceae). Am. J. Bot., 70 (3): 466-468.

Lewontin RC (1972). The apportionment of human diversity. Evol. Biol. 6: 381-398.

Lim W (2004). Population characterization of *Panax quinquefolium* with ginsenoside analysis and randomly amplified polymorphic DNA markers. Ph.D. dissertation. Cornell University, Ithaca, NY.

Miller MP (1997). AMOVA-PREP: A program for the preparation of input files for use in WINAMOVA. Department of Biological Sciences, Northern Arizona University, Flagstaff, AZ.

Miller MP (1998). TFPGA: A Windows® program for the analysis of allozyme and molecular population genetic data. Department of Biological Sciences, Northern Arizona University, Flagstaff, AZ.

Mooney HE (2007). Genetic and evolutionary consequences of harvest in American ginseng, *Panax quinquefolius* L. (Araliaceae). Ph.D. dissertation. West Virginia University, Morgantown, WV.

Nei M (1973). Analysis of gene diversity in subdivided populations. Proc. Nat. Acad. Sci., 70: 3321-3323.

Nei M (1978). Estimation of average heterozygosity and genetic distance from a small number of individuals. Genetics, 89: 583-590.

O'Hara M, Kiefer D, Farrell K, Kemper K (1998). A review of 12 commonly used medicinal herbs. Arch. Fam. Med., 7: 523-536.

Otero-Arnaiz A, Casas A, Hamrick JL, Cruse-Sanders J (2005). Genetic variation and evolution of *Polaskia chichipe* (Cactaceae) under domestication in the Tehuacán Valley, central Mexico. Mol. Ecol., 14: 1603-1611.

Pritts KD (1995). How to find, grow, and use America's forest gold. Stackpole Books. Mechanicsburg, PA.

Proctor JTA, Bailey WG (1987). Ginseng: Industry, botany, and culture. Hort. Rev., 9: 187-236.

Robbins CS (1998). American ginseng: The root of North America's medicinal herb trade. TRAFFIC North America Report. Number B347, Washington, DC.

Schlessman M (1985). Floral biology of American ginseng (*Panax quinquefolium*). Bul. Torrey Bot. Club, 112(2): 129-133.

Schluter C, Punja ZK (2002). Genetic diversity among natural and cultivated field populations and seed lots of American ginseng (*Panax quinquefolius* L.) in Canada. Int. J. Plant Sci., 163: 427-439.

Stewart Jr CN, Porter DM (1995). RAPD profiling in biological conservation: an application to estimating clonal variation in rare and endangered *Iliamna* in Virginia. Biol. Conserv. 74: 135-142.

Stewart Jr. CN, Excoffier L (1996). Assessing population genetic structure and variability with RADP data: Application to *Viccinium macrocarpon* (American Cranberry). J. Evol. Biol., 9: 153-171.

Wright S (1951). The genetical structure of populations. Annals of Eugenics, 15: 323-354.

Wright S (1978). Evolution and the genetics of populations, Variability within and among natural populations, University of Chicago Press, Chicago, IL. Vol. 4.

WVDOF (2008). http://www.wvforestry.com/ginseng.cfm?menucall=ginseng.

Yeh CF, Yang RC, Boyle T, Ye Z, Mao JX (1997). POPGENE. The user-friendly shareware for population genetic analysis. Molecular Biology and Biotechnology Center, University of Alberta, Edmonton, Canada.

Zohary D (2004). Unconscious selection and the evolution of domesticated plants. Econ. Bot., 58 (1): 5-10.

10

Feasibility studies on the potential of grafting and budding of frafra potato (*Solenostemon rotundifolius*)

K. Aculey[1], A. K. Quainoo[1]* and G. Mahanu[2]

[1]Department of Biotechnology, Faculty of Agriculture, University for Development Studies, Tamale, Ghana.
[2]Department of Horticulture, Faculty of Agriculture, University for Development Studies, Tamale, Ghana.

Frafra potato (*Solenostemon rotundifolius*) is a herbaceous root tuber that is grown mainly in the northern part of Ghana. It is used as a hunger crop and brewed for alcoholic beverages. In spite of the socio-economic importance of this crop, little research aimed at improving the crop and its small tuber size has been carried out. In an effort to improve upon the agronomic performance of frafra potato, a randomized complete block design was used to conduct an experiment to determine the possibility of grafting and budding the crop. Approach, Cleft and splice graftings and budding, using dormant and actively growing buds were used to bud the crop. With the exception of the approach grafting, all the grafted plants from cleft and splice grafting produced leaves and shoots. The splice grafting recorded the highest number of leaves and plant survival. The dormant buds used in budding recorded higher number of leaves, longer shoots and plant survival. Only the approach grafted plants established union between the stocks and the scion, while the cleft and splice grafting failed to establish a union. The experiment demonstrated the potential of mixing the genetic materials of different frafra potato materials through grafting. The experiment may be improved through the use of auxin to induce the union of stock and scion.

Key words: Frafra potato, small tuber size, grafting and budding.

INTRODUCTION

Due to the lack of proper storage and inadequate storage facilities, most farmers are not able to store enough food throughout the year. As such, food security has always been a common phenomenon especially during the period between planting and harvesting of succeeding crops (Weibe et al., 2000). As a result, crops that mature and are ready for consumption during these periods are normally cultivated before the main staple crops. One of such crops is the frafra potato, (*Solenostemon rotundifolius*). Frafra potato belongs to the family Labiatae (Tindall, 1983). It is a succulent crop with the stem square in cross-section (Dupriez, 1989). Frafra potato is suited for cultivation on marginal areas in the dry savannah regions with poor soils. It is propagated mainly by tubers. In Ghana, frafra potato is cultivated exclusively in the northern part of the country. Its tuber is used mainly as food. Among the frafra people, the tubers are presented to in-laws as a dowry. The tubers have also been used in the distillation of alcoholic beverage (Apabol, 1997). The leaves have been used in treating dysentery in some rural communities in Nigeria (Tindall, 1983).

Frafra potato is propagated mainly by tubers. However, vine cuttings of two to three nodes can also be used for its propagation (Kay, 1973). Notwithstanding the socio-economic importance of frafra potato, little research has been conducted to improve upon its tuber size which has been the main constrain to its cultivation as a main staple crop. The aim of the study was to determine the potential of grafting and budding frafra potato as a means of improving upon the agronomic performance of the crop.

MATERIALS AND METHODS

The experiment was conducted at the plant house of Nyankpala campus, University for Development Studies Ghana. The study area lies on latitude 9° 25 N and longitude 9° 58 W with altitude 183 m above mean sea level (SARI, 1997). Plastic containers of diameter and height 25 and 35 cm respectively were used as experimental pots for the experiment. The soil for growing the crop

*Corresponding author. E-mail: aquainoo@googlemail.com.

Figure 1. An insect pest feeding on frafra potato seedlings.

was prepared by mixing top soil, compost and river sand in the ratio 7:3:2. The pots were filled to 2/3 of volume with the prepared soil samples. Surgical blades were used to make various stock and scion cuttings of the experimental materials. Strips of gauze and polythene bag were used to hold the stock and scion together. A tailoring thread was used to tie the scion and the stock firmly together after grafting and budding. Staking sticks 90 cm long were used to keep the plants in an upward position. A randomized complete block design (RCDB) with five treatments and replicated five times were used. Five experimental treatments were carried out as follows: Treatment 1: approach grafting, Treatment 2: cleft grafting, Treatment 3: splice grafting, Treatment 4: budding using dormant buds and treatment 5: budding using actively growing buds.

Planting materials and protection of seedlings

Frafra potato tuber seeds were obtained from the seed bank of University for Development Studies. Healthy tuber seeds were selected and planted in July 2010. Frafra potato seedlings were sprayed with Karate to control insect pests feeding on the leaves (Figure 1). Unhealthy seedlings were uprooted and destroyed to prevent spread of diseases to the healthy seedlings.

Grafting and budding

The grafting and budding operations were carried out 12 weeks after germination of the frafra potato seeds. Cleft and splice grafting were employed with dormant and actively growing buds used for the budding operations.

Data collection and analysis

Data was collected every two weeks over a period of eight weeks; after grafting and budding on number of leaves on grafted and budded plants, length of plant shoots after grafting and budding and number of plants surviving eight weeks after grafting and budding. The data was subjected to analysis of variance (ANOVA), using the standard error difference (S.E.D) at significant level ($p < 0.005$) and the means separated using least significant difference (LSD).

RESULTS

Grafting of frafra potato

Number of leaves produced by grafted plants

The number of leaves produced by the grafted plants did not show any significant difference ($p > 0.05$) in Weeks 2 and 8. However, weeks 4 and 6 exhibited significant difference ($p < 0.05$) in the number of leaves produced by the grafted plants (Figure 2).

Length of plant shoot

Growth in the length of shoot produced by the grafted plants showed no significance differences at weeks 2 and 8. However, it showed significant differences ($p < 0.05$) at Weeks 4 and 6 (Figure 3).

Number of plants surviving grafting

There was no significant difference ($p > 0.05$) in the

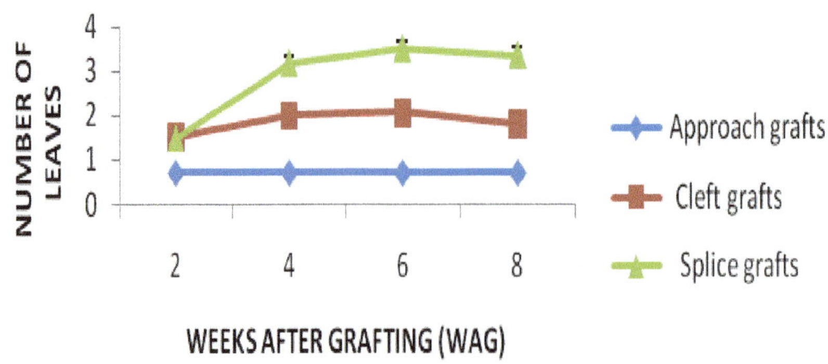

Figure 2. Mean number of leaves produced by grafted Frafra potato plants.

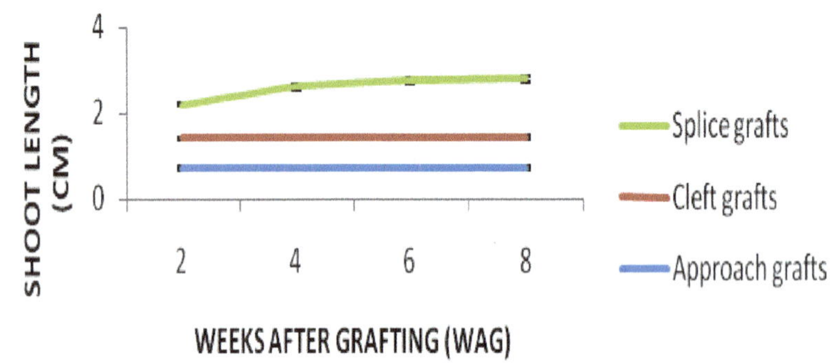

Figure 3. Mean shoot length of grafted Frafra potato plants.

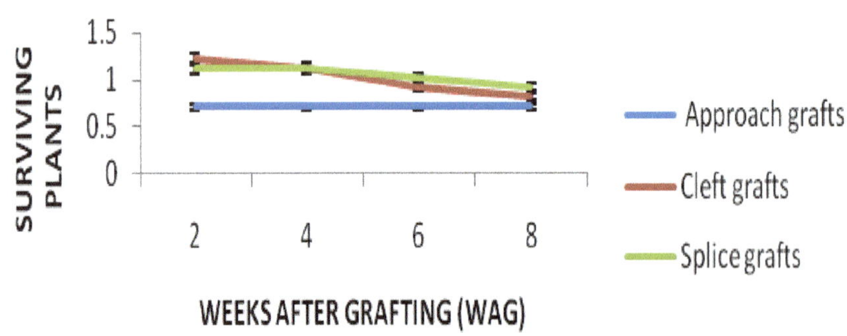

Figure 4. Mean number frafra potato plants surviving grafting.

number of plants that survived grafting at Week 8. However, at Weeks 2, 4 and 6, there were significant differences in the number of plants surviving grafting (Figure 4).

Budding of frafra potato

Number of leaves produced by budded plants

Significant differences ($P < 0.05$) were observed in the number of leaves produced by the dormant and active buds at Weeks 2 and 8. However, differences in the number of leaves produced by the dormant and active buds at weeks 4 and 6 were insignificant (Table 1).

Length of plant shoot

Significant differences ($p < 0.05$) were exhibited in the length of plant shoots generated by the budded plants using the dormant and active buds at weeks 2, 4 and 8.

Table 1. Mean number of leaves produced by budded plants.

Treatment	Week 2	Week 4	Week 6	Week 8
Dormant buds	1.600	2.980	3.290	3.580
Active buds	0.710	1.980	1.980	1.360
LSD	0.826	1.342	1.510	1.955
F. probability	0.041	0.107	0.074	0.034

Table 2. Average shoots length of budded frafra potato plants.

Treatment	Week 2	Week 4	Week 6	Week 8
Dormant buds	0.844	1.136	1.306	1.390
Active buds	0.7I0	0.804	0.982	0.828
LSD	0.133	0.308	0.439	0.403
F. probability	0.049	0.040	0.110	0.018

Table 3. Average number of budded frafra potato plants surviving.

Treatment	Week 2	Week 4	Week 6	Week 8
Dormant buds	1.220	1.220	1.220	1.220
Active buds	1.118	1.016	1.016	0.914
LSD	0.283	0.347	0.347	0.347
F. probability	0.374	0.178	0.178	0.070

However, at week 6, shoot length produced by the budded plants using the dormant and active buds showed no significant differences (Table 2).

Number of plants surviving budding

No significant differences ($P > 0.05$) were observed in surviving plant at Weeks 2, 4, 6 and 8. There were no significant differences in survival of plants generated from dormant and active buds (Table 3).

DISCUSSION

With the exception of the approach grafts, the growth pattern of leaves and length of shoots generated by the grafted and budded plants were similar. The number of leaves and shoots generated by the grafted plants initially increased rapidly but later increased at a decreasing rate. The splice grafts recorded the highest number of leaves and shoot lengths as compared with the approach and splice grafting (Figures 2 and 3).

Leaves and shoots length generated were also higher and longer respectively in plants budded with the dormant buds. Leaves and shoots growth of the budded plants decreased with time (Tables 1 and 2). The production of leaves and shoots in the grafted and budded plants was an indication of stock and scion/bud union (Hartmann et al., 1997). The decrease in the number of leaves as well as growth in shoot length by the grafted and the budded plants may be attributed to inadequate water through the limited conducting vessels as transpiration increased with increase in the number and size of leaves coupled with high temperatures. This situation, according to Hartmann et al. (1997) can lead to degeneration of the scion.

Survival of grafted plants was highest in the cleft grafted plants in the first two weeks after grafting followed by the splice grafted plants. However, at Week 8, average survival rate was highest in the splice grafts with the approach grafted plants recording the lowest (Figure 4). Low success rate in the initial approach grafting operation of uniting the stems of the frafra potato plants accounted for the low average number of leaves and shoots generated by the approach grafts.

Degeneration of either the stock or the scion was the cause of deaths in the grafted plants. The inability of the scion to obtain water and nutrients from the stock since the vascular connection between the stock and the scion has not been established may be attributed to the scion degeneration whilst degeneration of the stock may be attributed to the inability of the stock to obtain carbohydrates and other metabolites from the scion due to disruption of the phloem (Hartmann et al., 1997).

Apart from a single approach graft, none of the cleft and

Figure 5. Pictures of budded grafted frafra potatoes at various stages of development (a) frafra potato budded using a dormant bud at Week 2; (b) frafra potato budded using a dormant bud at Week 4; (c) a splice graft at Week 8; (d) a cleft graft at Week 8.

splice grafts as well as the budded plants was able to form a union between the stock and the scion/bud. This may be due to environmental conditions such as temperature and moisture and other factors such as plant species and craftsmanship.

CONCLUSION AND RECOMMENDATIONS

Generally, leaves and shoots were generated in most of the grafted and budded frafra potato plants. Moreover, most of the grafted and budded plants survived throughout the data collection period and although a union between the stock and the scion in almost all the grafted and budded plants was not successful, the results obtained give an indication that, with a controlled experiment, grafting and budding of frafra potato will be possible.

Since graft union was not formed in the cleft and splice grafts, as well as the budded plants, it is recommended that, auxin (IBA, NAA) and cytokine (BA) which was not used in the experiment should be applied during the grafting and budding operation since it has been found to enhance graft union formation in some plants.

It is also recommend, that the experiment should be conducted in a more controlled environment where temperature and moisture can be controlled to favour graft union formation. Other grafting methods can also tried in the grafting operation. Figure 5 depicts some grafted and budded frafra potatoes.

REFERENCES

Apabol RR (1997). Assessment of the performance of some frafra potato accession in Nyankpala area of Ghana. A dissertation submitted to the faculty of Agriculture, University for Development Studies, pp 1-55.

Dupriez H (1989). African Gardens and Orchards. Macmillan press limited. London and Basing stoke Translate from French into English by C. T. A., pp-316-317.

Hartmann HT, Kester DE, Geneve RL (1997). Plant propagation: principles and practices, 6th ed. pp 392-500.

Kay DE (1973). Root crops and products Digest 2. The tropical products institute, Foreign and Commonwealth Office (ODA) London, pp 76-245.

Savanna Agricultural Research Institute (SARI) (1997). In: Apabol RR 1997- a dissertation submitted to the faculty of Agriculture, University for Development Studies, pp 1-55.

Swiader JM; Ware GW and McCollum JP (1992). Producing vegetable crops, 4th ed. Interstate publishers, Inc, Danville. pp. 435-456

Tindall HD (1983). Vegetables in the Tropics. The Macmillan Press Limited, pp 242-245.

Weibe KD; Meredith JS; Clare N, Vince B (2000). Resource quality and Agriculture productivity: A multi country comparison. Selected paper presented at the annual meeting of the American Agricultural Economics Association, Tampa, FL. 31st July, 2000.

Up-scaling production of certified potato seed tubers in Kenya: Potential of aeroponics technology

Jane Muthoni*, Miriam Mbiyu and Jackson N. Kabira

Kenya Agricultural Research Institute (KARI), National Potato Research Centre, Tigoni, P. O. Box 338 00217, Limuru, Kenya.

Unavailability of certified seed tubers is a major constraint to potato production in Kenya. This compels most farmers to use planting materials from informal sources such as previous harvests, local markets and neighbours. Kenya Agricultural Research Institute (KARI), Tigoni has the national mandate to produce basic seed tubers but can only supply much less than 1% of the national seed requirements. In order to improve on the situation, aeroponics technique was introduced to enhance production of prebasic seeds from tissue culture plantlets. The aeroponics unit is still in experimental stages and is being evaluated for productivity and profitability. Some results show that aeroponics can increase production of prebasic minitubers. However, the aeroponic technology needs further and complete evaluation in terms of productivity, profitability and sustainability.

Key words: Aeroponics, certified seed, Kenya Agricultural Research Institute (KARI), Tigoni, minitubers, potatoes.

INTRODUCTION

Potato (*Solanum tuberosum* L.) is a crop of major economic importance worldwide (Tsegaw, 2005; FAO, 2008). On a global scale, potato is the fourth most cultivated food crop after wheat, rice, and maize (FAO, 2008). Potato is also the most important tuber crop ranking first in volume produced among root and tuber crops, with an annual production of approximately 300 million tonnes grown on about 19 million hectares (FAO, 2008), it is followed by cassava, sweet potato, and yam (FAO, 2004, 2008).

The world average potato production is about 17 t ha^{-1}, while direct consumption as human food is 31.3 kg per capita (kg yr^{-1}) (FAO, 1995, 2008). Worldwide, there are wide regional disparities in potato production (FAO, 2008). Asia and Europe are the world's major potato producing regions, accounting for more than 80% of world production while Africa is the least, accounting for about 5% (Table 1). North America is the clear leader in productivity at more than 40 t ha^{-1}, followed by Europe at 17.4 t ha^{-1} while Africa lags at about 10 t ha^{-1} (FAO, 2004; FAO, 2008). In Africa, the top ten potato producers in descending order are Egypt, Malawi, South Africa, Algeria, Morocco, Rwanda, Nigeria, Kenya, Uganda, and Angola (FAO, 2008). In Kenya, potato is the second most important food crop after maize (MoA/GTZ, 1998; CIP, 2006; MoA, 2008). It is grown by about 500 000 farmers, cultivating 120,000 hectares per season with an annual production of about 1 million tonnes in two growing seasons (MoA, 2008). National potato production ranges from 4.4 tha^{-1} to 15 tha^{-1} with an average of about 6.7 tha^{-1} although yields of 40 tha^{-1} are attainable under research conditions (Lung'aho et al., 1997; MoA, 2008). The low yields are attributed to production constraints such as low soil fertility, diseases such as bacterial wilt and late blight as well as high cost of inputs mainly certified seed tubers.

Certified potato seeds are produced only by Kenya Agricultural Research Institute (KARI), Tigoni, whose physical capacity is limited (Kaguongo et al., 2008).

Consequently, KARI can only produce 1% of the national certified seed requirement (Ayieko and Tschirley, 2006). Due to limited supply, the certified potato seeds are highly priced and the cost of seeds account for 42%

*Corresponding author. E-mail: jayney480@yahoo.com.

Table 1. Potato production by regions in 2007.

Region	Harvested area (,000 ha)	Production (,000 tons)	Productivity (t ha^{-1})
Africa	1,542	16,707	10.8
Asia/Oceania	8,733	137,344	15.7
Europe	7,474	130,224	17.4
Latin America	964	15,683	16.3
North America	616	25,345	41.2
World	19,328	325,302	16.8

Source: FAO, 2008.

Table 2. Amount of basic seed produced from KARI Tigoni and its subcentres from 2006 to 2009 (kg).

Season / Place	2006 Long rains	2006/2007 Short rains	2007 Long rains	2007/2008 Short rains	2008 Long rains	2008/2009 Short rains	2009 Long rains
Njambini sub station	5550	6250	4050	3000	6250	1500	16250
Marindas substation	14450	0	23650	0	8650	0	5050
Marimba substation	2050	2650	2650	4150	2650	3000	7500
Tigoni	13450	16500	20000	20000	16250	12600	9000
Total	35500	25400	50350	27150	33800	17100	37800

Source: KARI (2007).

of the total production costs (Kaguongo et al., 2008). Consequently, farmers depend on seed from informal sources which include farm-saved (self supply), local markets and neighbours (Kaguongo et al., 2008). Self-supply is the major source of potato seed tubers for most farmers (Kaguongo et al., 2008). This informal system leads to use of poor quality seeds and often accelerates the spread of seed-borne diseases (Ng'ang'a et al., 2003).

This paper reviews the certified potato seed production system in Kenya, the problems it faces, and the potential of aeroponic technology to improve on it.

PRODUCTION OF CERTIFIED POTATO SEED TUBERS IN KENYA

Conventionally, certified potato seed tuber production starts with *in vitro* plantlets from meristematic tissue in the tissue culture laboratory (KARI, 2007). The plantlets are multiplied and after hardening-off, they are planted in potted sand media in a screen house to produce pre-basic minituber seeds (generation 0). The prebasic minitubers are then multiplied for three generations in the field to produce generation 1, 2 and 3 basic seeds. The generation 3 basic seeds are planted for one generation at KARI, Tigoni and its substations, contracted certified seed farmers' fields as well as Agricultural Development Corporation (ADC) farms to produce certified seeds (KARI, 2007). Production of prebasic minituber seeds from *in vitro* plantlets is slow and this leads to a bottle-neck in certified seeds production chain (Muthoni et al., 2010). The multiplication rate is 6 to 8 minituber seeds/ plant (KARI, 2007). Table 2 shows amount of generation 3 basic seeds produced from KARI, Tigoni for the last seven years. Figure 1 shows a schematic system of seed potato production from pre-basic to basic as done at KARI, Tigoni

COMPARISON BETWEEN AEROPONICS TECHNOLOGY AND CONVENTIONAL METHODS

In a bid to compare the productivity of aeroponics technology and conventional methods, two most popular potato varieties *Tigoni* and *Asante* were evaluated for minituber production. The experiment was conducted between March 2009 and December 2009 (first season) while the second season was between March 2010 and December 2010.

MATERIALS AND METHODS

Aeroponics

The aeroponics unit at KARI, Tigoni was constructed in 2009. The

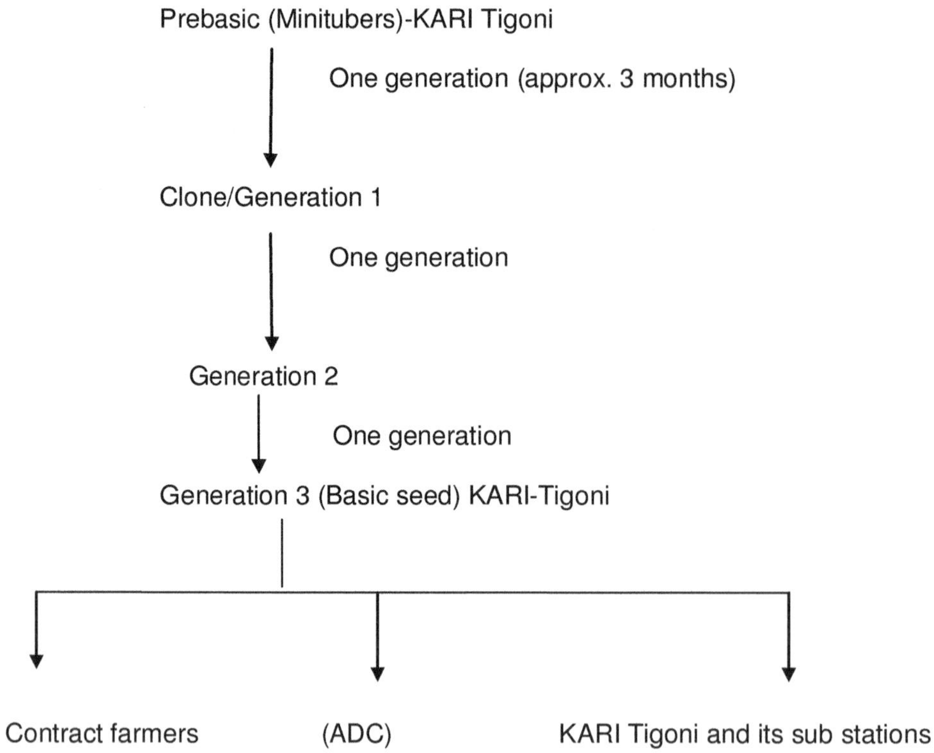

Figure 1. Schematic system of seed potato production from pre-basic to basic.
Source: KARI (2007).

Table 3. Composition of 500 liters nutrient solution.

Nutrient	Quantity (g)
KNO_3	2.2
NH_4NO_3	1.4
Ca superphosphate	0.8
$MgSO_4$	0.8
Fe(EDTA) Fe 6%	0.036
Micro(Fetrilon)	0.048
pH	6.5

Source: Farran and Mingo-Castel (2006).

unit consisted of a screen house measuring 30 m × 6 m. It was a metallic structure with translucent asbestos roof and the sides were made of white muslin shading net to lower the temperature inside. Inside the screen house were the growth chambers. Each chamber could accommodate 180 plants.

Each growth chamber consisted of a styrofoam box measuring 6 m × 1.8 m × 1.5 m. The sides were cut to ease minituber harvesting and monitoring of the fertigation system. The cut sides were covered with a black polythene paper which could be lifted to check the tuber growth. The fertigation system consisted of an underground plastic tank in which the nutrient solution was held. The nutrient solution was transmitted into growth chambers using pipes. The fertigation system was controlled by a timer and mists of nutrient solution were sprayed onto the dangling potato roots for 5 min after every 15 min. The nutrient solution was collected at the bottom of the boxes and pumped back into the tank. This system was powered by electricity, but a solar-powered generator was used in case of electricity failure. The composition of the nutrient stock solution used is given in Table 3. Each growth chamber represented a replication and each potato variety was grown in three such chambers giving a total of three replications for each potato variety. The experiment was laid out in a completely randomized design replicated three times.

Crop management

The plantlets from tissue culture laboratory were transplanted into growth boxes when they attained 5 cm tall with root length of about 2.5 cm and with 2 to 3 leaves. Small holes were made on the lids of the growth chambers and the plantlets were fitted into the holes at the crown with the roots hanging inside the box. The plantlets were held firmly in an upright position with the help of masking tape, this also ensured no light got inside the growth chamber. The lowest leaf of the plant was cut off every two weeks for the next two months. This allowed the plants to be pushed deep down into the growth chamber to promote stolon development.

The black polythene paper on the sides of the growth chamber was lifted to allow minituber harvesting. The minitubers were harvested when they attained a minimum of 8 g. This corresponded to a tuber diameter of 0.5 cm. This was done continuously to promote

Table 4. Average number of minitubers per replicate from aeroponics in the first season.

Minituber size	1	2	3	Total
Asante	7404	1856	470	9730
Tigoni	15298	3858	1146	20302

Minituber size (diameter): 1 = 0.5 to 1.5 cm, 2 = 1.51 to 2.5 cm, 3 = 2.51 to 3.5 cm.

Table 5. Average number of minitubers per replicate from aeroponics in the second season.

Minituber size	1	2	3	Total
Asante	7521	1820	521	9862
Tigoni	16620	4246	1059	21925

Minituber size (diameter): 1 = 0.5 to 1.5 cm, 2 = 1.51 to 2.5 cm, 3 = 2.51 to 3.5 cm.

Table 6. Average number of minitubers per replicate from pots in the first season.

Minituber size	1	2	3	Total
Tigoni	1155	761	164	2080
Asante	926	299	165	1390

Minituber size (diameter): 1 = 0.5 to 1.5 cm, 2 = 1.51 to 2.5 cm, 3 = 2.51 to 3.5 cm.

Table 7. Average number of minitubers per replicate from pots in the second season.

Minituber size	1	2	3	Total
Tigoni	1123	832	112	2067
Asante	912	322	153	1387

Minituber size (diameter): 1 = 0.5 to 1.5 cm, 2 = 1.51 to 2.5 cm, 3 = 2.51 to 3.5 cm.

promote tuber initiation. The first harvest was done five weeks after transplanting while subsequent harvesting was done weekly for the next 18 weeks. During each harvest, all the 180 plants in a growth chamber were harvested to give total replicate yield. The average replicate total yields for each variety were then calculated. The minitubers were then sorted out into three sizes based on diameter: Size 1 had a diameter of 0.5 to 1.5 cm, size 2 had diameter of 1.51 to 2.5 cm while size 3 had diameter of 2.51 to 3.5 cm. Harvested tubers were left to cure in wooden trays for 4 weeks before cold storage.

Conventional method

The experiment was laid out in a completely randomized design replicated three times. Each replication consisted of 180 pots. The plantlets from tissue culture laboratory were transplanted into pots when they were about 5 cm tall with root length of about 2.5 cm and with 2 to 3 leaves. Each plantlet was translated into its own pot. The pots were half filled with sterilised soil and put into screenhouse. The plants were watered regularly as need arose. Eight weeks after transplanting, all the plants were harvested. During each harvest, all the 180 plants in a replicate were harvested to give total replicate yield. The average replicate total yields for each variety were then calculated. The minitubers were then sorted out into three sizes based on diameter: Size 1 had a diameter of 0.5 to 1.5 cm, size 2 had diameter of 1.51 to 2.5 cm while size 3 had diameter of 2.51 to 3.5 cm. Harvested tubers were left to cure in wooden trays for 4 weeks before cold storage.

RESULTS AND DISCUSSION

Variety *Tigoni* produced more minitubers than *Asante* in both seasons for both aeroponics and conventional methods (Tables 4, 5, 6, and 7). For aeroponics, variety *Tigoni* produced an average total yield of 20,302 minitubers during the first season which translated to about 112 minitubers per plant (Table 4). In the second season, it produced 21,925 minitubers which translated to 122 tubers per plant (Table 5). By comparison, variety *Asante* produced an average total of 9,730 minitubers during the first season which translated to 54 minitubers per plant

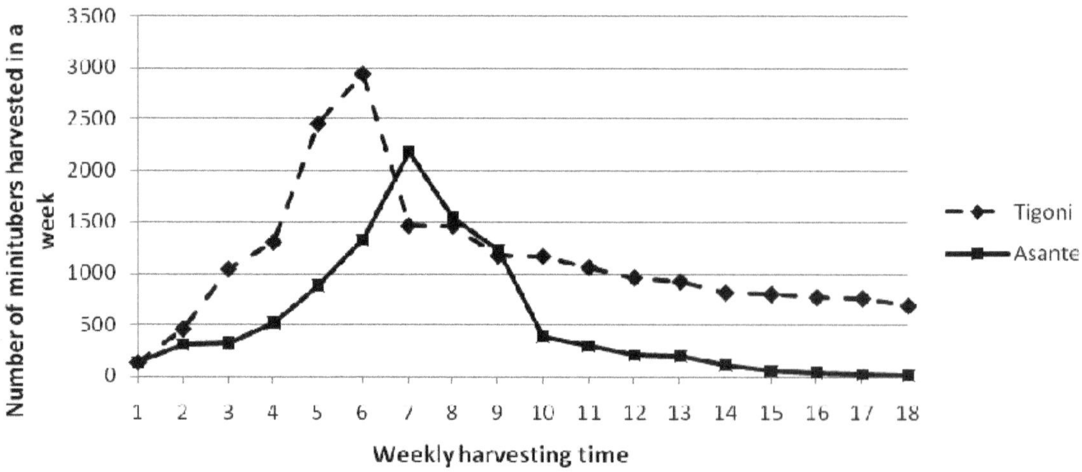

Figure 2. Average number of total minitubers harvested from each replication over a period of 18 weeks during the first season.

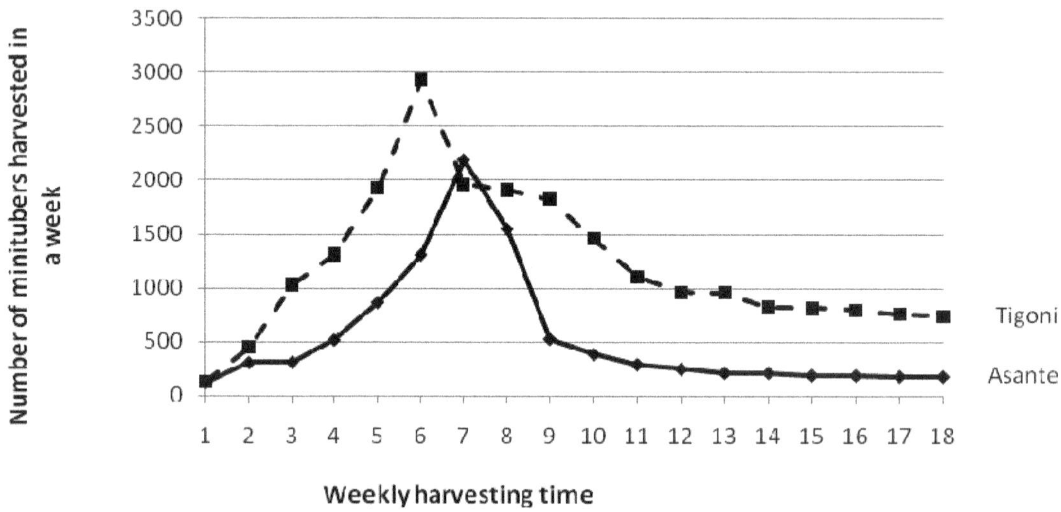

Figure 3. Average number of total minitubers harvested from each replication over a period of 18 weeks during the second season.

per plant (Table 4). During the second season, it yielded 9,862 minitubers which translated to 55 minitubers per plant (Table 5). This high multiplication rate was most likely due to longer harvesting period (Figures 2 and 3), which was made possible by the non-destructive harvesting method.

For the conventional pot method, variety *Tigoni* produced 2080 minitubers in the first season which translated into 12 minitubers per plant (Table 6). In the second season, it yielded 2067 minitubers which translated into 11 minitubers per plant (Table 7). By comparison, variety *Asante* produced an average total of 1,390 minitubers during the first season which translated to 8 minitubers per plant (Table 6). During the second season, it yielded 1,387 minitubers which translated to 8 minitubers per plant (Table 7).

For variety *Tigoni,* the minituber multiplication rate was 9.76 times higher for the aeroponics technology than the conventional pot method in the first season, in the second season, the minituber multiplication rate for aeroponics technology was 10.6 times higher than the conventional pot method. For variety *Asante,* the minituber multiplication rate was 7 times higher for the aeroponics technology than the conventional pot method in both seasons.

Previous study have indicated a multiplication rate 14 to 18 times higher for aeroponics technology than the conventional methods for variety *Tigoni* while for variety *Asante*, the multiplication rate was 7 to 9 times higher for aeroponics than the conventional pot method (CIP, 2010). In the same study, 67 minitubers/ plant were obtained from potato variety *Yungay*, 70 minitubers/ plant from variety *Canchán INIA*, and 69 minitubers/ plant from variety *Perricholi* when using aeroponic technique. These potato varieties produced an average of 5 to 10 minitubers per plant using conventional pot method (CIP, 2010).

In terms of minituber size distribution, the small size (diameter 0.5 to 1.51 cm) had the highest number of minitubers harvested in both seasons for both varieties (Tables 4 and 5). The largest size (diameter 2.51 to 3.5 cm) had the least number of minitubers (Tables 4 and 5). This has the advantage in that the small minitubers can be planted on a far much larger area than the large minitubers given the same mass.

CONCLUSIONS AND RECOMMENDATIONS

Aeroponics technology produced far more minituber than the conventional pot method. The aeroponics technology might offer some solution to the perennial problem of shortage of certified potato seeds in Kenya. However, the system needs further and complete evaluation in terms of productivity, profitability and sustainability. In addition, a comparative cost analysis is required between aeroponics and conventional methods in terms of unit cost per tuber.

ACKNOWLEDGEMENTS

The authors acknowledge International Potato Center (CIP) for putting up the aeroponics unit where the study was conducted, and for all the technical assistance without which the study could not have been conducted. The authors also acknowledge the technical staff at KARI, Tigoni who worked tirelessly in the aeroponics unit. The authors also acknowledge the assistance of the management KARI, Tigoni for facilitating the study.

REFERENCES

Ayieko M, Tschirley D (2006). Enhancing access and utilization of improved seed for food security in Kenya. Working paper No. 27/2006. Tegemeo Institute of agricultural policy and development, Egerton University, Kenya.

Centro Internacional de la Papa (2006). World potato atlas [Online]. Available at https://research.cip.cgiar.org/confluence/display/wpa/Kenya. (Accessed on 25 March 2010). CIP, Lima, Peru.

Centro Internacional de la Papa (2010). International Potato Centre, Press. Available at Releases.cip-webmaster@cgiar.org. Accessed on 25 April 2010). CIP, Lima, Peru.

Farran I, Mingo-Castel AM (2006). Potato mini tuber production using aeroponics: Effect of plant density and harvesting Intervals. American J. Potato Res., 2006.http://findarticles.com/p/articles/mi_qa4069/is_200601/ai_n17183789.

Food and Agriculture Organisation of the United Nations (1995). Potato in the 1990's. Situation and prospects of the world potato economy. A joint study by the Basic Foodstuffs Service. FAO Commodities and Trade Division, and Post Harvest Management Marketing Program. FAO, Rome.

Food and Agriculture Organisation of the United Nations (2004). Production statistics [Online] Available at http://www.apps.fao.org (Accessed 26 March 2010). FAO, Rome.

Food and Agriculture Organisation of the United Nations (2008). International year of the potato [Online]. Available at http://www.potato2008.org (Accessed 10 March 2010). FAO, Rome.

Kaguongo WP, Gildemache P, Demo P, Wagoire W, Kinyae P, Andrade J, Forbes G, Fuglie K, and Thiele G (2008). Farmer practices and adoption of improved potato varieties in Kenya and Uganda. Social Sciences Working Paper 2008-5. CIP, Lima, Peru.

Kenya Agricultural Research Institute (2007). Annual Report 2006. KARI. Nairobi, Kenya.

Kenya Agricultural Research Institute (2010). Annual Report 2009. KARI. Nairobi, Kenya.

Lung'aho C, M'makwa C, Kidanemariam HM (1997). Effect of source of mother plant, variety and growing conditions on the production of stem cuttings and subsequent yield of mini-tubers in the Kenyan potato programme. *In* Proceedings of the 4th Triennial Congress of the African Potato Association. 23-28 February, 1997. Pretoria, South Africa. African Potato Association, Kampala, Uganda.

Lutaladio NB, Ewell PT, Kidanemariam H (1995). Advances in potato research in eastern and central Africa. *In* M.O. Akoroda and I.J. Ekanayake (ed) Root crops and poverty alleviation: Proceedings of the Sixth Triennial Symposium of the International Society for Tropical Root Crops-Africa Branch (ISTRC-AB), Lilongwe, Malawi. 22-28 October, 1995. ISTRC-AB, Lilongwe, Malawi.

Ministry of Agriculture (2008). National policy on potato industry. Policy reforms to revitalize the potato industry. MoA, Nairobi, Kenya.

Ministry of Agriculture and Deutsche Gessellschaft fur Technische Zusammenarbeit (1998). Postharvest systems of potato and sweet potato in Kenya. Final report. MoA/GTZ, Nairobi, Kenya.

Muthoni J, Mbiyu MW, Nyamongo DO (2010). A review of potato seed systems and germplasm conservation in Kenya. J. Agric. Food Info., 11:157–167.

Ng'ang'a NM, Kinyae P, Walingo A, Wakahiu MW (2003). Establishing effective seed potato prices for small scale farmers in Kenya. Unpublished report. KARI. Nairobi, Kenya.

Tsegaw T (2005). Response of potato to paclobutrazol and manipulation of reproductive growth under tropical conditions. Ph.D. Thesis. University of Pretoria, South Africa.

Composting certain agricultural residues to potting soils

M. Saber[1], Badr-el-Din, S.[1], Awad, N.[1] and Mohammed, Z.[2]*,

National Research Center, Cairo, Egypt.
Faculty of Science, Cairo University Egypt.

Four agricultural residues, namely, sheath of peanut pods, flax shivers, red sawdust, and white sawdust were composted with different amendments for four months under greenhouse and field conditions. Results emphasized that composting proceeded at different rates and was principally influenced by the type of residues and/or added amendments. Initial bio-fortification of residues with cellulose decomposing microorganisms exhibited a recognized trend of stimulation and led to a higher rate of organic carbon loss. Percentage losses in organic carbon of the sheath of peanut pods after composting were higher in bio-fortified treatment. The organic accelerator associated with bio-fortification led to a loss in the organic carbon amounting to 15.61% of the initial value of flax shivers after 120 days of composting. Bio-fortification of red sawdust receiving either an organic or a chemical accelerator increased the amounts of bio-oxidized organic carbon to 9.13% in control and to 16.45 or 12.97% respectively in amended red sawdust. In white sawdust, the differences in the percentage of oxidized carbon were hardly detectable. The nitrogen (N) concentrations increased in the final compost, and the largest amount was found in the sheath of peanut pods, reaching 2.6%. The level of total N was more or less the same in composts prepared from either flax shivers, red or white sawdust where it was 1.65, 1.63 and 1.5%, respectively. Oxidation of organic matter associated with the increase in nitrogen concentrations, whether initially or during composting, led to a reduction in carbon-to-nitrogen (C/N) ratios in the final composts. Generally, all bio-fortified composts had narrower C/N ratios compared to the non-biofortified ones. The lowest organic carbon and total nitrogen contents and consequently C/N ratios were found in sheath of peanut pods compost compared to other residues. Initial neutral pH values of different composted residues were slightly shifted towards acidity before settling around neutrality. Total soluble salts in the final composts varied within the type of agricultural residues used. Nevertheless, values in all composts at the end of the composting did not exceed safe levels.

Key words: Potting soil, biofortification, composting, cellulose decomposing microorganisms, sawdust, carbon-to-nitrogen (C/N) ratios.

INTRODUCTION

Potting soil is usually prepared by composting various organic residues and used to grow certain horticulture plants in contained greenhouses. Most commercially available brands of potting soil have specific chemical and physical characters related to their bulk and particle density, size distribution and water holding capacity as well as carbon-to-nitrogen (C/N ratio, cation exchange capacity, buffering action, quantity and availability of nutrients and pH value. No doubt, microorganisms have a beneficial role both in composting and bio-furnishing potting soils. Saber (1996) confirmed that bio-fortification gives rise to composting of many agricultural residues to potting soils.

The present work aims to maximize the role of micro-organisms in composting certain agricultural residues to potting soil.

MATERIALS AND METHODS

Greenhouse trails

A completely randomized block design greenhouse experiment, in

*Corresponding author. E-mail: msaber1941@yahoo.com.

four replicates was carried out. Four agricultural residues, that is, shell of peanut pods, flax shivers, red sawdust (inner wood of red ceder tree) and white sawdust (inner wood of spruce tree), were composted to a potting soil in plastic bags under different experimental treatments. Agricultural residues were air-dried, sieved to pass through a 2 mm sieve, and moistened to 70% of their water holding capacity. Each residue was composted in association with either a chemical accelerator (7 kg superphosphate, 40 kg ammonium sulfate, and 35 kg calcium carbonate and 100 kg fertile soil per ton dry matter) or an organic accelerator (100 kg of chicken manure per ton dry matter). Control bags received neither chemical nor organic accelerator. Each treatment, as well as control, was divided into two sub-groups, the first was biofortified with a suspension of cellulose decomposing microorganisms, and the second was kept without biofortification. Treated and untreated agricultural residues were pocked in a sufficient number of net plastic bags containing 1 kg each, and were turned off every two weeks. The moisture content during 120 composting period was kept at a proper level throughout irrigation. Samples from each replicate were collected initially and after 15, 30, 60, 90 and 120 days for chemical and microbiological analyses.

Field trails

Two compost heaps, 500 kg each, using red or white sawdust were built in windrows. The agriculture residues were thoroughly mixed with a chemical accelerator at the rate of 40 kg ammonium sulfate, 7 kg superphosphate, 1.25 kg potassium sulfate, 35 kg calcium carbonate and 100 kg fertile soil per ton. The heaps were built in ten successive layers, and attention was paid to maintain moisture at a proper level for microbial proliferation. Both heaps were left for a 120 days composting course and were turned over and rebuilt again three times after 15, 30 and 60 days.

Chemical and physical methods

The dry matter was recorded after drying the plant samples at 70°C in a hot air oven until constant weight. Total nitrogen was determined by the semi-microkjeldahl method using H_2SO_4 and H_2O_2, sulfuric acid N/70 was used for titration with bromocresol green-methyl red mixture (Horowitz, 1980). Organic matter was determined by loss on ignition in a muffle furnace at 900 + 25°C for 1 h. (Michiels et al, 1979). Values of pH were determined in a paste 1:2.5 soil/water, after standing for 2 h, using the glass electrode method. Electrical conductivity (E.C.) was determined in soil using a conductivity bridge in 1:5 soil/water suspensions. Electrical conductivity of potting soil was determined by weighting 5 g compost in an Erlenmeyer 250 ml and adding 100 ml distilled water, shaking for 1 h, filter and measuring E.C. in the same way as explained for soil (Michiels et al., 1979). Soil water holding capacity was determined using Keen and Raezkovski box.

RESULTS

Chemical changes took place during composting of the four agricultural residues were followed up for four months. Initial incorporation of chicken manure as an organic accelerator to the different agricultural residues slightly raised their organic carbon contents. Biological oxidation of organic matter during the composting course led to a persistent decrease in the organic carbon contents in the different composted materials, yet at varied rates. Such decreases exhibit the efficiency of biomass in decomposing the raw materials. Data in Table (1) shows that biofortification of agricultural residues with cellulose decomposing microorganisms led to a higher rate of organic matter oxidation. The percentage losses in organic carbon of the sheath of peanut pods after 120 days composting period were higher in biofortified treatments associated with either chemical or organic accelerator. Nevertheless, the differences between the influences of the two accelerators were slight.The final organic carbon contents were 72 and 71% of their initial values in biofortified composts receiving respectively either chemical or organic accelerator, while it was 83% in control compost.

In case of flax shivers, biofortification with cellulose decomposing microorganisms exhibited the same effect, despite being at a lower rate. Generally, most biofortified composts lost more organic carbon during the composting course compared to non-biofortified ones. Results show that the organic accelerator associated with biofortification led to a loss in the organic carbon amounting to 16% of the initial value after 120 days of flax shivers composting. However, the chemical accelerator associated with biofortification resulted in slightly smaller losses in organic carbon, the differences were not appreciable.

Biofortification of red sawdust receiving either chicken manure as an organic accelerator or a chemical accelerator raised the amounts of bio-oxidized organic carbon to 9% in control and 17 or 13% respectively in both amended raw materials. The amounts of organic carbon lost in the absence of biofortification with cellulose decomposing microorganisms, followed the same trend, but were much smaller. White sawdust composting followed the same aforementioned trend, still, at varied rates. The differences in the percentage of oxidized carbon were small and were more or less within the same range. The rate of organic carbon oxidation during the composting period varied between the four raw materials tested. The greatest losses were reached in biofortified composts receiving chicken manure. The oxidation of organic carbon was greatest in case of sheath of peanut pods followed by red sawdust and flax shivers and was smallest in case of white sawdust. It was found that 71, 84 and 89% respectively of the initial values represented the amounts of organic carbon found after 120 days.

Results exhibited an initial increase in the total nitrogen contents of the different raw materials as a result of supplementation with either chemical or organic accelerators. The increases in total nitrogen contents proceeded at varying rates throughout the composting course due to the activation of free living nitrogen-fixing diazotrophs. Also, the decomposition in organic matter through bio-oxidation raised the percentage of total nitrogen in the composts compared to their initial values.

During the composting course of sheath of peanut pods the total nitrogen values in the different treatments increased by 112 and 127% of their initial values, respectively in non-biofortified and biofortified controls (Table 2).

Table 1. Effects of different treatments of on carbon percentage in the investigated agricultural residues (Oven dry basis).

Treatment	Time in days					
	0	15	30	60	90	120
Sheath of peanut pods						
Control	40.73	38.70	36.63	35.86	34.30	33.78
CDB	40.15	36.56	34.80	33.70	33.54	33.14
Chemical	39.17	35.75	34.39	33.36	33.03	32.31
Chemical + CDB	40.11	34.38	32.25	31.00	29.44	29.00
Organic	41.33	36.33	33.78	33.65	33.21	33.25
Organic + CDB	42.00	36.18	33.46	33.14	32.34	30.00
Flax shivers						
Control	48.05	44.70	43.62	42.92	41.84	41.84
CDB	50.50	47.00	45.99	45.19	44.37	43.98
Chemical	46.90	43.39	42.75	41.68	41.25	40.81
Chemical + CDB	48.82	45.14	44.47	42.98	42.51	42.28
Organic	48.12	43.95	43.18	42.36	41.86	40.56
Organic + CDB	51.53	46.83	46.06	45.29	44.79	43.49
Red sawdust						
Control	44.64	44.14	43.92	43.50	43.43	42.87
CDB	45.82	44.18	43.39	42.93	42.20	41.64
Chemical	45.63	43.81	43.15	42.72	42.15	41.37
Chemical + CDB	46.57	43.76	42.79	41.84	41.18	40.53
Organic	47.00	44.01	43.50	42.99	42.31	41.93
Organic + CDB	47.31	41.97	41.19	40.50	40.01	39.53
White sawdust						
Control	46.66	45.77	45.08	44.66	44.32	44.18
CDB	48.63	47.43	46.73	46.09	45.46	44.67
Chemical	48.63	47.43	46.73	46.09	45.46	44.67
Chemical + CDB	48.93	47.50	46.96	46.23	44.91	44.37
Organic	49.00	47.44	46.64	46.16	45.43	44.74
Organic + CDB	49.15	46.05	45.66	44.80	44.13	43.87

CDB= cellulose decomposing microorganisms; Chemical = chemical accelerator; Organic = organic accelerator

However, the increases recorded in total nitrogen contents in the chemically accelerated sheath of peanut pods, reached 168 and 174% from their initial values at the end of 120 days composting in non-biofortified and biofortified treatments respectively. The highest increase in total nitrogen was found in case of sheath of peanut pods amended with chicken manure as an organic accelerator. Total nitrogen increased by 194 and 213% from the initial values in non-biofortified and biofortified composts respectively.

Composting of flax shivers under the different treatments led to variable increases in total N contents. The greatest increases were recorded in flax shivers composts supplemented with the organic accelerator followed by those receiving a chemical accelerator. Generally, the increase in total N was greater in biofortified composts even in controls. Biofortification increased the total N contents in compost from the initial values after 120 days from 120 to 122% in controls, and from 174 to 183% in composts enriched with the chemical accelerator and from 206 to 209% in composts enriched with the organic accelerator.

Both red and white sawdust had low contents of total nitrogen, that is, 0.15 and 0.11% respectively. The initial treatments increased the total N contents several times to 0.47% in case of red sawdust and to 0.48% in case of

Table 2. Effects of different treatments of on nitrogen percentage in the investigated agricultural residues (Oven dry basis).

Treatment	Time in days					
	0	15	30	60	90	120
Sheath of peanut pods						
Control	1.10	1.07	1.10	1.20	1.20	1.23
CDB	1.10	1.00	1.20	1.31	1.33	1.40
Chemical	1.30	1.73	1.78	2.00	2.17	2.19
Chemical + CDB	1.32	1.80	2.00	2.20	2.30	2.30
Organic	1.16	2.10	2.30	2.30	2.31	2.25
Organic + CDB	1.25	2.28	2.43	2.53	2.59	2.66
Flax shivers						
Control	0.35	0.37	0.38	0.41	0.43	0.42
CDB	0.36	0.38	0.42	0.45	0.45	0.44
Chemical	0.69	0.79	0.98	1.10	1.20	1.20
Chemical +CDB	0.71	0.83	1.00	1.20	1.31	1.30
Organic	0.73	0.87	1.15	1.38	1.53	1.50
Organic + CDB	0.79	0.96	1.26	1.59	1.66	1.65
Red sawdust						
Control	0.15	0.17	0.23	0.25	0.26	0.25
CDB	0.16	0.19	0.25	0.29	0.33	0.40
Chemical	0.47	0.67	0.88	1.00	1.10	1.00
Chemical + CDB	0.47	0.69	0.91	1.31	1.37	1.35
Organic	0.43	0.71	0.95	1.23	1.29	1.20
Organic +CDB	0.46	0.75	0.99	1.68	1.65	1.63
White sawdust						
Control	0.11	0.12	0.13	0.15	0.17	0.18
CDB	0.12	0.15	0.21	0.25	0.27	0.25
Chemical	0.48	0.50	0.80	0.91	1.15	1.05
Chemical +CDB	0.49	0.53	0.90	1.00	1.20	1.20
Organic	0.46	0.58	0.87	1.00	1.46	1.43
Organic + CDB	0.46	0.60	0.99	1.30	1.50	1.50

CDB, cellulose decomposing microorganisms; Chemical, chemical accelerator; Organic, organic accelerator.

white sawdust. The organic accelerator led, more or less, to the same increases amounting to 0.43% in case of red sawdust and 0.46% in case of white sawdust. These increases markedly raised the total N contents at the end of sawdust composting over those recorded in other local composts. This however, might be ascribed to the initial low contents of total N. Such increases amounted to 167 and 250% from the initial values in non-biofortified and biofortified controls of red sawdust compost respectively. However, the recorded increases in total N contents in red sawdust composts, from the initial values, reached 213, 287 and 279 and 354% in non-biofortified and biofortified composts receiving either chemical or organic accelerator respectively.

More or less the same level of total N contents was recorded in case of white sawdust composts and the results followed the same trends displayed in case of red sawdust. The total N contents increased by 164 and 208%, from the initial values, in non-biofortified and biofortified control composts respectively. While in composts receiving either chemical or organic accelerator the increases in total N contents reached 219, 244 and 311, 326%, from the initial values, in non-biofortified and biofortified treatments respectively.

The N contents in the final compost increased according to the type of the different raw materials. The largest amount was found to be in the sheath of peanut pods reaching 2.6%. The level of total N was more or less the same in composts prepared from either flax shivers, red or white sawdust where it was 1.65, 1.63 and1.5%, respectively.

The C/N ratio in the final compost is usually used as an

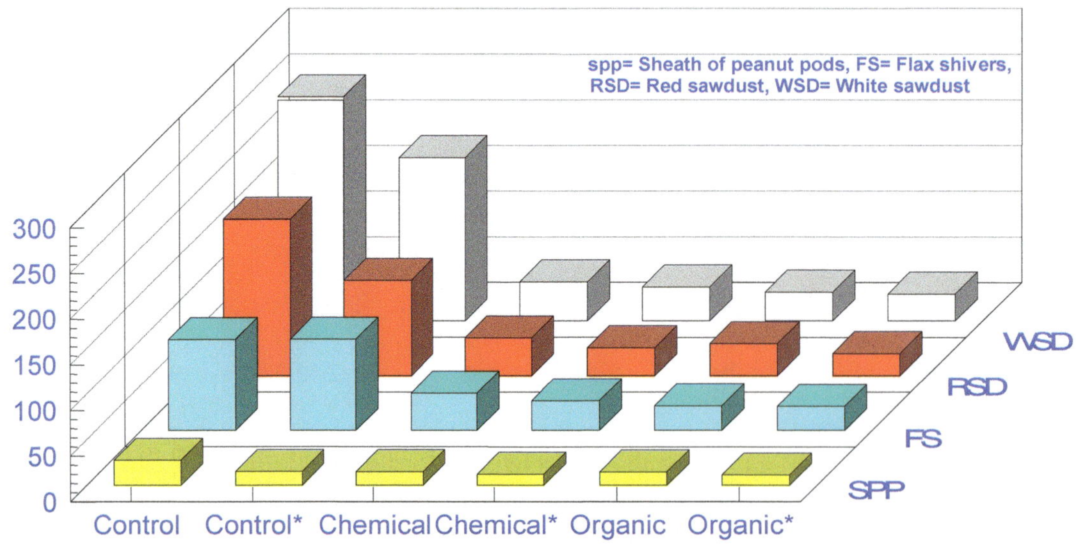

Figure 1. Changes in C/N ration of different raw materials using during composting. *Biofortified with cellulose decomposing microorganisms

Table 3. Effects of different treatments of on pH in the investigated agricultural residues (Oven dry basis).

Treatment	Sheath of peanut pods		Flax shivers		Red sawdust		White sawdust	
	I	F	I	F	I	F	I	F
Control	6.56	6.37	7.10	7.00	6.84	7.00	7.00	7.12
CDB	6.57	6.57	7.00	7.03	6.95	7.00	6.85	7.0
Chemical	6.38	6.38	6.92	7.11	6.70	7.03	6.50	7.00
Chemical + CDB	6.44	6.58	6.97	7.00	6.66	7.01	6.45	6.80
Organic	6.94	6.93	6.80	7.13	6.94	7.01	6.80	7.00
Organic + CDB	6.65	6.65	6.80	6.89	7.11	7.00	6.91	6.57

CDB, Cellulose decomposing microorganisms; Chemical, chemical accelerator; Organic, organic accelerator.

indicator for evaluating their suitability and maturity (Figure 1). In case of sheath peanut pods, the C/N ratios in the final composts ranged between 27.00 in non-biofortified controls and 11.28 in biofortified composts receiving organic accelerator. Nevertheless, the chemical accelerator led to more or less the same value of C/N ratio. The final C/N ratios in the flax shivers were 99.6 and 100 in non-biofortified and biofortified composts. On the other hand, the C/N ratios were calculated as 40.81, 32.52 and 26.88, 26.21 in case of biofortified and non-biofortified composts receiving respectively either chemical or organic accelerator. Incorporation of the chemical accelerator with either red or /and white sawdust resulted in a narrowing in C/N ratios in the final compost 41.37, 30.64 and 42.54, 36.97 in non-biofortified and biofortified red and white sawdust represented the calculated C/N ratios respectively. The influence of the organic accelerator, however, on composting of either red or white sawdust far exceeded that of the chemical one. The calculated C/N ratios in the final compost were 34.24, 24.25 and 31.28, 29.24 respectively in non-biofortified and biofortified red and white sawdust.

Data in Table 3 reveal that initial pH values of composted material were lowest in case of sheath of peanut pods (6.5) and highest in case of flax shivers (7.1). The incorporation of the chemical or organic accelerator in the presence and absence of biofortification with cellulose decomposing microorganisms did not show any marked changes in the recorded initial pH values. Nevertheless, the effect of biofortification was small in comparison with that of acceleration with either organic or chemical additives.

The initial values of total soluble salts expressed in terms of electrical conductivity units, given in Table 4, were increased as a result of treating the different plant materials with any of the chemical or organic accelerators. The added soluble salts through the chemical accelerator far exceed those added with the organic accelerator. Biofortification did not exert any initial influence on the amounts of total soluble salts. It is worth

Table 4. Effects of different treatments of on in EC (dS/cm^{-1}) in the investigated agricultural residues (Oven dry basis).

Treatment	Sheath of peanut pods		Flax shivers		Red sawdust		White sawdust	
	I	F	I	F	I	F	I	F
Control	1.58	1.51	0.47	0.47	0.60	0.59	0.50	0.53
CDB	1.40	1.38	0.65	0.63	0.50	0.50	0.50	0.49
Chemical	3.30	3.27	2.10	2.08	2.43	2.31	2.50	2.43
Chemical + CDB	2.93	2.90	2.26	2.13	2.57	2.40	2.20	2.20
Organic	1.70	1.70	1.08	1.00	1.00	0.93	1.00	1.00
Organic + CDB	1.63	1.65	1.20	1.20	1.00	1.00	1.00	1.00

CDB, cellulose decomposing microorganisms; Chemical, chemical accelerator; Organic, organic accelerator; I, initial; F, final.

mentioning that although the total soluble salts in sheath of peanut pods were at least three fold greater those measured in flax shivers, red sawdust or white sawdust,yet, they did not reach an injurious level (which is 6 dS cm^{-1}).

DISCUSSION

The most important parameters in relation to composting were changes in organic carbon, total nitrogen, C/N ratio and pH value. These changes reflect the extent of the biotransformations that took place in the different composts in chemical terms. Also, changes in the total soluble salts are of substantial importance in evaluating the quality of the potting soils compost (Fialho et al., 2010). Results obtained emphasized that the composting process proceeded at different rates and was principally influenced by the type of raw materials and/or the type of added amendments.

Biofortification with cellulose-decomposing microorganisms stimulated the rate of bio-oxidation of organic matter. Percentage losses in organic carbon after 120 days composting period were greater in biofortified treatments associated with either chemical or organic accelerators. Nevertheless, the differences between the influences of the two types of accelerator were slight.

Biofortification exerted substantial differences in C/N ratios in the final compost as achieved by Kostov et al. (1991) who found that sawdust had greater microbial biomass, greater CO_2 evolution, greater ammonification and more actinomycetes but less nitrification and fewer fungi compared with bark. All groups and activities were greater in sawdust and bark compared with the soil used as the substrate. They stated that cellulose decomposing strains of *Bacillus* sp., *Cephalosporium* sp. and *Streptomyces* sp. sometimes increased these activities but only marginally. Generally all biofortified composts had narrower C/N ratios compared to the non-biofortified ones. However, this holds true for all the four of the tested raw materials Otherwise, an immobilization of nitrogen and other macro- and/or micronutrient elements might be expected in soil with small amounts of available nitrogen, receiving compost having a wide C/N ratio (Zeng et al., 2010).

The oxidation of organic carbon was highest in case of sheath of peanut pods followed by red sawdust and flax shivers, and was lowest in white sawdust.

Atanasova et al. (1990) studied the mineralization of carbon and nitrogen during the composting of sawdust and bark associated with organic manure and found that decomposition of sawdust was more rapid than that of bark. Also, N'dayegamiye and Isfan (1991) carried comparative composting trials with wood shavings, sawdust and peat moss under field conditions. In each trial, the fresh material was mixed with cattle manure at 2:1 ratio by volume. The C/N ratio decreased after 36 month from 43 to 17 in WS pile and from 48 to 35 in peat moss pile. Tsai and Huany (1994) examined the use of chicken sawdust-rice straw, and dairy-sawdust-rice straw, under three aeration rates (0.0, 6.6 ± 0.5 and 25.9 ± 1.9 liter/min) and recorded a decrease in carbon content and an increase in N, P and K contents during composting

After the initial increase in total nitrogen contents a result of the amendments, total nitrogen continued to increase during composting due to the activity of free living nitrogen-fixing diazotrophs together with the decrease in organic carbon contents in the final composts (Tsai, 1994 and Bueno et al., 2008). The highest total nitrogen content in mature composts was recorded in sheath of peanut pods, 2.66%, however, total nitrogen contents in other composts ranged around 1.6%.

The C/N ratio is usually used as an indicator for maturity, suitability, potential nitrogen immabolization and stability of potting soils. Most of the published papers revealed that a C/N ratio of about 1:25 is proper for land application and characterizes a high grade of compost (Alexander, 1977 and Stevenson, 1982). Both organic carbon decreases and total nitrogen increases were reflected in narrowing C/N ratios in final composts as previously stated by Singh et al. (1992) and Tsai (1994). In order to prepare a potting soil from agricultural residue with a wide initial C/N ratio (>100), nitrogen is provided in the form of conventional organic manure and/or mineral fertilizers. The composting process results in softening initial differences in C/N ratios among various raw materials.

As a result of the bio-oxidation process during the different stages of composting, an obvious decrease in the pH values was observed in the case of sheath of peanut pods. Thereafter, the pH values under all treatments were probably affected by the high buffering capacity of the organic colloidal fraction present therein. Furthermore, the neutralizing effect of calcium ion, added through the chemical accelerator may play a marked buffering role, besides the decomposition of the organic acids that took place throughout the late stage of composting. Generally, the final pH values in all composted materials at the end of the experiment were around neutrality (Avnimelech et al., 1996) which is considered suitable for a high grade compost potting soils.

Most plants can survive at a wide range of substrate pH (4-8) without suffering significant physiological disorders as long all the nutrients are provided in available forms. However, growth rate and plant performance might be slowed down at extreme acid or alkaline conditions. The main effects of pH are exerted on nutrients availability, CEC and biological activity.

Total soluble salts in the final composts varied mainly within the type of raw materials used in composting. Recorded values for the total soluble salts in all composts at the end of the composting, however, did not exceed the safe level. It seems reasonable to state that both pH (Poole et al., 1981) and E.C. values in the final composts fulfilled the approved levels of a high-grade potting soil mentioned in the literature.

Peat usually satisfies these requirements and hence is widely used as potting soil. However, it should be examined carefully before use as peat is diverse in nature and may vary even within the same deposit (Hadar, 1986). Many organic materials have been tested as alternatives to peat according to the characteristics discussed above. Although peat seems to be the best of all the growth media tested, the ability of peat substitutes to promote the survival of targeted microorganisms justifies their development as an alternative to peat.

In conclusion, results show that the proper organic carbon and total nitrogen contents and consequently C/N ratios were found in sheath of peanut pods compost compared to other raw materials.

REFERENCES

Alexander M (1977). Introduction to Soil Microbiology. John Wiley & Sons Inc. (2nd ed.), 207p.

Atanasova G, Kostov O, Rankov V (1990). Microbiological processes during the composting of wastes from the wood-processing industry. Pochvoznanie. i. Agrokhimiya, 25(2): 64-72.

Avnimelech Y, Bruner M, Ezrony I, Sela R, Kochba M (1996). Stability indexes for municipal sold waste compost. Compost Sci. Utilization, 4(2): 12-20.

Bueno P, Tapias R, López F, Díaz M (2008) Optimizing composting parameters for nitrogen conservation in composting. Bioresour. Technol., 99(11): 5069-5077.

Fialho L, Silva W, Milori D, Simões M, Martin L (2010). Characterization of organic matter from composting of different residues by physicochemical and spectroscopic methods Bioresour. Technol., 101(6): 1927-1934

Hadar Y (1986). The role of organic matter in the introduction of biofertilizers and biocontrol agents to soil. In: The role of organic matter in modern agriculture. Eds. Y. Chen and Y. Avnimelech by Martins Nijhoff Pub., Dordrecht. Netherlands, pp. 169-176.

Horowitz W (1980). Official Methods of Analysis. 13th Eden-Association of Official Analytical Chemists, Washington, D. C., 1018p.

Kostov O, Rankov V, Atanacova G, Lynch JM (1991). Decomposition of sawdust and bark treated with cellulose-decomposing microorganisms. Biol. Fert. Soils, 11(2): 105-110.

N'dayegamiye A, Isfan D (1991). Chemical and biological changes in compost of wood shavings, sawdust and peat moss. Can. J. Soil Sci., 71(4): 475-484.

Poole RT, Conover CA, Joiner JN (1981). Soils and potting mixtures. In Foliage Plant Production. Ed. J.N. Joiner. Prentice-Hall Inc., Englewood Cliffs, N.J., pp. 179-202

Saber MSM (1996). Biofortified farming systems for sustainable agriculture and improved environment. Global Environmental Biotechnology Approaching the Year 2000. International Soc. for Environ. Biotechn., Third Intern. Symposium, July 15-20 1996, Boston, Massachusetts, USA. pp. 66-73

Singh S, Mishra MM, Sneh G, Kapoor KK (1992). Legume-cereal straw compost enriched with Mussoorie rock phosphate as a substitute of inorganic N and P fertilizers. Int. J. Trop. Agric., 10(3): 226-232.

Stevenson FJ (1982). Humus Chemistry. John Wiley & Sons Inc., 443p.

Tsai YF (1994). Production of compost from mushroom wastes. Bull. Taichung Dist. Agric. Improvement Station, 44: 13-21.

Tsai YF, Huany HC (1994). Effects of different organic wastes and aeration rates on the nutrients contents in composts. Bull. of Taichung District Agric. Improvment Station 43: 25-33.

Zeng G, Yu M, ChenY, Huang D, Zhang J, Huang H, Jiang R, Yu Z (2010) Effects of inoculation with Phanerochaete chrysosporium at various time points on enzyme activities during agricultural waste composting Bioresour. Technol., 101(1): 222-227

Impact of government financial incentives on peri-urban vegetable production in Botswana

M. E. Madisa and Y. Assefa*

Department of Crop Science and Production, Botswana College of Agriculture, Private Bag 0027, Gaborone, Botswana.

In its effort to increase agricultural productivity, the Government of Botswana has established a financial scheme to improve vegetable production. Citizen Entrepreneurial Development Agency (CEDA) and CEDA young farmers fund are encouraging farmers to get involved in vegetable production through financial and technical assistants. However, data on the impacts of these funds on vegetable production in the country are scarce and often unreliable. Very little is in fact known about the change in the quantity of vegetable produced, vegetable produce imported to the country and consumed after these funds are introduced. This paper discusses the impact of these government incentives on farm land allocated to vegetables, total vegetable production and import of vegetable in Botswana. The increasing interest in vegetable production, the change in demography and the trend in import of production inputs are also discussed.

Key words: Botswana, Citizen Entrepreneurial Development Agency (CEDA), financial incentives, government, urban, vegetable.

INTRODUCTION

The contribution of agricultural sector in Botswana has been declining since independence in 1966. This decline is largely attributed to the rapid development of the mining sector (Seleka, 1999; Anon, 2009). In its effort to reverse this trend and diversify the economy from mineral sector, the government of Botswana has introduced a number of interventions to promote agricultural development. The major strides in this regard are the financial schemes and import restriction systems developed starting from the late 1970s (MoA, 1977). Botswana's agricultural sector started getting stronger after the national development plan 9 (NDP 9) policy on critical sectoral issues and future strategy for development (MoA, 1991), which stated that agriculture should be strengthened and diversified. In line with this policy, the government of Botswana reviewed financial assistance policy (FAP) which was started in 1982 and lasted up to 1998. This project was then succeeded with a new financial scheme, citizen entrepreneurial development agency (CEDA) that is still running. In 2004, an additional financial scheme named CEDA young farmers fund (CEDA YFF) was initiated with a purpose of addressing the younger and educated farmers (CEDA 2006). In CEDA YFF, the government put an age restriction on beneficiaries and the project focused first on young farmers with age less than 35 years, which was later raised to 41 years to include some older farmers. These financial schemes provided financial assistance to several subsistence farmers in urban and peri-urban areas of the country. Several vegetable farmers using modern technologies and improved seeds have emerged and are actively involved in medium scale vegetable production. However, very little is in fact known about the change in the quantity of vegetable produced, vegetable produce imported to the country and consumed after these funds are introduced. This paper discusses the impact of these government incentives on farm land allocated to vegetables, total vegetable production and import of vegetable in Botswana. The increasing interest in vegetable production, the change in demography and the trend in import of production inputs are also discussed.

MATERIALS AND METHODS

Data collection and analyses

The data on area cultivated, vegetable produced, and number of vegetable projects, were obtained from Department of Crop Production, Horticulture Division Annual Reports. The authors

*Corresponding author. E-mail yosepharm@yahoo.com.

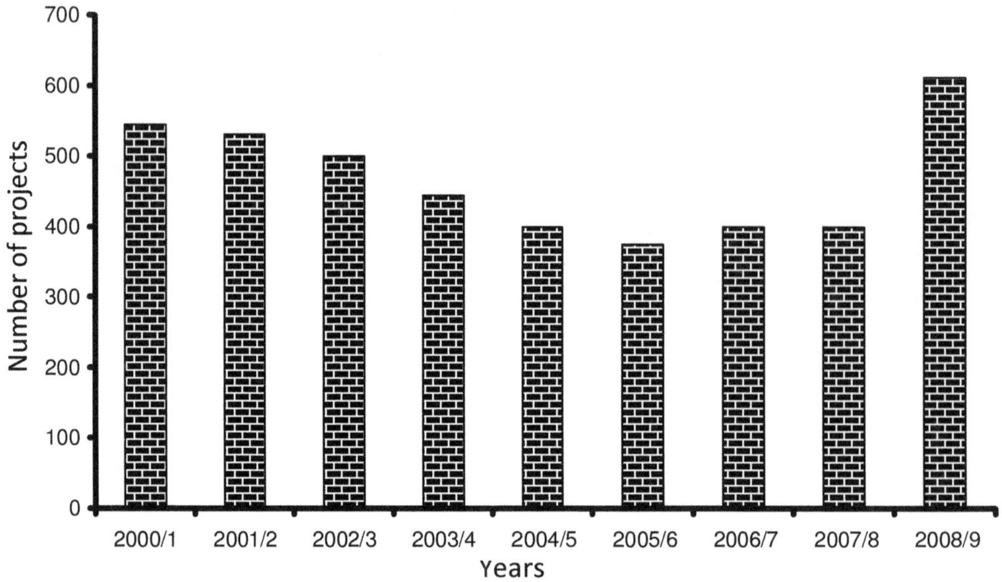

Figure 1. Variation in the number of vegetable projects funded by different financial schemes between 2000 and 2009 in Botswana. Source: Horticulture division annual reports.

participated in the collection and compiling of the annual reports of the department for the duration of the study. Data on vegetable import and consumers' demand were obtained from central statistics office (CSO). Proportion of vegetable farmers receiving fund from different schemes were calculated and graphs were produced using excel spread sheet.

Government financial assistance schemes and import restrictions

1. Financial assistance policy (FAP) was started in 1982 and ended in 1998. The objective of FAP was to promote employment creation, economic diversification (to reduce the economy's dependence on mining, the cattle industry and the public sector), rural development and active citizen, and women participation and ownership of productive ventures (Molokonmme, 1992; Rebaagetse, 1999). The majority of horticultural enterprises in Botswana were developed through grant money received from FAP. According to Rebaagetse (1999), FAP had funded 357 horticultural projects by 1998, about 167 of which were operational in 1999. The maximum repayment period for such loans was 5 years. In FAP, medium-scale vegetable projects were receiving loans ranging from P 150, 001 to P 2 million, at an annual interest of 7.5% and maximum repayment period of up to seven years (FAP, 1982).
2. Citizen entrepreneurial development agency fund (CEDA) was introduced in 2002 to promote the development of citizen business and citizen participation in the economy. Its specific objectives are; to support enterprises that which add value to the economy; to promote the development of citizen entrepreneurial culture and skills; to promote diversification; to promote the development of competitive and sustainable citizen-owned enterprises, to promote sustainable employment creation; to promote linkages of citizen-owned enterprises with primary industries; and to improve efficiency in business service delivery (CEDA guidelines, 2002).
3. CEDA young farmers fund (YFF) was established in 2006, under CEDA through which young people (aged between 18 and 35 years) will have improved access to finance and entrepreneurial training and engage in sustainable agricultural activities, better equipped with the requisite skills for running farming business; established to assist in reversing challenges of facing Botswana, such as unemployment, rural-urban migration and slow rural development and for the youth, they predominantly form Botswana's population, they are the most affected section of the society by varied challenges. The fund financed maximum loan of P 500.000.00 at an interest rate of 5% with payment period ranging from 5 to 10 years. According to CEDA 2009 annual report, 188 projects were approved under CEDA YFF at the value of P 76 million and horticulture accounted for 32 % of the total funding at P 24 million.
4. Import restriction: The government through Southern African Custom Union Agreement (SACUA) provisions have also put in place a mechanism to protect its horticultural producers through permit system. This mechanism was put in action by statutory instrument No.56 of 1977, control of Goods Act (Cap.43.07). Permits are issued for household consumption and commercial purpose (MoA, 1977). It is through this system that amount of fruits and vegetables that can be brought into the country at any one time are regulated.

RESULTS

Number of vegetable projects funded by government financial schemes

Figure 1 shows the number of projects in the country immediately after the termination of FAP. The vegetable project numbers declined from 550 in 2000 to 400 in 2005/2006. Thereafter, the projects grew slightly in 2006/2007 and 2007/2008. However, in 2008/2009, a dramatic increase in number of vegetable projects was recorded (from 400 projects to 742).

The proportion of vegetable projects and the contributions of government financial schemes are presented in Table 1. Out of the total number of projects, 414 (57.74%) were funded by individuals, while a total of

Table 1. Proportion of vegetable projects funded by government financial schemes in 2009.

Funding organization	Number of projects funded	Percentage
Individual	414	57.74
Financial assistance policy	166	23.15
CEDA	84	11.72
Government departments	39	5.44
NGO	14	1.95
Total	717	100

Source: Horticulture division annual reports.

Table 2. Cultivated area, vegetable production and productivity from 1997 to 2008.

Year	Area cultivated (ha)	% increase in area from previous year	Total Production (tonnes)	Productivity (tonnes/ha)
1997/1998	407.85	-	6,900	16.9
1998/1999	496.75	21.8	9,198	18.5
1999/2000	545.45	9.80	9,885	18.1
2000/2001	988.5	81.23	2,995	3.0
2001/2002	671.35	-32.08	8080	12.03
2002/2003	462.34	-32.13	13,406	29.0
2003/2004	546.0	18.09	16,000	29.3
2004/2005	670.96	22.89	18,000	26.8
2005/2006	698.0	4.03	24,076	34.5
2006/2007	1026.0	46.99	24,076	23.5
2007/2008	973.10	-5.16	31,985	32.9
2008/2009	830.0	-14.71	31,150	37.5

Source: Horticulture division annual reports.

287 (40.3%) of the vegetable projects were funded through government financial schemes (Table 1). FAP funded 166 vegetable projects (23.15%), CEDA and CEDA YFF supported 84 projects (11.72%), while other government departments funded 39 of the vegetable projects, accounting for 5.44%. There were relatively few non-governmental organizations in the country and they only supported 14 (1.95%) of the vegetable projects

Area allocated for vegetable production and productivity

Area allocated to vegetable production has fluctuated over the years (Table 2). In 1997/1998 season, the area under cultivation was 407.85 ha and this area increased to 496.75 ha in 1998/1999 season representing 21.8% increase. In 1999/2000 season, the area cultivated increased by only 9.80% to 545.45 ha. While in 2000/2001, there observed a drastic increase in land under cultivation to 988.5 ha (81.23 %). However, this increase was followed by a decline in area under cultivation by 32 % to 671.35 ha in 2001/02 season. This decline continued into 2002/2003 season, when the area under cultivation went down by 32.13% to 462.34 ha. In 2003/2004, the area increased to 546.0 ha (18.09% increase). In 2004/2005, the area increased by 22.89% to 670.96 ha. However, in the following year in 2005/2006, the area under cultivation increased to 698.0 ha (only by 4.03%). In 2006/7 another drastic increase in area under cultivation was recorded, as the cultivated area increased to 1026.0 ha (a 46.99% increase). After 2006/2007 season, a downward trend in area under cultivation was observed. In 2007/8 the area declined to 973.1 ha (a 5.16% reduction), while in 2008/2009, the area declined to 830.0 ha, that is, a decline of 14.71% from 2007/2008 season. However, it should be noted that the area under vegetable production increased from 407.85 ha in 1997/1998 to 830.0 ha in 2008/2009, representing a significant increase of more than hundred percent in a period of 12 years.

A general increase in productivity was observed over the years, even though there was no consistency (Table 2). In 1997/1998 season, 6,900 tonnes of vegetables were produced in the country at a production rate of 16.9 tonnes per hectare. While in 1998/1999 season, the total production increased to 9,198 tonnes and the average production also increased to 18.5 t/ha. The total production increased to 9885 tonnes in 1999/2000 with a slight reduction in the yield per hectare (18.1 t/ha). In 2000/2001 season the total production declined to 2,995 tonnes and the yield per hectare recorded for the year was also the lowest, averaging 3.0 t/ha. The following year, (in 2001/2002 season) total production increased to 8.080 tonnes and the productivity was 12.03 t/ha. In 2002/2003 total production was 13,406 with productivity of 29.0 t/ha. While in 2003/2004 season, the total production increased to 16, 000 tonnes without a change in productivity per hectare. In 2004/2005 season, total production increased to 18,000 tonnes with relatively lower productivity of 26.8 t/ha. In 2005/2006 and 2006/2007 season, total production increased to 24,076 tonnes with productivity of 34.5 and 23.5 t/ha respectively. In 2007/2008, total production increased to 31,985 tonnes with an average of 32.9 t/ha. In 2008/2009 season, total production stood at 31,150 tonnes with the highest productivity ever recorded of 37.5 t/ha. It is worth noting that total production increased four and half times from 6,900 tonnes in 1997/1998 to 31,150 tonnes in 2008/2009 and the productivity for the same period increased by more than two fold from 16.9 to 37.5 t/ha (Table 2).

Import of vegetables to Botswana and national vegetable demand

In 1997/1998, the national demand stood at 31033 tonnes and 77.77% (24,133 tonnes) of it was covered through vegetable imported from neighbouring countries (Table 3). While in 1998/1999, national demand declined slightly to 38,277 and quantities of imported vegetables increased to 29,079 tonnes, contributing 73.64% of national demand. Contrary to the previous year, national demand increased to 39,490 in 1999/2000 and imports increased slightly to 29,605 tonnes satisfying 74.97% of the demand. In 2000/2001, the national demand increased to 40,481 tonnes and the quantity of imported vegetables increased to 37,488 tonnes, with the contribution of import elevating to 92.61%.

After 2000/2001 season, the percent contribution of imports to national demand continued to decline. For instance, in 2001/2002, the national demand increased to 41,409, while imported vegetables declined to 33329 tonnes, satisfying only 80.49% of the nations demand for vegetables. Similarly in 2002/2003, the national demand rose to 43664 tonnes, while the quantity of imported vegetables decreased to 30258 tonnes, declining by 69.30%. Similarly, the national demand increased to 44667 tonnes in 2003/2004, while the quantity of imported vegetables declined to 28667, contributing only to 64.18% of the national vegetable demand.

By 2004/2005, the national demand stood at 46,000 tonnes, while the percentage contribution of imports to national demand was 28000 tonnes (60.87%). In 2005/2006, the national demand for vegetables was 46667 tonnes, whereas the quantity of imported vegetables was 22,591 tonnes. For the first time, the contribution of the imported vegetables went below fifty percent of the national demand at 48.41% in 2005/2006. National demand stood at 47333 tonnes in 2006/2007 at the same time as imports decreased to 23257 tonnes representing 49.13% of national demand. The national demand continued to grow, reaching 49,333 tonnes in 2007/2008, as the quantity of imported vegetables declined to 17348 tonnes, contributing 35.17% to the vegetable demand. The objective of Botswana Government desire for import substitution appear to be bearing fruits, as currently, the national demand stands at 50,000 tonnes, and the country imports 18,850 tonnes, with import's contribution falling to 37.70% to the national demand (Table 3).

DISCUSSION

Number of vegetable projects funded by government financial schemes

Despite the continuity in funding through government financial schemes, vegetable projects started declining from 550 in the year 2000/2001 to 400 in 2005/2006. This decline is mainly attributed to the collapse of FAP funded projects (Rebaagetse, 1999). As reported by Hovorka (2004), most FAP funded vegetable projects collapsed as they could not support themselves. Seleka at al. (2002) attributed the decline to low skill level of producers, inadequate labour (particularly skilled labour), the inability to identify and target support to farmers with potential for growth. Research results show that the process of adoption is highly dynamic and variable with farmers testing a farming practice and discontinuing or re-adopting them due to a whole range of factors, of which incentives and profitability could be among the reasons to be mentioned. Farmers getting involved in agriculture for such reasons may be termed 'pseudo-adopters' (Kiptot et al., 2007). The increase in number of vegetable projects observed in 2005/2006 may, therefore, be associated either with the increase in the re-adopting farmers or the newly adopting farmers funded by the newly introduced CEDA and CEDA YFF who got involved in vegetable production (CEDA, 2009). According to CEDA (2009), 188 projects were approved under CEDA YFF at the value of P 76 million and horticulture accounted for 32% of the total funding at P 24 million. This demonstrates the emphasis that government gave to the improvement of vegetable production in the country. The government's effort, however, has not resulted in continuous increase in the number of vegetable projects. The trend observed in the number of vegetable projects has some important implications. For urban and peri-urban vegetable production to be attractive to farmers, they must provide other tangible economic benefits besides financial assistance to commence vegetable farming. Detailed data on productivity, cost of production, harvest volume,

Table 3. Vegetable imported to the country from 1997 to 2008 and its contribution to satisfying the national demand.

Year	Vegetable imported to the country (tonnes)	National demand for vegetable (tonnes)	% contribution of import to satisfy demand
1997/1998	24133	31033	77.77
1998/1999	29079	38277	73.64
1999/2000	29605	39490	74.97
2000/2001	37488	40481	92.61
2001/2002	33329	41409	80.49
2002/2003	30258	43664	69.30
2003/2004	28667	44667	64.18
2004/2005	28000	46000	60.87
2005/2006	22591	46667	48.41
2006/2007	23257	47333	49.13
2007/2008	17348	49333	35.17
2008/2009	18850	50000	37.70

Source: CSO (2004).

wasted vegetable, and farmers' knowledge need to be collected and relevant correction measure taken to insure the sustainability of government funded vegetable projects.

Area allocated for vegetable production

The drive to increase vegetable production in the country has resulted in an increase in the area devoted to vegetable production by the farmers. This could be attributed to financial incentives from government that enabled them to acquire irrigation and other farm inputs, which otherwise they could not afford. Similar result was reported by Piotto et al. (2004), where the area devoted for agriculture responded to government and other incentives, with farmers dedicating portions of their land as they continue to receive government incentives. Remenyi (2000) stated that lack of access by urban poor to micro-finance services, especially credit for working capital, is an important constraint on greater, more varied and more innovative forms of participation in urban agriculture on wage employment or self-employment. Financial incentives such as CEDA and CEDA YFF have played a significant role in devoting more land to vegetable production in the country, leading to an increase in vegetable production. Good policies like this are reported to have a huge potential role in alleviating poverty and should be encouraged (Barbier, 2000).

Vegetable production and productivity

Botswana's effort to increase vegetable production is partially met as there is an observed gradual increase in the amount of vegetables produced in the country. In the 12 years considered in this study, the national vegetable production has doubled. The productivity has also shown a drastic improvement through time. The 16.9 t/ha reported for the year 1997/1998 has reached 37.5 t/ha in 2008/2009 season. This increase in vegetable production and productivity observed could be associated with an increase in area allocated for vegetable production and/or due to the use of production inputs like fertilizer and pesticides. In recent surveys conducted in vegetable producing urban and peri-urban areas of Botswana, Madisa et al. (2010) reported a significant change in the demography and educational background of vegetable farmers. Their observations on the educational background and age of vegetable farmers were found to be opposite to what is commonly known of small-scale farmers in Africa (Kamara et al., 2001; Lewu and Assefa, 2009) and what was previously reported from vegetable farms of Botswana (CSO, 2004). Most of the interviewed farmers in this study were literate and were at their active age. Existing literature show that improved crop production strategies require high levels of expertise from farmers in order to be implemented effectively (Crosby et al., 2000) and farmers who are better educated are generally more open to innovative ideas and new technologies that promote technical change (Lapar and Ehui, 2003). The educated farmers who get involved in vegetable production through incentives such as CEDA and CEDA YFF might have contributed to the observed increase in productivity as they have a better knowledge of the benefits of adopting improved vegetable production practices and are likely to have adopted the practices. Our view agrees with the rapid increase in the import of production inputs from the neighbouring countries (CSO, 2004).

Import of vegetables in Botswana

Though there is significant increase in production and productivity, Botswana is still heavily dependent on import to satisfy the nations demand for vegetable. In 1997/1998, Botswana national vegetable demand was 31033 tonnes and 77% was met by imports as the country was only producing 6900 tonnes or 22% of its national demand. The country managed to reduce the imports from 77 to 37.7% in 2008/2009. At one point (2001/2002), the country was importing 92% of its national vegetable demands. These significant results show that government's efforts on increasing vegetable production through financial incentives to farmers are bearing fruits. The increase in vegetable production saves the country a lot of foreign reserves, it also provide a lot of jobs to the nationals. Diversification into horticultural production as reported by Pumpkins et al. (2005) and Chand et al. (2008) is becoming attractive for many poor farmers around the world. The sector is reported to be profitable (Lumpkins et al., 2005) as farmers involved in horticultural production usually earn much higher farm incomes as compared to cereal producers per capita farm income. However, the continued dependence on import of vegetables to satisfy the national demand is negatively affecting the country. However, market oriented production requires the use of high yielding varieties and increase dependence on chemical fertilizer, pesticide and micro-nutrient addition (Brown and Shrestha, 2000) which ultimately increase production cost. The high level of operational costs incurred in small-volume-based vegetable production due to high input cost, unavailability of the inputs in nearby markets and lack of transport further soar the cost of production and hinders farmers from participating in these emerging markets (Machethe, 2004). Sooner or later, non-competitive segments of the market fall (the small-scale peri-urban farmers and traders) by the way side in due course. These issues are fundamental and need focusing. More efforts are needed to cushion the vulnerable groups, who have previously dominated the vegetable market, against the emerging big and medium commercial players who are currently in complete control of the wholesale, institution and retailer market. Hence, government incentive and import restriction need to be assisted with strategies that are required for tackling issues such as sustainability and viability and operational efficiency of the urban and peri-urban vegetable projects.

Responsible governmental institutes need to give a due consideration to and closely follow-up the vegetable production practices in urban and peri-urban areas and assist farmers with production, management and marketing of vegetables.

REFERENCES

Barbier EB (2000). The economic linkages between rural poverty and land degradation: some evidence from Africa. Agric. Ecosyst. Environ., 82(1-3): 355-370

Brown S, Shrestha B (2000). Market-driven land-use dynamics in the middle mountains of Nepal. J. Environ. Manage., 59 (3): 217-225.

CEDA (2002). Citizen Entrepreneurial Development Agency: Guidelines, Government Printer, Gaborone.

CEDA (2006). Citizen Entrepreneurial Development Agency; Young Farmers Fund Guidelines

Chand R, Raju SS, Pandey LM (2008). Progress and potential of horticulture in India. Indian J. Agric. Econ., 63(3): 299-309

Crosby CT, De Lange MM, Stimie CM, Van Der Stoep I (2000). A review of planning and design procedures applicable to small-scale farmer irrigation projects. WRC Report No. 578/2/00. Water Research Commission, Pretoria, South Africa.

CSO (2004). Population and Housing Census: Summary Statistics on Urban Areas Gaborone Government Printer, Vol. 1-3

Hovorka AJ (2004). Commercial Urban Agriculture in Greater Gaborone: Form and Function, Challenges and Prospects. Bot. J. Afr Stud.

Kamara A, Van Koppen B, Magingxa L (2001). Economic viability of small-scale irrigation systems in the context of state withdrawal: the Arabie Scheme in the Northern Province of South Africa. In Proceedings of the 2nd WARSFA / Waternet Symposium: Integrated water resources management: theory, practices, cases. Cape Town, 30–31 October.

Kiptot E, Hebinck P, Franzel, Richards S (2007). Adopters, testers or pseudo-adopters? Dynamics of the use of improved tree fallows by farmers in western Kenya. Agric. Syst., 94 (2): 509-519

Lapar MLA, Ehui S (2003). Adoption of dual-purpose forages: some policy impleications. Trop. Grasslands, 37: 284-291

Lewu FB, Assefa Y (2009). Farmers' knowledge in the cropping systems of Northern KwaZulu-Natal, South Africa: Current challenges and solutions for sustainable future food production. Afr. J. Agric Res. 4(11): 1148-1153

Lumpkin TA, Weinberger K, Moore S (2005). Increasing income through Fruit and vegetable Production Opportunities and Challenges. Consultative Group on international Agricultural Research Science Forum, CGIAR Priorities: Science for the poor. Marrakech, Morocco.

Machethe CL, Mollel NM, Ayisi K, Mashatola MB, Anim FDK, Vanasche F (2004). Smallholder irrigation and agricultural development in the Olifants River Basin of Limpopo province: management, transfer, productivity, profitability and food security Issues. WRC Report No. 1050/1/04. Water Research Commission, Pretoria, South Africa.

Madisa. ME, Assefa Y, Obopile M (2010). Assessment of Production Constraints, Crop and Pest management Practices in Peri-urban Vegetable Farms of Botswana. Egypt. Acad. J. Biol., Sci. 1(1): 1-11.

Minstry of Agriculture (MoA) (1977). statutory instrument No.56 of 1977, control of Goods Act (Cap.43.07). Permits are issued for household consumption and commercial purpose. Division of Agricultural TradeDepartment of Agribusiness Promotion. Gaborone, Botswana

Ministry of Agriculture (MoA) (1991). Botswana's Agricultural Policy; Critical Sectoral Issues and Future Strategy for Development. Information section MoA, Gaborone, Botswana.

Molokomme IM, (1992). Report on FAP Service Sector Study. Prepared for Botswana Government. Ministry of Finance and Development Planning, Gaborone, Botswana.

Piotto D, Montagnini F, Kanninen M, Ugalde L, Viquez E (2004). Forest plantations in Costa Rica and Nicaragua: Performance of species and preferences of farmers. J. Sustainb. For., 18(4): 59-77

Rebaagetse S (1999). Overview of the performance of Agricultural Projects Funded through the financial Assistance Policy (FAP) for the period 1982-1998. Ministry of Agriculture, Gaborone, Botswana.

Remenyi J (2000). Poverty Reduction and Urban Renewal through Urban Agriculture and Microfinance: A Case Study of Dhaka, Bangladesh. jvr. uaBerlin.

Seleka TB, Malope P, Madisa ME (2002). Situational analysis on tomato production, marketing and post harvest activities in Botswana, Gaborone.

14

Effect of auxins on adventitious root development from nodal cuttings of *Saraca asoka* (Roxb.) de Wilde and associated biochemical changes

Goutam Kumar Dash[1], Sunil Kumar Senapati [2]* and Gyana Ranjan Rout[3]

[1]Plant Biotechnology Division, Regional Plant Resource Centre, Bhubaneswar- 751 015, Orissa, India.
[2]MITS School of Biotechnology, Plant Biotechnology Division, Infocity, Patia, Bhubaneswar-751 024, India.
[3]Department of Agricultural Biotechnology, College of Agriculture, Orissa University of Agriculture and Technology, Bhubaneswar-751 003, Orissa, India.

An attempt was made to induce rooting from nodal cuttings of *Saraca asoca* under controlled conditions and study its biochemical changes during rooting. The nodal cuttings were pretreated with different concentrations of IAA, NAA and IBA and kept in a growth chamber [25±2°C, 16 h photoperiod (55 $\mu molm^{-2}s^{-1}$) with cool, white fluorescent lamps and 65% relative humidity] for 12 h. Among the three auxins used for pretreatment, IBA showed more positive response on rooting as compared to IAA and NAA within 4 weeks of transfer to growing medium under mist condition. Among four concentrations of IBA tested, 500 ppm gave maximum percentage of rooting, number of roots and root length. Therefore, IBA was used further in experiments for biochemical investigation. The adventitious rooting was obtained in three distinct phases that is induction (0 to 10 d), initiation (10 to 20 d) and expression (20 to 30 d). IAA-oxidase activity of IBA-treated cuttings increased slightly as compared to control. The activity was found to decrease during induction and initiation phases and increase during expression phase. The peroxidase activity in IBA-treated cuttings increased up to initiation phase and declined at the expression phase. Polyphenol oxidase activity increased both in IBA-treated and control cuttings during induction and initiation phase but declined slowly during expression phase. Total phenolic content was higher in IBA-treated cuttings, particularly in initiation and expression phases and it also correlated with peroxidase activity. Phenolics might be playing key role for induction of adventitious rooting, and phenolic compounds can be used as rooting enhancer in *S. asoca*, an important medicinal plant.

Key words: Auxins, enzyme activity, nodal cuttings, *Saraca asoca*, propagation.

INTRODUCTION

Saraca asoka (Roxb.) de Wilde (Family Leguminosae) is an important medicinal plant used by uterine tonic. It grows in the central and Eastern Himalays as well as in the West Coast of India. Bark of the plant has medicinal properties. Bark is also used as a uterine tonic and reported to cure biliousness, dyspepsia, dysentery, colic piles and pimples. It contains tannins and glycosides. Alkaloids leucopelargonidin and leucocyanidin have been extracted from the bark. Leaves are known to have blood purifying properties and flowers are used to treat diabetes. Most of the pharmaceutical companies are collecting the bark from the plants growing in the forest. In recent years, many of the plant genetic resources have been lost due to unplanned construction and destruction of natural forest. Conventionally, this plant propagated by seeds and the seed viability is very short. Lepidopterous caterpillars attack the seeds in the early stage of fruit (pod) formation. Commercial exploitation of the cuttings is limited by heavy losses during rooting and hardening procedures. Normal cuttings without treatment of plant growth regulators are also difficult for rooting. Adventitious root formation from the stem cutting is influenced

*Corresponding author. E-mail: sunilsenapati2005@gmail.com.

by a number of internal and external factors (Davis et al., 1988; Cheniany et al., 2010). Different researchers have recognized different phases of adventitious root formation in plants based on physiology (DeKlerk et al., 1999; Gaspar et al., 1994; Kevers et al., 1997; Kulka, 2008; Cheniany et al., 2010). Auxin has been known to be intimately involved in the process of adventitious root formation (Wiesmann et al., 1988) and the interdependent physiological stages of the rooting process are associated with changes in endogenous auxin concentrations (Gaspar et al., 1997; Satisha et al., 2008). In several cases, high levels of IAA are associated with the promotion of adventitious rooting. In some cases, root development also occurred by decreasing the endogenous level of IAA (Hausman, 1993).

Root initiation has also occurred without any changes in IAA levels in the root occurring region (Nordstrom and Eliasson, 1991). In addition, several studies have emphasized that polyamines and auxins played major roles during the induction of rooting (Davis et al., 1988; Nag et al., 2001). The formation of adventitious roots involves the process of redifferentiation in which predetermined cells switch from their morphogenetic path to act as mother cells for the root primordia (Aeschbacher et al., 1994). Among these changes, the process of lignification in the cell wall, catalyzed by a particular peroxidase, may occur during rooting (Sato et al., 1993). It is well known that adventitious root formation can be stimulated by application of auxins exogenously but the mechanism of this physiological response is still a matter of debate. This plant is going to extinct in near future. There is a need to conserve *S. asoca* for *ex Situ* conservation. The aim of the present study is to establish the rooting efficiency from cuttings of *S. asoca* under mist conditions and to investigate the biochemical changes during rooting process.

MATRIALS AND METHODS

Collection of plant material

Freshly growing twigs (15 cm long) of *S. asoca* were collected from Champagarh reserve forest at Khurda district of Orissa, India and brought to the laboratory in a water cool box. Each twig was further segmented into smaller segments resulting into 3 to 4 node with leaf in each segment. The cuttings were used for rooting experiments.

Pretreatment of cuttings

The cuttings were put in 500 ml glass beakers containing 50 ml MS (Murashige and Skoog, 1962) nutrient solution without organics and sucrose. The basal region of the nodal cuttings was dipped into the solution. The pH of the nutrient solution was maintained at 5.8. Different concentrations of IAA, NAA and IBA (0, 100, 200, 300, 500 and 700 ppm) were used for pretreatment. Nutrient solution without growth regulator was used as control. The pretreatment of cuttings was maintained in a controlled growth chamber (25 ±2°C), 16 h photoperiod (55 $\mu mol m^{-2} s^{-1}$) with cool, white fluorescent lamp and 65% relative humidity] for 12 h. After 12 h, the cuttings were planted into rooting medium.

Rooting media and growth conditions

Pre-treated cuttings were planted in the nursery bed containing sterile sand and soil in the ratio of 1:1. Hundred cuttings were planted in each bed under mist condition with 75% relative humidity for 4 weeks. Misting was made at every 30 min interval.

Scoring of data in rooting experiment

Rooting percentage, number of roots per cutting, root length was recorded at every week interval up to 4 weeks. 50 cuttings were used per treatment. The experiment was repeated thrice.

Biochemical investigation

After recording the information from the rooting experiment, an independent experiment was conducted for biochemical studies. For biochemical experiment, 20 cuttings were pretreated with nutrient solution with or without 500 ppm IBA for 12 h and subsequently planted in the rooting medium. The experiment was repeated 3 times. The growth condition was used as earlier. On the basis of the preliminary observation, it was decided to collect samples from 5 to 30 days after pretreatment. The basal part (about 0.5 cm of the rooting zone) of the cuttings was taken after 5, 10, 15, 20, 25 and 30 days of transferring to the medium for biochemical analysis.

Extraction for enzyme assay

1 g of fresh tissue was ground in mortar and pestle with 10 ml of 50 mM potassium phosphate buffer (pH 6.0). The homogenate was centrifuged for 20 min at 20,000 g. After centrifugation, the pellet was discarded and the supernatant was mixed with cold acetone to a final concentration of 70% and centrifuged at 20, 000 g for 15 min. The precipitate was resuspended in the same buffer and centrifuged at 20,000g for 20 min and the supernatant was used for enzyme assay. Protein concentrations were determined according to Bradford (1976) and bovine albumin (BSA) was used as the standard.

Peroxidase assay

The reaction mixture was made by mixing 0.1 ml enzyme extract, 0.01 ml 20 mM Guaiacol, 0.1 ml 50 mM H_2O_2, and 0.76 ml 2.5 mM 3,3-dimethylglutaric acid (3,3-DGA)-NaOH at pH 6.0. Peroxidase activity was determined spectrophotometrically by monitoring the formation of tetraguaiacol at 470 nm after 10 min incubation at 30±0.5°C (Beffa et al., 1990). Each value is the mean of 3 replicates. One unit of peroxidase activity corresponds to a ΔA_{470} of 1.0 for 1 mg of protein in 10 min.

IAA oxidase assay

The reaction mixture was made by mixing the 0.2 ml enzyme extracts, 0.78 ml of 50 mM potassium-phosphate buffer (pH 6.0), 0.01 ml of 5 mM $MnCl_2$, 0.01 ml of 5 mM 2,4-dichlorophenol and

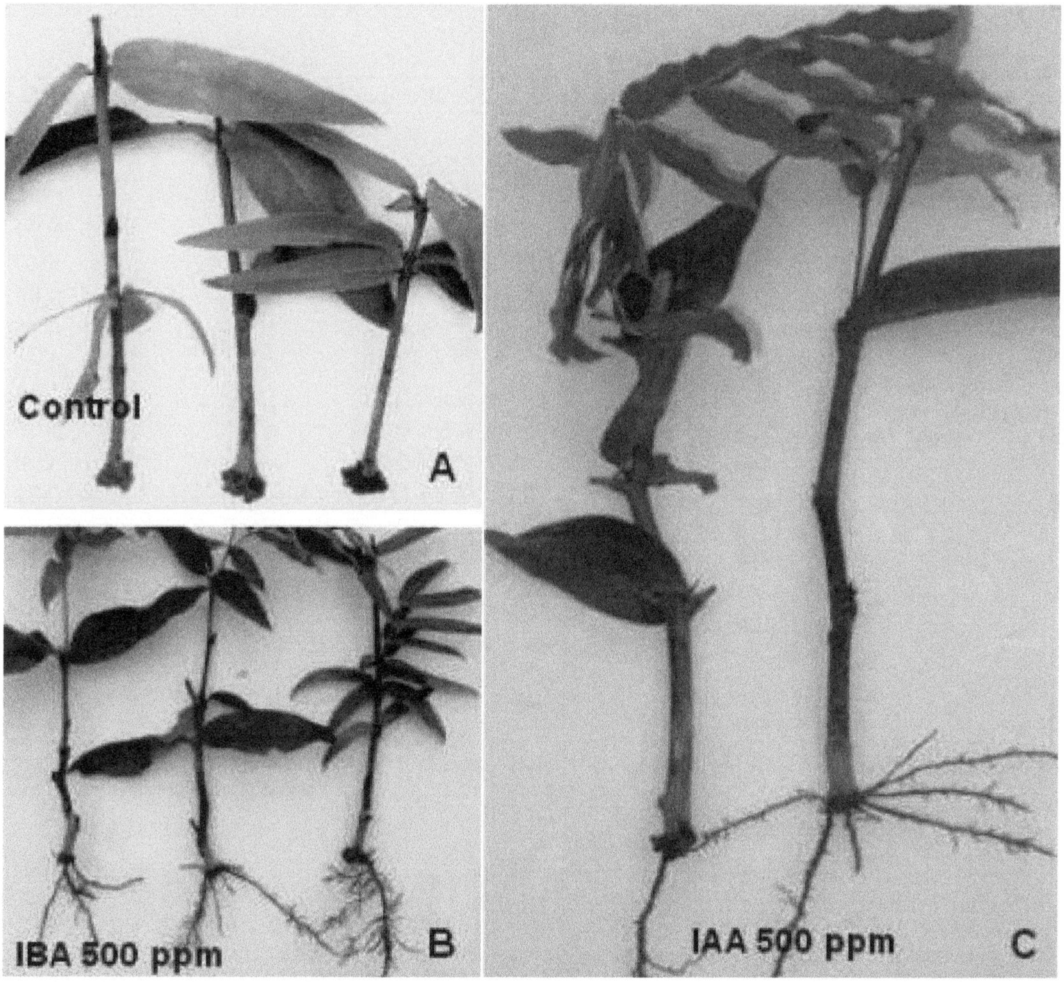

Figure 1. A) Cuttings pretreated with nutrient solution after 3 weeks transferred to growing medium under mist condition. B) Cuttings pretreated with 500 ppm IBA after 4 weeks transferred to potting medium. C) Cuttings pretreatment with 700 ppm IAA after 4 weeks transferred to potting medium.

0.02 ml of 2.5 g/L IAA. Assay was conducted at 25± 0.5°C for 30 min. The Salkowski reagent (2 ml) was then added and the destruction of IAA was determined by measuring the absorbance at 535 after 30 min (Beffa et al., 1990). Each value was the mean of 3 replicates. One unit of IAA oxidase activity is equivalent to a ΔA_{535} of 1.0 for 1 mg of protein in 30 min.

Polyphenol oxidase assay

Polyphenol-oxidase enzyme assay was made by using pyrogallol as the substrate (Kar and Mishra, 1976). The activity was determined in terms of enzyme activity/mg. protein/min. Each value was the mean of 3 replicates.

Total phenol content

Total phenol was extracted and estimated by Folin-Phenol reagent (Bray and Thorpe, 1954) and expressed as mg/gm fresh weight. Each value was the mean of 3 replicates.

RESULTS AND DISCUSSION

Effect of IAA-, IBA- and NAA-pretreatments on rooting

Cuttings treated with nutrient solution without growth regulators did not show any response on root development after 3 weeks on transfer to potting medium (Figure 1A). Root initiation was developed after 20 days of transfer to potting medium in most of the cuttings treated with growth regulators. The nodal cuttings pretreated with IBA showed higher response on rooting than NAA and IAA pretreatments (Table 1). The cuttings treated with 500 ppm IBA showed maximum percentage of rooting and maximum number of roots per cutting (Figure 1B). Among the different concentrations of IAA used for pre-treatment of cuttings, the maximum percentage (54.5%) of rooting (Figure 1C) was obtained in cuttings treated with 500 ppm IAA. Among the different concentrations of

Table 1. Effect of different pretreatment of IAA, IBA and NAA on rooting from cuttings of *Saraca asoca* after 4 weeks transferred to potting medium.

Different treatments	% of rooting (mean ± S.E)*	No. of roots/cutting (Mean ± S.E)*	Av. root length (cm) (mean ± S.E)*
Control (H_2O)	0	0	0
IAA 100 ppm	16.4 ± 0.8^{b}	1.2 ± 0.12^{a}	0.38 ± 0.10^{a}
IAA 200 ppm	20.5 ± 0.6^{d}	2.4 ± 0.30^{b}	0.74 ± 0.22^{d}
IAA 300 ppm	38.2 ± 0.9^{g}	3.6 ± 0.71^{d}	0.91 ± 0.25^{e}
IAA 500 ppm	54.5 ± 0.7^{l}	3.1 ± 0.61^{d}	1.01 ± 0.41^{f}
IAA 700 ppm	42.8 ± 0.5^{i}	2.1 ± 0.4^{b}	0.88 ± 0.32^{e}
IBA 100 ppm	24.8 ± 1.0^{e}	2.4 ± 0.34^{b}	0.76 ± 0.12^{d}
IBA 200 ppm	42.7 ± 1.1^{l}	2.1 ± 0.2^{b}	0.89 ± 0.14^{e}
IBA 300 ppm	56.6 ± 1.0^{m}	3.6 ± 0.42^{d}	1.16 ± 0.33^{g}
IBA 500 ppm	76.8 ± 1.4^{n}	4.5 ± 0.53^{e}	1.36 ± 0.52^{h}
IBA 700 ppm	52.4 ± 0.8^{k}	2.4 ± 0.31^{b}	0.45 ± 0.2^{b}
NAA 100 ppm	18.2 ± 0.5^{c}	1.1 ± 0.26^{a}	0.32 ± 0.08^{a}
NAA 200 ppm	30.6 ± 0.4^{f}	2.3 ± 0.32^{b}	0.62 ± 0.11^{c}
NAA 300 ppm	48.2 ± 0.9^{j}	2.7 ± 0.45^{c}	1.01 ± 0.31^{f}
NAA 500 ppm	56.8 ± 0.7^{m}	2.2 ± 0.62^{b}	1.12 ± 0.33^{g}
NAA 700 ppm	40.6 ± 0.8^{h}	1.1 ± 0.21^{a}	0.72 ± 0.08^{d}
IAA 200 + IBA 200 ppm	42.4 ± 1.1^{i}	1.1 ± 0.45^{a}	0.75 ± 0.06^{d}
IAA 200 + IBA 500 ppm	57.2 ± 0.7^{m}	2.1 ± 0.32^{b}	0.86 ± 0.04^{e}
IAA 500 + IBA 500 ppm	15.5 ± 0.6^{a}	1.2 ± 0.25^{a}	0.30 ± 0.02^{a}

*Mean having the same letter in a column were not significantly different by Duncan`s multiple comparison test ($P < 0.05$). 35 cuttings/treatment; repeated twice.

NAA used for pretreatment, the maximum percentage of rooting (56.8%) was observed in cuttings treated with 500 ppm NAA within 4 weeks transfer to potting medium. However, the cuttings pretreated with 500 ppm IBA showed higher percentage of rooting (76.8%) as well as higher number of roots per cutting within 30 days of transfer to the medium. Therefore, this concentration was used further for biochemical studies during rooting process. The formation of adventitious roots from *in vitro* raised shoots of different plant species promoted by different auxins has been reported by various researchers (Sharma et al., 1999; Kulka, 2008). Sharma et al. (1999) achieved root initiation (71.6%) from *in vitro* grown tea micro-shoots on pretreatment with 500 ppm IBA for 30 min.

Subsequently, Gunasekare and Evans (2000) reported that the root initiation from microshoots with a pretreatment of IBA (50 ppm for 3 h) followed by culture of shoots in auxin-free half-strength MS liquid medium with continuous agitation.

Biochemical changes during rooting

Biochemical make up plays a major role during rooting process in plant. The biochemical studies were conducted from 5 to 30 days at every 7 days intervals during the rooting process. Generally, the adventitious rooting occurred in three distinct phases that is induction, initiation and expression, as proposed for poplar (Hausman et al., 1997). The results showed that the IAA-oxidase activity of the IBA-treated cuttings decreased during induction (0 to 10 d) and initiation phases (10 to 20 d) and increased during expression phase (20 to 30 d). The IAA-oxidase activity was higher in control treated cuttings as compared to IBA-treated one (Figure 2). The low IAA-oxidase activity during the induction period in IBA-treated cuttings appears to be responsible for better development of adventitious roots with IBA, possibly serving as the source of free auxin (Nag et al., 2001). Kulka (2010) reported that auxin produced by the plantlets induces root development. Hartmann et al. (1993) reported that the adventitious root formation initially occurs in two phases: an auxin-sensitive phase and an auxin-insensitive phase. At the expression phase, the IAA-oxidase activity was high which might be related to low IAA content (Nag et al., 2001). Figure 3 indicates higher peroxidase activity in IBA-treated cuttings as compared to control. The activity of peroxidases in IBA-treated cuttings clearly increased up to the expression phase, whereas in the case of control cuttings it showed stabilization or modest increase during the initiation and expression phases (Figure 3). The increase in peroxidase activity observed in IBA-treated cuttings during induction and initiation phase

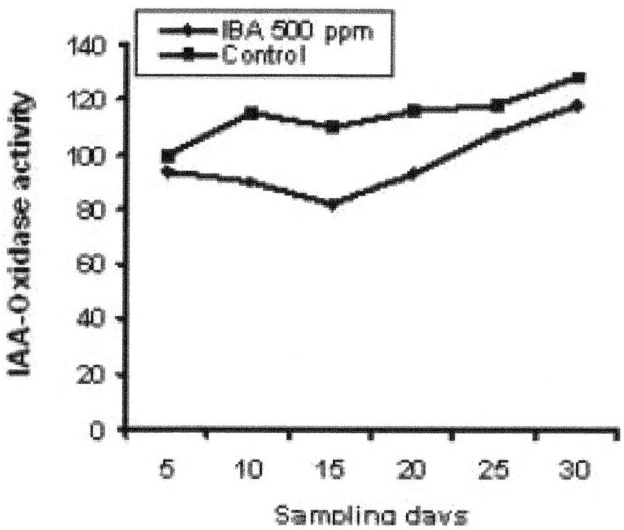

Figure 2. Changes of IAA-oxidase activity in different time period of root development from cuttings of *Saraca asoca* pretreated with or without IBA (500 ppm). One unit of IAA-oxidase activity is equivalent to a ΔA_{535} of 1.0 for 1 mg of protein in 30 min. Values are means ± SE.

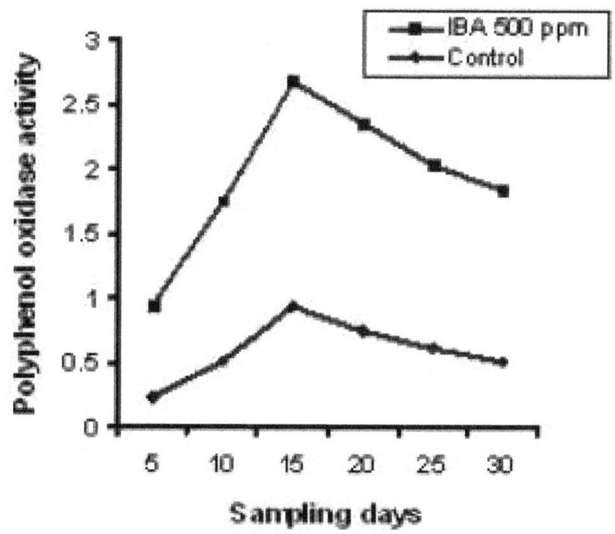

Figure 4. Changes of polyphenoloxidase activity in different time period of root development from cuttings of *Saraca asoca* pretreated with or with out IBA (500 ppm). The activity was determined

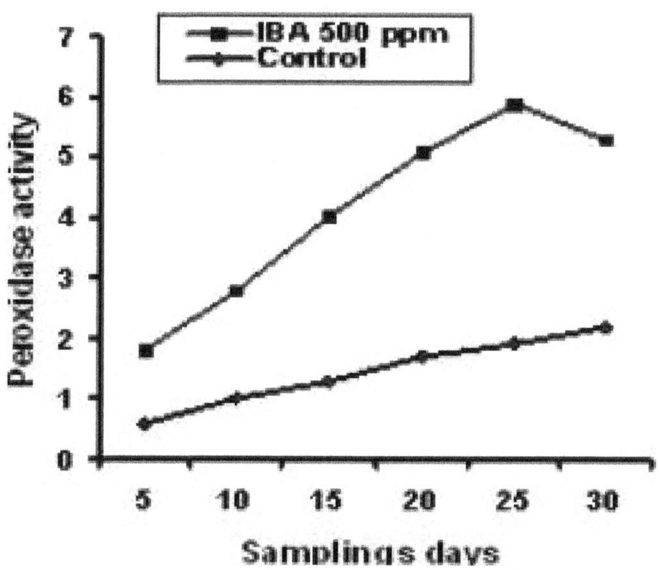

Figure 3. Changes of peroxidase activity in different time period of root development from cuttings of *Saraca asoca* pretreated with or without IBA (500 ppm). One unit of peroxidase activity is equivalent to a ΔA_{470} of 1.0 for 1 mg of protein in 10 min. Values are means ± SE

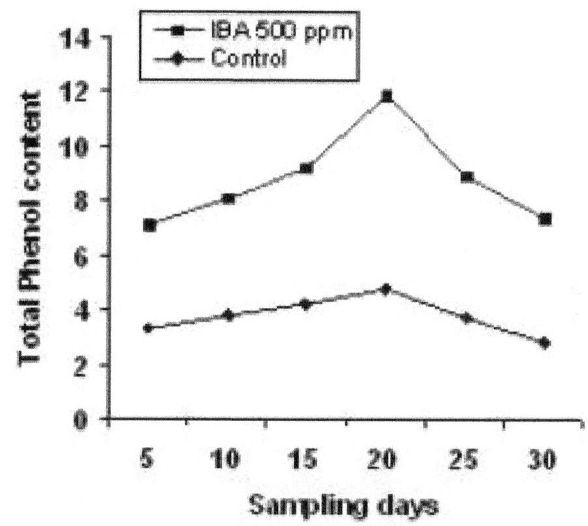

Figure 5. Total phenol content in different time period of root development from cuttings of *Saroca asoca* pretreated with or with out IBA (500 ppm). The phenol content was expressed as mg/gm. fresh weight. Values are means ± SE.

might serve as a good marker for rooting ability in cuttings (Moncousin and Gaspar, 1983). Plant peroxidases are known to be involved in auxin metabolism as well as lignification processes in the cell wall in the presence of phenol, and the present results showed that peroxidase activity increased later than IAA-oxidase activity which indicates that peroxidase activity is more involved in cell wall synthesis at the later phase and obligatory step in root formation (Sato et al., 1993).

An auxin-induced change in peroxidase and IAA-oxidase during the rooting process has also been reported (Mato et al., 1988; Fett-Neto et al., 1992; DeKlerk, 1996).

Polyphenoloxidase activity increased both in IBA-treated and control-treated cuttings during induction and initiation phase but declined slowly during expression phase (Figure 4). The present result also showed that total phenol content was maximum in IBA-treated cuttings, particularly in initiation and expression phases (Figure 5). It might be possible that phenolics play key role for induction of adventitious rooting. Polyphenol oxidase activity has also significant role during the rooting of hardwood cutting as reported (Satisha et al., 2008; Yilmaz et al., 2003). De Klerk et al. (1999) indicated that the wounding related compounds and successive application of auxins leads to higher percentage of rooting. They found that the phenolic compounds act as antioxidants, thereby protect-ting IAA from oxidation and plant tissue from oxidative stress due to wounding. Santos et al. (2009) reported that root induction linked to oxidative stress and that activity of stress induced alternative oxidase (AOX) is importantly involved in adventitious rooting. Cheniany et al. (2010) reported that peroxidase and polyphenol oxidase acti-vities are involved in induction of rooting in *Juglans regia*.

In conclusion, IBA pre-pretreatment helps adventitious root formation from cuttings of *S. asoca*. Among all the biochemical parameters evaluated, the more distinct between IBA-treated and control cuttings were IAA-oxidase (which includes a subgroup of peroxidases, and whose activity was lower in rooting cuttings during the induction phase) and total phenolic content (higher in rooting cuttings during the initiation phase) (Figures 2 and 5). Peroxidase activity was higher in IBA-treated cuttings up to expression phase whereas polyphenol oxidase activity was higher up to initiation phase and thereafter declined (Figures 3 and 4). Phenolic compounds as modifiers of the activities of peroxidase and polyphenol oxidase, as both inhibitors and stimulators, may influence the emergence of roots from cuttings of *S. asoca*. Thus, this investigation will be useful for mass-scale propagation and conservation of important medicinal plants; future experiments should examine the validity of these results with recalcitrant and easy-to-root woody germplasm.

ACKNOWLEDGEMENTS

The authors wish to acknowledge the National Medicinal Plants Board, Ministry of Health and Family Welfare (Sanction No. GO/OR-03/2008), Government of India for financial assistance to conduct this experiment.

REFERENCES

Aeschbacher RA, Schiefelbein JW, Benfey PN. (1994). The genetic and molecular basis of root development. Ann. Rev. Plant Physiol. Plant Mol. Biol., 45:25-45.

Bradford MM (1976). A rapid and sensitive method for the quantification of microgram quanties of protein utilizing the principle of protein-dye binding. Anal. Biochem., 72:248-254.

Beffa R, Martin HV, Pilet PE. (1990). *In vitro* oxidation of indoleacetic acid by soluble auxin-oxidases and peroxidases from maize roots. Plant Physiol., 94: 485-491.

Bray HG, Thorpe WV (1954). Analysis of phenolic compounds of interest in metabolism. In: Glick, D. (Ed.) Methods of biochemical analysis. Vol. 1, Intersci Publ., New York, pp. 27-52.

Cheniany M, Ebrahimzadeh H., Masoudi-nejad A, Vahdati K, Leslie C. (2010). Effect of endogenous phenols and some antioxidant enzyme activities on rooting of Persian walnut (*Juglans regia* L.). Plant Sci., 4 (12):479-487.

Davis TD, Haissig BE, Sankhla N. (1988). Adventitious root formation in cuttings. In:Dudley, T.R. (Ed.) Advances in plant sciences series, Vol. II, Portland, Oregon, Dioscorides Press, pp.128-142.

DeKlerk GJ. (1996). Marker of adventitious root formation. *Agronomie,* 16: 609-616.

DeKlerk GJ, Krieken WVD, Jong JC (1999). The formation of adventitious roots:New concepts, new possibilities. *In Vitro* Cell Dev. Biol.Plant, 35:189-199.

Fett-Neto AG, Teixeira SL, Dasilva EAM, Sant Annna R. (1992). Biochemical and morphological changes during *in vitro* rhizogenesis in cuttings of *Sequoia sempervirens* (D.Don) Endl. J. Plant Physiol., 140:720-728.

Gaspar T, Kevers C, Hausman JF, Ripetti V. (1994). Peroxidase activity and endogenous free auxin during adventitious root formation. In: Lumsden PJ, Nicholas JR, Davis WJ. (Eds.) Physiology, growth and development of plants in culture. Kluwer Acad. Publ., Dordrecht, The Netherlands.

Gaspar T, Kevers C, Hausman JF. (1997). Indissociable chief factors in the inductive phase of adventitious rooting. In: Altman A and Waisel M. (Ed.). Biology of root formation and development. Plenum Press, New York.

Gunasekare MTK, Evans PK. (2000). *In vitro* rooting of microshoots of tea (*Camellia sinensis* L.). SriLanka J. Tea Sci., 66 (1/2): 5-15.

Hartmann HT, Kester DE, Davies FT. (1993). Plant propagation-principles and practices. 5th Edition. Prentice Hall of India Pvt. Ltd., pp.199-255.

Hausman JF. (1993). Changes in peroxidase activity, auxin level and ethylene production during root formation by poplar shoots raised *in vitro*. Plant Growth Regul., 13: 263-268.

Hausman JF, Evers D, Kevers C, Gaspar T. (1997). Internal controls of root induction in poplar shoots raised *in vitro*. Angew. Bot., 71:104-107.

Kar M, Mishra D. (1976). Catalase, peroxidase and polyphenol oxidase activities during rice leaf senescence. Plant Physiol., 57:315-319.

Kevers C, Hausman JF, Fairre-Rampant O, Evers D, Gaspar T. (1997). Hormonal control of adventitious rooting: progress and questions. J. Appl. Bot., 71:71-79.

Kulka RG. (2008) Hormonal control of root development on epiphyllous plantlets of *Bryophyllum* (Kalanchoe) *marnierianum*: role of auxin and ethylene. J. Exp. Bot., 59 (9):2361-2370.

Mato MC, Rua ML, Ferro E (1988). Changes in levels of peroxidases and phenolics during root formation in *Vitis* cultured *in vitro*. Physiol. Plantarum, 72:84-88.

Moncousin CH, Gaspar T (1983). Peroxidase as a marker for rooting improvement in *Cynara scolymus* L. cultivated *in vitro*. Biochem. Physiol. Pflanzen, 178:263-271.

Murashige T, Skoog F (1962). A revised medium for rapid growth and bioassay with tobacco tissue cultures. Physiol. Plantarum, 15: 473-497.

Nag S, Saha K, Choudhuri MA (2001). Role of auxin and polyamines in adventitious root formation in relation to changes in compounds involved in rooting. J. Plant Growth Regul., 20:182-194.

Nordstrom AC, Eliasson L (1991). Levels of endogenousindole-3-acetic acid and indole-3-acetylaspartic acid during adventitious root formation in pea cuttings. Plant Physiol., 82: 599-605.

Santos ME, Cardoso HG, Hernandez A, Peixe AA, Polidoros A, Satisha J, Raveendran P, Rokade ND. (2008). Changes in polyphenol oxidase activity during rooting of hardwood cuttings in three grape rootstocks under Indian conditions. S. Afr. J. Entomol. Vitic., 29(2)

:94-97.
Sato Y, Sugiyama M, Gorecki RJ, Fukuda H, Komamine A (1993). Interrelationship between lignin deposition and the activities of peroxidase isozymes in differentiating tracheary elements of *Zinnia*. Planta, 189: 584-589.
Sharma M, Sood A, Nagar PK, Prakash Om, Ahuja PS (1999). Direct rooting and hardening of tea microshoots in the field. Plant Cell. Tiss. Org. Cult., 58 (2): 111-118.
Wiesman Z, Riov J, Epstein E (1988). Characterization and rooting ability of indole-3-butyric acid conjugates formed during rooting of mung bean cuttings. Plant Physiol., 91: 1080-84.
Yilmaz H, Taskin T, Otludil B (2003) Polyphenol oxidase activity during rooting in cuttings of grapes (*Vitis vinifera* L.) varieties. Turk. J. Bot., 27:495-498.

Comparative assessment of floristic diversity in a buffer zone community forest and a community forest of Barandabhar Corridor, Chitwan, Nepal

Rishi Ram Dhakal, Gandhiv Kafle* and Jay N. Yadava

Institute of Forestry, Tribhuvan University, Hetauda, Nepal.

This research was carried out in the Bandevi buffer Zone Community Forest (BZCF) and Satkanya Community Forest (SCF) of Barandabhar corridor area in Chitwan district of Nepal to assess and compare the status of floristic diversity in buffer zone community forest and community forest in Barandabhar corridor in Chitwan district of Nepal, managed under different rules and regulations. Primary data were collected from reconnaissance survey, direct observation, forest inventory, interviews with forest user group members and key informant interview. Secondary data were collected from the forestry stakeholders working in community forestry and buffer zone sectors. The floral diversity was assessed by using Simpson's Diversity Index (SI), Shannon Weiner Diversity Index (WI) and Margalef Species Richness Index (MI). Information on management practices were assessed by field observation, key informant interview and review of operational plan of respective forests. Diversity index (SI=0.9367 and WI =3.3714) and species richness index of (MI=10) of BZCF were found higher than the diversity index (SI=0.8749 and WI =3.0099) and species richness index (MI=9.0491) of SCF. We conclude that floral diversity is higher in BZCF than the CF outside the buffer zone under similar edaphoclimatic conditions.

Key words: Community forest, buffer zone community forest, floristic diversity, diversity index, species richness index, Nepal.

INTRODUCTION

Nepal, a small Himalayan country of Asia represents one of the unique places of the world, which boasts of high bio-diversity is under continual disturbance by the local people. Plant species richness of Nepal comprises 465 species of lichens, 1,822 species of fungi, 687 species of algae, 853 species of bryophytes, 534 species of pteridophytes, 27 species of gymnosperms; and 5,856 species of angiosperms (GoN, 2009). The biological diversity contained in the Terai and Siwalik Hills (lowlands) ecosystems are of international importance both in view of the number of globally threatened species of fauna and flora as well as the diversity of ecosystems in these area. There are 1,885 species of angiosperms, 61 species of bryophytes and 81 species of pteridophytes from the Terai plain and Siwalik hill (BPP, 1995). The community forests are national forests handed over to the forest user group (FUG) under section-25 of forest act for management and utilization for the collective benefit of the community (HMG, 1995). The buffer zone community forests are national forests handed over to the buffer zone user committee (BZUC) under section-21 of National park and wildlife conservation Act -1972, with amendment to manage, utilize the forest product and conserve the biodiversity for the collective benefit of the buffer zone community (HMG, 1999). Buffer zone community forest is one of the participatory forest management initiatives within the buffer zone management program which has a crucial role to improve the park-people relationships.

With the shift to active forest management in community forestry, several types of silvicultural operations such as cleaning and weeding, thinning, pruning, coppicing, selective felling, singling, collecting litter, grass, and dry twig, grazing, establishing and monitoring of trial plots, harvesting and removing dead and logged trees along with planting new species have been

*Corresponding author. E-mail: rishi.saveflora@gmail.com.

designed and undertaken by community forest user groups (CFUGs) (Dhital et al., 1998; Ojha and Bhattarai, 2001; Khadka and Schmidt-Vogt, 2008; Acharya, 2003). Practices such as seedling plantation, controlling wildlife hunting, forest fire and grazing, regulating forest encroachment, protecting soil erosion prone area and water resource area assist biodiversity conservation, paradoxically other practices such as species selection, removal of unwanted species during silvicultural activities, leaf litter collection, elite dominance in decision making, and traditional knowledge depletion have detrimental impact on biological diversity and ecosystem function of community managed forest (Shrestha et al., 2010). Silviculture affects these three attributes of forest ecosystem in different scale and intensity if not carried out with caution; it can threaten biodiversity.

The community forestry management approach offers an attainable means to conserve the biodiversity of Nepal. The programme is successful in rehabilitating degraded hills and thereby in increasing the biodiversity (Shrestha et al., 2010). However, the prevalent management approach in community forestry indicates increasing threats to the conservation of biodiversity. In addition, existing legal and administrative base favors for the management approach, which threats biodiversity and argued that the prevalent forest management approach in community forestry recognizes biodiversity conservation as secondary issue and there is evidences that biodiversity has been either decline or has been altered in community managed forest (Acharya, 2003).

On other hand, it is perceived that it has contributed to biodiversity conservation (Adhikari et al., 2004; Kijtewachakul et al., 2004). Maintaining compositional, structural and functional attributes of forest ecosystem is one of the important approaches of biodiversity conservation. Pokharel et al. (2005) claimed that community forests have improved overall forest conditions including biodiversity. Pandey (2007) found comparatively higher tree species diversity on community-managed forest stands than the national parks and government managed forests. However, more rigorous studies are necessary to understand whether the current management practices in community forests have been ameliorating or aggravating or bringing no change in the forest biodiversity. In this context, this study was carried out to find out and compare the current status of floral diversity in two forest categories: 1) a buffer zone community forest and 2) a community forest; managed under different management rules and regulations in Barandabhar corridor area of Terai region of Nepal.

MATERIALS AND METHODS

Study area

Two community forests of Barandabhar corridor area, Chitwan district of Nepal (Figure 1) were selected for the present study with a criterion: Satkanya community forest (SCF) is managed under Forest Act and Regulation (1993, 1995), and Bandevi Barandabhar buffer zone community forest (BZCF) is managed under National Parks and Wildlife Conservation Act and Regulation (1972, 1996). Both these forests are part of important Barandabhar corridor area of Nepal.

BZCF lies at 27°38' N latitude and 84°26' E longitude in the buffer zone area of Chitwan National Park. The area has alluvial soil with shallow ground water table with 5 to 15% gentle slope. The forest covers an area of 167 ha. There are 1500 households in the user's group of this buffer zone community forest. The forest was handed to local community in 1995. It is one part of Barandabhar corridor area of Nepal and an important place from biodiversity conservation point of view. This forest is dominated by Sal (*Shorea robusta)* associated with Asna (*Terminalia tomentosa*), Barro (*Terminalia belerica*), Kyamun (*Syzygium cerasoides)*, Jamun (*Syzygium cumini*) and Botdhanyero (*Lagerstroemia parviflora*). SCF lies at 27° 40' N latitude and 84°33' E longitude. The area has alluvial soil with shallow ground water table. The forest covers an area of 72 ha. There are 384 households in the user's group of this community forest. The forest was handed to local community in 1997. It is one part of Barandabhar corridor area of Nepal and an important place from biodiversity conservation point of view. This forest is dominated by Sal associated with Asna, Barro, Kyamun, Jamun and Botdhanyero.

Data collection and analysis

Management practices of both forests were studied from their operational plan, direct observation of practices on-site, reviewing the government rules and regulation for provisions on the management and utilization of the forest resources of respective forest types according to management authority and objective of management. Systematic random sampling with square nested sample plots was employed to collect quantitative information on floral diversity with 1% sampling intensity for tree species (≥30 cm diameter at breast height (DBH)), 0.5% for shrubs and saplings (≤10 cm DBH and more than 1 m height), 0.001% for herbs and seedlings (30 to 100 cm height) in each forest. The plot size for surveying tree and poles (pole has 10 to 29.9 cm DBH) was 10×10 m; 5×5 m for shrubs and sapling; and 1×1 m for herbs and seedling (Oosting, 1956; in Gysel and Lyon, 1980; in Sutherland, 1996; Rayamajhi, 1994). 18 sample plots were laid in SCF and 13 were located in BZCF. The number of species and their abundance with respect to their diameter were recorded in a standard data sheet. The plant species were identified with the help of standard literature of plant identification in Nepal and visual inspection by taxonomist and knowledgeable local informants.

Secondary data were gathered from published literatures, maps, and related line agencies such as District Forest Office, Terai Arc Landscape Programme Office, Chitwan National Park, Community Forest User Group Office, Federation of Community Forest User Groups in Nepal, Forest Action. These literatures were reviewed in order to have better understanding, interpretation and analysis for the present study. Formal and informal interviews were organized among community forest user groups, concerned buffer zone user committee, buffer zone community forest users, key informants, district forest officer, national park's warden and buffer zone management committee members about current management practices in respective forests and legal status of management.

Data collected from the field survey were processed to calculate species richness and diversity function of the study area, as well as density and relative density, frequency and relative frequency of the species, diversity indices and species richness index using the standard formulas in Magurran (2004). Secondary data were systematically reviewed and information on forest management practices; provisions of existing rules and regulations for management and utilization of forest resources in CF and BZCF;

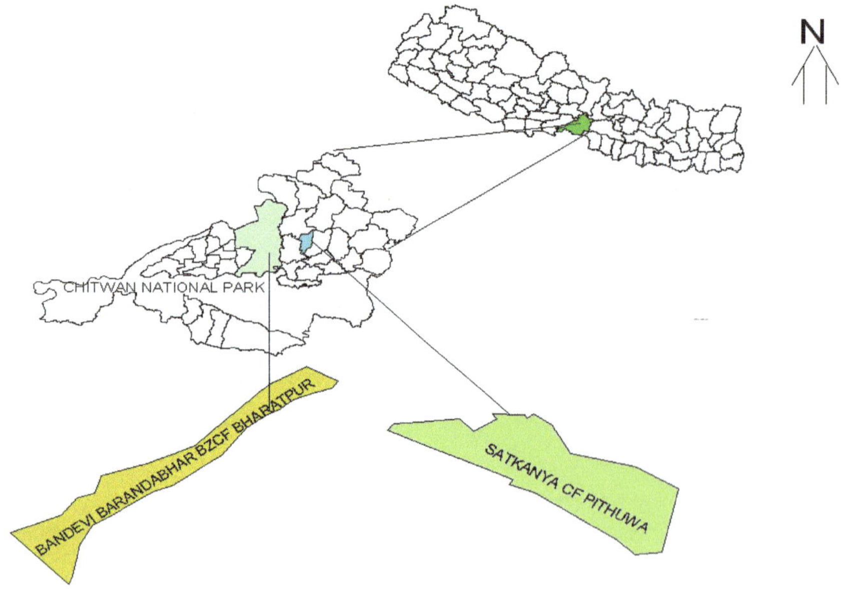

Figure 1. Map showing research sites: Study area.

and research methodology were extracted qualitatively.

RESULTS AND DISCUSSION

Historical perspectives on forest management

The review of operational plan of the community forest showed that until the early 1970s, Barandabhar forest was covered by dense vegetation and was a good habitat for Tiger, Rhino and other wildlife species. After 1972, a large number of people migrated to the study area from the adjacent hills. This resulted in clearing, degradation and fragmentation of the forest due to encroachment. During this time, the forest area was under the jurisdiction of the district forest office, and some conservation initiatives had already been initiated by the government. The degradation of forest continued till the early 1990s. The scarcity of forest resources became severe in the area and the pressure on the forest area increased. In the mean time some conservation initiatives were started, such as fencing around the forest, management of grasslands, plantation of fodder and timber species. After the declaration of the buffer zone, the area came under the jurisdiction of the authority of Chitwan National Park and was considered as a buffer zone forest and it was later handed over to the community as buffer zone community forest. Now, this corridor forest is being managed mainly in two types of management categories that is 1) government managed forest management type and 2) community managed forest management type. Community forests are also managed in two different ways under two acts and rules with different objectives.

Patrolling, forest fire control, control of encroachment, illegal felling, grazing and hunting, and soil conservation and wildlife conservation are common forest protection and conservation practices in SCF and BZCF. BZCF is fenced along the boundary. Soil and water source conservation activities in wildlife habitat, wetland management, grass land management, species conservation, bird and insect conservation and conservation education, conservation awards and awareness are the major biodiversity conservation activities in BZCF. Cleaning, singling, pruning, thinning, retention of mother tree for regeneration, felling as silvicultural operation and dead wood removal are the common silvicultural operations in SCF and BZCF. In SCF, forest nursery development, plantation, check dam and forest road construction and NTFP identification and promotion also exist as forest development activities. Plot fixing, marking, tree felling, depot of timber, distribution and marketing of timber, pole extraction, firewood collection, leaf litter collection, long bole (lingo) collection, NTFP collection by users and supply of timber outside from user group boundary by bidding (tender) are the activities of forest product collection, utilization and distribution in SCF.

In BZCF, timber of only fallen trees were collected, poles were collected from branch and tending materials, firewood, leaf litter, charcoal and NTFPs were collected according to the collection plan. Forest products are not supplied outside the buffer zone area according to Buffer Zone Management Regulation. Agroforestry programme is promoted to fulfill the insufficient forest products demanded by local community. Road gravelling, construction of culvert, drinking water supply and support for schools are the activities of community development. Group empowerment, training, demonstration plot

Table 1. Summary of diversity and species richness index by forest management types.

Management types	Number of spp.	Maximum number of ind./ spp.	Minimum number of ind./spp.	Total number of ind. of all spp.	Richness index	Diversity index	
	S	Ni max	Ni min	N	MI	SI	WI
SCF	58	196	1	556	9.0491	0.8749	3.0099
BZCF	71	184	1	987	10.00	0.93675	3.3714

Ind. = Individuals, spp = Species, SI=Simpson's diversity index, WI=Shannon Weiner diversity index, MI=Margalef species richness.

establishment, forest based income generation (NTFP cultivation, broom grass farming, mushroom farming, goat farming, candle manufacturing training) are major pro-poor focused activities as forest management and livelihood enhancement practices in SCF and BZCF.

Overall budget of the SCF was allocated 20% in community development, 15% in administrative cost, 35% for livelihood programme, 25% for forest development programme and 5% as saving. Total budget allocated for different components in BZCF are: community development (23%), administrative cost (15%), income generation (5%), wildlife damage compensation (2%), institutional development (5%), forest conservation and management (35%), education and conservation education (5%), and eco-tourism management (10%).

Diversity and species richness of flora by forest management types

A total of 71 species of flora were recorded in BZCF and 58 in SCF. Shannon Weiner species diversity indices (WI) of BZCF and SCF were 3.3714 and 3.0099 respectively. Similarly, the Simpson's diversity indices (SI) of BZCF and SCF were 0.93675 and 0.8749 respectively. Jha and Acharya (2008) found the Shannon diversity index of 1.63 in some CFs of mid-hills of Nepal which were already handed over to CFUGs 5 years ago with a diversity index of 1.22.

Kharal (2000) found Shannon diversity index of 1.8 in rural farmlands of Chitwan district, which has similar physiography, elevation and climatic condition with Barandabhar area. Most of the tree species were represented by less number of individuals. The trend of species richness index (Margalef) in the study area was found almost similar to that of Shannon diversity index. The species richness index of whole BZCF (MI=10) was found higher than that of SCF (MI=9.0491) (Table 1).

Diversity and species richness of flora by plant life forms

Both Shannon and Simpson's diversity indices of tree/pole sapling, tree seedling and herb were higher in BZCF than SCF while both diversity indices of seedlings of shrub and established shrubs were higher in SCF than BZCF. The species richness index of tree and shrub categories were lower in BZCF than that of SCF while species richness index of herb was found higher in BZCF than in SCF. The comparative diversity and species richness index are illustrated in Table 3.

Status of plant life forms and abundance

The number of trees and poles in the BZCF (711/ha) was higher than that of SCF (677/ha). Similarly, the number of tree seedlings (79444/ha), shrub (1444/ha) and herbs (228333/ha) in BZCF was also found higher than the SCF (70769/ha, 454/ha and 113846/ha respectively). The number of saplings (3844/ha) and shrub seedlings (40000/ha) in BZCF were lower than the saplings (4308/ha) and shrub seedlings (43846/ha) in SCF (Figure 2). It reflects that the BZCF is dense than SCF, in terms of the number of plant individuals.

Distribution of individuals of different plant life forms

The percentage of tree and shrub species in SCF is higher as compared to BZCF while percentage of herb species is less in SCF area. Regarding the whole area, number of individuals of tree species is more than that of shrub as well as herb species. It was found that the abundance of shrub species is lower than that of other plant categories in both BZCF and SCF. A total of 58 and 71 plant species were recorded in SCF and BZCF area respectively. The range of species richness found within a sample plot varied between 3 and 23 in BZCF; whereas, it was 2 and 18 species in SCF. The number of species of tree, shrub and herb was 22, 13 and 41 in BZCF respectively; whereas it was 21, 12 and 25 in SCF.

The proportion of herb species was higher in BZCF than in CF but proportion of tree and shrubs were higher in SCF than BZCF. Sigdel (2008) found a total of 147 species (36 trees, 37 shrubs and 74 herbs) in Shivapuri National Park in mid-hill of Nepal (1000 to 2000 m elevation) where the proportion of herbs were

Table 2. Abundance distribution of different life forms of plants in sampled area.

Plant life form	Category of forest									
	BZCF					SCF				
	No. of spp.	Total no. of individuals	% of ind.	Max. no.	Min. no.	No. of spp.	Total no. of individuals	% of ind.	Max. no.	Min. no.
Tree /Pole	14	128		14	1	9	88		4	1
Sapling	17	173	45	10	3	17	140	59	8	1
Seedling	14	143		9	2	10	92		7	1
Established shrub	8	65	14	6	1	6	18	14	5	1
Shrub seedling	6	72		4	1	10	57		9	2
Herb	39	411	41	15	3	25	148	27	13	5
Total	71	992	100	23	3	58	543	100	18	2

Ind. = Individuals, spp. = Species.

Table 3. Diversity and species richness index by plant life forms.

Plant life forms	Management type							
	BZCF				SCF			
	Diversity index		Richness index	No. of spp.	Diversity index		Richness index	No. of spp.
	SI	WI	MI	S	SI	WI	MI	S
Tree	0.7707	2.01639	2.6793	14	0.55878	1.23999	4.4773	9
Sapling	0.8248	2.1315	3.1048	17	0.62477	1.61557	3.2378	17
Tree seedling	0.7621	2.0006	2.4559	14	0.7396	1.45098	2.6538	10
Established shrub	0.7351	1.46825	1.6769	8	0.8497	1.69202	1.7299	6
Shrub seedling	0.2766	1.436193	1.1619	6	0.90226	2.229526	4.043	10
Herb	0.8939	2.79706	6.3138	39	0.8605	2.694898	4.6025	24
Total	0.9367	3.2126	10	71	0.8749	3.0099	9.0491	58

SI=Simpson's diversity index, WI=Shannon Weiner diversity index, MI=Margalef species richness, S=Total number of species.

higher as that of BZCF. The detailed information on growth form of plant species found in the study area is presented in Table 2.

Conclusion

Diversity of tree species, measured in terms of Shannon diversity index, Simpson's diversity index, Margalef species richness index in BZCF and SCF areas as a whole were found to be

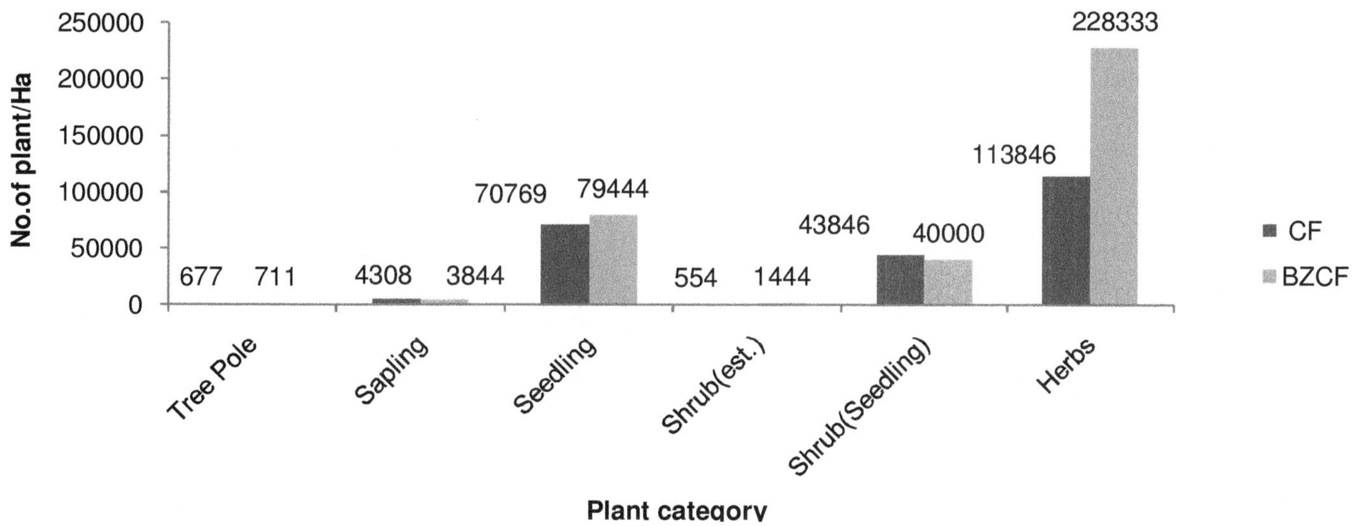

Figure 2. Abundance of plant categories.

higher than those values in CFs of mid-hills of Nepal and also to be higher than those in farmlands of similar geographical region of Nepal. The index values are found to be higher in BZCF than SCF area. The major cause influencing both index values in both management types was due to the higher abundance of individuals of single species especially *S. robusta*. A total of 71 plant species were found in BZCF area, while 58 species were recorded in SCF area. The distribution of individuals of herbs and seedlings was found more than 80% in both management types (CF and BZCF) while that of tree /pole forms were less than 1%.

The number of trees per ha and species number in community forest was found to be lower than that of BZCF. One of the major causes of that difference in number of individuals is management activities and silvicultural operation. Species preference, selection, ignorance and removal of lower plants, insufficient knowledge on biodiversity, overgrazing, forest fire and forest encroachment were mainly responsible for lowering more species in SCF of corridor area. The distribution of individuals of tree, shrub and herb forms in BZCF was found to be higher while those of shrub seedling and tree sapling forms were found too low than in CF. Similar trend was found for species diversity and species richness index in both management types. The diversity index of tree, tree sapling, seedling and herb forms were found to be higher in BZCF than of SCF but diversity index of shrub forms is high in SCF than in BZCF. Percentage of individual number of tree species in SCF sampled area was higher than the sampled area of BZCF.

Herb and shrub species were found higher in BZCF sampled area than the SCF sampled area. The Simpson's diversity index (SI) of seedling and shrub seedling of SCF was found higher than in BZCF whereas the same index of tree; shrub and herb were found higher in BZCF than in SCF. Shannon Weiner diversity index (H') of shrub and shrub seedling of SCF was found higher than in BZCF whereas the same index of tree, sapling, seedling and herb shrub and herb were found higher in BZCF than in SCF. It is concluded that floral diversity and species richness was found higher in BZCF than in CF even though these areas have the same edhaphoclimatic condition. It gives general clue that community forest management practice under the NPWC Act 1972; was favorable for floral diversity conservation than the management practice under the Forest Act 1993; however this clue needs further research for deriving specific conclusion.

ACKNOWLEDGEMENTS

We are grateful to the Community Based Natural Forest and Tree Management in Himalaya (ComForM) for supporting this research. We are especially indebted to the local people for taking part in this research during data collection. We are highly grateful to two anonymous reviewers for providing critical comments to enhance the manuscript.

REFERENCES

Acharya KP (2003). Conserving biodiversity and improving livelihoods: the case of community forestry in Nepal. Paper presented in International Conference on Rural Livelihood, Forests and Biodiversity, Bonn, Germany.

Adhikari B, Di Falco S, Lovett JC (2004). Household characteristics and forest dependency: evidence from common property forest management in Nepal. Ecol. Econ. 48: 245-257.

BPP (1995). Biodiversity profile of Terai and Siwalik physiographic zone. Biodiversity Profile Project, Ministry of Forest and Soil Conservation, Kathmandu Nepal.

Dhital RM, Pokhrel MP, Poudel LR, Dahal SP (1998). Creative silviculture by the community forest user groups in the Hills of Nepal. Paper presented in international seminar on Cultivating Forests: Alternative Forest Management Practices and Techniques for community forestry. RECOFTC, Bangkok, Thailand.

GoN (2009). Fourth National Report to the Convention on Biological Diversity. Government of Nepal, Ministry of Forests and Soil Conservation, Kathmandu, Nepal.

HMG (1995). The Forest Act 1993 and the Forest Regulations 1995. Kathmandu, Law Books Management Board, FDP/USAID/HMGN, Kathmandu Nepal.

Jha PK, Susanta A (2008). Impact of Community Forestry in Forest Condition Change: Evidences from Four CFs from the Mid -hills of Nepal. Livelihood and Forestry Program, Kathmandu, Nepal.

Khadka SR, Schmidt VD (2008). Integrating biodiversity conservation and addressing economic needs: An experience with Nepal's community forestry. Local Environ., 13: 1-13.

Kharal DK (2000). Diversity and Dynamic of Tree Species and Its Sustainability in rural farmland: a case study in Chitwan district, central Terai of Nepal. A M.Sc. thesis. Available at http://himaldoc.icimod.org/record/263/files/207.pdf. Date accessed: 26 May 2011.

Kijtewachakul N, Shivakoti GP, Webb EL (2004). Forest health, collective behaviors and management. Environ. Manage., 33: 620-636.

Magurran AE (2004). Measuring biological diversity. Oxford: Blackwell Publishing.

Ojha HR, Bhattarai B (2001). Understanding community perspectives of silvicultural practices in the middle hills of Nepal. Forests, Trees People Newslett., 40: 55-61.

Pandey SS (2007). Tree species diversity in existing community based forest management systems in central mid-hills of Nepal. Uppsala, Sweden: Swedish Biodiversity Centre. Swedish University of Agricultural Sciences.

Pokharel BK, Stadtmuller T, Pfund JL (2005). From degradation to restoration: An assessment of the enabling conditions for CF in Nepal. Nepal Swiss CF Project (NSCFP).

Sigdel S (2009). Altitudinally coordinated pattern of plant community structure in the Shivapuri National Park, Nepal. Banko Janakari, Nepal. Available at: <http://www.nepjol.info/index.php/BANKO/article/view/2161>. Date accessed: 26 May 2011.

Household tree planting and its related constraints in meeting woodfuel production in Kiambu, Thika and Maragwa Districts of Central Kenya

Githiomi J. K.[1]*, Mugendi D. N.[2] and Kung'u J. B.[2]

[1]Forest Products Research Centre, Kenya Forestry Research Institute P.O. Box 64636-00620 Nairobi, Kenya.
[2]Department of Environmental Sciences, Kenyatta University, P.O. Box 43844-00100 Nairobi, Kenya.

The objective of this study was to assess household tree planting for fuelwood production and investigate current constraints to tree establishment in Kiambu, Thika and Maragwa Districts of Central Kenya. The study was undertaken through household survey in Kiambu, Thika and Maragwa Districts of Central Kenya. The districts were purposively sampled on the basis of various factors which include diverse ecological conditions and population densities among others. Sampling of the households was done using multistage stratified random sampling technique where stratification was based on the weights in socio-economic and climatic activities as indicators. The results of the study showed that over 90% of the households in the three districts have planted trees in their farms and tree planting was found to be positively correlated to household farm size. Boundary tree planting was the most preferred as compared to woodlots which had the least preference. *Grevillea robusta* and *Eucalyptus* species were the most preferred tree species. Inadequate land was the leading major obstacles to tree planting in the three districts with 74, 60 and 57% of the household respondents followed by scarcity of seedlings with 16, 33 and 28% in Kiambu, Thika and Maragwa districts, respectively. The third most important constraint was tree establishment cost. The household land size ownership varied among the districts with a range of 1.9 to 3.6 acres. The small household land holdings indicate the need to integrate woodfuel production with farming systems as agricultural sector has a key role in supplementing wood production. The study was concluded by recommending development of decentralised woodfuel planning with site specific implementation strategies in the study area as there were varying tree planting parameters among the districts. Establishment of tree nurseries was also recommended for a sustainable seedling production.

Key words: Household land size, species preference, woodfuel planning, sustainability.

INTRODUCTION

In developing countries, woodfuel is the major source of cooking and heating where about 2 billion people rely solely on fuel wood for cooking (FAO, 2005). This figure demonstrates the critical importance of wood energy in meeting energy requirements in these countries. It is estimated that about 90% of Kenyan rural households use woodfuel either as firewood or charcoal (Ministry of Energy, 2002; Theuri, 2002; Kituyi, 2008). Wood energy provides 70% of Kenya's national energy needs and it is expected to continue as the country's main source of energy for the foreseeable future (Republic of Kenya, 2002a). Besides being the standard cooking fuel for the majority of Kenyan households, fuelwood is also an important energy source for small-scale rural industries like tea factories. A comprehensive biomass study undertaken in Kenya in 2000 revealed that the principal sources of fuel wood are the farm lands with a production of 84% of the total woodfuel requirement (NEMA, 2004).

*Corresponding author. E-mail: josephgithiomi@ymail.com.

Table 1. Proportion of land sizes (acres) for Kiambu, Thika and Maragwa districts.

District	Land size (acres)											
	<1		1.1-2		2.1-3		3.1-4		> 4.1		% total (n)	
	%	n	%	n	%	n	%	n	%	n	%	n
Kiambu	46	92	22	43	8	15	6	12	18	36	100	198
Thika	41	63	20	30	13	20	7	11	18	28	100	152
Maragwa	44	86	24	48	19	38	5	9	8	16	100	197

Kenya has 3.467 million ha of forest cover which is equivalent to 5.9% of land area out of which 1.417 million ha comprises of indigenous closed canopy forests, mangroves and plantations (Kenya Forest Service Strategic Plan, 2009). Much of the closed canopy forest has been depleted due to internal and external influences. Continued losses of forests and associated resources have had far reaching negative effects on the country's economy and welfare of Kenyans. Some of the consequences include inadequate supply of woodfuel and timber which lead to overharvesting of trees leading to environmental degradation and loss of biodiversity among others (Nellie and Githiomi, 2009).

The Kenya Forest Service (KFS) strategic plan indicates that 10.385 million ha of land is covered with trees on farmlands with wood stocking of about 9.7 m^3/ha. The moratorium on tree harvesting from public forests in Kenya was imposed in 1999 and complete ban in 2002 which precipitated a shortage of sawn timber and wood products. This shortage lead to increase of timber prices which encouraged farm tree planting. The integration of trees in agriculture systems is a road map to follow to enhance sustainability (Schuren and Snelder, 2008). This integration has been made easier through research on farm forestry systems which are more diverse, efficient and easily adopted to the local condition (Adensinu and Chianu, 2002). The Government of Kenya has been involved in promoting tree planting at the farm level with the aim of increasing tree cover to 10% by the year 2030 (Republic of Kenya, 2007). There have been successful tree planting programs involving rural communities in Kenya led by government rural forest extension services and various non-governmental organizations (Githiomi et al., 2012). Lack of sustainable woodfuel production planning strategy has lead to scarcity and over-exploitation of natural resources and environmental degradation as supported by past studies by Akinga (1980) and Ministry of Energy (2002), which despite being two decades apart showed a widening gap between supply and demand in woodfuel.

The deficit in woodfuel was due to higher tree cutting rate than replenishment. The study aimed at determining the factors influencing tree planting for woodfuel production in Kiambu, Thika and Maragwa districts of central Kenya and suggest possible interventions strategies to ensure self sufficiency.

MATERIALS AND METHODS

The study was carried out in Kiambu, Thika and Maragwa districts of central Kenya. The districts were purposively sampled on the basis of various factors which include diverse ecological conditions and population densities, presence of agro based industries like tea factories among others. The selected districts have rainfall range of 800 to 1400, 500 to 2500 and 900 to 2700 mm for Kiambu, Thika and Maragwa districts, respectively (Republic of Kenya, 1997a, 1997b, 2002b). Sampling of the households was done using multistage stratified random sampling technique beginning with stratification sampling procedures as outlined by Lee-Ann and Martin (1997). Each of the three districts under study was stratified according to the weights in socio-economic and climatic activities as indicators. Using this procedure, at least 40% of the divisions with relatively homogeneous characteristics in each district were sampled to form a stratum. This ensured heterogeneity was well captured and represented. Similar procedure was followed for selection of administrative locations and sub-locations. Based on these sampling procedures, Kiambu, Thika and Maragwa districts were stratified into four, three and two strata, respectively. Each stratum in the district consisted of one or more divisions. A total of 200 households were sampled from each district.

Allocation of the sampling units in each sampled division was done proportionally to the total number of household obtained from 1999 census (Republic of Kenya, 2001). Data collection was done using structured and semi-structured household questionnaires. The information collected included proportions of land size per household, household land area under trees, tree planting practices, tree preferences and tree planting obstacles. The generated data was coded and entered in the computer using spread sheet of Ms-Excel. Statistical package for social sciences (SPSS) was used in analyzing the data.

RESULTS

The results showed that over 40% of the household respondents within the three districts had less than 1 acre of land (Table 1) and the proportion of land size ownership category varied significantly between the districts. ($\chi^2 = 21.489$, d.f = 8, $p = 0.006$). The mean land size holdings for Kiambu, Thika and Maragwa were 3.6, 3.2 and 1.9 acres with s.e.d of 0.282, respectively, which was significantly different ($p = 0.033$). Further, multiple comparison of estimated land sizes among the three districts showed that there were significant differences ($p<0.016$) in land size between Maragwa and Kiambu as well as Maragwa and Thika ($p<0.05$), but no significant difference ($p = 0.649$) between Kiambu and Thika districts (Table 2). The significant differences in land size

Table 2. Pairwise comparisons of land sizes (acres) between Kiambu, Thika and Maragwa districts.

Comparison of district	Mean difference	s.e	p-value	95% confidence Interval for difference	
Kiambu versus Thika	0.345	0.759	0.649	-1.145	1.835
Kiambu versus Maragwa	1.719	0.709	0.016	.326	3.112
Maragwa versus Thika	-1.374	0.706	0.050	-2.760	0.012

Table 3. Mean land holdings of sampled division per district.

District of survey	Division of survey	Mean	Std. deviation	Range (acres)	N
Kiambu	Kiambaa	1.7	1.25	4.5	70
	Lari	2.7	2.72	13.8	64
	Ndeiya	2.3	3.50	14.5	16
	Limuru	7.8	18.82	84.9	48
Thika	Gatundu	2.4	2.31	14.0	68
	Kakuzi	5.1	6.32	40.0	48
	Thika Municipality	2.4	7.44	45.0	37
Maragwa	Makuyu	1.9	1.37	6.8	84
	Kigumo	1.9	1.67	9.8	113

Table 4. Proportion of household respondents who had planted trees on farm in Kiambu, Thika and Maragwa districts.

District	Proportion of households who planted trees on their farms					
	Planted		Not planted		Total	
	%	n	%	n	%	n
Kiambu	92	182	8	15	100	197
Thika	92	145	8	12	100	157
Maragwa	96	184	4	7	100	191

Table 5. Proportion of land planted with trees in Kiambu, Thika and Maragwa districts.

	Land size currently under trees													
	<0.5		0.5-1		1-2		2-3		3-5		>5		Total	
District	%	n	%	n	%	n	%	n	%	n	%	n	%	n
Kiambu	65	121	22	42	7	14	3	6	0	0	2	4	100	187
Thika	84	121	13	19	2	3	1	1	0	0	0	0	100	144
Maragwa	85	167	12	24	2	20	1	1	1	1	0	0	100	196

noted in this study were clearly manifested within the divisions sampled where large variations were observed (Table 3). For example, Limuru division in Kiambu district had a large standard deviation implying that there was a high range of land sizes.

Proportion of household land under trees

The results showed that over 90% of the households sampled in the three districts had planted trees on their farms and there was no significant difference ($\chi^2 = 2.82$, d.f = 2, p = 0.242) in tree planting among Kiambu, Thika and Maragwa districts (Table 4). There were significant differences ($\chi^2 = 30.406$, d.f = 2, p<0.01) between the proportion of land under trees in the three districts. Majority of the households (over 65%) had less than a half an acre of land under trees cover (Table 5). There was a positive correlation (r = 0.55, p<0.01) between land size per household and area under trees (Table 6) across

Table 6. Association between household land size and proportion under trees across Kiambu, Thika and Maragwa districts.

Land size	Farm proportion in acres currently under trees													
	<0.5		0.5-1		1-2		2-3		3-5		>5		Total	
	%	n	%	n	%	n	%	n	%	n	%	n	%	n
<1	99	219	1	3	0	0	0	0	0	0	0	0	100	222
1-2	79	90	20	23	1	1	0	0	0	0	0	0	100	114
2.1-3	72	51	24	17	4	3	0	0	0	0	0	0	100	71
3.1-4	42	14	48	16	9	3	0	0	0	0	0	0	100	33
> 4.1	34	26	32	25	17	13	10	8	1	1	5	4	100	77

Table 7. Tree planting technologies adopted on farms in Kiambu, Thika and Maragwa districts.

District	The planting practices									
	Boundary/fence		Scattered trees in crop land		Home compound		Woodlot		Total	
	%	n	%	n	%	n	%	n	%	n
Kiambu	73	138	11	21	13	25	3	6	100	190
Thika	85	126	13	19	1	2	1	1	100	148
Maragwa	95	185	2	3	3	6	1	1	100	195

Table 8. Preferred tree species on farms in Kiambu, Thika and Maragwa districts.

District	Grevillea		Croton		Eucalyptus		Cypress		Pines		Others		Total	
	%	n	%	n	%	n	%	n	%	n	%	n	%	n
Kiambu	84	164	1	2	10	20	1	2	1	2	3	6	100	196
Thika	89	160	4	7	7	13	0	0	0	0	0	0	100	180
Maragwa	92	184	4	9	3	7	0	0	0	0	1	2	100	199

* Others: *Callistemon citrinus*, *Dovyalis caffra*, *Prunus*, *Olea* and *Terminalia brownie*.

all the three districts. Most of the households with less than 1 acre of land had planted less than 0.5 acres of their land with trees. There was also a significant Pearson correlation ($r = 0.54$, $p< 0.01$) between the land size and the number of trees planted.

Tree planting practices

The results revealed significant differences ($\chi^2 = 51.211$, d.f = 6; $p<0.01$) among districts in tree planting practices adopted by farmers on their farms (Table 7). Boundary planting was the most preferred practice with 73, 85 and 95% of the household respondents in Kiambu, Thika and Maragwa district, respectively, while woodlots had the least preference.

Tree planting preferences

The results showed that exotic tree species were the most preferred for planting on the farms with *Grevillea robusta* being the most popular species followed by *Eucalyptus* spp. There was a significant difference ($\chi^2 = 47.906$, d.f = 20, $p<0.01$) on tree preference among the districts (Table 8).

Tree planting obstacles

Household respondents identified various obstacles to tree planting as shown in Table 9, where inadequate land was the leading major obstacles, followed by scarcity of seedlings. These obstacles were found to be significantly different ($\chi^2 = 151.2$, d.f = 54, $p = 0.003$) among the districts.

DISCUSSION

The reported mean land sizes between 1.9 to 3.6 acres in the three districts were higher than what was reported by a study in Kakamega district in western part of Kenya where average farm size was 1.44 acres (Van Gelder and Kekholf, 1984). The small land sizes in all the three districts indicate that tree planting has to be integrated to

Table 9. Obstacles to tree planting in Kiambu, Thika and Maragwa districts.

District	Obstacles encountered in woodlot establishment									
	Scarcity of seedlings		Inadequate land		High cost of establishment		Others*		Total	
	%	n	%	n	%	n	%	n	%	n
Kiambu	16	32	74	144	2	4	8	8	100	188
Thika	33	60	60	110	1	1	6	11	100	182
Maragwa	28	54	57	112	2	4	14	25	100	195

* Include poor weather, lack of interest, pest and diseases.

the household farming systems. The increase in frequency of small land sizes is likely to be intensified in future due to continued increase in population resulting to fragmentation of land holdings into smaller farm sizes (Kamau, 1998). The high variability in land sizes among the districts was due to large tea and coffee plantations that were found in Limuru, Kakuzi and Thika Municipality divisions within the districts. The high rate of household tree planting in the study area (90%) compares well with an earlier study done in Kakamega district in western Kenya, where 80% of the rural household had planted trees on 25% of their farms despite the small household land sizes in the district (van Gelder and Kerkhof, 1984). The high tree planting rate in the three districts indicates that there was high awareness on tree planting in all districts as it had been shown in an earlier study by Mercer in 2004 that tree growing awareness through extension services related positively to tree growing in the fields.

The study in the three districts revealed that trees on farms have assumed an important place as one of the many smallholder land use options. Similar observations had been reported in a study done earlier in the neighboring district of Muranga in 1995, where it was observed that despite the pressure of land, trees were grown in 5 to 10% of the agricultural land (Dewees, 1995). The high rate of tree planting among the rural smallholder farmers is likely to have been contributed by its low labor input and minimal annual operating cost which is coupled by greater risk resistance in case of rain deficiency. The significant difference between the proportions of land in the three districts justifies development of area specific wood energy plan with area specific implementation strategies under each of the district development committees (DDC). The high level of sensitization on tree planting gives a positive starting point to development of sustainable wood energy plan through increasing the farmlands area under trees. Taking into account that over 40% of the household had less than 1 acre of land, the tree planting of even less than a quarter acre by a household is a substantial contribution to wood production. The positive correlation between land size per household and area under trees implies that the larger the land size, the bigger the area under trees on households farms. This was further evidenced by significant Pearson correlation between the land size and the number of trees planted.

Similar observation was reported in Zimbabwe, Philippines and Ethiopia where land size of household was positively correlated with number of trees planted by individual household (Price and Campbell, 1998; Nick and Jungho, 2004; Zenebe et al., 2010). This positive relationship can be used in developing area specific wood energy plan where the proposed increase of the area under trees is taken proportionately to the household farm size. The reason for high boundary tree planting in the districts was due to the fact that land sizes were small. Traditionally, boundary tree planting was done to mark demarcation between sub-clan lands (Leakey, 1997). However, the practice has been adopted even when dividing different cropping systems in the same farmland as was observed in the study area. In all districts, *G. robusta* was the most preferred tree species due to its ability to intercrop in agroforestry systems. *Eucalyptus* species were the second preferred species due to their fast growth with coppicing ability and also being a good timber species. Similar observation was also reported in a study done in Kilosa District, Tanzania where another agroforestry species *Senna* spp were the most popular followed by Eucalyptus species (Aalbaek, 2000). An earlier study in Kakamega district in Western Kenya also found *Eucalyptus* tree species to be the most preferred species within that district due to similar characteristics (Van Gelder and Kerkhof, 1984). Eucalyptus has also been reported to have over 20% returns to investment by farmers in Northern Ethiopia which makes it economically viable to plant (Jagger and Pender, 2003).

Fast growth and utilization potential of species are important factors to be considered in future woodfuel development. The obstacles to tree planting were found to be significantly different among the districts where inadequate land size was more pronounced in Kiambu district, whereas scarcity of seedling was more evident at Thika than in Maragwa and Kiambu districts. Similar observation was reported in a study by Aalbaek (2000) who found that about 28% of surveyed farmers throughout Tanzania stated that land scarcity was one of the top constraints to tree planting. Given the importance of land as a household asset, the agricultural land use activities and poverty reduction strategies need to take into account the declining farm sizes. The other obstacles included poor weather, lack of tree planting interest and termite attack. These factors need to be considered while promoting establishment of trees at the farm level. The variation of the obstacles in tree planting and tree preferences among the districts justifies the development of area specific strategies for wood energy development in each district. This will avoid the generalized strategies that may not apply uniformly across all the regions.

The interventions suggested to increase farm tree planting should be compatible with the existing farming practices for easy adoption by farmers.

CONCLUSIONS AND RECOMMENDATION

The proportion of land under trees by the households was found to be correlated to their land size which is an important factor to consider in future development of wood energy plan. The household land size categories, tree preference and constraints were found to vary between the three districts which justify the decentralized micro-level wood energy planning with site specific implementation strategies. The main obstacle to household tree planting was inadequate land for planting, unavailability of seedling and the high cost of tree establishment. The most preferred tree species planted were *G. robusta* followed by *Eucalyptus* species which was based on their growth and utilization potential. The culture of tree planting was found to be eminent in the three districts where over 90% of the household had planted trees in their farms with boundary tree planting being the most prevalent as a result of small land sizes. The study recommended promotion of tree planting in the study area through assisting in establishment of tree nurseries for seedling production.

Integrations of tree planting to farming systems were also recommended as agricultural sector has a key role in supplementing wood production.

ACKNOWLEDGEMENT

Authors would like to thank the district forest officers of Kiambu, Thika and Maragwa districts for assisting in data collection and KEFRI management for supporting the study.

REFERENCES

Aalbaek A (2000). Farmer tree planting in Tanzania. Proceedings of a workshop held in Arusha, Tanzania. The African Academy of Sciences (AAS). pp. 156 – 170.

Adensinu AA, Chianu J (2002). Determinants of farmers'adoption and adaptation of alley farming technology in Nigeria, Agro for. Syst., 55: 99-112

Akinga AW (1980). Woodfuel Survey in Kenya. Forestry Department, Ministry of Natural Resources report, Nairobi, Kenya. p. 42.

Dewees PA (1995). Social economic incentives for smallholder tree growing: A case study from Muranga district Kenya. Retrieved on 6/8/09 from: http://www.fao.org//docrep/u899e/u899 e00.html.

Food and Agriculture Organization (2005). The State of the Worlds Forests. Accessed from: http://www.fao.org//docrep/007/y557e/y557e00.html.

Githiomi JK, Mugendi DN, Kung'u JB (2012). Analysis of household energy sources and woodfuel utilisation technologies in Kiambu, Thika and Maragwa districts of Central Kenya. J. Hortic. For., 4(2): 43-48.

Jagger P, Pender J (2003). The role of trees for sustainable management of less favoured land: The case of Eucalyptus in Ethiopia. Forest Policy and Economics.

Kamau FK (1998). A Kenyan experience on research and development effort linking crop and livestock improvement, Natural resource management and human health. Workshop hld in ILRI Adis Ababa, Ethiopia 11-15 May.

Kenya Forest Service (2009). Kenya Forest Service Strategic Plan 2009/10 – 2013/14.

Kituyi E (2008). Policy Proposal for Sustainable Consumption and Production of Energy in Kenya: A Memorandum to the National Environment Policy Committee. Renewable Energy Technology Assistance Programme, Policy Brief Series No. 2, 2008. Nairobi, Kenya. p. 8.

Leakey LS (1997). The Southern Kikuyu before 1903. Academic press. London, UK.

Lee-Ann CH, Martin AB (1997). Surveying Natural Population. Columbia University Press, New York. pp.113-152.

Mercer DE (2004). Adoption of agroforestry innovations in the tropics: A review in Vistas Agroforestry. A compendium for first World congress of Agroforestry, Kluwer, Dordretch, Netherlands.

Ministry of Energy (2002). Study on Kenya's Energy Demand, Supply and Policy Strategy for Households, Small scale Industries and Service Establishments. Kamfor Consultants, Nairobi, Kenya. p. 133.

Nellie O, Githiomi J (2009). Forest Energy for Livelihoods; A paper presented in the 4th Forestry Society of Kenya Scientific Conference November, pp. 15-34.

NEMA (2004). State of Environment Report by National Environment Management Authority, Kenya. www.nema.go.ke

Nick E, Jungho S (2004). Social-economic factors affecting smallholder tree planting and management intentions in Leyte Province, Philippines. Small- scale Forest Economics, Management and Policy, Queensland, Australia, 3: 257-271.

Price L, Cambell B (1998). Household tree holdings: A case study in Mutoko communal area, Zimbabwe. Agroforestry Systems, 39: 205-210.

Republic of Kenya (1997a). Kiambu District Development Plan. Government Printers, Nairobi, Kenya, p. 156.

Republic of Kenya (1997b). Thika District Development plan 2002-2008. Government Printers, Nairobi, Kenya. p. 165.

Republic of Kenya (2001). 1999 Population and Housing Census: Counting our people for development. Government Printers, Nairobi, Kenya. p. 157.

Republic of Kenya (2002a). National Development Plan 2002-2008: Effective Management for Sustainable Economic Growth and Poverty Reduction. Government Printers, Nairobi, Kenya. p. 79.

Republic of Kenya (2002b). Maragwa District Development Plan 2002-2008: Effective Management for Sustainable Economic Growth and Poverty Reduction. Government Printers, Nairobi, Kenya. p. 59.

Republic of Kenya (2007). Ministry of Planning and National Development ; Kenya integrated household budget survey 2005/06 Revised Edition, Nairobi, Kenya. p. 96.

Schuren DJ, Snelder RD (2008). Tree growing on farms in Northern Luzon (Philipines): Small holders motivations and other determinants for adopting agroforestry systems.

Theuri K (2002). Woodfuel Policy and Legislation in Kenya; a paper presented during the Regional Workshop on Woodfuel Policy and Legislation held in ICRAF, Nairobi, Kenya. p. 8.

Van Gelder, B, Kerkholf P (1984). The agroforestry survey in Kakamega District. Kenya Woodfuel Development Programme, working paper no.3. Nairobi, Kenya. Workshop held in Arusha, Tanzania. The African Academy of Sciences (AAS). pp. 156 - 170.

Zenebe G, Menalek K, Gunamark K (2010). Household in Tigrai, Northern Ethiopia: Tree species, purpose and determinants. hht://www.efdinitiative.org.research/publication– epository/household-tree.

17

Using different nitrogen and compost levels on lettuce grown in coconut fiber

A. A. A. Farag[1], M. A. A. Abdrabbo[1] and E. M. Abd-Elmoniem[2]

[1]The Central Laboratory for Agricultural Climate, 6 Michail Bakhoum St., Dokki, Giza, Egypt.
[2]Soils Department, Faculty of Agriculture, Ain Shams University, Shoubra El-khema, Cairo, Egypt.

The present investigation was conducted for two seasons (2006 and 2007) into white polyethylene container filled with coconut fiber, to study the effect of three nitrogen levels (50, 100 and 150 ppm) applied by fertigation system and three compost levels (0, 2 and 4% by volume). Lettuce plant (*Lactuca sativa L.*) with two types' iceberg and romaine lettuce were transplanted in October to study the effect of treatments on vegetative growth, nutrients content (nitrogen, phosphors, potassium, calcium, magnesium and nitrate) as well as total yield. The experimental design was a split- plot with three replications. The obtained results indicated that the increasing nitrogen level up to 150 ppm significantly increased plant height, number of leaves per plant, dry weight and yield. Compost at 4% gave the highest vegetative growth and yield comparing with the other treatments. The highest nitrogen level (150 ppm) combined with compost 4% increased significantly vegetative growth and total yield of lettuce. On the other hand, 50 ppm nitrogen gave the highest nitrogen use efficiency (NUE) and the least value of nitrate content in leaves followed by 100 ppm.

Key words: Lettuce, polyethylene container, nitrogen use efficiency, coconut fiber, nitrate.

INTRODUCTION

Vegetables are the group of foods which makes the greatest contribution to nitrate consumption and leafy crops such as lettuce and spinach are particularly likely to contain high levels of nitrates. An EU regulation has set limits of 4,500 mg·kg^{-1} of nitrate in the fresh product for lettuce harvested from November 1 to March 31 and 3,500 mg·kg^{-1} from April 1 to October 31(Byrne et al., 2001).

Lettuce (*Lactuca sativa* L.) is a plant of considerable agricultural and economic interest but as a leafy vegetable it accumulates large quantities of nitrate especially when grown in high NO_3-N availability and low radiation. The accumulation of nitrate in plants depends on their genetic characteristics as well as on many environmental factors such as nitrogen supply for example replacement a part of NO_3-N concentration by urea or methods of application, light intensity, photoperiod, temperature o Large inputs of mineral fertilizer nitrogen (N) are routinely used to maintain the yield and quality of crops. Nitrogen not recovered by crops can pollute adjacent r water supply (Maynard andBarker, 1972; Zhu et al., 2000). terrestrial and aquatic habitats. This can cause a reduction in biodiversity since species that are characteristic of infertile habitats do not tend to thrive following nutrient-enrichment (Grime, 2001).

Nitrogen is one of the most important mineral nutrients determining plant growth and crop yield. Its effects are

Table 1. Chemical and physical properties of the coconut fiber of the experiment analyzed before cultivation.

Salinity (dS m^{-1})	pH	Moisture content, *Wm* (% m/m)	Dry Matter, *DM* (%nm/m)	Organic matter, *Wom* (% m/m)	Bulk density, *DB* (g L^{-1})	Water capacity (g/100 g DM)	Total pore space, *PS* (% v/v)	Water capacity, *WV*(% v/v)	Air capacity, *AV* (% v/v)
1.2	6.5	17	83	95	40	230	96	10	88

associated with leaf area growth and photosynthetic rate (Pinheiro Henriques and Marcellis, 2000; Pons and Westbeek, 2004).

The particular structure of coconut fibers and their physical and chemical properties, make them suitable for container media purposes (Batra, 1985). In fact the use of coconut fiber in European greenhouse production is well accepted as new technology. Coir contains equal portions of lignin and cellulose and is rich in potassium and the micronutrients Fe, Mn, Zn and Cu. Due to the high potassium content of the media a reduction in potassium fertilization has been shown to produce beneficial results (Savithri et al., 1993).

Compost contains many essential nutrients andimproves soil physical and chemical properties. It without a doubt is a valuable product as compost improves soil organic matter content, nutrient availability, soil aeration, and water holding capacity, and reduces soil bulk density. Compost, if properly prepared, is beneficial to the productivity of field and container crops (Brinton, 1997).

MATERIALS AND METHODS

The two experiments were carried out in the two successive seasons in 2006 and 2007 at Dokki location, Agricultural Research Center, Giza governorate Egypt. Each experiment included three nitrogen levels 50,100 and 150 ppm applied by fertigation, which were the combination with three compost levels 0, 2 and 4% volume of medium. Seeds of iceberg lettuce (" *Lactuca sativa*" var. capitata L.) cv. 'Vanity' and romaine lettuce (" *Lactuca sativa*" var. longifolia L.) cv. 'Balady' sown were at 22 and 26 August. Dates of transplanting were October 5 and 9 of 2006 and 2007, for the first and the second seasons, respectively. Lettuce grown in white polyethylene container of 0.60 m Length × 0.25 m Width × 0.25 m height, filled with 30 L coconut fiber. Each container contained three plants. Experiment design was a split plot with three replications where nitrogen fertilizer levels were distributed in the main plot and compost levels allocated in sub plots. Analysis of data was done by computer, using ANOVA program for statistical analysis. The differences among means for all traits were tested for significance at 5% level according to Waller and Duncan (1969). All other agriculture practices of cultivation were performed as recommended by the Ministry of Agriculture. Fresh and dry weights of individual heads were taken at 55- day interval on samples of three plants randomly harvested. Total dry weight was determined after oven-drying the samples at 75°C for 48 h. N, P and K were determined in leaf. Samples which were dried at 70°C in an air forced oven for 48 h. Dried leaves were digested in H_2SO_4 and the fallowing mineral contents were estimated (phosphorous "and" potassium) in the acid digested solution by colorimetric method (ammonium molybdate) using spectrophotometer and flame photometer (Chapman and Pratt, 1961). Nitrogen uptake was derived from dry weight in which total nitrogen concentration was determined by a micro-Kjeldahl method (Bremner and Mulvaney, 1982). Nitrogen use efficiency (NUE) was calculated as the yield obtained from the N(Yn) fertilized plot minus control (Yc), divided by a unit weight of the applied fertilizer (Nw for nitrogen):

NUE = (Yn-Yc)/Nw

RESULTS AND DISCUSSION

Vegetative growth

The effect of different nitrogen levels on vegetative growth characters of different compost levels in Tables 1 and 2. Regarding the effect of different nitrogen treatments, data showed that using 150 ppm nitrogen level increased lettuce plant height, number of leaves and dry weight significantly followed by 100 ppm treatment. The lowest vegetative growth was obtained by 50 ppm treatment during the two studied seasons.

The compost levels had same response on vegetative growth indicated that 4% compost had the highest vegetative growth followed by 2% compost with significant difference between them with the two lettuce types. Regarding the interaction effect between different nitrogen levels and compost levels, data illustrated that the highest vegetative growth characters obtained by using 150 ppm nitrogen level combined with 4% compost followed by 100 ppm nitrogen level with 4% compost. On the other hand, the lowest vegetative growth characters were obtained by using 50 ppm nitrogen level treatment with 0% compost with the two lettuce types.

Compost improves coconut fiber properties water holding capacity and CEC and that reflect to improve plant mineral uptake that led to improve plant growth. Nitrogen is one of the most important mineral nutrients determining plant growth; this result agree with Broadley et al. (2000) and Pons and Westbeek (2004) which found that removing N supply to lettuce considerably reduced plant growth. Reductions in the plant growth of N-limited plants correlate with decreases in leaf area, fresh and dry weight a reallocation of C from shoots to roots and a reduction in net C assimilation.

Yield

The effect of different treatments on lettuce yield is

Table 2. Chemical properties of the compost used in the experiment analyzed before cultivation.

EC	pH	Cations (Meg/L)				Anions (Meg/L)				N	C	C/N	O.M.
		Ca^{2+}	Mg^{2+}	Na^{+}	K^{+}	CO^{3-}	HCO_3^{-}	Cl^{-}	$SO4^{2-}$	(%)	(%)	(%)	(%)
7.95	5.39	19.0	51	48	1.05	0.0	3.1	34	38.15	0.32	4.94	15.41	9.88

Table 3. Effect of different applied nitrogen and compost levels on plant height, and number of leaves for iceberg lettuce plants during the 2006 and 2007seasons.

Nitrogen levels (%)	2006 season				2007 season			
	50	100	150	Mean A	50	100	150	Mean A
	ppm				ppm			
Hybrids		**Plant height**				**Plant height**		
0	12.6^{g}	13.7^{f}	15.4^{d}	13.9^{c}	13.1^{e}	14.3^{d}	16.1^{c}	14.5^{c}
2	13.8^{f}	17.6^{c}	18.8^{b}	16.7^{b}	14.4^{d}	18.3^{b}	18.1^{b}	16.9^{b}
4	15.1^{d}	18.9^{b}	20.1^{a}	18.0^{a}	14.5^{d}	18.2^{b}	19.3^{a}	17.3^{a}
Mean B	13.8^{c}	16.7^{b}	18.1^{a}		14.0^{b}	16.9^{aB}	17.8^{a}	
		Number of leaf				**Number of leaf**		
0	29.2^{f}	32.5^{e}	35.7^{d}	32.4^{c}	30.4^{g}	33.9^{e}	37.2^{d}	33.8^{c}
2	31.2^{eF}	40.6^{c}	44.2^{bC}	38.6^{b}	32.5^{f}	42.3^{c}	42.8^{bC}	39.2^{b}
4	34.4^{dE}	45.5^{b}	48.7^{a}	42.8^{a}	33.1^{eF}	43.9^{b}	46.9^{a}	41.3^{a}
Mean B	31.6^{c}	39.5^{b}	42.8^{a}		32.0^{c}	40.0^{b}	42.3^{a}	

Mean A: The mean value of the compost level. Mean B: The mean value of the nitrogen level. Means followed by the same letter within column are not significantly different ($P<0.05$).

Table 4. Effect of different applied nitrogen and compost levels on plant height, and number of leaves for romaine lettuce plants during the 2006 and 2007 seasons.

Nitrogen levels (%)	2006 season				2007 season			
	50	100	150	Mean A	50	100	150	Mean A
	ppm				ppm			
Hybrids		**Plant height**				**Plant height**		
0	22.4^{h}	28.3^{f}	33.8^{d}	28.2^{c}	21.6^{h}	27.3^{f}	32.6^{e}	27.2^{c}
2	26.4^{g}	33.6^{d}	38.4^{c}	32.8^{b}	25.4^{g}	35.8^{d}	40.9^{c}	34.0^{b}
4	31.68^{e}	40.3^{b}	46.0^{a}	39.3^{a}	33.7^{de}	43.9^{b}	49.6^{a}	42.4^{a}
Mean B	26.8^{c}	34.1^{b}	39.4^{a}		26.9^{c}	35.7^{b}	41.0^{a}	
		Number of leaf				**Number of leaf**		
0	11.9^{g}	18.2^{e}	23.1^{d}	17.7^{c}	11.4^{h}	17.5^{f}	22.3^{de}	17.1^{c}
2	15.3^{f}	22.1^{de}	27.2^{b}	21.5^{b}	14.7^{g}	23.5^{d}	29.6^{b}	22.6^{b}
4	18.7^{e}	25.5^{c}	35.7^{a}	26.6^{a}	19.9^{e}	27.2^{c}	38.1^{a}	28.4^{a}
Mean B	15.3^{c}	21.9^{b}	28.7^{a}		15.3^{c}	22.7^{b}	30.0^{a}	

Mean A: The mean value of the compost level. Mean B: The mean value of the nitrogen level.Means followed by the same letter within column are not significantly different ($P<0.05$).

presented in Tables 3 and 4. Referring the effect of different nitrogen levels, data showed that using 150 ppm nitrogen level increased had weight significantly followed by 100 ppm nitrogen level comparing with 50 ppm nitrogen level. Regarding the effect of different compost levels on yield, data showed that the highest yield weight obtained by 4% compost followed by 2% compost with significant difference between them. The lowest yield weight was obtained by 0% compost with the two lettuce types. Regarding the interaction effect between different

Table 5. Effect of different applied nitrogen and compost levels on head fresh and dry weight of leaves for iceberg lettuce plants during the 2006 and 2007 seasons.

Head	2006 season				2007 season			
Nitrogen levels	50	100	150	Mean A	50	100	150	Mean A
(%)	ppm				ppm			
Hybrids	Head fresh weight (g)				Head fresh weight (g)			
0	230^{h}	275^{f}	344^{d}	283^{c}	240^{d}	287^{c}	359^{bc}	295^{c}
2	261^{g}	351^{d}	382^{c}	331^{b}	272^{c}	366^{b}	368^{b}	335^{b}
4	291^{e}	422^{b}	442^{a}	385^{a}	280^{c}	407^{ab}	426^{a}	371^{a}
Mean B	260^{c}	349^{b}	389^{a}		264^{c}	353^{b}	384^{a}	
	Head dry weight (g)				Head dry weight (g)			
0	15.0^{i}	23.9^{h}	26.8^{f}	21.9^{c}	15.6^{g}	24.9^{f}	27.9^{d}	22.8^{c}
2	24.5^{g}	32.9^{d}	35.9^{c}	31.1^{b}	25.5^{ef}	34.3^{c}	34.6^{c}	31.4^{b}
4	27.3^{f}	39.6^{b}	41.5^{a}	36.1^{a}	26.3^{e}	38.2^{b}	40.1^{a}	34.8^{a}
Mean B	22.2^{c}	32.1^{b}	34.7^{a}		22.4^{c}	32.46^{b}	34.2^{a}	

Mean A: The mean value of the compost level. Mean B: The mean value of the nitrogen level.Means followed by the same letter within column are not significantly different ($P<0.05$).

Table 6. Effect of different applied nitrogen and compost levels on head fresh and dry weight of leaves for romaine lettuce plants during the 2006 and 2007 seasons.

Romaine	2006 season				2007 season			
Nitrogen levels	50	100	150	Meana	50	100	150	Mean A
(%)	ppm				ppm			
Hybrids	Plant fresh weight (g)				Plant fresh weight (g)			
0	234^{g}	365^{d}	497^{c}	365.3333	225^{f}	352^{de}	479^{c}	352
2	295^{f}	505^{c}	616^{b}	472	284^{e}	538^{b}	657^{ab}	493
4	347^{e}	623.9^{ab}	636.2^{a}	535.7	370^{d}	665^{ab}	678^{a}	571
Mean B	292	497.9667	583.0667		293	518.3333	604.6667	
	Plant dry weight (G)				Plant dry weight (G)			
0	21.1^{f}	32.8^{d}	44.7^{bc}	32.9^{c}	20.3^{g}	31.6^{ef}	43.1^{d}	31.7^{c}
2	26.5^{e}	45.4^{b}	55.4^{ab}	42.4^{b}	25.5^{f}	48.4^{c}	59.6^{b}	44.5^{b}
4	31.3^{de}	56.1^{a}	57.2^{a}	48.2^{a}	33.3^{e}	59.8^{b}	61.5^{a}	51.5^{a}
Mean B	26.3^{c}	44.8^{b}	52.4^{a}		26.4^{c}	46.6^{b}	54.7^{a}	

Mean A: The mean value of the compost level. Mean B: The mean value of the nitrogen level. Means followed by the same letter within column are not significantly different ($P<0.05$).

nitrogen levels and compost, data showed that using 150 ppm nitrogen level combined with 4% compost increased yield weight significantly followed by 100 ppm nitrogen level combined with 4% compost with iceberg lettuce, but there were no significantly differences between them with romaine lettuce. The lowest yield was obtained by 50 ppm nitrogen level with 0% compost with the two lettuce types. The increases in yield might be due to the increased utilization of N fertilizer in stimulating meristimatic activities. The accumulation of synthesized metabolites resulted in high dry matter accumulation and finally high yield. The obtained results are a good line with those reported by Pinheiro and Marcellis (2000). Yuri et al. (2004) conclude that the use of 56.0 t ha^{-1} of organic compost applied in pre plant provides an increase in yield and commercial quality of crisp head lettuce.

Nitrogen use efficiency (NUE)

Data in nitrogen use efficiency which is defined as the economic production obtained per unit of nitrogen applied is presented in Tables 5 and 6. It is evident from the data obtained that the highest NUE was obtained by 50 ppm nitrogen level followed by 100 ppm nitrogen level with

Table 7. Effect of different applied nitrogen and compost levels on nitrogen use efficiency of leaves for iceberg lettuce plants during the 2006 and 2007 seasons.

Head	2006 season				2007 season			
Nitrogen levels	50	100	150	Mean A	50	100	150	Mean A
(%)	ppm				ppm			
Hybrids	NUE				NUE			
0	38.1^{c}	22.7^{fg}	18.9^{h}	26.6^{c}	39.7^{b}	23.6^{e}	19.7^{f}	27.7^{c}
2	43.5^{b}	29.1^{e}	21.0^{g}	31.2^{b}	45.2^{ab}	30.3^{d}	20.3^{f}	31.9^{b}
4	48.2^{a}	34.9^{d}	24.3^{f}	35.9^{a}	46.4^{a}	33.6^{c}	23.4^{e}	34.5^{a}
Mean B	43.3^{a}	29.0^{b}	21.5^{c}		43.9^{a}	29.2^{b}	21.2^{c}	

Mean A: The mean value of the compost level. Mean B: The mean value of the nitrogen level.Means followed by the same letter within column are not significantly different (P<0.05).

Table 8. Effect of different applied nitrogen and compost levels on nitrogen use efficiency of leaves for romaine lettuce plants during the 2006 and 2007seasons.

Romaine	2006 season				2007 season			
Nitrogen levels	50	100	150	Mean A	50	100	150	Mean A
(%)	ppm				ppm			
Hybrids	NUE				NUE			
0	39.0^{e}	30.2^{g}	27.5h	32.2^{c}	37.5^{d}	29.0^{e}	26.5^{f}	31.0^{c}
2	48.9^{c}	41.9^{d}	34.1^{f}	41.6^{b}	47.1^{c}	44.6^{cd}	36.3^{de}	42.7^{b}
4	57.7^{a}	51.7^{b}	35.1^{f}	48.2^{a}	54.9^{a}	50.6^{b}	37.4^{d}	47.6^{a}
Mean B	48.5^{a}	41.2^{b}	32.2^{c}		46.5^{a}	41.4^{b}	33.4^{c}	

Mean A: The mean value of the compost level. Mean B: The mean value of the nitrogen level.Means followed by the same letter within column are not significantly different (P<0.05).

significant difference between them. The lowest NUE was obtained by 150 ppm nitrogen level. On the other hand, 50 ppm nitrogen level combined with4% compost gave the highest NUE followed by 50 ppm nitrogen level combined with 2% compost with iceberg lettuce, but with romaine lettuce 50 ppm nitrogen level combined with 4% compost gave the highest NUE followed by 100 ppm nitrogen level combined with 2% compost with significant difference between them.

This result agree with (Abu- Rayyan et al. 2004) who found that maximizing the nitrogen-use efficiency of crop production can be achieved (i) by optimizing the supply of N to meet the requirements of a crop during growth and development or (ii) by growing N-efficient crop genotypes. There is therefore much advantage to be derived from using economic rate that will enhance higher nitrogen use efficiency and maximize production, based on the value of increased nitrogen input resulted in a decline observed in NUE (Kogbe and Adediran, 2003).

Mineral contents

According to the effect of nitrogen levels (Tables 7 and 8), it was shown that using 150 ppm nitrogen led to increase N, P, K, Ca and Mg% significantly in dry weight followed by 100 ppm nitrogen with significant difference between them. The lowest N, P and K% was obtained by 50 ppm nitrogen. Data showed the effect of different compost levels on N, P, K, Ca and Mg% in dry weight. The highest N, P, K, Ca and Mg% was observed by 4% compost followed by 2% compost with significant difference between them. The lowest N, P, K, Ca and Mg% was obtained by 0% compost. The interaction effect between nitrogen levels and compost showed that using 150 ppm nitrogen combined with 4% compost increased N, P, K, Ca and Mg%. The lowest N, P, K, Ca and Mg% proceeded by 50 ppm nitrogen with 0% compost with the two lettuce types.

These results are in agreement with that of Economakis and Koleilat (1997) they reported that the uptake of N, P and K increased with increasing N concentration in the nutrient solution. Souza et al. (2005) reported that the protein, phosphorus, potassium, calcium and magnesium contents in lettuce leaves increased with the doses of organic compost.

Nitrate contents

Data in Tables 9 and 10 indicate NO_3-N in dry weight and

Table 9. Effect of different applied nitrogen and compost levels on N, P, K, and Ca, Mg and NO_3 of leaves for iceberg lettuce plants during the 2006 and 2007seasons.

Head	2006 season				2007 season			
Nitrogen levels	50	100	150	Mean A	50	100	150	Mean A
(%)	ppm				ppm			
Hybrids	**N (%)**				**N (%)**			
0	1.91^{f}	2.44^{d}	2.83^{bc}	2.36^{c}	1.96^{h}	2.53^{e}	2.94^{c}	2.47^{c}
2	2.14^{e}	2.63^{c}	2.90^{b}	2.55^{b}	2.17^{g}	2.75d	3.13^{b}	2.68^{b}
4	2.32^{d}	2.81^{bc}	3.15^{a}	2.73^{a}	2.34f	2.92^{c}	3.36^{a}	2.87^{a}
Mean B	2.12^{c}	2.62^{b}	2.93^{a}		2.16^{c}	2.73^{b}	3.14^{a}	
	P (%)				**P (%)**			
0	0.33^{h}	0.36^{g}	0.38f	0.35^{c}	0.34^{g}	0.37f	0.39^{e}	0.36^{c}
2	0.35^{gh}	0.41^{e}	0.46^{c}	0.41^{b}	0.36f^{g}	0.42d	0.44^{c}	0.41^{b}
4	0.44^{d}	0.48^{b}	0.55^{a}	0.49^{a}	0.38^{g}	0.46^{b}	0.53^{a}	0.45^{a}
Mean B	0.37^{c}	0.41^{b}	0.46^{a}		0.36^{c}	0.41^{b}	0.45^{a}	
	K (%)				**K (%)**			
0	1.77^{h}	1.89^{g}	2.05f	1.90^{c}	1.84i	1.97^{h}	2.14^{g}	1.98^{c}
2	2.15^{e}	2.27^{d}	2.40^{c}	2.27^{b}	2.24f	2.36^{c}	2.31d	2.30^{b}
4	2.35^{cd}	2.61^{b}	2.73^{a}	2.56^{a}	2.26^{e}	2.51^{b}	2.63^{a}	2.46^{a}
Mean B	2.09^{c}	2.25^{b}	2.39^{a}		2.11^{c}	2.28^{b}	2.36^{a}	
	Ca (%)				**Ca (%)**			
0	0.51^{f}	0.59^{e}	0.66d	0.59^{c}	0.53^{g}	0.61f	0.68^{e}	0.61^{c}
2	0.72^{cd}	0.74^{c}	0.79^{bc}	0.75^{b}	0.75d	0.77^{c}	0.76^{c}d	0.76^{b}
4	0.81^{b}	0.85^{ab}	0.89^{a}	0.85^{a}	0.78^{bc}	0.80^{b}	0.85^{a}	0.81^{a}
Mean B	0.68^{a}	0.72^{b}	0.78^{c}		0.68^{c}	0.72^{b}	0.76^{a}	
	Mg (%)				**Mg (%)**			
0	0.18^{f}	0.21^{e}	0.24d	0.21	0.18^{e}	0.21^{d}	0.25^{c}	0.21^{c}
2	0.26^{cd}	0.27^{c}	0.28^{b}	0.27	0.27^{bc}	0.28^{b}	0.27^{bc}	0.27^{b}
4	0.29b	0.31^{a}	0.32^{a}	0.306667	0.27^{bc}	0.29^{ab}	0.30^{a}	0.29^{a}
Mean B	0.24^{c}	0.26^{b}	0.28^{a}		0.24^{b}	0.26^{ab}	0.27^{a}	
	NO_3 (ppm)				**NO_3 (ppm)**			
0	527f	723^{d}	1055^{b}	768^{b}	495^{h}	712^{e}	1060^{c}	764^{b}
2	544^{e}f	738^{cd}	1120^{ab}	800^{ab}	549g	731^{de}	1105^{b}	791^{ab}
4	581^{e}	774^{c}	1155^{a}	828.0^{a}	589f	767^{d}	1148^{a}	821^{a}
Mean B	541^{c}	745^{b}	1110^{a}		536^{c}	736^{b}	1104^{a}	

Mean A: The mean value of the compost level. Mean B: The mean value of the nitrogen level.Means followed by the same letter within column are not significantly different ($P<0.05$).

NO_3-N as a percentage from total nitrogen, as affected by nitrogen levels and compost levels in lettuce plants. Nitrate contents in dry weight recorded the least values with 50 ppm nitrogen (366 to 540 D.W) ppm and 0% compost level (586 to 825 D.W) ppm in both type's iceberg and romaine of lettuce plant. The interaction between nitrogen and compost recorded the last value in 0% compost with 50 ppm nitrogen in both types of lettuce plant.

Accumulation of nitrates in lettuce has been shown to be affected by the soil texture, amount of fertilizer-N, the timing of fertilizer-N release, the light intensity and duration. In most types of lettuce, including the romaine type had the highest concentration of nitrates and is normally observed in the external leaves (Santamaria et al., 1999; Abu-Rayyan et al., 2004).

Table 10. Effect of different applied nitrogen and compost levels on N, P, K, and Ca, Mg and NO_3 of leaves for romaine lettuce plants during the 2006 and 2007seasons.

Romaine	**2006 season**				**2007 season**			
Nitrogen levels	**50**	**100**	**150**	**Mean A**	**50**	**100**	**150**	Mean A
(%)	**ppm**				**ppm**			
Hybrids	**N (%)**				**N (%)**			
0	2.12^{f}	2.68^{d}	3.13^{bc}	2.64^{c}	2.06h	2.66^{e}	3.09^{c}	2.60c
2	2.35^{e}	2.91^{c}	3.24^{b}	2.83^{b}	2.28^{g}	2.89^{d}	3.29^{b}	2.81^{b}
4	2.57^{de}	3.14^{bc}	3.47^{a}	3.06^{a}	2.46f	3.06^{c}	3.52^{a}	3.01^{a}
Mean B	2.35^{c}	2.91^{b}	3.28^{a}		2.26^{c}	2.87^{b}	3.30^{a}	
	P (%)				**P (%)**			
0	0.36^{f}	0.40^{de}	0.42^{d}	0.39^{c}	0.35^{g}	0.38f	0.41^{e}	0.38c
2	0.39^{e}	0.45^{cd}	0.51^{bc}	0.45^{b}	0.37^{fg}	0.48^{d}	0.54^{c}	0.46^{b}
4	0.49^{c}	0.53^{b}	0.61^{a}	0.54^{a}	0.52^{cd}	0.57^{b}	0.65^{a}	0.58^{a}
Mean B	0.41^{c}	0.46^{b}	0.51^{a}		0.41^{c}	0.48^{b}	0.53^{a}	
	K (%)				**K (%)**			
0	0.26^{h}	0.31^{g}	0.34^{f}	0.303333	0.25^{h}	0.30^{g}	0.33^{f}	0.29^{c}
2	0.38^{de}	0.39^{d}	0.42^{c}	0.396667	0.36^{e}	0.41^{d}	0.44^{cd}	0.40^{b}
4	0.37^{e}	0.44^{b}	0.47^{a}	0.426667	0.45^{c}	0.47^{b}	0.52^{a}	0.48^{a}
Mean B	0.34^{c}	0.38^{b}	0.41^{a}		0.35^{c}	0.39^{b}	0.43^{a}	
	Ca (%)				**Ca (%)**			
0%	0.57^{h}	0.67^{g}	0.74^{f}	0.66	0.55^{h}	0.64^{g}	0.71^{f}	0.63^{c}
2%	0.81^{e}	0.83^{de}	0.88^{c}	0.84	0.78^{e}	0.89^{d}	0.94^{cd}	0.87^{b}
4%	0.85^{d}	0.95^{b}	1.06^{a}	0.953333	0.97^{c}	1.02^{b}	1.07^{a}	1.02^{a}
Mean B	0.74^{c}	0.87^{b}	0.89^{a}		0.77^{c}	0.85^{b}	0.91^{a}	
	Mg (%)				**Mg (%)**			
0	0.20^{g}	0.24^{f}	0.27^{e}	0.24^{c}	0.20^{h}	0.23^{g}	0.26f	0.23^{c}
2	0.29^{de}	0.30^{d}	0.32^{c}	0.30^{b}	0.28^{e}	0.32^{d}	0.34^{cd}	0.31^{b}
4	0.33^{bc}	0.34^{b}	0.36^{a}	0.34^{a}	0.35^{c}	0.37^{b}	0.39^{a}	0.37^{a}
Mean B	0.27^{c}	0.29^{b}	0.32^{a}		0.28^{c}	0.31^{b}	0.33^{a}	
	NO_3 (ppm)				**NO_3 (ppm)**			
0	549i	772^{f}	1160^{c}	845^{b}	560^{h}	799^{f}	1122^{c}	817^{b}
2	588^{h}	812^{e}	1232^{b}	880^{ab}	586^{h}	785^{e}	1191^{b}	851^{ab}
4	639^{g}	852^{d}	1270^{a}	910^{a}	654^{g}	864^{d}	1242^{a}	880^{a}
Mean B	596^{c}	819^{b}	1221^{a}		576^{c}	792^{b}	1180^{a}	

Mean A: The mean value of the compost level. Mean B: The mean value of the nitrogen level.Means followed by the same letter within column are not significantly different ($P<0.05$).

REFERENCES

Abu-Rayyan A, Kharawish BH, Al-Ismail K (2004). Nitrate content in lettuce (Lactuca sativa L.) heads in relation to plant spacing, nitrogen form and irrigation level. J. Sci. Food Agric., 84: 931–936.

Abu-Rayyan A, Kharawish BH, Al-Ismail K (2004). Nitrate content in lettuce (Lactuca sativa L.) heads in relation to plant spacing, nitrogen form and irrigation level. J. Sci. Food Agric. 84:931-936.

Batra SK (1985). Other long vegetable fibers. Handbook of Fiber Science and Technology. p. 4.

Bremner JM, Mulvaney CS (1982). Nitrogen - Total. In: Methods of Soil Analysis (A. L. Page et al., ed.). Agronomy Monograph 9, Part 2, 2nd ed. American Society of Agronomy, Madison, WI. pp. 595-624.

Brinton WF (1997). Proc. 6th annual conference on composting. October 11-13, Beltsville, MD.

Broadley MR, Escobar-GutieÂrrez AJ, Burns A, Burns IG (2000). What are the effects of nitrogen deficiency on growth components of lettuce? New Phytologist. 147:519-526.

Byrne C, Maher MJ, Hennerty MJ (2001). Reducing the nitrogen content of protected lettuce crops. Irish J. Agric. Food Res. 39(3):491. Cairo, Egypt.

Chapman HD, Pratt PF (1961). Methods of analysis for soils, plants and waters. Division of Agricultural Sciences, University of California, Riverside.

Economakis CD, Koleilat R (1997). Effect of nitrogen concentration on growth, water and nutrient uptake of lettuce plants in solution culture.

Acta Hort. 2:449,

Grime JP (2001). Plant strategies, vegetation processes, and ecosystem properties, 2nd edn. Chichester, UK: John Wiley.

Hardwick RC (1987). The nitrogen content of plants and the self thinning rule of plant ecology: a test of the core-skin hypothesis. Ann. Bot. 60:439-446.

Kogbe JOS, Adediran JA (2003), Influence of nitrogen, phosphorus and potassium application on the yield of maize in the savanna zone of Nigeria. Afr. J. Biotechnol. 2(10):345-349.

Maynard DN, Barker AV (1972). Nitrate Content of Vegetable Crops. HortScience **7**:224–226.

Pinheiro Henriques AR, Marcelis FM (2000). Regulation of growth at steady-state nitrogen nutrition in lettuce (*Lactuca sativa* L.): interactive effects of nitrogen and irradiance. Ann. Bot. 86:1073-1080.

Pons TL, Westbeek MHM (2004). Analysis of differences in photosynthetic nitrogen-use efficiency between four contrasting species. Physiol. Plant. 122:68-78.

Santamaria P, Elia A, Serio F, Todaro E (1999). A survey of nitrate and oxalate content in fresh vegetables. J. Sci. Food Agric. 79:1882–1888.

Savithri P, Murugappan V, Nagarajan R (1993). Possibility of economizing K fertilization by composted coir peat application. Fert. News. 38:39-40.

Souza PA, de Negreiros, MZ, de Menezes JB, Bezerra Neto F, Souza GLFM, Carneiro, CR, Queiroga RCF de (2005). Chemical evaluations of lettuce leaves grown under the residual effect of soil fertilized with organic compost. Hortic. Bras. 23(3):754-757.

Waller RA, Duncan DB (1969). Ways for the symmetric multiple comparison. Problem Am. Stat. Assoc J.19:1485-1503.

Yuri JE, Resende GM. De, Rodrigues JJ (2004). Effect of organic compost on crisp head lettuce production and commercial characteristics. Hortic. Bras. 22(1):127-130.

Zhu Z, Gerendas J, Sattelmacher B (2000). Influence of nitrate and urea supply on growth and nitrate accumulation in hydroponically grown lettuce (*Lactuca sativa*). Xth International Colloquium for the Optimization of Plant Nutrition. April 8–13 2000, Cairo Sheraton,

18

Functional distances of woody species, proxy of ecological elasticity and provision for forest management

Shiferaw Abebe[1,3,4]*, Kåre Arnstein Lye[1] and Diress Tsegaye[1,2]

[1]Department of Ecology and Natural Resource Management, Norwegian University of Life Sciences, P. O. Box 5003, N-1432 Ås, Norway.
[2]Department of Biology, University of Oslo, P. O. Box 1066 Blindern, N-0316 Oslo, Norway.
[3]Forestry Agricultural Research Center, Ethiopian Institute of Agricultural Research, P. O. Box 2003, Addis Ababa, Ethiopia.
[4]Norwegian Institute for Nature Research (NINA), Gaustadallèen 21, No-0349 Oslo, Norway.

The natural forests in north western Ethiopia have been dwindling at an alarming rate. However, church forests are found as a spot in each community serving as unique and conspicuous *in situ* conservation of indigenous species. Five church forests with adjacent degraded forests were surveyed to investigate the functional distance and diversity. Fifty woody species were identified both from the church and adjacent degraded forests and have 34 species in common (S_j = 0.68). Eleven new species were recorded in the adjacent degraded forest, but five species which were found in the church forest disappeared. The average functional distances between woody species were 22.3 ± 14.4 (mean ± SD) and 23.8 ± 17.0 for church and adjacent degraded forests, respectively. Dominant woody species were functionally dissimilar. Functional distances between 10 minor of the church and 10 dominant of adjacent degraded forest species were significantly different (χ^2 = 12.24, P = 0.01), suggesting that minor species have offered insurance. Passenger species and some of the dominant species in the church forest have also offered insurance. Functional distance which is a cumulative effect of functional attributes is a proxy for ecological elasticity to disturbances and helps in forest management decisions.

Key words: Church forests, degraded forests, ecosystem, functional attributes, relative abundance, resilience.

INTRODUCTION

The natural forests in north western Ethiopia has been dwindling at an alarming rate (Mamo et al., 2006; Gole et al., 2008) due to conversion to arable lands, overgrazing, and excessive utilization for fuelwood and construction triggered by increasing population growth (Haugen, 1989; Taddese, 2001; Zeleke and Hurni, 2001). However, church forests are found as a spot in a landscape in each community (that is, the lowest administrative unit in the region) and mostly represent the original natural forest of the area (Wassie et al., 2005). The degraded natural forests may be composed of a combination of species from the original natural forest and additional new species. The church forests are unique and conspicuous spots serving as an *in situ* conservation, mainly for indigenous trees and shrubs in north western parts of Ethiopia. Though, the churches are mainly built as congregational sites of worship, burials, meditating and religious festivals, they often provide exclusive, secured habitats for plants, animals and microorganisms (Wassie, 2002; Wassie et al., 2005). The reason why church forests survive the intensive deforestation in the region is that, as many people remain adherents to the Orthodox faith which espouses protection of woodland within its

*Corresponding author. E-mail: shiferab@yahoo.com, shiferaw.abebe.zeleke@nina.no.

territory, they are reluctant to degrade such areas. Nevertheless, some disturbances such as spot clearing for burial purpose and selective logging for the maintenance of the churches are occasionally carried out. Due to differences in intensity of disturbances, species composition in church and degraded forests may thus be different even if they are located at short distances. Disturbance is thus one of the key factors that shape the structure of vegetation (Rusch et al., 2003; Westoby et al., 2002). Although, a few traits have a general association with responses to disturbance, some studies have documented that plant attributes are specific to a particular disturbance (Girme, 2001; Walker et al., 1995).

The functional attributes of species have pivotal roles for the ecosystem function of a given species assembly (Delarum et al., 2010) and help to predict changes in vegetation and biodiversity in response to changes in environmental factors and land-use shifts at regional and global scales (Rusch et al., 2003). It is now generally accepted that the structure of species assemblages may be as important as the concept of species richness (Naeem and Wright, 2003; Petchey and Gaston, 2006; Petchey et al., 2007; Griffin et al., 2009). This approach signifies the importance of functional diversity than biodiversity (Dalerum et al., 2010). There is thus an increasing interest in quantifying functional diversity. Holling (1973) has introduced the perception of ecosystem resilience as a means to investigate ecological processes. Such ecological resilience can be regarded as a measure of how much disturbance an ecosystem can tolerate without altering its original state (Gunderson, 2000) or amount of disturbance that a species assembly can take up without varying its ecological utility (Walker, 1995). Dominant species of a natural forest are few in number, and they are responsible for performing the bulk of photosynthesis, transpiration and nutrient uptake (Peterson et al., 1998). Dominant species would be functionally dissimilar to each other (Walker et al., 1999). Minor species, which constitute a smaller proportion of the species abundance in the ecosystem are similar to the dominant species in ecological functioning, but react in a different way to environmental requirements and disturbance; as a result they offer insurance to the vegetation after a certain interruption (Walker, 1995; Walker et al., 1999; Bengtsson et al., 2000).

Identifying dominant and minor species and functional distances between the dominant and minor species in a vegetation, therefore, helps to predict the possible structures of a forest and the likelihood of such occurrence (Archibold, 1995; Peterson et al., 1998; Walker et al., 1999). Despite the potential of the church forest in understanding the history of the original ecosystem, consequences of disturbance and resilience mechanism studies are rare (Wassie et al., 2002, 2005, 2010) and absent particularly in relation to functional diversity. Thus, the objectives of the study were to: 1) identify dominant and minor woody species and their functional attributes both in the church and degraded forests, 2) estimate functional diversity of the forests and functional distances between pairs of woody species, and 3) examine how minor species in the church forests have offered resilience for the adjacent degraded forest.

MATERIALS AND METHODS

Study area

The study was conducted in Awabel district of East Gojam administrative zone in Amhara national regional state, Ethiopia. Each community (the lowest local administrative unit) has its own church which is mostly surrounded by natural forests. We selected five churches with natural forest that have adjacent degraded forests for this study: Yeleb Lideta (10° 09' N, 37° 59' E), Yeterbina Gebreal (10° 09' N, 38° 02'), Tsid Mariam (10° 08' N, 37° 58' E), Mehir Abo (10° 06' N, 37° 59' E) and Mekides Mariam (10° 07' N, 38° 00' E). The elevation of the study area ranges from 1100 to 3200 m above sea level. The annual rainfall and mean annual temperature is about 1100 to 1400 mm and 21°C, respectively.

Sampling

Five churches together with five adjacent degraded forests were randomly selected. The five selected churches were: Yeleb Lideta, Yeterbina Gebreal, Tsid Mariam, Mehir Abo and Mekides Mariam. Twelve circular plots with a radius of 5.64 m (100 m^2 area) (approximately 10% of the total area of each church forest) were randomly distributed in each of the church and adjacent degraded forests, a total of 60 plots were established in each of the forest types. Circular plots were used to avoid edge effects (Magurran, 2004). All woody species were recorded. However, seedlings were recorded in a subplot (1 m radius) established in each of the circular plots.

Estimating functional distance and functional attributes

As shown in Table 1, the most important plant functional attributes such as life form, height, leaf area, leaf life time, and fruit type (Archambault, 1998; Walker, 1995; Walker et al., 1999) were used to estimate the functional diversity of the two forests and functional distance between pair of woody species. Some of the attributes play vital role in performing photosynthesis and biomass accumulation by receiving more light for longer periods. Fruit type influences seed dispersal, storage and germination potential and mechanism which in turn have great impact on the regeneration and dominancy of a species (Ngulube et al., 1997; Teketay, 1997; Vanclay et al., 1997; Eriksson et al., 2003). These functional attributes and botanical names for each species were identified in the field according to "Flora of Ethiopia and Eritrea" (Edwards et al., 2000; Hedberg and Edwards, 1989; Hedberg et al., 2006). Specimens were collected for species which were not identified in the field and identified at the National Herbarium at Addis Ababa University. The functional distance was estimated following Walker (1995) and Walker et al. (1999) equation:

$$ED = \left[\sum_{i=1}^{I} \left(A_{ij} - A_{ik} \right)^2 \right]^{1/2}$$

Table 1. Functional attributes and values attached to each attribute.

Functional value	Height (m)	Leaf area (cm^2)	Growth form	Leaf life time	Fruit type
1	0-5	0-10	Shrub	Deciduous	Berry
2	5-10	10-20			Drupe
3	10-15	20-30	Shrub/tree	Semidecidous	Nut
4	15-20	30-40			Capsule
5	> 20	> 40	Tree	Evergreen	Pod

Source: Walker et al. (1999) and Rutina et al. (2005). Note that the two conifer species lack fruits. *Cupressus lusitanicus*, which has dry seeds, is in the table classified as having nut, while *Juniperus excelsa*, which has fleshy seeds, is in the table classified as having a berry.

Where A_{ij} and A_{ik} are the attribute values of species j, k for attribute i, and l is the total number of attributes being considered.

We also classified functional similarity intervals using ecological distance (ED) as functionally similar ($0 \leq ED \leq 6$), similar to average ($7 \leq ED \leq 14$), average to dissimilar ($14 \leq ED \leq 33$) and dissimilar ($ED > 33$) following Walker et al. (1999). Jacard's similarity coefficients (Magurran, 2004) were used to determine the similarities of woody species based on their abundance. The relative abundance of each species (Krebs, 1999; Smith and Smith, 2001) in both church and adjacent degraded forests was compared using a Z test of two proportions in Minitab software version 16.0. Then, a Chi-square test was used to compare the functional distances among the ten most dominant species of the church forest and between the ten least dominant (minor) of the church forest and the ten most dominant species of the adjacent degraded forest.

RESULTS

Relative abundance and similarity of woody species

Fifty woody species have been identified from both the Church and adjacent degraded forests. Fifteen woody species were significantly more abundant in the church forests than in the degraded forests ($P < 0.05$). Similarly, seven woody species were significantly more abundant in the adjacent degraded forests than in the church forest. Five woody species, *Cordia africana* Lam, *Dracaena steudneri* Engel., *Jasmium abyssinicum* Hochst. ex DC., *Malva verticilata* L. and *Olea europaea* L. ssp. *Cuspidata* which were found in the church forest have disappeared in the adjacent degraded forest. Eleven new woody species, *Acalypha fruitcosa* Forssk., *Clausena anisata* (Willd.) Benth., *Combretum molle* R.Br. ex G.Don, *Dodonaea angustifolia* L.F., *Ekebergia capensiss* Sparrm., *ficus plamta* Forssk., *ficus vasta Forssk.*, *Myrsine Africana* L. and *Rhus glutinosa* A. Rich. which were not found in the church forest have been recorded in the adjacent degraded forest (Table 2). Most of the woody species which were dominant in church forests had reduced in abundance in adjacent degraded forests. The similarly between the church and the adjacent forests as a whole was the highest ($S_j = 0.68$) of all indexes (Table 3).

On average, similarity index was higher among church forests ($S_j = 0.52$) compared to adjacent degraded natural forest ($S_j = 0.34$). Mekdes church forest and its adjacent degraded forest were the most dissimilar ($S_j = 0.18$).

Functional distance and diversity

The functional distances between woody spices ranges from 0 to 72 and 0 to 81 and church and adjacent degraded forests, respectively (Appendix 1 and 2). The total number of functional distances between woody species in the church and adjacent forest was 737 and 926, respectively. The average functional distances between woody species were 22.3 ± 14.4 (mean ± SD) and 23.8 ± 17.0 for church and adjacent degraded forests, respectively (Appendix 1 and 2). Out of the total functional distances in the church forests, 14% (106) pairs were functionally similar, 26% (193) pairs were similar to average, 38% (277) pairs were average to dissimilar and 22% (161) pairs were dissimilar for the church forests (Figure 1A). In the degraded adjacent forests, 17% (154) pairs were functionally similar, 23% (215) pairs were similar to average, 33% (304) pairs were average to dissimilar and 27% (253) pairs are dissimilar (Figure 1B). We found significant variation among the four functional distance groups in the distribution of functional distances of the 10 most dominant species in the church forest ($\chi^2 = 21.40$, $P < 0.001$) and between 10 minor of the church and 10 dominant adjacent degraded forest species ($\chi^2 = 12.24$, $P = 0.01$) (Table 4). The average dissimilar ($14 \leq ED \leq 33$) and dissimilar ($ED \geq 33$) functional groups combined represent 87% of the total functional distances in the church forest (Table 4).

Only the rest 20% of the functional distances among the ten most dominant woody species of the church forest lies in similar to average ($6 < ED \leq 14$) and similar ($0 \leq ED \leq 6$) functional ranges (Table 4). Whereas, the functional groups for average to dissimilar ($14 \leq ED \leq 33$) and dissimilar ($ED \geq 33$) combined have captured only 66% ($\chi^2 = 12.240$, $P = 0.01$) of the functional distances between the ten minor species of the church forest and the ten most dominant woody species of the adjacent degraded forests. Five woody species (*Carissa spinarum*

Table 2. Woody species identified and their relative abundance and functional attributes.

Species	Relative abundance (%) in 120 plots		Mature leaf area (cm^2)	Height at maturity (m)	Growth form	Leaf life time	Fruit type
	Church	Adjacent					
Acacia abyssinica Hochst. ex Benth.	15^a	28^a	0-5	20	Tree	Semi-deciduous	Pod
Acacia nilotica (L.) Willd. ex Del.	17^a	20^a	0-5	2-14	Shurb/tree	Evergreen	Pod
Acalypha fruticosa Forssk.	0^a	3^a	>40	2-5	Shrub	Semi-deciduous	Capsule
Acanthus pubescens (Oliv.) Engl.	32^a	30^a	>40	0-5	Shrub	Deciduous	Capsule
Albizia gummifera (J.F. Gmel.) C.A.Sm.	8^a	18^a	0-5	15	Tree	Evergreen	Pod
Asparagus racemosus Willd.	0^b	7^a	0-5	7	Shrub	Semi-deciduous	Berry
Bersama abyssinica Fresen.	2^b	15^a	>40	2-5	Shrub	Semi-deciduous	Capsule
Buddleja polystachya Fresen.	15^a	8^a	>40	4-12	Tree	Semi-deciduous	Capsule
Calpumia aurea (Ait.) Benth..	58^a	18^b	0-5	2-5	Shrub	Semi-deciduous	Capsule
Capparis micrantha A.Rich.	53^a	13^b	0-5	0-5	Shrub	Semi-deciduous	Capsule
Carissa spinarum L.	58^b	83^a	5-10	0-5	Shrub/tree	Evergreen	Berry
Clausena anisata (Willd.) Benth.	0^b	15^a	0-5	0-5	Shrub	Evergreen	Capsule
Clerodendrum myricoides (Hochst.) Vatke	18^a	3^b		0-5	Shrub	Evergreen	Capsule
Combretum molle R.Br. ex G.Don	0^a	2^a	20-30	2-5	Shrub	Evergreen	Capsule
Cordia africana Lam.	13^a	0^b	300	15-20	Tree	Deciduous	Drupe
Croton macrostachyus Del.	32^a	27^a	150	25	Tree	Deciduous	Nut
Cupressus lusitanica Lindl. ex Parl.	8^a	7^a	0-5	35	Tree	Evergreen	Dry seed
Cussonia holstii Harms ex Engl.	3^a	7^a	20-30	0-5	Shrub	Evergreen	Nut
Dodonaea angustifolia L.f.	0^a	5^a	5-10	0-5	Shrub	Evergreen	Capsule
Dombeya torrida (J.F.Gmel.) P. Bamps	8^a	17^a	100	12-15	Shrub/tree	Semi-deciduous	Capsule
Dracaena steudneri Engl.	3^a	0^a	>40	5-10	Shrub/tree	Evergreen	Capsule
Ehretia cymosa Thonn.	8^a	7^a	15	0-5	Shrub	Semi-decidous	Capsule
Ekebergia capensis Sparrm.	0^a	3^a	>40	20-30	Tree	Evergreen	Capsule
Eucalyptus globulus Labill.	3^a	2^a	30-40	30	Tree	Evergreen	Capsule
Euclea racemosa Murr. ssp. *schimperi*	3^a	8^a	10-15	3-5	Shrub/tree	Evergreen	Drupe
Euphorbia abyssinica J.F.Gmel.	35^a	3^b	0-5	10	Shrub	Evergreen	Capsule
Ficus palmata Forssk.	0^a	5^a	>40	20-30	Tree	Deciduous	Berry
Ficus vasta Forssk.	0^b	8^a	>40	>20	Tree	Deciduous	Fig, berry
Grewia ferruginea Hochst. ex A. Rich.	60^a	7^b	>40	0-5	Shrub	Evergreen	Capsule
Hibiscus micranthus L.f.	12^a	2^b	6-30	0-5	Shrub	evergreen	Capsule
Jasminum abyssinicum Hochst. ex DC.	18^a	0^b	20-30	5-10	Shrub	Semi-deciduous	Nut
Juniperus excelsa M.Bieb.	8^a	2^a	0-5	40	Tree	Evergreen	Fleshy seed
Justicia schimperiana (Nees) T. Anders.	40^a	5^b	36-40	0-5	Shrub/tree	Evergreen	Capsule
Malva verticillata L.	7^a	0^b	20-30	2-5	Shrub	Evergreen	Capsule

Table 2. contd.

Maytenus senegalensis (Lam.) Exell	58^a	40^b		0-5	0-5	Shrub	Evergreen	Berry
Myrsine africana L.	0^a	5^a		0-5	0-5	Shrub	Evergreen	Capsule
Ocimum urticifolium Roth	8^a	7^a		0-5	1-2	Shrub	Semi-deciduous	Capsule
Olea europaea L. ssp. *cuspidata*	17^a	0^b		30	10-20	Tree	Evergreen	Nut
Olinia rochetiana A. Juss.	0^b	8^a		0-5	0-5	Shrub	Evergreen	Capsule
Osyris quadripartita Decn.	12^a	18^a		0-5	0-5	Shrub	Evergreen	Drupe
Phytolacca dodecandra L'Hérit.	23^a	3^b		0-5	2-5	Shrub	Evergreen	Capsule
Pittosporum viridiflorum Sims	30^a	13^b		20-30	0-5	Shrub/tree	Semi-deciduous	Pod
Prunus africana (Hook.f.) Kalkm.	3^a	7^a		0-5	0-5	Shrub	Evergreen	Pod
Protea gaguedi J.F.Gmel.	25^a	30^a		0-5	2-15	Shrub	Evergreen	Pod
Pterolobium stellatum (Forssk.) Brenan	48^a	18^b		50	40	Tree	Evergreen	Drupe
Rhus glutinosa A.Rich.	0^b	12^a		24	3-10	Shrub/tree	Evergreen	Drupe
Rosa abyssinica Lindley	2^a	10^a		0-5	0-5	Shrub	Evergreen	Drupe
Schefflera abyssinica (A. Rich.) Harms	5^a	13^a		>40	10-30	Tree	Evergreen	Capsule
Terminalia schimperiana Hochst.	5^a	15^a		20-30	2-5	Evergreen	shrub	Capsule
Vernonia amygdalina Del.	8^a	15^a		>40	3-10	Shrub/tree	Evergreen	Capsule

Figures with different superscripts indicates that the relative abundance of a species is significantly different ($P<0.05$) in church and degraded forests.

L., *Maytenus senegalensis* (Lam.) Exell, *Calpurnia aurea* (Ait.) Benth., *Justicia schimperiana* (Nees) T. Anders and *Osyris quadripartite* Decn. out of the most dominant 10 woody species of the adjacent degraded forest had similar ecological distance (Appendix 1 and 2).

DISCUSSION

The similarity index shows that church forests are more similar to each other than adjacent degraded forests, suggesting that disturbance can create more dissimilar forests. Our result revealed that functional diversity of the degraded natural forest had increased only by 6% after disturbance considering the church forest as a reference. However, Abebe et al. ('under review') has reported higher species diversity for church forest compared to adjacent degraded forest using the same plots of these forests. A slightly higher mean and standard deviation of functional diversity for the adjacent degraded forest compared to the church forest indicates that disturbance may some how increase functional diversity. However, this occurs at moderate level of disturbance as reported in other studies (Delarum et al., 2010). This study indicates that functional diversity is less sensitive to disturbance than species diversity. This study thus contributes to the increasing interest in quantifications of functional diversity (Petchey et al., 2007; Bracken et al., 2009; Dalerum et al., 2009). The findings of this study must be cautiously interpreted not to diminish the role of the church forest in maintaining intact natural forest of indigenous species. The relative abundance of the species confirms that rare species are many in number and few in abundance and dominant species are few in number but covers most of the area. It agrees with the ecological concept which claims that dominant species are few in number and minor or tail species are many in number, but less dominant (Peterson et al., 1998; Walker et al., 1999).

The result reveals that dominant species are functionally dissimilar to each other, supporting previous findings which claimed that differences in functional niches allow these species to be co-dominant (Walker et al., 1999; Peterson et al., 1998). This implies that less disturbed natural forests have high functional diversity that can play a vital role in avoiding competition among species. This also signifies the advantage of natural forest or mixed plantations compared to monoculture plantations. The functional distances between the

Table 3. Similarity in species abundance among the possible pairs of church forests (Chf) and adjacent degraded forests (Af).

Chf	Adf	Yeleb		Yeterbina		Tsid Mariam		Mekides		Mehir	
	0.68	**Chf**	**Af**	**Chf**	**Af**	**Chf**	**Af**	**Chf**	**Af**	**Chf**	**Af**
Yeleb	Chf	1.00	0.63 (18)	0.63 (20)	0.49 (17)	0.47 (15)	0.38 (11)	0.53 (17)	0.22 (9)	0.49 (17)	0.35 (13)
	AF		1.00	0.52 (20)	0.38 (13)	0.31 (10)	0.35 (9)	0.33 (11)	0.25 (9)	0.31 (11)	0.41 (13)
Yeterbina	Chf			1.00	0.47 (16)	0.50 (15)	0.36 (10)	0.48 (15)	0.30 (11)	0.46 (16)	0.37 (13)
	AF				1.00	0.36 (12)	0.36 (10)	0.52 (16)	0.41 (14)	0.42 (15)	0.33 (12)
Tsid Mariam	Chf					1.00	0.65 (10)	0.62 (16)	0.23 (14)	0.48 (15)	0.34 (11)
	AF						1.00	0.46 (11)	0.37 (7)	0.31 (10)	0.29 (8)
Mekdes	Chf							1.00	0.18 (7)	0.50 (16)	0.36 (12)
	AF								1.00	0.40 (14)	0.39 (12)
Mehir	Chf									1.00	0.50 (13)
	AF										1.00

minor species of the church forest and the dominant species of the adjacent forests suggests that minor species of the church forest have shown tendency of offering insurance in the adjacent degraded forest. The result also confirms that new comer (passenger species) have offered insurance after disturbance. However, some species (*C. spinarum*) which are dominant in the church forest have still remained dominant in the adjacent forest. On the other hand, five species (*C. africana*, *D. steudneri*, *J. abyssinicum*, *M. verticillata* and *O. europaea* ssp. *cuspidate*) which were rare in the church forest were totally absent in the adjacent degraded forest. Their timber quality and sensitivity to browsing made these species extinct locally due to animal and human disturbances. In sum, dominant species and rare species may be functionally similar, but may not be necessarily different in responding to disturbance. Minor species, which offer insurance regenerate in different ways, that is, by seed banks, seedling banks, seed rain or underground stumps (Teketay, 1997; Argaw et al., 1999; Eriksson et al., 2003; Wassie and Teketay, 2006; Wang et al., 2009. About 82% of the species, which are not found in the church forests, have capsule fruits and the rest 12% have berry fruits.

The seed germination strategy and dispersal methods might have contributed to the existence of those species. Tiny seeds from capsule are highly dependent on wind for dispersal and wind dispersal is more limited in the church forests compared to the open adjacent degraded forests. The main disturbances observed in the area were combinations of human disturbances mainly for wood products (that is, fuelwood and construction) and intensive browsing by animals. However, other disturbances such as fire, mineral and water stress, shed, wind, pest, disease and other natural disturbances such as climate change could also be possible causes (Vanclay et al., 1997; Meers et al., 2010). Most of the dominant species in the adjacent degraded forest, except *C. spinarum* are not preferred by animals as a forage or by people for different uses (Bekele-Tesemma et al., 1993). *C. spinarum* which is preferred as browse for animals and for its excellent fire wood is abundant both in church and adjacent degraded forests. Its unique potential and adaptation has helped to perfectly respond to the disturbances and offer insurance. Its sharp thorns are important for resisting browsing pressure. Its amorphous big underground root also helps to produce multiple sprouts after grazing or cutting by people or damage by animals. This suggests that a single species with unique ability in responding to disturbances can also offer insurance for the ecosystem after disturbances.

The result lends hand to the general understanding that disturbance is among the most

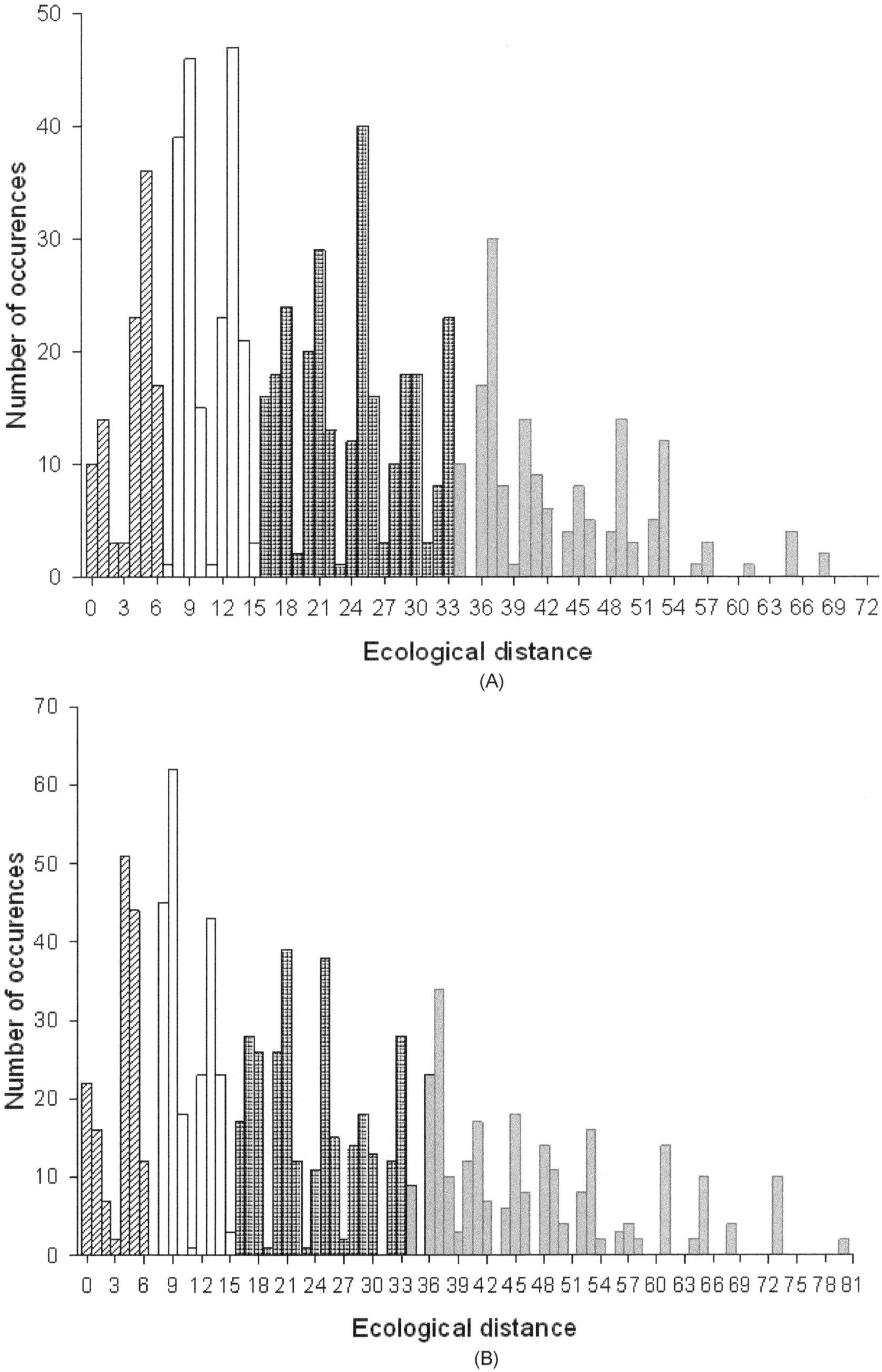

Figure 1. Frequencies of ecological distances for all species pairs in the church forest. Groupings reflect apparent clusters of frequent ecological distances (light hashed, white bars, striped, and grey bars); the lower the ecological distance, the more functionally similar are the two species.

Table 4. Functional distances between 10 most abundant species in the church forest, and between 10 minor church and 10 dominant adjacent degraded forest species.

Functional similarity	Functional distance pairs between species	
	Church dominant	Church minor versus adjacent dominant
Similar (0 ≤ ED ≤ 6)	3	14
Average to similar (6 < ED ≤ 14)	4	20
Average to dissimilar (14 < ED < 33)	19	37
Dissimilar (ED ≥ 33)	19	29
Total pairs	45	100
χ^2	21.40	12.24
P-value	< 0.001	0.01

important key factors shaping the ecologies of the vegetation and attributes that confer success under disturbance are central (Rusch et al., 2003; Girme, 2001; Westoby et al., 2002). Functional diversity which is a cumulative effect of functional attributes is a proxy for ecological elasticity to disturbances. Thus, the concept of functional distances between species may be used as criteria to determine the functional distances between species for plantation and conservation purposes. It can also be used to predict changes in vegetation driven by anthropogenic disturbance such as land-use shifts and climate changes at local, regional and global scales.

ACKNOWLEDGMENTS

The authors are grateful to the Norwegian University of Life Sciences and the Norwegian State Educational Loan Fund (lånekassen) for providing financial assistance.

REFERENCES

Archambault L, Morissette J, Bernier-Cardou M (1998). Forest succession over a 20-year period following clear cutting in balsam fir-yellow birch ecosystems of eastern Québec, Canada. Forest Ecol. Manag., 102: 61-74.

Archibold OW (1995). Ecology of World vegetation. Chapman and Hall, London, UK. pp. 510.

Argaw M Teketay D Olsson M (1999). Soil seed flora, germination and regeneration pattern of woody species in an Acacia woodland of the Rift Valley in Ethiopia. J. Arid Environ., 43: 411-435.

Bekele-Tesemma A, Birnie A, Tengnäs B (1993). Useful trees and shrubs for Ethiopia. identification, propagation and management for agricultural and pastoral communities. Regional Soil Conservation unit (RSCU), Swedish International Development Authority, Nairobi. p. 474.

Bengtsson J, Nilsson SG, Franc A, Menozzi P (2000). Biodiversity, disturbances, ecosystem function and management of European forests. Forest Ecol. Manag., 132: 39-50.

Bracken MES, Friberg SE, Gonzalez-Dorantes CA, Williams SL (2009). Functional consequences of realistic biodiversity change in a marine ecosystem. Proc. Natl. Acad. Sci., USA 105: 924–928.

Dalerum F, Cameron EZ, Kunkel K, Somers MJ (2010). Interactive effects of species richness and species traits on functional diversity and redundancy. Theor. Ecol. DOI 10.1007/s12080-010-0104-y.

Edwards S, Hedberg I (EDS.) (1989). Flora of Ethiopia. The National Herbarium, Addis Ababa University, Addis Ababa and Department of Systematic Botany, Uppsala University, Uppsala. 3: 659.

Edwards S, Tadesse M, Sebsebe D, Hedberg I (eds) (2000). Flora of Ethiopia and Eritrea Vol. 2 Part 1, *Magnoliaceae* to *Flacourtiaceae*. The National Herbarium, Addis Ababa, Ethiopia; Uppsala, Sweden. p. 532.

Eriksson I, Teketay D, Granström A (2003). Response of plant communities to in an Acacia woodland and a dry Afromontane forest, southern Ethiopia. Forest Ecol. Manag., 177: 39-50.

Girme JP (2001). Plant strategies, vegetation process, and ecosystem properties. 2nd.ed. Wiley, Chichester, UK. p. 456.

Gole TW, Borsch T, Denich M, Teketay D (2008). Floristic composition and environmental factors characterizing coffee forests in southwest Ethiopia. Forest Ecol. Manag., 255: 2138-2150.

Griffin JN, Mendez V, Johnson AF, Jenkins SR, Foggo A (2009). Functional diversity predicts over yielding effects of species combinations on primary productivity. Oikos 118: 37–44.

Gunderson LA (2000). Ecological resilience in theory and application. Ann. Rev. Ecol. Syst., 31: 425–439.

Haugen T (1989). Woody vegetation of Gidoloe and Konso, Gamogofa region, South Ethiopia. Lidia 2: 9-32.

Hedberg I, Ensermu K, Edwards S, Sebsebe D, Persson E (eds) (2006). Flora of Ethiopia and Eritrea, Gentianaceae to Cyclocheilaceae. The National Herbarium, Addis Ababa, Ethiopia; Uppsala, Sweden. 5: 690.

Holling CS (1973). Resilience and stability of ecological systems. Ann. Rev. Ecol. Syst., 123.

Krebs CJ (1999). Ecological Methodology. Harper and Row Publisher, New York. p. 620.

Magurran AE (2004). Measuring Biological Diversity. Blackwell Science, Oxford. p. 264.

Mamo N, Mihretu M, Fekadu M, Tigabu M, Teketay D (2006). Variation in seed and germination characteristics among Juniperus procera populations in Ethiopia. Forest Ecol. Manag., 225: 320-327.

Meers TL, Kasel S, Bell TL, Enright NJ (2010). Conversion of native forest to exotic Pinus radiata plantation: Response of understorey plant composition using a plant functional trait approach. Forest Ecol. Manag., 259: 399-409.

Naeem S, Wright JP (2003). Disentangling biodiversity effects on ecosystem functioning: deriving solutions to a seemingly insurmountable problem. Ecol. Lett., 6: 567-579.

Ngulube MR, Hall JB, Maghembe JA (1997). Fruit, seed and seedling variation in Uapaca kirkiana from natural populations in Malawi. Forest Ecol. Manag., 98: 209-219.

Petchey OL, Evans KL, Fishburn IS, Gaston KJ (2007). Low functional diversity and no redundancy in British avian assemblages. J. Anim. Ecol., 76: 977–985.

Petchey OL, Gaston KJ (2006). Functional diversity: back to basics and looking forward. Ecol. Lett., 9: 741–758.

Peterson G, Allen CR, Holling CS (1998). Ecological resilience, biodiversity and scale. Ecosystems 1: 6-18.

Rusch GM, Pause JG, Leps J (2003). Plant Functional Types in relation to disturbance and land use Introduction. J. Veg. Sci., 14: 307-310.

Smith RL, Smith TM (2001). Ecology and Field Biology, Sixth Edn.

Benjamin Cummings, an Imprint of Addison, Wesley Longman. p. 843.
Taddese G (2001). Land degradation: a challenge to Ethiopia. Environ. Manag., 27: 815–824.
Teketay D (1997). The impact of clearing and conversion of dry Afromontane forests into arable land on the composition and density of soil seed banks. Acta. Oecol., 18: 557-573.
Vanclay JK, Gillison AN, Keenan RJ (1997). Using plant functional attributes to quantify site productivity and growth patterns in mixed forests. Forest Ecol. Manag., 94: 149-163.
Walker B (1995). Conserving biological diversity through ecosystem resilience. Conserv. Biol., 9: 747–752.
Walker BH, Kinzig A, Langridge J (1999). Plant attributes diversity, resilience, and ecosystem function: The nature and significance of dominant and minor species. Ecosystems 2: 95-113.
Wang J, Ren H, Yang L, Li D, Guo Q (2009). Soil seed banks in four 22-year-old plantations in South China: Implications for restoration. Forest Ecol. Manag., 258: 2000-2006.
Wassie A (2002). Opportunities, Constraints and Prospects of the Ethiopian Orthodox Tewahido Churches in Conserving Forest Resources: The Case of Churches in South Gonder, Northern Ethiopia. M.Sc. thesis, Swedish University of Agricultural Sciences, Skinnskatterberg, Sweden.
Wassie A, Sterck FJ, Bongers F (2010). Species and structural diversity of church forests in a fragmented Ethiopian Highland landscape. J. Veg. Sci. 21: 938-948.
Wassie A, Teketay D (2005). Soil seed banks in church forests of northern Ethiopia: Implication for the conservation of woody plants. Flora, 201: 32-45.
Wassie A, Teketay D (2006). Soil seed banks in church forests of northern Ethiopia: Implications for the conservation of woody plants. Flora Morp. Dist. Fun. Ecol. Plants, 201: 32-43.
Westoby M, Falster DS, Moles AT, Vesk PA, Wright IJ (2002). Plant ecological strategies: Some Leading Dimensions of Variation Between Species Annu. Rev. Ecol. Syst., 33:125–59.
Zeleke G, Hurni H (2001). Implications of land use and land cover dynamics for mountain resource degradation in the north-western Ethiopia highlands. Mt. Res. Dev., 21: 184-191

Appendix 1. Functional distance between species and diversity of the church forest.

Species	A.e	A.n	A.p	A.g	B.a	B.p	C.a	C.m	C.s	C.m	C.a	C.m	C.l	C.h	D.t	D.s	E.c	E.g	E.r	E.a	G.f	H.m	J.a	J.e	J.s	M.v	M.s	O.u	O.e	O.q	P.d	P.v	P.g	
Acacia abyssinica																																		
Acacia nilotica	9																																	
Acanthus pubescens	46	41																																
Albizia gummifera	5	4	53																															
Bersama abyssinica	42	29	4	41																														
Buddleia polystachya	18	25	24	21	20																													
Calpurnia aurea	26	13	20	25	16	36																												
Capparis micrantha	26	13	20	25	16	36	0																											
Carissa spinarum	34	21	38	25	26	30	18	18																										
Clerodendrum myricoides	34	13	20	25	8	28	8	8	14																									
Cordia africana	38	49	20	45	24	12	40	40	30	40																								
Croton macrostachys	25	44	33	40	37	9	53	53	49	53	17																							
Cupressus lusitanica	9	12	65	8	53	25	37	37	25	37	49	32																						
Cussonia holstii	33	20	9	32	5	25	5	5	13	5	25	40	40																					
Dombeya torrida	22	21	12	25	8	4	24	24	26	16	16	13	29	13																				
Dracaena steudneri	29	18	21	22	9	9	25	25	19	9	25	30	30	14	5																			
Ehretia cymosa	27	14	13	26	9	29	1	1	17	5	33	46	38	2	17	18																		
Eucalyptus globulus	15	18	49	14	37	9	45	45	33	33	37	18	10	38	13	14	40																	
Euclea racemosa	27	14	33	18	21	25	13	13	1	9	29	46	22	10	21	14	12	28																
Euphorbia abyssinica	9	6	49	2	37	21	21	21	15	21	37	42	10	26	25	20	22	18	10															
Grewia ferruginea	46	25	16	37	4	24	20	20	22	4	36	49	49	9	12	5	13	33	17	33														
Hibiscus micranthus	34	13	20	25	8	28	8	8	14	0	40	53	37	5	16	9	5	33	9	21	4													
Jasminum abyssinicum	28	17	10	29	6	22	6	6	14	6	26	33	33	1	10	13	3	31	11	25	10	6												
Juniperus excelsa	5	8	65	4	53	25	37	37	37	37	57	36	4	44	29	30	38	10	30	10	49	37	37											
Justiia schimperiana	39	18	17	30	5	25	13	13	17	1	37	50	42	6	13	6	8	32	12	26	1	1	7	42										
Malva verticillata	34	13	20	25	8	28	8	8	14	0	40	53	37	5	16	9	5	33	9	21	4	0	6	37	1									
Maytenus senegalensis	45	24	41	36	29	49	13	13	5	13	49	68	36	12	37	30	14	50	6	26	25	13	13	48	18	13								
Ocimum urticifolium	26	13	20	25	16	36	0	0	18	8	40	53	37	5	24	25	1	45	13	21	20	8	6	37	13	8	13							
Olea europaea	12	13	46	9	34	10	34	34	18	26	30	21	5	29	14	13	31	3	15	9	30	26	24	9	27	26	33	34						
Osyris quadripartita	38	17	36	29	24	44	8	8	6	8	48	65	33	9	32	25	9	45	5	21	20	8	10	41	13	8	1	8	30					
Phytolacca dodecandra	30	9	32	21	20	40	4	4	14	4	52	65	33	9	28	21	5	41	9	17	16	4	10	33	9	4	9	4	30	4				
Pittosporum viridiflorum	18	13	12	17	8	12	8	8	14	8	16	29	29	5	8	9	5	25	9	13	12	8	6	29	9	8	21	8	18	16	12			
Protea gaguedi	29	8	33	20	21	41	5	5	21	5	57	68	36	12	29	22	6	42	14	18	17	5	13	32	10	5	16	5	33	9	1	13		
Pterolobium stellatum	21	4	37	16	25	37	9	9	25	9	61	56	24	16	25	22	10	30	18	18	21	9	13	20	14	9	20	9	25	13	5	17	4	
Prunus africana	23	26	53	22	41	13	49	49	25	37	33	18	10	38	17	18	44	4	24	22	37	37	31	18	36	37	42	49	3	41	45	29	50	5
Rosa abyssinica	26	13	40	17	28	32	12	12	2	12	36	53	21	13	28	21	13	33	1	9	24	12	14	29	17	12	5	12	18	4	8	12	13	12

Appendix 1.contd.

Schefflera abyssinica	22	25	48	21	36	8	52	52	38	36	36	17	17	41	12	13	45	1	33	25	32	36	34	17	33	36	57	52	6	52	48	28	49	4
Terminalia schimperiana	34	13	20	25	8	28	8	8	14	0	40	53	37	5	16	9	5	33	9	21	4	0	6	37	1	0	13	8	26	8	4	8	5	16
Vernonia amygdalina	29	18	21	22	9	9	25	25	19	9	25	30	30	14	5	0	18	14	14	20	5	9	13	30	6	9	30	25	13	25	21	9	22	5

Appendix 2. Functional distance between species and diversity of the adjacent degraded forest.

Species	A.e	A.n	A.p	A.g	B.a	B.p	C.a	C.m	C.s	C.m	C.a	C.m	C.l	C.h	D.t	D.s	E.c	E.g	E.r	E.a	G.f	H.m	J.a	J.e
Acacia abyssinica																								
Acacia nilotica	9																							
Acanthus pubescens	46	41																						
Albizia gummifera	5	4	53																					
Bersama abyssinica	42	29	4	41																				
Buddleia polystachya	18	25	24	21	20																			
Calpurnia aurea	26	13	20	25	16	36																		
Capparis micrantha	26	13	20	25	16	36	0																	
Carissa spinarum	34	21	38	25	26	30	18	18																
Clerodendrum myricoides	34	13	20	25	8	28	8	8	14															
Cordia africana	38	49	20	45	24	12	40	40	30	40														
Croton macrostachys	25	44	33	40	37	9	53	53	49	53	17													
Cupressus lusitanica	9	12	65	8	53	25	37	37	25	37	49	32												
Cussonia holstii	33	20	9	32	5	25	5	5	13	5	25	40	40											
Dombeya torrida	22	21	12	25	8	4	24	24	26	16	16	13	29	13										
Dracaena steudneri	29	18	21	22	9	9	25	25	19	9	25	30	30	14	5									
Ehretia cymosa	27	14	13	26	9	29	1	1	17	5	33	46	38	2	17	18								
Eucalyptus globulus	15	18	49	14	37	9	45	45	33	33	37	18	10	38	13	14	40							
Euclea racemosa	27	14	33	18	21	25	13	13	1	9	29	46	22	10	21	14	12	28						
Euphorbia abyssinica	9	6	49	2	37	21	21	21	15	21	37	42	10	26	25	20	22	18	10					
Grewia ferruginea	46	25	16	37	4	24	20	20	22	4	36	49	49	9	12	5	13	33	17	33				
Hibiscus micranthus	34	13	20	25	8	28	8	8	14	0	40	53	37	5	16	9	5	33	9	21	4			
Jasminum abyssinicum	28	17	10	29	6	22	6	6	14	6	26	33	33	1	10	13	3	31	11	25	10	6		
Juniperus excelsa	5	8	65	4	53	25	37	37	37	37	57	36	4	44	29	30	38	10	30	10	49	37	37	
Justiia schimperiana	39	18	17	30	5	25	13	13	17	1	37	50	42	6	13	6	8	32	12	26	1	1	7	42
Malva verticillata	34	13	20	25	8	28	8	8	14	0	40	53	37	5	16	9	5	33	9	21	4	0	6	37
Maytenus senegalensis	45	24	41	36	29	49	13	13	5	13	49	68	36	12	37	30	14	50	6	26	25	13	13	48
Ocimum urticifolium	26	13	20	25	16	36	0	0	18	8	40	53	37	5	24	25	1	45	13	21	20	8	6	37
Olea europaea	12	13	46	9	34	10	34	34	18	26	30	21	5	29	14	13	31	3	15	9	30	26	24	9

Appendix 2. contd.

Osyris quadripartita	38	17	36	29	24	44	8	8	6	8	48	65	33	9	32	25	9	45	5	21	20	8	10	41
Phytolacca dodecandra	30	9	32	21	20	40	4	4	14	4	52	65	33	9	28	21	5	41	9	17	16	4	10	33
Pittosporum viridiflorum	18	13	12	17	8	12	8	8	14	8	16	29	29	5	8	9	5	25	9	13	12	8	6	29
Protea gaguedi	29	8	33	20	21	41	5	5	21	5	57	68	36	12	29	22	6	42	14	18	17	5	13	32
Pterolobium stellatum	21	4	37	16	25	37	9	9	25	9	61	56	24	16	25	22	10	30	18	18	21	9	13	20
Prunus africana	23	26	53	22	41	13	49	49	25	37	33	18	10	38	17	18	44	4	24	22	37	37	31	18
Rosa abyssinica	26	13	40	17	28	32	12	12	2	12	36	53	21	13	28	21	13	33	1	9	24	12	14	29
Schefflera abyssinica	22	25	48	21	36	8	52	52	38	36	36	17	17	41	12	13	45	1	33	25	32	36	34	17
Terminalia schimperiana	34	13	20	25	8	28	8	8	14	0	40	53	37	5	16	9	5	33	9	21	4	0	6	37
Vernonia amygdalina	29	18	21	22	9	9	25	25	19	9	25	30	30	14	5	0	18	14	14	20	5	9	13	30

Production efficiency of yam in Zing Local Government Area of Taraba State, Nigeria

Musa, Y. H.[1], Onu, J. I.[2], Vosanka, I. P.[1]* and Anonguku, I.[3]

[1]Department of Agricultural Extension and Management, Taraba State College of Agriculture, Jalingo, Nigeria.
[2]Department of Agricultural Economics and Extension, Federal University of Technology Yola, Adamawa State. Nigeria.
[3]Department of Agricultural Extension and Communication, University of Agriculture, Makurdi, Nigeria.

The research was carried out to analyze the efficiency of yam production in Zing Local Government Area of Taraba State. Data were collected from 103 respondents using multistage sampling techniques, and analyzed by means of descriptive statistics, multiple regression and profitability analysis. Findings indicated that 95% of the respondents were males, 38.8% fall within 31 to 40years of age. 84.4% were married and about 72.8% have one formal educational level or the other. 53.4% operate only 1 to 2 hectares of farm land. The double log function with R^2 of 0.745 was the best fit. The study established that farm size, yam seed and fertilizer were positively related to yam output and significant at 1 and 5% levels respectively. The marginal analysis of input utilization revealed that farm size, yam seed and labour (family and hired) were rationally used but not at optimal levels. Therefore, more than 97, 66.55 and 36% increase in MVP respectively were required. Average gross income was found to be N241,800.00 with average total cost of N125,320.00 per hectare. Yam seed constituted the greatest share of the total cost representing 66.8% with an average cost of N85 per kg. Operating ration of 0.43 indicates moderate total revenue over total variable cost. RRI was 92.9% and profitability index of 48.2%. Issuing of micro soft loan at bearable interest rate, technical research and the provision of improved practices was therefore recommended.

Key word: Zing, economic analysis, yam, production, efficiency.

INTRODUCTION

Yam is an annual tuber and monocot plant. It belongs to the genus "Dioscorea" and the family "dioscoreacea". The food plant comprises of 600 species out of which ten species produces edible tubers and only six are cultivated in Africa. The current estimate of world yam production is far from reliable. The latest FAO statistics of 1989 to 1990 still do not include Asia, India, China or Indonesia. Moreover, the structure of production system that includes yams clearly creates an obstacle to making estimate. In this regard, yam production is under estimated, but it is not known by how much. Yam production trend is to be at least around 25 million tons in 1992; it is thus the fourth major root crop in the world after potatoes, cassava and sweet potatoes (Degras,1993; Asiedu, 1999). Yam productions continue to increase from 6% in Central America to 15% in Asia and yield raises to 20% (National Root Crop Research Institute, 2004). Amegbeto and Asiedu (2000) reported that the international trade of yam originates from Jamaica as the leading exporter in Central America, Brazil lead in South America while Japan leads the production in Asia. Asomugha and Ujoku (2007) stressed that West and Central Africa accounts for about 93% of the world total production (38 million tons). The dominant yam production zones stretches from Coted'ivoire through Ghana, Togo, Benin, Nigeria, Cameroon, Gabon, Central African Republic and the Western part of the democratic Republic of Congo. Ethiopia and Sudan are major yam producers in East Africa. IITA (1991, 1993) reported that Nigeria remains the principal yam producer

*Corresponding author. E-mail: vosanka@yahoo.com

with the world production percentage of 71.9% in recent years. Coted'ivore, the second largest producer had production percentage increase from 9.3 to 10.6% with its yield increase by 7.6% with approximately the same percentage of cultivated surface area of 10.4%. Ghana, the third largest yam producer increase its production almost by 20% (Asumugha and Njoku, 2007). Yam production in Nigeria is quite high.

The annual production in the country is estimated at 26.59 million metric tons (FAO, 2006). The annual growth rate for the same period was 6% for the yield and 10% for the area planted. With growing demand, yam has assumed great importance in Nigeria. The nation produces about 31.5 million metric tons of yam annually. CBN (2003) and FAO (2002) reported that Nigeria accounts for 71% (26 million tons) of the total world production of yam harvested from 2,760.00 hectares. On the basis of quality of root and tuber crops produced in Nigeria, yam ranks second only after cassava (NBS, 2007). The edible varieties of yam are important food crop and serve as an important carbohydrate staple for millions of people in both the tropical and sub-tropical countries in West Africa, Caribbean, the Northern and central part of South-East Asia including some part of China, Malaysia, Japan and Oceania (Adetuyi et al., 2010). In Nigeria, yam is becoming more expensive and relatively unaffordable in urban areas as production growth has not kept pace with population growth leading to demand exceeding supply (Kushwaha and Polycarp, 2000). Production of yam in Nigeria is believed to be constrained mostly by high cost of seed. It is observed that the defect of traditional farming of the people of Zing, in Taraba State has tremendously undermined the high production trend of yam in the area. However, not all farmers' can allocate resources efficiently for yam production in the study area. The specific objectives of this study were to:

i) Describe the socio-economic characteristics of yam producers;
ii) Estimate the efficient use of yam production resources, and;
iii) Determine the profitability status of yam production.

METHODOLOGY

Study area

Zing Local Government Area is one of the 16 LGAs in Taraba State. It is bounded by Yorro LGA in the south, in the North-East and West respectively by Adamawa State. The area lies between longitude 10° and 11°E and latitude 9° and 10°N of the equator with estimated population of about 115,384 (NPC, 2006). The area falls within the transitional belt of savanna in north eastern Nigeria. It has good climatic conditions and rich in agricultural opportunities with the temperature ranging from 28 to 34°C, the mean annual rainfall of the area is 1500 mm. The study area is endowed with abundant natural resources including, streams, natural grassland and economic trees.

Data collection and sampling techniques

Primary data was employed for this study. The primary data were obtained through the use of structured questionnaires which were administered by trained enumerators under the supervision of the researchers. Information on yam farmers' socio-economic characteristics, production activities and cost and return on yam production were among the bulk of data collected. Yam farmers were the population from which samples were drawn. From the list of the districts, political wards, villages and yam farmers, a sampling frame of 520 yam farmers were obtained for the study. Accordingly, purposive and multistage random sampling techniques were employed to draw respondents namely: The first stage was the selection of 10 political wards in the local government area. The selection of 30% of the villages (66) in each of the council wards to give a total of 20 villages was the second stage. Finally, the third stage was the drawing of 25% of the farmers in each of the selected villages to give a total sample size of 130 respondents. In each of the stages, random sampling technique (Lottery method) was employed to draw the sample units.

Method of data analysis

Simple descriptive statistic, multiple regression and profitability (gross margin) analysis were used to analyze the data obtained for the research objectives. The descriptive statistic involved the use of frequencies, and percentages to describe the socio-economic characteristics of the yam farmers. Multiple regression analysis was employed to examine the efficient utilization of variable inputs on the output of the local yam farmers in the study area. The general form of the equation is stated as:

$$Y = f(X_1, X_2, X_3, X_4, X_5, X_6, U_i) \quad (1)$$

Where:

Y = Output of yam (kg)
X_1 = Farm size (ha)
X_2 = Quantity of yam seed in numbers (100 tubers)
X_3 = Family labour in man days
X_4 = Hired labour in man-days
X_5 = Quantity of fertilizer used (kg)
X_6 = Farming experience (years)
U_i = Error term

Four different functional forms (linear, exponential, semi-log and double log) were employed for the analysis, out of which the double log function was chosen as the best fit. The selection of double log function as the lead equation was based on the magnitudes and appropriateness of signs of the estimated regression coefficients, standard errors of estimates, magnitudes of the coefficient of multiple determination (R^2) significance of the estimated coefficients and F- value. The model is stated as follows:

$$\text{Log } Y = \beta_0 + \beta_1 \log X_1 + \beta_2 \log X_2 + \beta_3 \log X_3 + \beta_4 \log X_4 + \beta_5 \log X_5 + \beta_6 \log X_6 + Ui$$

Where,

Log = Natural logarithm
Y, X_1 ---- X_6 are as defined in Equation 1
B_1 ----- B_6 are parameters to be estimated
B_0 = Constant term
Ui = Error term.

Marginal analysis of input utilization was used to determine the

Table 1. Distribution of respondents by socio-economic characteristics (n = 103).

Variable	Frequency	Percentage
Gender		
Male	98	95.0
Female	05	5.0
Age (year)		
21- 30	32	31.1
31 – 40	40	38.8
41 – 50	23	22.3
51 – Above	08	7.8
Marital status		
Married	87	84.4
Single	12	11.7
Widowed	03	2.9
Divorced	01	1.0
Educational attainment		
Non-formal education	28	27.2
Primary education	23	22.3
Secondary education	29	28.2
Tertiary education	23	22.3
Farming experience (years)		
1 - 10	20	19.4
11 - 20	50	48.5
21 - 30	19	18.5
31 - Above	14	13.6
Farm size (hectares)		
Less than 1	07	6.8
1 - 2	55	53.4
Above 2	41	39.8

Source: Field survey (2010).

resource use efficiency of the inputs used by the farmers. The values of marginal product were estimated using the regression coefficient of each input and the arithematic mean value of farm output and inputs. Fallowing Shehu et al. (2006), the MVP of each resource was measured. The acquisition cost of each resource was used as the marginal factor cost (MFC). This was based on the assumption that farmers operate in a pure competitive input markets (Olukosi and Ogungbile, 1989). The MVP was compared with MFC of the inputs to determine the efficiency of use of the inputs. The decision rule used to ascertain whether an input is over, under or optimally utilized was interpreted as described by Iheanacho et al. (2000) that when the efficiency unit (r) is unity, it indicates resource use optimization, negative unit means over utilization of resources while positive unit suggest under utilization of resources. The gross margin is the difference between gross farm income and the total variable cost of production. It was used to estimate the profitability level of yam production in the area.

Gross margin (profitability) analysis is used to evaluate the efficiency of an individual business (Olukosi and Erhabor, 1998; Idowu, 2009) while the net farm income is the difference between the gross margin and the total cost of production less the sum of fixed variable cost. The gross margin model states as follows:

$$GM = GI - TVC \quad (2)$$

$$NFI = GM - TFC \quad (3)$$

Where,

GM = Gross margin per hectare (N)
GI = Gross income per hectare (N)
TVC = Total variable cost per hectare (N)
NFI = Net farm income per hectare (N)
TFC = Total fixed cost per hectare (N)

RESULTS AND DISCUSSION

Socio-economic characteristics of respondents

The summary of the results on social economic

Table 2. Double-log regression estimates for yam production.

Variable	Coefficient	Std. error	T- value
Farm size (X_1)	0.091	0.072	15.184 ***
Yam seed (X_2)	0.118	0.065	1.820**
Family labour (X_3)	0.011	0.035	0.321NS
Hired labour (X_4)	-0.025	0.032	-0.793NS
Fertilizer (X_5)	0.035	0.013	2.827 ***
Farming exp. (X_6)	-0.073	0.067	-1.077 NS
R^2		0.745	0.164
F-value	46.674***	--	--
Constant	3.499***	--	--

Source: Survey data (2010); *** Significant at 1%; ** Significant at 5%; NS (not significant).

Table 3. Marginal analysis of input used by yam farmers.

Variable	MPP	MVP	MFC	MVP/MFC
Farm size (X_1)	8766.31	745,136.35	15,675	4.07
Yam seed (X_2)	0.9624	81.0884	27.077	2.9
Family labour (X_3)	0.6997	59.475	38.211	1.56
Hired labour (X_4)	5.8100	-493.85	191.200	-2.58
Fertilizer (X_5)	5.6177	477.504	475.4	1.00

Source: Survey data (2010).

characteristics are presented in Table 1. This indicated that 95% were males, while only 5% were females. This means that men do most of the yam production activities than women probably do due to their ownership of farmland. This result is also supported by Ani (2004). Age brackets of the respondents in the area shows that 38.8% were within the ages of 31 to 40 years as reported by Fadeji (2006) that the average age of 49% of the respondents falls between 31 to 40 years. This is most obvious as more strenuous yam operations such as heap making are mostly done by young men that are active and energetic. The educational status of the respondents indicated that 28.2% attended secondary school, 22.3% attended tertiary educational level and 27.2% had no formal education. Majority of the respondents represen-ting 60.2% had farm size of less than or equal to 2 hectares, only about 39.8% had above 2 hectares with farming experience between 11 to 20 years. This therefore, reveals that the yam farmers in the study area are really small scale farmers. Ibrahim (2004) observed that small scale farmers are those that cultivate farm land not more than 2 hectares.

Resources use relationship in yam production

The results of the regression analysis as reflected in Table 2 revealed that the coefficient of multiple deter-mination (R^2) was 0.745. This indicates that about 75% of the variation in output is accounted for by the variations in the independent variables used for the production. The remaining 25% may be attributed to variations in other factors not included in the model. Farm size, yam seed and fertilizer were significant and have positive relation-ship with the output. This result agreed with that of Sulumbe et al. (2010), who established that age, family size, income and extension contact were positively related to output and significant at 1% level. The coef-ficients of farm size (X_1) and fertilizer (X_5) were positive and significant at 1% level. This means that a unit increase in these variables, under static condition of other explanatory variables result in increased output level. This result is in conformity with Shehu et al. (2009) that increase in farm size implies more output is expected. The coefficient of yam seed (X_2) was also positive but significant at 5% level. Also, the coefficient of family labour (X_3) was positive but not statistically significant.

Marginal analysis of resource efficiency utilization

The result shows that increase in one hectare of farm size would result in 8,766.31 (Table 3). This implies that increase in farm size by one hectare would result to extra 8766 kg of yam. Also, an extra use of 1 kg of yam seed would give an additional 0.9624 kg of yam and extra use of one man day of family labour would give extras of0.6997 kg of yam. Increase in fertilizer usage by 1 kg

Table 4. Absolute value of the required adjustment in MVPS (in percentage) for optimal allocation of variable input.

Variable	Adjustment required
Farm size (X_1)	97.43
Yam seed (X_2)	66.55
Family labour (X_3)	35.90
Hired labour (X_4)	61.24
Fertilizer (X_5)	100.00

Source: Survey data (2010).

Table 5. Average costs and returns of yam production.

Variable	Unit	Unit cost (N)	Quantity	Value (N)	Percentage (%)
Yam output (YO)	Kg	120	2,015	241,800	
Variable costs (VC)					
Yam seed	Kg	85	985	83,725	66.8
Labour	Man-days	125	34	4,250	3.4
Fertilizer	Kg	70	84.2	5,894	4.7
Transportation	Kg	5	1545	7,725	6.2
Storage	Kg	2	778	1, 556	1.2
Total variable cost (TVC)	-	-	-	103,150	
Fixed costs (FC)					
Land lease	Ha	8300	2.4	19,920	15.9
Utilities	-	-	-	2,250	1.8
Total fixed cost (TFC)	-	-	-	22,170	
Total cost (TC)				125,320	
Gross margin	-	-	-	138,650	
Net farm income	-	-	-	116,480	

Source: Field data (2010).

would increase output of yam by 5.6177 kg. Economic theory states that a firm maximizes profit with respect to an input if the ratio of its MVP to its MFC is unity (Kay, 1986). Comparison of the MVP to MFC ratios shows resources such as farm size and yam seed are greatly more than unity. This implies that they are largely underutilized while family labour is slightly underutilized. Fertilizer ratio is exactly unity (1) suggesting that it was efficiently utilized. Hired labour ratio was negative signifying that it was over utilized during the growing period of yam production.

Optimum resource allocation adjustment requirement for mvp$_s$

The required adjustment allocation for optimum resources in yam production in the study area is hereby presented in Table 4. The results indicate that for optimal allocation of farm size, more than 97% increase in MVPS was required. Also, more than 66.55 and 40% increase in MVPS were required for yam seed and family labour respectively. For hired labour about 61.24% resource application need be reduced to obtain the optimum requirement.

Average costs and returns of yam production

The result of the average costs and returns per hectare is hereby presented in Table 5. The average gross income that accrued from the sales of yam at the average cost of N120 per tuber (1 kg equivalent) was found to be N241,800.00. The average total cost was N125,320.00. Yam seed constituted the greatest share of the total cost representing 66.8% with an average cost of N85 per kg followed by land lease with 15.9% which has the average rentage cost of N8,300/ha. This study agreed with Abubakar et al. (2005) that yam seed accounted for the greater portion (60.00%) of the total variable cost. This

Table 6. Profitability ratio analysis of yam production.

Parameter	Value	Percentage
Profitability index (PI)	0.482	48.2
Rate of return on investment (RRI)	0.929	92.9
Rate of returns on variable cost (RRVC)	2.129	212.9
Operating ration (OR)	0.426	42.6

Source: Field data (2010).

could be attributed to the scarcity of yam tubers as well as seed yam during planting period. The total variable cost constituted 82.3% (N103,150.00) of the total cost while the total fixed cost was 17.7% (N22,1700) of the total cost. The average gross margin and net farm income of the production per hectare were N138,650.00 and N116,480.00 respectively. This can be deduced that yam production in the area is profitable.

Profitability ratios of yam production

The profitability ratio of yam production was computed to establish the profitability level of yam production. Result of Table 6 revealed that the profitability index (return to scale) (PI), rate of returns on investment (RRI), rate of returns on variable cost (RRVC) and operating ratio (OR) were ₦0.482, ₦0.929, ₦2.129 and ₦0.426. PI of 0.482 indicates that for every naira invested, about 48 kobo is returned as net profit to the farmer. Also, the farmer earns about 93 kobo as net income on every naira spent on yam production on the average. The rate of returns on variable cost of ₦2.129 indicates the farmers' returns on every naira expended. Finally, an operating ratio that is less than one indicates a good, efficient and profitable venture. Therefore, the OR of 0.43 indicates moderate total revenue over total variable cost, which is good for the venture.

Conclusion

The result on the socio-economic characteristics indicated that most (95%) of the farmers are males and only 5% are females. 13.8% were between 31 to 40 years of age, signifying that the farmers are in their youthful age. Also, the educational status of the farmers indicated that they had formal education up to secondary school level. Their farming experience was between 11 to 20 years and majority of 53.3% had farm holdings of between 1 to 2 hectares. The result of the regression analysis revealed that farm size and fertilizer were positive and significant at 1% level, while yam seed was also positive and significant at 5% level. This implies that increasing these variable inputs would result in an increased output level of yam. The result of marginal analysis of resource efficiency utilization shows that increase in one hectare of farm size, 1 kg of yam seed, extra use of one man-day of family labour and increase fertilizer usage by 1 kg would result in the corresponding increase in extra 8,766, 0.9624, 0.6997 and 5.6177 kg of yam respectively. Result of the average cost and returns indicated that GI, GM and NFC were ₦241,800.00, ₦138,650.00 and ₦116, 480.00 respectively. It is therefore recommended that the farmers should increase their farm size, yam seed and fertilizer usage. Also, government in their effort in improving yam production should enhance technical research on yam and timely provision of credit facilities to farmers in order to boost their yam production abilities.

REFERENCES

Abubakar I, Iheanacho AC, Abdullahi AB (2005). Determinants of profitability in yam production in Northern part of Taraba State, Nigeria. J. Sustain. Dev. Agric. Environ., 1: 24-29.

Adetuyi FO, Aladekoyi G, Adetuyi OO (2010). The nutritional composition of cut surface of yam (Disocorea rotundata). J. Sustain. Dev. Agric. Environ., 5(1): 108-116.

Amegbeto K, Asiedu R (2000). Improvement of Yam Based Production Systems. IITA Project 5: 1-4.

Ani AO (2004). Women in Agriculture and Rural Development. (1 ed), priscaquilla publishers, Kaduna.

Asiedu JJ (1999). Processing Tropical Crops by Technical Approach, Macmillan Publishers Limited London, pp. 149-161.

Asomugba GN, Ujoku ME (2007). Minisett Techniques of Seed Yam Production in Two Major Yam producing states In Nigeria.

CBN (2003). Annual Report and statement of Account Various Issues. Central Bank of Nigeria (CBN) Abuja.

Degras L (1993). "The Yam", A Tropical Root Crop: Macmillan with CTA Publishers, pp. 42-52.

Fadeji TO (2006). Improved Maize Cultivation in the rural communities of Kaduna State, Nigeria: The Problems and Prospects. Proceedings, 11th Annual Conference of Agricultural Extension Society of Nigeria (AESON), pp. 172-183.

FAO (2002). Root Statistics, Proceeding of the Expert Consultation of Root Crop.

FAO (2006). Food and Agricultural Organisation Database.

Iben MS (1984). The Nigeria Farmer and Agricultural Institutions. An Assessment by Nigerian Institute of Social and Economic Research, pp. 12-45.

Ibrahim A (2004). Analysis of production Efficiency of yam in Northern part of Taraba State, Nigeria. M.Sc dissertation, Department of Agricultural Economics and Extension, University of Maiduguri, Borno State, Nigeria, p. 76.

Idowu AO (2009). Economics of poultry production in Egba Division of Ogun State, Nigeria. J. Agric. Res. Policie, 4(1) 85 – 90.

Iheanacho AC, Olukosi JO, Ogungbile AO (2000). Economic Efficiency of Resource use in Millet-Based Cropping Systems in Borno State, Nigeria J. Trop. Agric., (2): 18-29

IITA (1991). Production Cost in the Yam Based Cropping System of South-Eastern Nigeria, RCMP Monograph Ibadan, Nigeria, 6: 2-26.

IITA, (1993). Production of Yams, Present and Future Perspective; IITA Research Guide, 46: 5.
Kay RD (1986). Farm Management Planning, control and Implimentation. Mc-Graw-Hill Book Company, New-York, p. 350.
NBS (2007). National Bureau of Statistics Annual Agricultural Survey Report: 1994/95-2005/06.
Olukosi JO, Erhabor PO (1988). Introduction to farm management Economics: Principles and Applications, AGITAB publishers, Zaria, p. 112.
Shehu JF, Mshelia SJ, Tashikalma AK, Gabdo BH (2009). Economic of small-scale Rain-fed upland Rice Production in Adamawa State, Nigeria. J. Res. Agric. 6(2):14 -21.
Sulumbe MI, Iheanacho AC, Mohammed ST (2010). Profitability Analysis of cotton production Under Sole-cropping system In Adamawa State, Nigeria. J. Sustain. Dev. Agric. Environ., 5 (1): 7-13.

Effect of methyl bromide alternatives on seedling quality, nematodes and soilborne fungi at the Blenheim and Trenton nurseries in South Carolina: 2008 to 2009

S. A. Enebak*, T. E. Starkey and M. Quicke

School of Forestry and Wildlife Sciences, Southern Forest Nursery Management Cooperative, Auburn University, AL 36849, USA.

Fumigation with methyl bromide (MBr) has been the most commonly used method for producing high quality, pest-free forest tree seedlings in the southeastern United States. Of the 1 billion seedlings produced annually, approximately 95% of seedlings grown for reforestation purposes are loblolly pine. Growth and survival of seedlings after outplanting is strongly linked to high quality, pest-free forest-tree seedlings produced in the nursery. This large scale study compares seven soil fumigants and a non-fumigated treatment using operational fumigation methods and standard operating nursery management practices over two growing seasons at two nurseries in South Carolina. Soilborne fungi, nematode and weed control was dependent upon nursery and soil fumigant tested. When examining overall seedling characteristics, the best MBr alternatives tested were chloropicrin, Pic+ and Chlor 60, all of which resulted in similar seedling densities to that of MBr. Information gathered from this study should be used by nursery managers in the southern U.S. to choose an MBr alternative that would be useful in the production of forest-tree seedlings in their nurseries based on their individual nursery soil conditions and pest pressures.

Keywords: Soil fumigation, chloropicrin, soilborne fungi, nematodes, loblolly pine.

INTRODUCTION

While fumigation with methyl bromide (MBr) has been the most commonly used method for producing high quality, pest-free forest-tree seedlings in the southeastern United States, concerns on the continued reliance of MBr in these production systems has resulted in new soil fumigant chemistries. As part of a long-term, continuing effort to identify and evaluate soil fumigants as an alternative to methyl bromide (MBr), a program funded by the USDA – ARS Area-wide Pest Management Project for Methyl Bromide Alternatives – South Atlantic Region, was undertaken in 2006. This 5-year program was to look at the efficacy of a number of soil fumigants against common nursery pests and their effect on seedling quality of forest-tree seedlings in the southern United States under normal nursery management practices over a two year cropping cycle in forest-tree nurseries that varied by ownership, production methods and soil types. While studies were previously conducted in Georgia, these studies compare seven soil fumigants at the ArborGen Super Tree Nursery in Blenheim, SC and the South Carolina Forestry Commission's Nursery in Trenton, SC, for the production of loblolly pine. Information from these studies should be used by nursery managers in the southern U.S. to choose an MBr alternative that would be useful in the production of forest-tree seedlings in their nurseries.

*Corresponding author. E-mail: enebasa@auburn.edu.

MATERIALS AND METHODS

Soil fumigation treatments at these two nurseries included two MBr rates and five currently available alternatives selected based on results of small plot studies previously conducted by the Southern Forest Nursery Management Cooperative (Table 1). The

Table 1. Soil fumigants and rates used in the 2008 to 2009 USDA Areawide demonstration plots at the ArborGen Super Tree Nursery in Blenheim, SC and the South Carolina Forestry Commission's Taylor Tree Nursery in Trenton, SC.

Soil fumigant treatment	Rate used	Chemical components
methyl bromide + chloropicrin #1	392 kg/ha (400 lb/a)	98% MBr and 2% chloropicrin
methyl bromide + chloropicrin #2	265 kg/ha (235 lb/a)	98% MBr and 2% chloropicrin
dimethyl disulfide + chloropicrin	107 l/ha (70 gal/a)*	79% DMDS and 21% chloropicrin
MBrC 70/30	448 kg/ha (400 lb/a)	70% MBr (98/2) and 30% Solvent A
Pic+	336 kg/ha (300 lb/a)	85% chloropicrin + 15% Solvent A
chloropicrin	336 kg/ha (300 lb/a)	100% chloropicrin
Chlor 60	448 kg/ha (400 lb/a)	60% chloropicrin and 40% 1,3-dichloropropene (Telone®)

* Compound is now labeled under the name Paladin®. Solvent A is a proprietary compound of Hendrix and Dail approved by EPA for use in forest-tree nurseries and under full label considerations.

Table 2. Site information for Trenton, SC and Blenheim, SC fumigation.

Application parameter	Trenton, SC	Blenheim, SC
Fumigation date	October 2, 2007	April 3, 2008
Fumigation type	Broadcast/Shank Injected	Broadcast/Shank Injected
Area in trial	2 ha	2 ha
Air temperature range	16 - 28°C	6 to 9°C
Wind speed	5 – 17 km/h	8 – 18 km/h
Soil moisture	5%	7%
Soil series	Wagram sand – (sandy loam)	Autryville sand – (loamy sand)
Plastic in place	10 days	7 days
Soil particle size	75:15:10 (sand: silt: clay)	87:6:7 (sand: silt: clay)

experiment occupied approximately 2 ha out of a total 35 production ha within each nursery and for each, the soil fumigants were shank-injected to a depth of 20 to 25 cm using a Raven Flow Meter Control System® (Sioux Falls, SD) and immediately covered with 1 mm High Density Polyethylene Tarp (Cadillac Plastics Inc.) under the weather and soil conditions listed in Table 2. At the Trenton Nursery, the nursery manager allowed a non-fumigated control treatment (no soil fumigation) to be installed. Soil fumigants were applied in October 2007 (Blenheim) and April 2008 (Trenton) and were laid out in nursery sections that contained nine seedling beds between irrigation pipelines (20 m), with each treatment plot approximately 100 m long × 4 m wide. The experimental design was a randomized complete block replicated four times over the five nursery sections. Each 9-bed nursery section contained three soil treatments. Both the Blenheim and Trenton nurseries sowed a single family of loblolly pine (*Pinus taeda* L.) seed in late April 2008 with the first seedling crop lifted in early December 2008. The area was fallow during the winter of 2008 and in March 2009 the beds were prepared for sowing. The second seedling crop was sown in April 2009 with seedlings lifted in December 2009. With respect to seedling production over the two growing seasons, standard nursery production systems used at each nursery (e.g. irrigation, fertilization, weed control, insect control, etc.) were allowed to continue within the experimental area.

Seedling quality

To determine the effects of the various fumigants on seedling quality, seedling and soil samples were collected from the middle seedling bed of each 3-bed treatment plot within each section.

Seedling densities were assessed in four subplots per treatment plot at 7 weeks post sowing, mid-summer (15 weeks post sowing) and just prior to lifting in the fall (26 weeks post sowing) in both production years. To determine the effect of each soil fumigant on seedling quality, at the end of each growing season in 2008 and 2009 at both nurseries, 25 seedlings per subplot and returned to the laboratory for measurements. Each seedling was measured for root collar diameter (RCD), shoot height and seedling dry weight (biomass). In addition, root morphology which included total root length, root surface area, average root diameter and the number of root tips was measured on ten seedlings per subplot using WinRhizo® software by Regents Instruments Inc. Quebec, Canada.

Soil Trichoderma and nematodes

Throughout the two cropping seasons, soil samples were collected from the center seedling bed of each 3-bed plot: at pre-sowing, post-sowing, mid-summer and just prior to seedling lifting in November of each season. Soils were returned to Auburn University where half of each soil sample was plated onto *Trichoderma*-selective media (Elad et al., 1981) and the remaining half was sent to the Soils Laboratory at Auburn University for a quantitative assessment of nematode populations.

RESULTS

Seedling quality

At the end of the first growing season, there were no

Table 3. Loblolly pine seedling density (No./m^2) at the end of each cropping season at Blenheim, SC and Trenton, SC.

Treatment	Blenheim, SC		Trenton, SC	
	Dec 2008	Dec 2009	Oct 2008	Oct 2009
MBr #1	236^a	247^a	182^a	118^a
MBr #2	226^a	247^a	169^a	115^a
Chloropicrin	247^a	247^a	170^a	115^a
Chlor 60	247^a	247^a	161^a	114^a
MBrC 70/30	236^a	236^a	175^a	117^a
DMDS+Chloropicrin	247^a	258^a	173^a	114^a
Pic+	236^a	247^a	174^a	121^a
Control	-	-	175^a	116^a
$lsd_{(0.05)}$	22	22	24	21

Within column means followed by the same letter do not differ at P = 0.05 level using Duncan's Multiple Range Test. Target seedling density is 236 seedlings/m² at Blenheim, SC and 247 seedlings/m^2 at Trenton, SC.

Table 4. Loblolly pine seedling root collar diameter (mm) at the end of each growing season at Blenheim, SC and Trenton, SC.

Treatment	Blenheim, SC		Trenton, SC	
	Dec 2008	Dec 2009	Oct 2008	Oct 2009
MBr #1	4.37^{ab}	4.38^{cd}	4.02^b	4.51^{abc}
MBr #2	3.93^c	4.21^d	3.97^b	4.52^{abc}
Chloropicrin	4.46^a	4.43^{bcd}	4.08^{ab}	4.61^{ab}
Chlor 60	4.22^{ab}	4.52^{bc}	4.00^b	4.65^{ab}
MBrC 70/30	4.18^b	4.32^{cd}	4.10^{ab}	4.79^a
DMDS + Chloropicrin	4.43^a	4.81^a	3.89^b	4.61^{ab}
Pic+	4.32^{ab}	4.65^{ab}	4.34^a	4.48^{bc}
Control	-	-	3.08^c	4.33^c
$lsd_{(0.05)}$	0.25	0.28	0.32	0.29

Within column means followed by the same letter do not differ at P = 0.05 level using Duncan's multiple range test.

significant differences for seedling densities among the soil fumigants tested at either the Blenheim or Trenton nurseries (Table 3). With respect to meeting seedling production targets, Blenheim was at or above their production target while Trenton was far below the target of 247 seedlings per m^2 for both growing seasons at an average of 172 and 116 seedlings per m^2 for 2008 and 2009, respectively (Table 3).

At Blenheim in 2008, seedlings growing in soils treated with either Chloropicrin or DMDS + Chlor had significantly larger RCD than seedlings from soils treated with a lower rate of MBr (265 kg/ha) and the MBrC 70/30 (Table 4). In 2009, seedlings grown in soils fumigated with DMDS+Chlor and Pic+ had larger RCDs than either of the MBr treatments. In the 2008 growing season at Trenton, the root collar diameter (RCD) of loblolly pine seedlings growing in the non-fumigated treatment was significantly smaller than all other soil fumigants. Of the soil fumigants tested at this nursery, seedlings growing in Pic+ were significantly larger than seedlings growing in MBr, DMDS + Chlor and Chlor 60 treated soils (Table 4). Not unexpected, the effect of soil fumigants producing larger seedlings was lessened and, in 2009, there was no significant difference in seedling RCD between MBr, Pic+ and the non-fumigated treatments.

The proportion of seedlings for each grade in 2008 was variable across all soil treatments examined with none of the soil fumigants standing out at Blenheim (Table 5). In contrast, Chlor 60 had the highest proportion of seedling culls when compared to the other soil fumigants tested in Blenheim soils. In 2009, the proportion of Grade 1 seedlings increased for all soil treatments except 100% chloropicrin. Despite the low seedling densities at the Trenton nursery, the proportion of seedlings produced in 2008 for each grade was similar for all soil fumigants tested: 22% Grade 1, 60% Grade 2 and 20% Cull, except the non-fumigated soil which had the greatest percent of culls with 68% (Table 6). In 2008, soils treated with Pic+

Table 5. Loblolly pine seedling grade by soil fumigation in 2008 and 2009 at Blenheim, SC.

Soil treatment	2008				2009			
	Total	Grade 1	Grade 2	Culls	Total	Grade 1	Grade 2	Culls
MBr #1	236^{a}	64^{ab}	161^{ab}	11^{a}	247^{a}	75^{b}	150^{a}	21^{a}
MBr #2	226^{a}	32^{b}	172^{ab}	21^{a}	247^{a}	63^{b}	139^{a}	22^{a}
Chloropicrin	247^{a}	96^{a}	139^{b}	11^{a}	247^{a}	86^{b}	150^{a}	11^{a}
Chlor 60	247^{a}	43^{b}	129^{b}	7^{a}	247^{a}	96^{ab}	139^{a}	11^{a}
MBrC 70/30	236^{a}	43^{b}	182^{a}	11^{a}	236^{a}	75^{b}	139^{a}	21^{a}
DMDS + Chlor	247^{a}	86^{a}	150^{b}	11^{a}	258^{a}	118^{a}	129^{a}	11^{a}
Pic+	236^{a}	75^{a}	150^{b}	11^{a}	247^{a}	107^{a}	129^{a}	11^{a}
lsd $_{(0.05)}$	22	27	29	14	22	23	21	12

Within column means followed by the same letter do not differ at P = 0.05 level using Duncan's Multiple Range Test. Numbers are seedlings per square meter of nursery bed. Target seedling density was 236 seedlings/m².

Table 6. Loblolly pine seedling grade by soil fumigation in 2008 and 2009 at Trenton, SC.

Soil treatment	2008				2009			
	Total	Grade 1	Grade 2	Culls	Total	Grade 1	Grade 2	Culls
MBr #1	182^{a}	39^{b}	107^{a}	36^{b}	118^{a}	53^{a}	54^{a}	11^{ab}
MBr #2	169^{a}	34^{b}	101^{ab}	34^{b}	115^{a}	47^{a}	53^{a}	15^{a}
Chloropicrin	170^{a}	36^{b}	110^{a}	24^{bc}	115^{a}	48^{a}	51^{a}	16^{a}
Chlor 60	161^{a}	32^{b}	98^{ab}	28^{bc}	114^{a}	53^{a}	56^{a}	5^{b}
MBrC 70/30	175^{a}	45^{ab}	99^{ab}	31^{b}	117^{a}	59^{a}	49^{a}	9^{ab}
DMDS + Chlor	173^{a}	24^{bc}	109^{a}	40^{b}	114^{a}	54^{a}	49^{a}	11^{ab}
Pic+	174^{a}	62^{a}	92^{b}	17^{c}	121^{a}	49^{a}	62^{a}	10^{b}
Control	175^{a}	9^{c}	49^{c}	117^{a}	116^{a}	39^{a}	58^{a}	19^{a}
lsd $_{(0.05)}$	24	25	12	12	21	17	18	9

Within column means followed by the same letter do not differ at P=0.05 level using Duncan's Multiple Range Test. Numbers are seedlings per square meter of nursery bed. Target seedling density was 247 seedlings/m²

had a greater proportion of Grade 1 seedlings than all the other soil fumigant treatments examined. In 2009, the proportion of Grade 1 seedlings increased for all soil treatments; with 44% Grade 1, 46% Grade 2 and 9% Cull. In 2009, the higher proportion Grade 1 and Grade 2 seedlings resulted in a lower proportion of cull seedlings compared to 2008. As expected, the non-fumigated soils produced the highest number of cull seedlings. Lower seedling densities typically results in a higher seedling RCD (less seedling to seedling competition), yielding more Grade 1 seedlings per square meter. Thus, the higher number of Grade 1 seedlings in the second season (2009) is due to the lower seedling densities in 2008 compared to 2009.

Except for average root diameter, the overall seedling root architecture and root morphology was less on seedlings during the 2009 growing season than on seedlings produced during the 2008 growing season at Blenheim (Table 7). Of the soil fumigants tested at this nursery, Pic+ and Chlor 60 consistently resulted in the best root morphology of the MBr alternatives. An interesting point in quantifying root systems is that total seedling root length in these trials ranged from 181 cm to 439 cm of total fine roots per seedling. Like that at Blenheim, the overall seedling root architecture and root morphology was less in 2008 than in 2009 at the Trenton nursery (Table 8). Of the soil fumigants tested, DMDS + Chlor and Chloropicrin consistently resulted in the best root morphology of the MBr alternatives to produce a fibrous root system which increases in seedling survival in the field after outplanting.

Soil Trichoderma and nematodes

At the end of the first growing season in 2008 at Blenheim, soils treated with Chlor 60 had significantly higher levels of the soil borne fungus *Trichoderma* than soil fumigation treatments that contained some amount of MBr (Table 9). By the end of the second growing season in 2009 the *Trichoderma* levels within the soil fumigants tested were similar to MBr. Within the two MBr rates, the higher rate of MBr (448 kg/ha) had higher *Trichoderma* levels than MBr at (265 kg/ha). At Trenton, soils treated with MBr had significantly lower levels of *Trichoderma* than treatments that contained Chloropicrin in 2008 (Table 9). By the end of the second growing season in 2009, *Trichoderma* levels within the soil fumigants

Table 7. Loblolly pine seedling root morphology at the end of each growing season, Blenheim, SC.

Treatment	Root length (cm)		Root surface area (cm²)		Avg root dia (mm)		No. root tips	
	2008	2009	2008	2009	2008	2009	2008	2009
MBr #1	420^{ab}	217^{ab}	79^{a}	74^{a}	1.09^{a}	0.59^{a}	826^{ab}	474^{abc}
MBr #2	346^{b}	209^{ab}	62^{a}	72^{a}	1.10^{a}	0.57^{a}	724^{b}	449^{abc}
Chloropicrin	419^{ab}	181^{b}	76^{a}	58^{a}	1.05^{a}	0.58^{a}	830^{ab}	400^{bc}
Chlor 60	416^{ab}	229^{a}	77^{a}	78^{a}	1.09^{a}	0.59^{a}	838^{ab}	502^{ab}
MBrC 70/30	408^{ab}	205^{ab}	75^{a}	69^{a}	1.09^{a}	0.59^{a}	832^{ab}	418^{bc}
DMDS+Chlor	439^{a}	198^{ab}	80^{a}	70^{a}	1.14^{a}	0.58^{a}	850^{a}	397^{c}
Pic+	411^{ab}	224^{ab}	75^{a}	76^{a}	1.12^{a}	0.58^{a}	801^{ab}	529^{a}
$lsd_{(0.05)}$	86	45	19	21	0.11	0.56	122	105

Within column means followed by the same letter do not differ at P=0.05 level using Duncan's multiple range test.

Table 8. Loblolly pine seedling root morphology at the end of each growing season, Trenton, SC.

Treatment	Root length (cm)		Root surface area (cm²)		Avg root dia (mm)		No. root tips	
	2008	2009	2008	2009	2008	2009	2008	2009
MBr #1	239^{bc}	233^{a}	90^{ab}	48^{b}	1.25^{bc}	0.67^{ab}	547^{a}	518^{abc}
MBr #2	248^{ab}	262^{a}	98^{ab}	54^{ab}	1.22^{c}	0.69^{ab}	706^{a}	527^{abc}
Chloropicrin	234^{bc}	251^{a}	106^{ab}	54^{ab}	1.36^{a}	0.73^{a}	661^{a}	490^{bc}
Chlor 60	273^{ab}	255^{a}	100^{ab}	59^{ab}	1.26^{bc}	0.69^{ab}	652^{a}	567^{ab}
MBrC 70/30	272^{ab}	215^{a}	83^{b}	58^{ab}	1.25^{bc}	0.68^{ab}	553^{a}	599^{ab}
DMDS+ Chlor	258^{ab}	267^{a}	112^{a}	58^{ab}	1.34^{ab}	0.72^{a}	662^{a}	533^{abc}
Pic+	296^{a}	240^{a}	93^{ab}	64^{a}	1.24^{c}	0.70^{a}	629^{a}	616^{a}
Control	189^{c}	266^{a}	106^{ab}	36^{c}	1.27^{bc}	0.62^{b}	660^{a}	446^{c}
$lsd_{(0.05)}$	58	56	26	14	0.09	0.07	211	121

Within column means followed by the same letter do not differ at P = 0.05 level using Duncan's multiple range test.

Table 9. Post-fumigation recovery of *Trichoderma spp.* from nursery soils in Blenheim, SC and Trenton, SC nurseries in 2008.

Treatment	Blenheim, SC			Trenton, SC		
	7 weeks	15 weeks	26 weeks	7 weeks	15 weeks	26 weeks
MBr #1	47^{b}	28^{ab}	114^{a}	53^{a}	69^{a}	83^{ab}
MBr #2	20^{b}	15^{b}	40^{b}	54^{a}	74^{a}	101^{ab}
Chloropicrin	56^{ab}	33^{ab}	65^{ab}	69^{a}	76^{a}	61^{b}
Chlor 60	99^{a}	55^{a}	76^{ab}	57^{a}	57^{a}	67^{ab}
MBrC 70/30	24^{b}	18^{b}	79^{ab}	69^{a}	71^{a}	96^{ab}
DMDS+Chlor	31^{b}	17^{b}	62^{ab}	62^{a}	74^{a}	77^{ab}
Pic+	92^{a}	37^{ab}	68^{ab}	49^{a}	54^{a}	67^{ab}
Non-Control	-	-	-	69^{a}	69^{a}	124^{a}
$lsd_{(0.05)}$	48	37	63	39	32	60

Within column means followed by the same letter do not differ at P=0.05 level using Duncan's multiple range test. There were no non-fumigated plots at Blenheim, SC. Numbers are in colony forming units (cfu's) per mg of soil.

examined were similar to those soils treated with MBr. The non-fumigated soils (6+ years) had the highest levels of *Trichoderma* out of all the soil fumigants at this nursery.

Over the course of the 2-year study, nematodes were assayed five times for both the number and species within the soil/seedling interface. Nematode populations within the soil are rarely uniformly distributed throughout the soil and these studies had relatively low numbers for all soil fumigants used at both nurseries (Table 10). Thus, there is no data to suggest that one soil fumigant is better than another in controlling nematodes as they were

Table 10. Soil nematode levels at the end of each growing season in Blenheim, SC and Trenton, SC.

Treatment	Nematode	Blenheim, SC		Trenton, SC	
		2008	2009	2008	2009
MBr #1	Stunt	0	0	2	24
	Root knot	0	0	0	1
	Ring	0	1	0	0
MBr #2	Stunt	1	0	1	8
	Root knot	0	0	1	0
	Ring	0	0	1	0
Chloropicrin	Stunt	0	0	27	69
	Root knot	0	0	0	1
	Ring	0	0	0	0
Chlor 60	Stunt	1	0	21	62
	Root knot	0	0	0	3
	Ring	0	2	0	0
MBrC 70/30	Stunt	0	0	2	9
	Root knot	0	0	0	1
	Ring	0	2	0	0
DMDS+Chlor	Stunt	2	0	8	96
	Root knot	0	0	0	0
	Ring	0	1	0	0
Pic+	Stunt	0	0	12	136
	Root knot	0	0	0	0
	Ring	0	1	1	2
Non-fumigated	Stunt	0	0	49	91
	Root knot	0	0	0	0
	Ring	0	0	11	0

Due to the non-uniform distribution of nematodes there were no differences observed for any treatment x nematode x year, P = 0.05. Numbers are the average of 5 replicates per treatment recovered in 100cc of soil.

equally as effective as MBr. However, at the Trenton nursery, the stunt nematode (*Tylenchorhynchus claytoni*), which can cause problems in seedling nurseries, appeared in all soil fumigant treatments during the second cropping season. It is interesting to note that Chlor 60, which contains Telone,® (a good soil nematicide), had more nematode species recovered from the soil samples than all other fumigants tested (Table 10).

DISCUSSION

The primary objective of the USDA Areawide MBr Alternative program is to identify possible alternatives to MBr using large-scale, multi-year trials in a number of different soils and growing conditions throughout the southern U.S. One of the unique aspects of MBr as a soil fumigant is its ability to consistently control weeds, insects, nematodes and fungi across many different growing conditions. Because of the cropping system unique to forest-tree nurseries, evaluations of MBr alternatives are conducted over two growing seasons because the true test of a soil fumigant is its performance during the second growing season where treatment differences usually begin to appear. These trials bear that out as the various MBr alternatives behaved differently at each of the nurseries tested.

When MBr is no longer available, those soil fumigants

with chloropicrin appear to be most useful in controlling pests and producing high-quality seedlings at Blenheim, SC. MBr alternatives that could be considered for Blenheim are Pic+, DMDS+Chlor and Chlor 60. While there were good seedling characteristics with the soil fumigant DMDS+Chlor (Paladin®), this compound has a significant odor problem that persisted into the summer growing season which will limit its acceptance as an alternative. Of those treatments tested at Blenheim, Chlor 60 had more nematodes (numbers and species) despite having Telone®, and the largest percent of cull seedlings in the first growing season. The large percentage of culls with this soil fumigant during the first growing season is unexplainable but may be due to seedling sowing densities. Taken together, of the soil fumigants tested, Pic+ performed the best for seedling production at Blenheim, SC.

Similarly, at the Trenton nursery, those soil fumigants with chloropicrin appear to be the most useful in controlling pests and producing high-quality seedlings. DMDS + chloropicrin resulted in adequate RCD and root morphology characteristics and soil-borne *Trichoderma* levels. However, the strong odor of garlic to propane turned nursery personnel away from this treatment. By far the best MBr alternatives tested were Chloropicrin and Chlor 60, with both soil fumigants controlling nematodes and producing quality seedlings. If buffer zone restrictions that come into force in 2012 limit the use of 100% chloropicrin, then Chlor 60, with 60% chloropicrin would be the next best alternative at Trenton, SC.

One of the unique aspects of soil fumigants currently being tested in southern forest nurseries is that they do not completely eliminate beneficial fungi which are needed for seedling growth. Previous Nursery Cooperative research has shown that *Trichoderma* is not as sensitive to MBr as other soil fumigants (Carey et al., 2005; Starkey et al., 2006). In these two nurseries, the soil fumigants tested did not completely eliminate all soilborne fungi and the population levels of non-target soilborne fungi rebounded quickly with all soil fumigants. Not surprising, the highest levels of *Trichoderma* were in the non-fumigated plots at Trenton that had not been fumigated in over 6 years.

One of the aspects of determining the effects of MBr on root architecture is that a fibrous root system increases the chance of seedling survival in the field (Hatchell and Muse, 1990; Frampton et al., 2002; Davis and Jacobs, 2005). All of the soil fumigants tested were similar in their root morphology characteristics when compared to the standard MBr. Not surprisingly, the effects of soil fumigation on seedling quality can be observed in the second growing season. In all treatments, root morphology was less in the second season when compared to the first growing season. For this reason, most forest-tree nurseries limit their production to 2 seedling crops per soil fumigation as the soilborne pests become too severe to produce Grade 1 and 2 seedlings.

The final decision when selecting an MBr alternative needs to take into consideration the ability of the soil fumigant to work under individual nursery soil conditions and the impact of the new EPA reregistration eligibility decision (REDs) on each individual nursery. While it would be useful for nursery managers and researchers to continue to use MBr to grow forest-tree seedlings, MBr will eventually be unobtainable and each nursery manager will need to identify the best alternative for their nursery. For these trials, Pic+ was the better soil fumigant at Blenheim, while 100% chloropicrin would be the alternative of choice at Trenton.

REFERENCES

Carey WA, McGraw D, Enebak SA (2005). Seedling production by seed treatment and fumigation treatment at the Glennville Regeneration Center in 2004. Auburn Univ. South. For. Nursery Manage. Coop. Res. Rep. 05-04, Auburn, AL, 5p.

Davis AS, Jacobs DF (2005). Quantifying Root System Quality of Nursery Seedlings and Relationship to Outplanting Performance. New For., 30:295-311.

Elad Y, Chet I, Henis Y (1981). A selective medium for improving quantitative isolation of *Trichoderma* spp. from soil. Phytoparasitica. 9:1 59-67.

Frampton J, Isik F, Goldfarb B (2002). Effects of Nursery Characteristics on Field Survival and Growth of Loblolly Pine Rooted Cuttings. South. J. Appl. For., 26:207-213.

Hatchell GE, Muse HD (1990). Nursery Cultural Practices and Morphological Attributes of Longleaf Pine Bare-Root stock as Indicators of Early Field Performance. USDA For. Serv. Res. Paper. SE. 277.

Starkey TE, Enebak SA, McGraw D (2006). Seedling quality and weed control with dazomet, methyl bromide and methyl iodide at the Glennville Regeneration Center 2005-2006. Auburn Univ. South. For. Nursery Manage. Coop, Auburn, AL. Res. Rep., 06-05, 6p.

Analysis of household energy sources and woodfuel utilisation technologies in Kiambu, Thika and Maragwa districts of Central Kenya

Githiomi J. K.[1]*, Mugendi D. N.[2] and Kung'u J. B.[2]

[1]Forest Products Research Centre, Kenya Research Institute, P.O. Box 64636-00620 Nairobi, Kenya.
[2]Department of Environmental Sciences, Kenyatta University, P.O. Box 43844-00100 Nairobi, Kenya.

This study was carried out in Kiambu, Thika and Maragwa districts of Central Kenya. Its objective was to analyze household energy sources and utilization technologies used. Primary data was collected from households using structured and non structured questionnaires. Trees on farm were found to be the major supply of the woodfuel energy where firewood was the main source of household energy followed by charcoal. Traditional three stones stoves were the most commonly used with 76 59 and 65 (household respondents) in Kiambu, Thika and Maragwa districts respectively. Improved charcoal stoves were the second commonly used while only a very negligible percentage used kerosene stoves and gas burners. Over 70% of the respondents were aware of the improved stoves but their adoption was less than 29%. The low adoption of improved stoves was due to their high cost as noted by the respondents. Over 90% of the households had the opinion that woodfuel sources were decreasing and there was a need to develop strategies for its future sustainability. The study recommended integration of woodfuel production to local farming systems and establishment of fuelwood plantations by Kenya Forest Service to substitute on farm sources. It also recommended promotion of improved stoves with higher efficiency to reduce the woodfuel used as well as improve on environmental pollution.

Key words: Farmlands, Central Kenya, improved stoves, firewood, charcoal.

INTRODUCTION

It is estimated that about 90% of Kenyan rural households use woodfuel either as firewood or charcoal (MoE, 2002) and it provides income to over 3 million people (ESDA, 2005). A significant portion of the rural population is employed in wood energy trade and it constitutes a major source of energy in Kenya (Republic of Kenya, 2002a). The Government of Kenya has been involved in promoting tree planting at the farm level with the aim of increasing tree cover to 10% by the year 2030 (Republic of Kenya, 2007). There have been successful tree planting programs involving rural communities in Kenya led by government rural forest extension services and various non governmental organizations (NGOs). Green Belt Movement is among the most active NGOs which have assisted planting of over 45 million trees in different parts of Kenya for the last three decades. Besides being the standard cooking fuel for the majority of Kenyan households, fuelwood is also an important energy source for small-scale rural industries such as tobacco curing, tea drying, brick making, fish smoking, and bakeries, among others. However, despite its importance, the available data is scarce and uncertain which is mainly due to the fact that it is handled in the informal sector and does not pass through monetized economy like in the case of liquefied petroleum gas (LPG), kerosene and electricity which are alternatives to wood energy. The energy sector concentrates more on national energy planning for conventional fuels while forestry sector focus more on planning for commercial wood supply and conservation of protected areas (Republic of Kenya, 2002a). These two key sectors deny

*Corresponding author. E-mail: josephgithiomi@ymail.com.

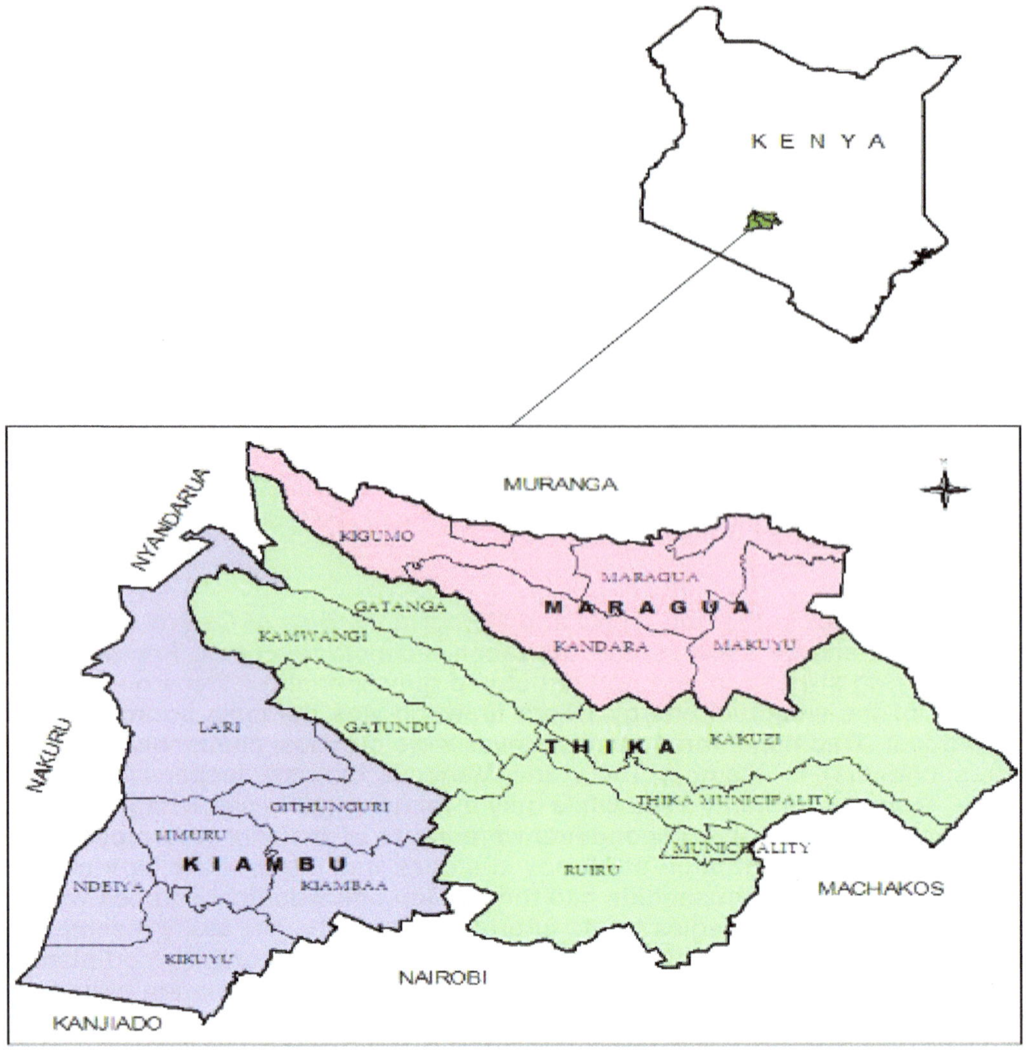

Figure 1. Map of the study area showing the Kiambu, Thika and Maragwa districts.

biomass energy the comprehensive consideration it deserves due to the role it plays in supply of energy to majority of Kenyans.

This study was undertaken between 2006 and 2007 and it was aimed at establishing household energy sources and its utilization technologies in Kiambu, Thika and Maragwa districts of Central Kenya. The data obtained is useful in policy formulations in these three districts.

METHODOLOGY

This study was carried out in Kiambu, Thika and Maragwa districts of Central Kenya as shown in Figure 1. The districts were purposively sampled on the basis of various factors which include diverse wood production systems (farmlands, plantations, and natural forests), diverse ecological conditions and population densities among others. The selected districts had rainfall range from 800 to 1400 mm, 500 to 2500 mm and 900 to 2700 mm for Kiambu, Thika and Maragwa districts respectively (Republic of Kenya, 1997a, b, 2002b). Sampling of the households was done using multistage stratified random sampling technique beginning with stratification sampling procedures as outlined by Lee-Ann and Martin (1997). Each of the three districts under study was stratified according to the weights in socio-economic and climatic activities as indicators. Following this procedure, at least 40% of the divisions with relatively homogeneous characteristics in each district were sampled. This ensured that heterogeneity was well captured and represented. Similar procedure was followed for selection of administrative locations and sub-locations. Based on these sampling procedures, Kiambu, Thika and Maragwa districts were stratified into four, three and two strata respectively. Each stratum in the district consisted of one or more divisions. A total of 200 households were sampled from each district. Allocation of the sampling units in each selected division was done proportionally to the total number of household obtained from the 1999 census (Republic of Kenya, 2001).

The data collection was done using structured and semi-structured household questionnaires which captured data on the source of woodfuel, supply trend, the type of energy sources used, cooking technologies and awareness on the improved stoves. The generated data was coded and entered in the computer using spread sheet MS-Excel. Statistical package for social sciences (SPSS) was used in analyzing the data.

Table 1. Sources of woodfuel in Kiambu, Thika and Maragwa districts.

District	Percentage responses on supply sources of woodfuel													
	Indigenous trees scattered on farm		Exotic trees scattered on farm		Planted woodlots		Purchase from vendors		Gazetted forests		Others		Total	
	%	n	%	n	%	n	%	n	%	n	%	n	%	n
Kiambu	4	8	24	49	4	9	57	117	9	18	2	4	100	205
Thika	26	50	12	22	3	6	57	109	1	1	3	3	100	191
Maragwa	40	80	31	63	21	42	5	10	2	4	1	1	100	200

Table 2. Responses on supply trend of woodfuel sources in Kiambu, Thika and Maragwa districts.

District	Source of woodfuel	Supply trend of woodfuel from different sources					
		Increase		Decrease		Total	
		%	n	%	n	%	n
Kiambu	Existing scattered indigenous trees on farms	33	2	67	4	100	6
	Existing exotic scattered trees on farm	7	3	93	39	100	42
	Planted woodlots	33	1	67	2	100	3
	Purchase from sellers	3	2	97	67	100	69
	Gazetted forests	0	0	100	18	100	18
	Others	0	0	100	3	100	3
Thika	Existing scattered indigenous trees on farms	13	6	87	41	100	47
	Existing exotic scattered trees on farm	14	3	86	18	100	21
	Planted woodlots	50	3	50	3	100	6
	Purchase from sellers	37	29	63	50	100	79
	Gazetted forests	0	0	100	12	100	12
	Others	0	0	100	3	100	3
Maragwa	Existing scattered indigenous trees on farms	4	3	96	70	100	73
	Existing exotic scattered trees on farm	19	11	81	47	100	58
	Planted woodlots	2	1	98	40	100	41
	Purchase from sellers	10	1	90	9	100	10
	Gazetted forests	0	0	100	11	100	11
	Others	0	0	100	2	100	2

RESULTS

The results on the woodfuel supply sources in the three districts under study (Table 1) showed that majority of the respondents from Kiambu and Thika districts (57%) purchased woodfuel from vendors whereas others obtained woodfuel directly from the farm. The woodlots were taken as an area in the farm set aside for tree planting while the other trees in the farm were either referred to as scattered indigenous or exotic. Maragwa district had the highest number of households sourcing woodfuel from indigenous trees scattered on farm. In Maragwa and Thika districts, there were less than 2% of the households sourcing woodfuel from the gazetted forests. Charcoal was the main product purchased from vendors as firewood was mainly sourced directly from the farms. Overall, there was a significant difference ($\chi2=157.852$; d.f=2; $p<0.01$) among districts in terms of supply sources of wood energy.

The households' response to the trend of future supply sources which included existing scattered indigenous trees on farms, existing exotic scattered trees on farm, planted woodlots, gazetted forests and purchases from vendors showed a decrease in all woodfuel supplies sources (Table.2). Majority of the household respondents in Kiambu and Thika districts indicated that the purchase of woodfuel from sellers will decrease as compared to Maragwa who had the highest number (73) indicating the highest decrease was to be in existing scattered indigenous trees on farms which was also their main source of

Table 3. Sources of energy used by households in Kiambu, Thika and Maragwa districts.

District	Commonly used sources of energy											
	Firewood		Charcoal		Kerosene		Gas		Crop residue		Total	
	%	n	%	n	%	n	%	n	%	n	%	n
Kiambu	87	181	12	25	1	1	1	2	0	0	100	209
Thika	80	159	15	30	4	7	1	2	1	1	100	199
Maragwa	96	192	1	2	2	4	1	1	1	1	100	200

Table 4. Cooking technology devices used in Kiambu, Thika and Maragwa districts.

District	Response on each technology used													
	Three stone fire		Traditional metal stove		Improved charcoal stove		Improved firewood stove		Kerosene stove		Gas burner		Total	
	%	n	%	n	%	n	%	n	%	n	%	n	%	n
Kiambu	76	155	4	8	18	36	0.5	1	0.5	1	1.0	2	100	203
Thika	59	116	2	4	30	59	2	3	5	9	2	3	100	198
Maragwa	64	130	6	12	24	49	3	7	1	1	2	4	100	203

Table 5. Awareness of improved technology in Kiambu, Thika and Maragwa districts.

District	Aware		Not aware		Total	
	%	n	%	no	%	n
Kiambu	79	162	21	42	100	204
Thika	70	136	30	59	100	195
Maragwa	89	177	11	23	100	200

Table 6. Reasons for not using improved firewood stoves in Kiambu, Thika and Maragwa districts.

Reasons for not using improved firewood stoves	Respondents					
	Kiambu		Thika		Maragwa	
	%	n	%	n	%	n
Cost	31	56	25	31	57	74
Non availability	23	41	35	43	16	21
Not interested	34	62	22	27	14	18
Never heard of it	7	12	15	19	12	16
Very slow in cooking	5	10	3	3	1	1
Total	100	181	100	123	100	130

source of woodfuel. The results showed that there were significant differences ($p<0.01$) in the sources of energy used by households within the districts studied.

Firewood was commonly used as shown by 87, 80 and 96% of the households' respondents in Kiambu, Thika and Maragwa districts respectively followed by charcoal. Crop residue, gas and kerosene were the least used (Table 3). The average weekly firewood and charcoal consumption per household in the three districts was found to be 32 and 10.4 kg respectively.

The results of woodfuel utilisation technologies are shown in Table 4. Inferential analysis of the data showed that there were significant differences ($p<0.01$) among technologies used within each district where three stones was the most commonly used by the household respondents' with 76, 59 and 64% followed by improved charcoal stove with 18, 30 and 24% in Kiambu, Thika and Maragwa districts respectively. A very negligible household percentage used kerosene stoves and gas burners.

On the awareness of the improved technology, the results (Table 5) showed that over 70% of the respondents were aware of improved cooking technologies. This significantly varied among the districts ($\chi^2=21.162$; d.f=2; $p<0.00$) where majority were from Maragwa district followed by Kiambu. The adoption of improved charcoal and fire wood stoves was 19, 24.5 and 28% in Kiambu, Thika and Maragwa districts, respectively.

A number of reasons were cited for not using the improved stoves with cost of purchasing improved stoves being regarded as main reason for its non-adoption (Table 6). This was significantly different ($p<0.05$) across the three districts but it was highly pronounced in Maragwa followed by Kiambu district. Other factors that cause non adoption of these improved stoves are their non availability, and lack of awareness by potential users.

DISCUSSION

The significant difference among districts in terms of supply sources of wood energy implied that each supply sources of woodfuel had different weights in each district.

For example, Maragwa got its woodfuel mainly from scattered indigenous trees on farm as compared to Thika and Kiambu who purchase mainly from vendors. The study indicated that farmlands form a critical part of wood energy resource supply as observed in all districts and only a small proportion of household were sourcing woodfuel from the gazzeted forests. Similar observations were reported by Ministry of Energy's study on energy supply and demand for household, small scale industries and service establishment in Kenya where most woodfuel was found to be from the farmlands (MoE, 2002). Encouraging farmers to plant more trees could increase the supply of woodfuels and other wood products considerably within the study area. However, there is limitation in promoting woodfuel production as it is taken as a by-product of other activities (like timber) which are more valued.

The decreasing trend of major supply sources in the three districts was caused by scarcity of land for continuous tree planting, deforestation of gazetted and on farm forests and slow maturity of indigenous tree species. These constraints can be addressed through planting of fast growing species and adopting planting technologies that maximizes the woodfuel production in agroforestry systems. The decreased supply of woodfuel supports the need for wood energy planning in each of the study districts to avoid eminent deficit.

The high number of household residents using firewood as source of energy (over 80%) had also been observed in a study by Ministry of Energy on energy demand and supply for households where 89% of rural households were found to be using firewood countrywide (MoE, 2002). This also compares well with another study by Kenya National Bureau of Statistics which indicated that 87.7% of households in Kenya were using firewood (Ministry of Planning and National Development, 2007). On-farm tree planting has a significant role to play in woodfuel production in Kenya. The government of Kenya has promoted tree planting on farm and in gazetted forests through the Kenya Forestry service and several other projects by development partners and NGOs. Tree reforestation has been done in degraded areas like Mau forest and other Kenyan water towers.

The three stone stoves which were commonly used in the three districts despite having low efficiency of about 10% and high health hazards, however, are popular with the households as they are cheap and they also contribute to house warming. The reported percentages (59 to 76%) of the households' respondents using three stone stoves are lower than what was reported in a study by Mugo (1999a) on rural households, but the variation could be due to the fact that this current study considered both rural and urban centres.

The proportion of households aware of the improved cooking technology was over 70% in all the districts. However, despite this high awareness, majority of the respondents were using the three stone stoves and only a small proportion of less than 28% of the household respondent had adopted the improved charcoal and firewood stoves which was much lower than 47% reported in a study on Kenya's energy demand and supply for households by Ministry of Energy (MoE, 2002). The difference in adoption among the districts could be due to variation in household income among the three districts. In Maragwa district, the improved stove awareness was higher than the other districts as there was an NGO which had been promoting improved stoves. This could be the reason why the district also had relatively higher rate of adoption of improved stoves compared to the other two districts. It is known that three stone stoves consume a lot of firewood as they are not efficient and they emit several serious pollutants which cause health problems (Kirk, 1993; Pandey, 1997). The improved stoves need to be promoted as they use less woodfuel and have higher combustion efficiency which gives greater amount of heat and less smoke. The use of improved stoves will lead to reduced environmental related diseases as outlined in Kenya vision 2030. In addition, the use of improved stoves will also reduce the rate of deforestation and environmental degradation (Mugo, 2001). The use of less fuelwood by improved stoves also reduces the time used in fuelwood collection, which can be re-directed into other income generating activities. There is need for further sensitization on the importance of using improved stoves as most of those who were aware about them were not using them. The main reason for the low adoption of these stoves was their higher cost which can be reduced through facilitation of their production at the local level.

Some of the improved stoves that have been adopted in Kenya and the neighboring countries of East and Central Africa include Kenya Ceramic Jiko (KCJ). This is a simply traditional charcoal metal stove fitted with ceramic liner which increases its efficiency several times as compared to the traditional all metal alternative. The other stove which has been successfully produced by local women groups in western Kenya is the Upesi stove which also has simple molding of clay liner.

CONCLUSIONS AND RECOMMENDATIONS

The woodfuel supply trend indicated that it was decreasing and therefore there is need to integrate its production with local farming systems as agricultural sector has a key role to play in supplementing woodfuel. This can be supported through development of on-farm management regimes for use at the farm level to ensure sustainability of woodfuel production. The Kenya Forest Service should also develop plantations for woodfuel production as a national priority as it does with timber production. The firewood plantations should be established with appropriate fast growing trees which match specific environmental and ecological conditions for maximum productivity.

The conservation of wood energy should be given a priority through promotion of improved stoves with higher efficiency as it was found that over 70% of the households in the study area use three stone stoves which were inefficient. The improved stoves to be promoted for adoption should consider users needs which include cooking comfort, convenience, health and safety. The improved stoves should also be of affordable prices as it was noted in this study that low adoption of improved woodfuel cooking devices was contributed by their high cost. This can be achieved through building capacity of making improved stoves at local level.

ACKNOWLEDGEMENTS

The authors would like to thank the District Forest Officers of Kiambu, Thika and Maragwa districts for assisting in data collection and KEFRI management for supporting the study.

REFERENCES

ESDA (2005). National Charcoal Survey: Exploring the potential for sustainable charcoal industry in Kenya. Energy for Sustainable Development Africa, Nairobi, Kenya. p 79.

Kirk RS (1993). Wood energy health effects in developing countries. Paper presented at seminar on "policy instruments for wood energy development" held in May 1993, Bankok, Thailand. p 4.

Lee-Ann CH, Martin AB (1997). Surveying Natural Population. Columbia University Press, New York, pp.113-152.

Ministry of Energy (2002). Study on Kenya's Energy Demand, Supply and Policy Strategy for Households, Small scale Industries and Service Establishments. Kamfor Consultants, Nairobi, Kenya, pp. 133

Ministry of Planning and National Development (2007). Kenya integrated household budget survey 2005/06 Revised Edition. KNBS, Nairobi, Kenya, pp. 96

Mugo F (1999). The effect of fuelwood demand and supply characteristics, land factors and gender on tree planting and fuelwood availability in highly populated rural areas of Kenya. A PhD dissertation, Cornell University, New York, USA, pp. 7.

Mugo FW (2001). The role of woodfuel conservation in sustainable supply of the resource: The case for Kenya. Paper presented at charcoal stakeholders workshop held at World Agroforestry Centre (ICRAF), 15th October, 2001, Nairobi, Kenya, pp. 4.

Pandey MR (1997). Women, Wood Energy and Health. Wood Energy News, Food and Agriculture Organization-Regional Wood Energy Development Programme in Asia (FAO-RWEDP), Bangkok, Thailand, 12(1):3-5

Republic of Kenya (1997a). Kiambu District Development Plan. Government Printers, Nairobi, Kenya, pp. 156.

Republic of Kenya (1997b). Thika District Development plan 2002-2008. Government Printers, Nairobi, Kenya, pp. 165.

Republic of Kenya (2001). 1999 Population and Housing Census: Counting our people for development. Government Printers, Nairobi, Kenya, pp. 157.

Republic of Kenya (2002a). National Development Plan 2002-2008: Effective Management for Sustainable Economic Growth and Poverty Reduction. Government Printers, Nairobi, Kenya, pp. 79.

Republic of kenya (2002b). Maragwa District Development Plan 2002-2008: Effective Management for Sustainable Economic Growth and Poverty Reduction. Government Printers, Nairobi, Kenya, pp. 359.

22

Community consultations on use and management of preferred NTFPs: A case study of the four ecological zones of Swaziland

Cliff S. Dlamini*, Coert J. Geldenhuys

Department of Forest and Wood Science, Faculty of Agrisciences, University of Stellenbosch, P/B X1, Matieland 7602, South Africa.

Traditional forest management systems were part of daily life in the natural forest and woodland resources in olden days Swaziland. The objective of the study was to embark on community consultations to collect information on their perception of preferred edible and medicinal non-timber forest products (NTFPs), their direct uses, the existing management strategies, the threats to forest biodiversity and the domestication and commercialization initiatives. Group discussions, individual perceptions and review of national forest policy were undertaken. The findings indicated that local communities lack knowledge of the existing policies and legislation that safeguard the sustainable use of NTFPs in the adjacent natural forests and woodlands, and further stated that there are no existing traditional local-level NTFP management systems. Uncontrolled trade in NTFPs, by non-resident collectors in Swaziland has been seen to be one of the predominant threats to forest biodiversity. This reaffirms the weak and ineffective national policies and legislation, and shows that the existing policies and legislation are not implementable. The positive side is that local communities have identified potential threats to forest biodiversity. All communities are willing to participate in the conservation and sustainable use of the adjacent natural forests and woodlands. Most local communities already have initiatives towards selection of top priority species for domestication and commercialization. The institutional, cultural, socio-economic, ecological/environmental and policy issues raised by local communities are crucial and essential elements for the formulation and development of guidelines for local-level sustainable management and development of NTFPs. Proper and innovative policies and legislation need to be put in place to cope with the current challenges.

Key words: Natural forests, natural woodlands, forest products, sustainability, national forest policies, forest management, community participation, forest biodiversity.

INTRODUCTION

In Swaziland, natural forests and woodlands are facing great loss of biodiversity due to uncontrolled, unsustainable forest products extraction. This shows that resource management and research programmes are not effective. This is mainly because ecological, economic and social planning and research are uncoordinated and done in isolation from each other (Geldenhuys, 2002, 2003, 2004; Dlamini, 2007, 2010a). Furthermore, it is clear that the current control mechanisms, policy and legislation are failing to prevent forest degradation, and one option to save this situation could be to embark on more comprehensive ecological research for a sound scientific understanding of natural forests and woodlands in order to be able to design successful sustainable forest management strategies (Geldenhuys, 2002; Dlamini, 2007).

New approaches to forest sustainability anywhere have developed strong partnerships between academia, government, the environmental community and the

*Corresponding author. E-mail: csdlamini@uniswa.sz.

industry and this has proved to be the most successful remedy (Abubakr et al., 1997; Crafter et al., 1997; GOS, 2002; Dlamini, 2007). Sustainable forestry is achievable when the needs of landowners and managers are balanced with the ecological capacity of the forest ecosystem. In view of the complexity and dynamic nature of social and natural systems, sustainable forestry must be flexible and adaptable necessitating adaptive management (Abubakr et al., 1997; Crafter et al., 1997; Geldenhuys, 2002, 2003; Dlamini, 2007). University researchers, industry and natural resource agencies will have time to develop long-term partnerships for evaluating the criteria and indicators of sustainable forest management (Abubakr et al., 1997; Crafter et al., 1997; 2002a; Dlamini, 2007)

All stakeholders in the forestry sector have a role to play in long-term sustainability of the resources, and these include local communities (adjacent to the natural forests and woodlands), forest industries (in commercial forestry), municipalities (involved with urban forestry/ greening), government (policy makers) and other interested parties (Abubakr et al., 1997; Crafter et al., 1997; Prasad, 1999; GOS, 2002; Nibbering and Samyn, 2002; Sreedharan, 2002; Dlamini and Geldenhuys, 2011a).

Community participation in forest management is the trend in Africa and beyond (Crafter et al., 1997; Wily, 2002; Geldenhuys, 2003). A recent overview showed that participatory forest management (PFM) in Africa is sufficiently widespread and effective to be recognized as a significant route towards securing and sustaining forests (Wily, 2002). While each state is striving for more participatory approaches to especially natural forest management, broad commonalities among the processes and paradigms are notable. The issue of PFM is in short democratization in the forestry sector, which is an important socio-political transformation towards more inclusive norms in the governance of society and its natural resources (Grundy and Breton, 1998; Prasad, 1999; Nibbering and Samyn, 2002; GOS, 2002; Sreedharan, 2002; Dlamini, 2007).

Non-timber forest products are important to joint participatory forest management (JPFM) in the following ways: they are integral to the lifestyle of the forest dependent communities, as they fulfill basic requirements; non-timber forest products (NTFPs) have an advantage over timber in terms of the time needed to achieve significant volumes of commercially valuable production (they become available even in the early stages of a rehabilitation of degraded forest areas). In countries like India, about 50% of forest revenue and about 70% of forest export revenue comes from NTFPs (Campbell et al., 1997). NTFP management and development therefore have clear ecological, social and economic benefits. It is necessary to understand how rural communities think about participatory approaches to resource use and management.

The specific objective of the study was to embark on community consultations to gather information on the communities' perception of preferred edible and medicinal NTFPs, their direct uses, the existing management strategies, the threats to forest biodiversity and the domestication and commercialization initiatives. The study focused on Swaziland.

The specific research questions were:

1. Are the local people conversant with existing policies and legislation that affect the NTFP sector? If not, what are the factors leading to this and what strategies can be developed to ensure optimization of these policies for equitable and sustainable development of NTFPs at the local level?
2. What are the potential threats to forest biodiversity that can lead to the loss of these NTFPs?
3. Are there existing local level traditional management plans for the natural forests and woodlands and national policies and legislation that safeguard the sustainable extraction of preferred edible and medicinal NTFPs?
4. Are the local people really aware of the opportunity cost of the surrounding natural forests and woodlands?
5. Are the local people willing to put some effort into the conservation of natural forest and woodlands for the sustainable supply of timber and NTFPs?
6. What domestication and commercialization initiatives are in place for preferred edible and medicinal NTFPs?

The hypothesis to be tested was:

There are no existing traditional forest management plans that can complement the national policies

METHODS

Selection of the study sites

The study sites were selected according to two key criteria:

1. They had to cover a broad spectrum of sites to allow calculation of variance. This was ensured by covering the four ecological zones (within the six physiographic zones) of the country (Godoy et al., 1993; FAO, 2001; Dlamini, 2007; Dlamini and Geldenhuys, 2011b) (Figure 1). This makes it possible to use data for comparison and generalization and the full range of Swaziland's major forest types with their associated variability in climatic and socio-economic conditions (Hassan et al., 2002);
2. The selected villages had to be part of communities that live adjacent to natural forests and woodlands and harvest, extract or collect and utilize NTFPs from the neighbouring natural forests and woodlands (Appasamy, 1993; Godoy and Bawa, 1993; Hall and Bawa, 1993; Hedge et al., 1996; Shackleton, 1996; Campbell et al., 1997; Crafter et al., 1997; Qureshi and Kumar, 1998; Shackleton and Shackleton, 2000; Shackleton et al., 2002; Dovie, 2003b; Dlamini, 2007; Dlamini and Geldenhuys, 2011b). Only rural communities/villages were included in the study due to the low dependence of urban populations on direct harvesting of NTFPs from natural forests and woodlands (Hassan et al., 2002).

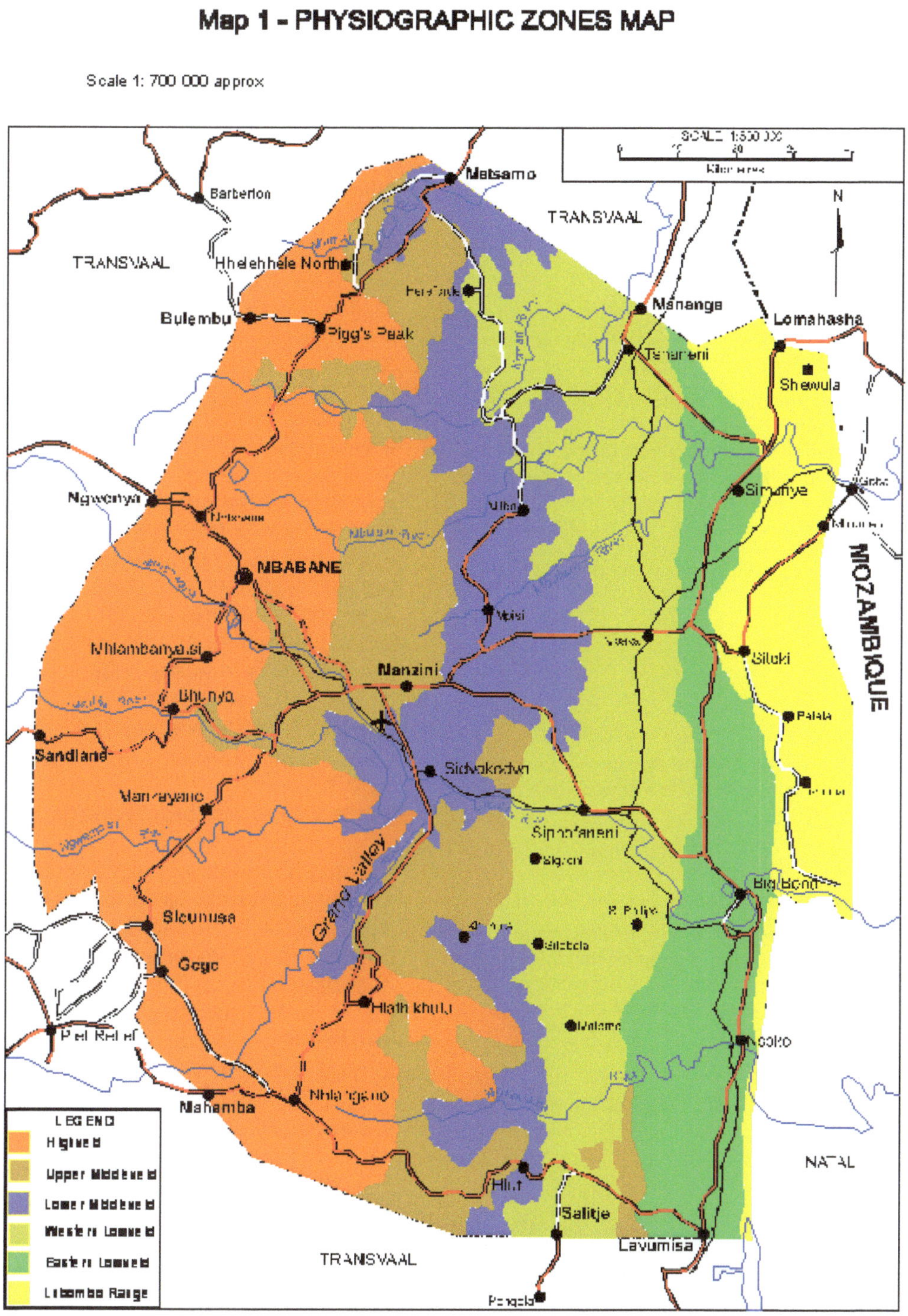

Figure 1. A map of the six physiographic zones of Swaziland [showing Grand Valley (Middleveld), Hhelehhele North (Highveld), Shewula (Lubombo) and Siphofaneni (Lowveld)].

Community profiles

The communities studied comprised mainly illiterate representatives, who came from relatively poor rural households. Most of the households practice livestock and crop farming at the subsistence level and harvest NTFPs. Most of the communityrepresentatives were old and unemployed people. From each area 20 men and 20 women were included in the survey.

Sampling procedure

Four communities were studied and only two villages were selected under each.

The sampling procedure was guided by the following:

1. The District Forestry Officer in each study area, in consultation with the District Agricultural Extension Officers, were requested to

select only two villages appropriate for this study in accordance with selection criterion 2;
2. A total of 40 community representatives in each study area were shortlisted from 40 households to participate in the community consultation meetings by the District Forestry Officer in close cooperation with the District Agricultural Extension Officers. The criteria used in choosing community representatives, amongst other factors, included their track record in attending other community development meetings and active participation in rural development programmes, gender (50% men and 50% women), age and harvesting of NTFPs. In addition, at least two community leaders were invited as observers in Box 1

Data collection

Community consultations were held in the following centres: Shewula Mountain Camp Conference Centre (Shewula-Lubombo); Siphofaneni RDA (Siphofaneni-Lowveld); Church of the Nazarene (KaKholwane-Middleveld); and Hhelehhele North Umphakatsi (Hhelehhele North-Highveld). A total of 4 meetings per study site were conducted. There were three programmes (group work, individual responses and discussion of National Forest Policy) undertaken in each study area and a compilation of questions raised is presented in Box 2.

Data analysis

Group discussions

The issues and comments raised by the group of community representatives for each question asked were compiled for each study area.

Individual perceptions

The data collected on all the responses was compiled, sorted and coded, stored in Excel and analyzed. The data was of binomial type and the incidence of people (denoted as percentage) was subjected to analysis of variance (4-factor factorial experiment in a one-way classification design), using SAS version 8.2 (SAS, 1999). The statistical model used, was:

$$Y_{ijkln} = \mu + \alpha_i + \beta_j + \gamma_k + \delta_l + \alpha\beta_{ij} + \alpha\gamma_{ik} + \alpha\delta_{il} + \beta\gamma_{kl} + \beta\delta_{jl} + \gamma\delta_{kl} + \varepsilon_{ijkln}$$

Where:

μ =population mean
α_i = product effect (site or species or community)
β_j= site effect
γ_k = species
δ_l = gender
$\alpha\beta_{ij}$ = product * Site effect
$\alpha\gamma_{ik}$ = product * species
$\alpha\delta_{il}$ = product *gender
$\beta\gamma kl$ = site *species
$\beta\delta_{jl}$ = site * gender
$\gamma\delta_{kl}$ =species * gender
ε_{ij} = error
The higher order interaction were used as part of error (ε_{ijkln})

The Shapiro-Wilk test was performed to test for non-normality (Shapiro and Wilk, 1965). In some cases where evidence of non-normality was found, it was due to high kurtosis and not skewness. According to Glass et al. (1972) the analysis is valid. Student's t-least significant difference was calculated at the 5% confidence level to compare main factor means (for example, products, sites, gender and species).

Review of current national forest policy

The issues raised by communities on local-level sustainable forest management, after the discussion of the National Forest Policy (NFP) were summarized for the entire country, and the full range of original submissions from individual study sites was also kept for reference on specific issues of concern by site through the four ecological zones of Swaziland.

RESULTS

Group discussions

Four full-day meetings were conducted per study site, as planned. It was remarkable to note that the target participants and observers were in attendance with a 100% turn-up in all the four study sites. However, it was also observed that due to unforeseen circumstances and social commitments the composition of the participants was changing in every subsequent meeting, but the total number and gender still remained at 20 men and 20 women as earlier planned.

Threats to forest biodiversity

The communities in the various study areas reported a number of threats to forest biodiversity. Responses per study area are presented in Table 1. Most threats were reported only in one of the four study areas. The responses depict the number of community representatives that raised and supported specific issues (that is, threats). For example 40/40 community respondents pointed out that "Alien Invasive Species" as a threat to forest biodiversity at Hhelehhele North, Grand Valley and Shewula). The most common threats were alien invasive plants and drought.

Traditional species management systems

Traditional authorities (village leadership structures) are responsible for the local-level management of natural forests and woodlands, which entails the sustainable harvesting of forest foods, medicines and other timber and NTFPs. Over the years there was a steady decline in the supply of NTFPs from natural forests and this is attributed to the issues highlighted in Box 3.

National Policies and legislation

The communities are not versed with any functional and

Box 1. Brief description of the four ecological zones of Swaziland.

Ecological zone and characteristics
Highveld:
The Swaziland Highveld (altitude: 900 to 1400m) is the upper part of an overall escarpment, comprising complex steep slopes between low and high levels, dissected plateaux, plateau remnants, and associated hills, valleys and basins. Mean annual rainfall is 850 to 1400 mm. Characterized by short grassland with evergreen forest patches
Middleveld:
The Upper Middleveld (altitude: 600 to 800m) consists of strongly eroded plateau remnants and hills at intermediate level of the overall escarpment. It also has structurally defined basins in relatively protected positions, which are only weakly eroded. The Lower Middleveld (altitude: 400 to 600m) is a piedmont zone of the escarpment, with generally strongly eroded foot slopes. The slopes are mostly moderate and the zone classifies at the first level as a plain. Mean annual rainfall is 650 to 1000 mm. Dominated by tall grassland with scattered trees and shrubs and broad-leaved savanna
Lowveld:
The Lowveld plain comprises sedimentary and volcanic Karroo beds as opposed to the igneous and metamorphic rocks of the Highveld and Middleveld. The Lowveld is subdivided into the higher Western Lowveld (altitude: 250 to 400m) on sandstone or claystone and the lower Eastern Lowveld (altitude: 200 to 400m) on basalt. Mean annual rainfall is 550 to 725 mm. There is a combination of Mixed savanna and Acacia savanna
Lubombo:
The Lubombo Range (altitude: 250 to 600m) is a cuesta with a steep escarpment bordering the Eastern Lowveld and a gradual dip slope of about 5% descending east. As a major landform, the Lubombo qualifies as a plateau. Mean annual rainfall is 700 to 825 mm. Usually has hillside bush and plateau savanna

Source: Dlamini (2007).

nominal national and international policies and legislation that govern the extraction of natural forest products in the adjacent natural forests and woodlands. About 47.5, 37, 100 and 100% of the responses were that there are no policies and legislation governing the use of NTFPs, in Hhelehhele North, Grand Valley, Siphofaneni and Shewula respectively.

This is regardless of the series of national policies and legislation that government has developed and the international conventions that government has signed and ratified. In the Grand Valley area, 22.5% of the responses mentioned that misunderstanding and strict enforcement of the Grass Fire Act of 1955 has lead to dense natural woodlands and displacement of grasses, which is a sign of a disturbed ecosystem. However, all the communities studied wished there were effective policies and laws to govern the sustainable use of natural forest and woodlands to prevent the gradual loss of forest bio-diversity. They expressed 100% support to save the dwindling natural woodlands for the remaining species of timber and NTFPs.

Opportunity cost of the forest

Some local people are aware of the economic value of the adjacent natural forests and woodlands. Hhelehhele North, Siphofaneni and Shewula had 72, 57.5 and 92.5% responses admitting that people are aware of the opportunity cost of the forest, while 35% responses in Grand Valley suggested that people needed to be educated about the concept.

Domestication and commercialization of indigenous non-timber forest products (NTFPs)

Only Shewula community has established a pilot nursery for the domestication and propagation of indigenous plants of socio-economic importance. The Hhelehhele North communities have established a community woodlot for valuable indigenous hardwoods that include *Breonadia salicina, Brachylaena ilicifolia* and other unidentified species. In addition, the communities are also in the process of establishing orchards for indigenous edible species. All these initiatives are under the Chief's management (a chief being the highest person in the village leadership).

The other communities in the Grand Valley/ KaKholwane and Siphofaneni study sites have priority species (*Syzygium cordatum* and *Berchemia zeyheri*) for domestication and mass propagation but do not have any

Box 2. Questions raised to participants.

Group discussions

The following questions were posed to the group for submission of group perceptions on the issues raised:

(a) What are the major land uses in the area?

(b) Are there national policies and laws governing the adjacent natural forests and woodlands?

(c) What are the current traditional species utilization and management programmes?

(d) What are the threats to forest biodiversity?

(e) Are the local people willing to participate in natural forest conservation?

(f) Are the local people aware of the opportunity cost of the adjacent natural forest and woodland resources?

(g) Is there a domestication and commercialization potential for NTFPs in the area?

(h) Which natural forests and woodlands can be considered or nominated for research and development?

Individual perceptions

The following set of questions was formulated and posed to the community representatives on an individual basis, in each study site:

(a) What are the most preferred edible NTFPs?

(b) What are the most preferred medicinal NTFPs?

(c) Which indigenous edible and medicinal NTFPs are threatened?

(d) Give a list of the top priority desired edible and medicinal NTFPs that are worthy of immediate domestication and commercialization (Refer to Appendix 11 for a specimen of the data collection form)

Review of the current National Forest Policy

Step 1: the SiSwati version of the NFP was read and discussed during a community consultation meeting in all the study sites.

Step 2: the policy issues and statements on Natural Forests and Woodlands were given special consideration and the communities were to develop guidelines for the management of their natural forests and woodlands in relation to those in the new NFP.

Step 3: summaries of their views were fully captured and recorded. The community consultations were conducted between July 2003 and October 2003. This was during the winter and spring seasons when farmers were not busy

Box 3. Existence of traditional species management in the various communities.

Hhelehhele North-Highveld: the traditional authorities are responsible but communities do not cooperate

Grand Valley/Kakholwane-Middleveld: there were stringent policies and laws against the felling/cutting of edible trees, felling of trees for fuel wood, woodcrafts, and medicinal purposes. These are now outdated. They are not documented and the youth challenge them and violate them

Siphofaneni-Lowveld: none at present, heard about the Forest Act of 1910 and traditional authorities used to enforce, but now it is outdated and there is nothing protecting forests

Shewula-Lubombo: traditional policies and laws are in place except that the youth violate them, but penalties are imposed on offences of misuse of forests

formal domestication programmes yet. It is encouraging that they are willing to embark on such programmes soon. All the communities are willing to embark on massive domestication and commercialization programmes. Hhelehhele North, Grand valley, Siphofaneni and Shewula had 62.5, 77.5, 47.5 and 67.5% positive responses for implementing NTFP domestication initiatives.

Resource use

The Siphofaneni site in the lowveld has the highest

Table 1. The reported threats to forest diversity from community consultations in the four ecological zones of Swaziland.

Threats	% responses (out of a total of 40 representatives per session) in study sites over the four ecological zones			
	Hhelehhele North	Grand Valley	Siphofaneni	Shewula
Alien Invasive Plants	100	100	0	100
Biopiracy	22.5	0	0	0
Desertification	0	0	47.5	0
Drought	100	0	0	100
Encroaching species	0	85	0	0
Farm structures	0	0	57.5	0
Fire arms	0	12.5	0	0
Forest fires	72.5	0	0	0
Fuel wood	0	0	67.5	12.5
Ignorance of laws	0	80	0	0
Overstocking	0	72.5	0	0
Population explosion	0	0	80	0
Sugarcane farming	0	0	92.5	0
Traditional healers	37.5	0	0	0
Warthogs	12.5	0	0	0
Wood crafts	0	0	0	30

Encroaching species in this regard refers to *Dichrostachys cinera.*

Table 2. Reported number of species in the various study sites - community consultations.

Product group	Grand Valley	Hhelehhele North	Shewula	Siphofaneni
Preferred edible plant species	38	16	16	20
Preferred edible animal species	4	0	5	7
Preferred medicinal plant species	20	20	13	9
Threatened edible animal species	0	0	0	4
Threatened medicinal plant species	7	5	5	0
Threatened edible plant species	2	1	5	9
Top priority species	5	7	10	15

number of harvested edible and medicinal products (mammals, fruits and berries, vegetables, mushrooms, bee honey, insects, caterpillars, wax and medicines), followed by the Shewula site (fruits and berries, vegetables, mushrooms, bee honey, insects, caterpillars and medicines) and the Grand Valley/KaKholwane site (fruits and berries, vegetables, mushrooms, bee honey, wax, caterpillars and medicines). The Hhelehhele North site is last with just fruits and berries, vegetables and medicines (for more details refer to Table 1).

Individual perceptions

The communities from the various study sites were further requested to develop lists of species under seven product groups. The product groups were as follows: preferred edible plants species, preferred edible animal species, preferred medicinal plants, threatened edible animal species, threatened medicinal plant species (plants facing extinction), threatened edible plant species and top priority species (most desirable or preferred) for immediate domestication and commercialization (Table 2). The Grand Valley site listed the highest number of edible plant species and the Siphofaneni the highest number of edible animals, while the Hhelehhele North and Grand Valley sites reported the highest number of medicinal plants. The various study sites have between 5 and 15 top priority species targeted for domestication and commercialization.

Analysis of responses on preferred non-timber forest products (NTFPs)

The results were analyzed for products (7 groups), site (4), species (129) and gender (2). The product groups were edible animals, edible plants, medicinal plants,

Table 3. The variation in the average number of responses and number of NTFP species reported during community consultations between product categories.

Level of product	Number of NTFPs species reported (including repetitions)	Mean number of responses out of a possible 40	Standard deviation
Medicinal plants	124	28.1 a	1.3
Top priority species	74	27.4 a	1.2
Threatened edible animals	8	26.5 a	1.8
Threatened edible plants	34	25.6 ab	0.7
Edible plants	180	25.1 ab	1.9
Threatened medicinal plants	34	21.6 b	1.4
Edible animals	32	13.4 c	1.3

Means with same letter indicates no differences.

Table 4. The variation in the number of responses and number of NTFP species reported during community consultations between the various study sites.

Level of site	Number of NTFP species reported (including repetitions)	Mean number of responses out of a possible 40	Standard deviation
Siphofaneni	128	27.7 a	1.1
Grand Valley	152	26.2 a	1.0
Shewula	108	23.4 b	1.2
Hhelehhele North	98	22.6 b	1.4

Means with same letter indicates no differences.

Table 5. Analysis of variance for community consultations showing the statistical differences (depicted by the p-values) in responses between products, sites, gender, species and the various interactions across study sites.

Source of variation	Degrees of freedom	Mean square	P-values
Product	6	1058.6	<0.0001**
Site	3	1004.5	<0.0001**
Product*site	13	936.5	<0.0001**
Species	127	436.8	<0.0001**
Site*species	61	147.2	<0.0008**
Product*species	32	249.9	<0.0001**
Gender	1	227.4	0.813ns
Site*gender	3	1108.3	<0.0001**
Product*gender	6	1263.3	<0.0001**
Species*gender	127	190.5	<0.0001**
Error	106	73.4	
Corrected Total	485		

P<0.05, statistically significant; P>0.05, not statistically significant at 95% confidence Interval; **, highly significant; *, Significant; ns, not-significant.

threatened edible animals, threatened edible plants, threatened medicinal plants and top priority species. A breakdown of the number of species, mean number and standard deviation of responses for product groups is presented in Table 3 and for sites presented in Table 4. The detailed analysis of variance (ANOVA) across sites is shown in Table 5, and for individual sites in Table 6.

Graphical representation of the percentage responses of community representatives for the various product groups within each of the study areas is shown in Figures 2, 3, 4 and 5. Please note the dependent variable here in this analysisanalysis is % people and not % species and that top priority species is a product class as per the people's perceptions and not statistically validated.

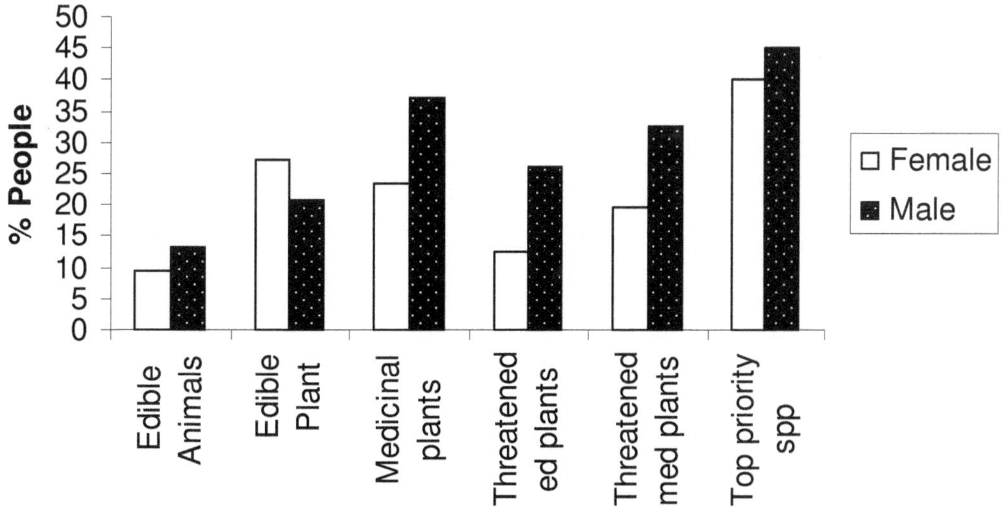

Figure 2. Graphical representation of the percentage esponses of community representatives for the various product groups at Grand Valley area.

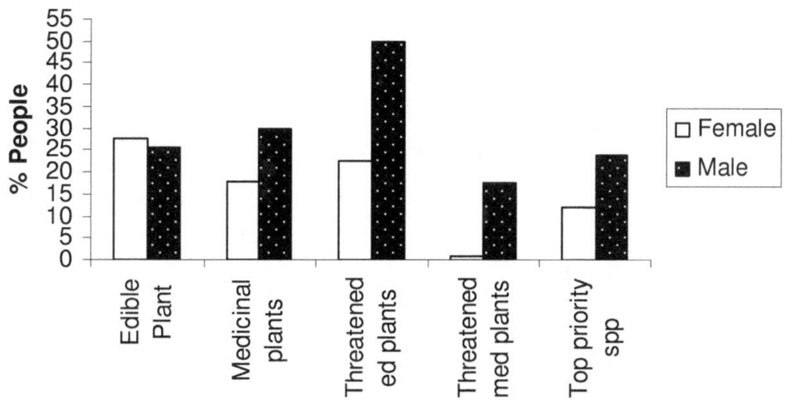

Figure 3. Graphical representation of the percentage responses of community representatives for the various product groups at Hhelehhele area.

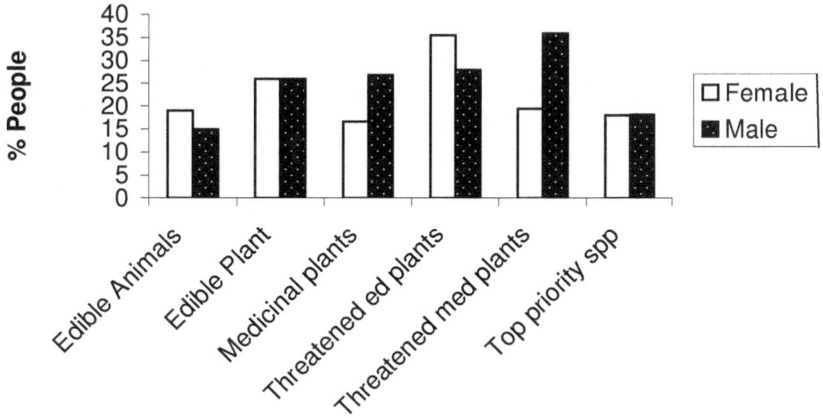

Figure 4. Graphical representation of the percentage responses of community representatives for the various product groups at Shewula Nature Reserve

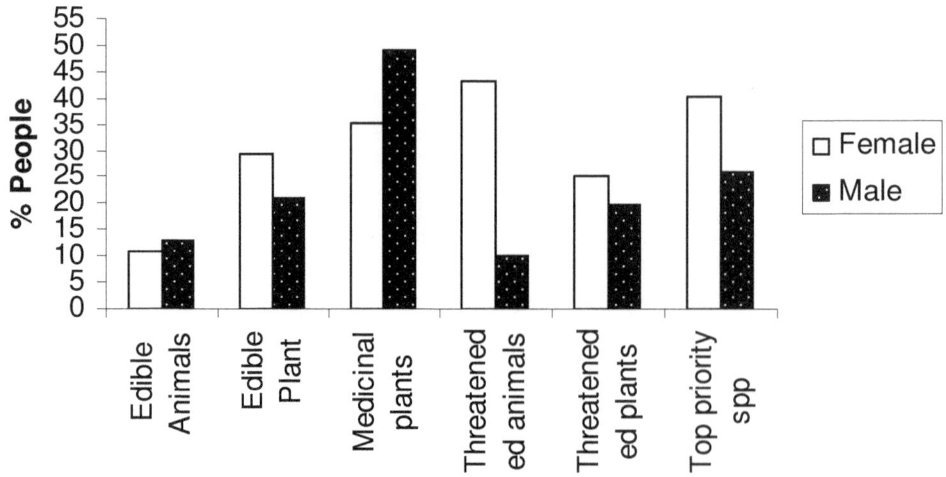

Figure 5. Graphical representation of the percentage responses of community representatives for the various product groups at Siphofaneni area.

The ANOVA of overall community consultations across sites shows highly significant differences between products (p<0.0001), between sites (p<0.0001), in the interaction between products and sites (p<0.0001), between species reported (p<0.001), in interaction between sites and species reported (p<0.008), in the interaction between products and species reported (p<0.0001), in the interaction between site and gender (p<0.0001), in the interaction between species and gender (p<0.0001), and in the interaction between products and gender (p<0.0001). There were no significant differences between gender (p<0.813)(Table 5).

The Grand Valley site-Middleveld

This site reported 66 of the 129 total species. No threatened edible animals were reported. Gender response was balanced between men and women representatives. In only edible plants were responses by women higher than by men.

The Hhelehhele North site-Highveld

This site reported 41 of the 129 total species. No threatened edible animals and edible plants were reported. Gender response was balanced in some cases.

The Shewula site-Lubombo Plateau

This site reported 38 of the 129 total species. No threatened edible animals were reported. Gender response was balanced in certain cases.

The Siphofaneni site-Lowveld

This site reported 51 of the 129 species. No threatened medicinal plants were reported. Gender response was balanced in some cases.

In overall, commonly reported preferred edible NTFPs include the following: *Sclerocarya birrea, B. zeyheri, Strychnos madagascariensis, Strychnos spinosa, Englerophytum natalense, Aloe maculata, Syzygium cordatum, Vangueria infausta, Psidium guajava, Psalliota campestris*, edible caterpillars and bee honey. The commonly preferred medicinal NTFPs included the following: *S. birrea, P. guajava, B. zeyheri, A. maculata, Rotheca hirsuta, Pittosporum viridiflorum, Drimia delagoensis, Peltosporum africanum, Aloe* spp., *Ekerbegia capensis,* and *Hypoxis* spp. Several species are used for both food and medicines, and categorized as multi-purpose species. Some of the preferred species are alien invader species, such as *P. guajava.* This shows that some invasive alien species contribute positively to household health and food needs.

Review of current National Forest Policy

Concerns with regard to the current status of forest and woodland resources in Swaziland revolve around sustainable management issues. Participants in the various sites made submissions of twenty one (21) issues that need urgent consideration in order to under pin the national efforts and programmes to combat degradation of natural resources and the environment (Table 7).

DISCUSSION

The results of the study show that the major land uses in the four study sites are more inclined to traditional

Table 6. Analysis of variance for community consultations showing the statistical differences (depicted by the p-values) in responses between products, sites, gender, species, and the various interactions among individual study sites.

Grand Valley Site (Middleveld)			
Sources of variation	**Degrees of freedom**	**Mean square**	**P-values**
Product	5	1125.5	0.0001**
Species	64	360.9	0.0016**
Product*Species	6	184.7	0.0163*
Gender	1	243.7	0.0231*
Product*Gender	5	632.7	0.0007**
Species*Gender	64	114.6	0.0352*
Error	6	26.5	
Corrected Total	151		
Hhelehhele North Site (Highveld)			
Product	4	753.3	0.0217*
Species	40	426.8	0.0473*
Product*Species	4	22.0	0.8616ns
Gender	1	1653.1	0.0088**
Product*Gender	4	357.6	0.0754ns
Species*Gender	40	144.8	0.2638ns
Error	4	72.3	
Corrected Total	97		
Shewula Site (Lubombo Plateau)			
Product	5	427.1	0.0058**
Species	36	370.3	0.0022**
Product*Species	12	288.8	0.0121*
Gender	1	244.5	0.0920ns
Product*Gender	5	258.4	0.0336*
Species*Gender	36	191.5	0.0379*
Error	12	72.9	
Corrected Total	107		
Siphofaneni Site (Lowveld)			
Product	5	1758.1	<0.0001**
Species	48	228.3	0.0037**
Product*Species	10	333.5	0.0016**
Gender	1	1411.1	0.0002**
Product*Gender	5	812.7	<0.0001**
Species*Gender	48	163.2	0.0137*
Error	10	43.0	
Corrected Total	127		

P<0.05, Statistically significant; P>0.05, not statistically significant at 95% confidence Interval; **, highly significant; *, significant; ns, not-significant.

subsistence and pastoral farming, where there is communal grazing of livestock from various homesteads. The literacy levels of the participants were low, as most people, especially the old men and women, never went to school. All the community members who participated in the community consultation meetings were unemployed, and only earned a living through subsistence farming and harvesting of forest products from the adjacent forests and woodlands. These products include indigenous medicines, wild food resources, woodcarvings and construction material. This is the part of the country where about 75% of the entire population reside (GOS, 1999;

Table 7: Issues raised in Community Consultations

Major Issues	Associated Issues
Institutional issues	-Community-based natural resources management committees
Socio-cultural issues	-Modern agro-forestry systems -Empowerment of chiefs -Incentives for chiefs
Environmental and Ecological issues	-Youth environmental programmes -Invasive alien plants strategy -Forest fires -Species re-introduction and replanting programmes
Socio-economic issues	-Eco-tourism -Alternative forms of livelihoods
Policy issues	-National and international policies and legislation -Forest reserves -Harvesting monitoring -Education and awareness -Matching resource use with resource availability -Alternative sources of energy -Forest act -Development projects versus natural woodlands -Gender, biodiversity and local knowledge systems -Resettlement programmes -Sustainable natural resources management briefs

Dlamini, 2007). In similar studies, Dovie et al. (2001), Dovie (2003b), Shackleton (1996), Shackleton and Shackleton (2000, 2002, 2004, 2006) and Shackleton et al. (2000, 2002) reported that the majority of South Africans reside in the rural areas characterized by limited infrastructure and lack of employment opportunities. Many therefore, make extensive use of NTFPs as part of their daily livelihoods. Rapid population growth and increasing poverty of the population has caused tremendous pressure on natural forests and woodlands to meet its subsistence needs and generate income as stated by Nibbering and Samyn (2002), Dlamini (1997, 1998, 2007, 2010b).

The threats to forest biodiversity and recommendations mentioned are similar to those in the National Forest Policy, National Environmental Policy and the National Biodiversity Strategy and Action Plan. Threats to forest biodiversity in Swaziland include alien invasive species, encroachment, browsing by livestock, uncontrolled fires, frequent drought and unsustainable exploitation of natural forest and woodlands. Land conversion to agriculture, infrastructure and uncontrolled resettlement are other threats to forest biodiversity (GOS, 1999, 2001, 2002; Dlamini, 2007). There is a strong need for development programmes to address these issues. Such programmes and projects should include the control of invasive species, overexploitation of endangered species, the control of forest fires and the reduction of overstocking and overgrazing. These programmes need to be developed in conjunction with the National Biodiversity Strategy and Action Plan (NBSAP). Also important is the Environmental Impact Assessment (EIA) and Natural Resources Accounting (NRA) to monitor the authenticity of land conversion and alleviate loss of forest biodiversity. The NRA will give the communities/resource managers a good picture of the opportunity cost of the adjacent natural forest and woodlands, while the EIA will give a good picture of the biodiversity status of the natural forests and woodlands. Similar observations were made by Godoy et al. (1993, 2000) based on 24 studies on NTFPs analyzed at the Harvard Institute for International Development.

The status of domestication and commercialization in the study sites shows that the country is behind the rest of the Sothern Africa Development Community (SADC) member countries. Most of the neighbouring countries have embarked on massive domestication and commercialization programmes and projects. Swaziland might benefit from the CP Wild project (CP Wild, 2004; Dlamini, 2007), which is seeking to establish a commercialization and domestication initiative that focuses on South Africa, Namibia, Botswana, Zimbabwe, Malawi, Mozambique, Swaziland and Lesotho. Their main aim is to develop the use of natural forest and woodland resources for socio-economic benefit in the SADC region of Africa. The willingness of the communities to participate in the sustainable management and conservation of natural forests and woodlands is a positive step considering the contemporary participatory forest management approach towards Sustainable Forest Management (SFM), where forest goods and services are managed for present and future generations. This is seen in the 100% attendance and participation of community representatives reported earlier.

Forests and woodlands play a significant role in supplying fuelwood for energy and about 70% of rural households in Swaziland rely on fuelwood, and this percentage is expected to remain high for some time (GOS, 2002; Dlamini and Geldenhuys, 2009). Fuelwood consumption estimates are consistent but there are clear indications that local shortages prevail, especially in the Upper Middleveld and parts of the Lower Middleveld and Lowveld, particularly in dense settlements and arable areas (Lasschuit, 1994, 1995). Supplementary sources of fuelwood are the private and communal wattle forests, though not always on a sustainable basis. Issues of other sources of affordable energy need to be promoted and these are addressed by the draft National Energy Options Paper. Trade in fuelwood and charcoal from indigenous species has exacerbated the over-exploitation of these natural resources leading to forest degradation. According to Dlamini (2007), the new National Forest Policy outlines other sources of fuelwood and ways to increase supply and energy efficiency that include the following:

1. Increased and improved distribution systems of wood waste from commercial plantations;
2. More efficient use and management of wattle forests;
3. Enrichment planting of selected fast growing natural tree species in the degraded natural forests and woodlands;
4. Buffer-zone planting of suitable exotic species to protect natural forests and woodlands;
5. Establishment of multipurpose woodlots;
6. Increased individual tree growing and Agro-forestry;
7. Agricultural residues, industrial wood waste and baggase from the sugar industry;
8. Charcoal in specific locations, such as from Highveld wattle forests and bush encroachment on farms;
9. Increased energy efficiency through improved wood stoves; and
10. Eradication programmes of harmful alien invasive plant species.

The aforementioned options have different potentials and need further evaluation.

The communities from the various study sites reported several preferred species of edible and medicinal NTFPs, the threatened species under each product group, and the top priority with regard to immediate domestication and commercialization. Studies in the Republic of South Africa done by Shackleton and Shackleton (1997, 2004, 2006) and Shackleton et al. (2002), and in Swaziland by Dlamini (2011a) and Dlamini and Geldenhuys (2009, 2011a, b) also highlighted several NTFP resources that are utilized by rural communities, that is, fuelwood, construction wood, edible fruits, edible herbs, edible insects, medicinal plants, bush meat, bee honey, reeds for weaving and grass hand brushes. Most of the NTFP species reported in South Africa are also extensively utilized in Swaziland as well, according to the list of species reported in the community consultations.

The Grand Valley site reported a significantly higher number of edible species. This area falls within the mixed woodland zone, which is the richest in botanical diversity in the entire country (Hess et al., 1990; DANCED, 1999; Dlamini, 2007). The findings of this study are in full agreement with the previous national forest inventories. All the sites reported more than five top priority species that are candidates for domestication and comercialization.

This means local, national, regional and international initiatives and programmes towards domestication and commercialization of indigenous NTFPs will be greatly appreciated and will receive full cooperation of the communities. There were differences in the information reported by men and women, and that confirms that both genders may not have the same practical experience with the surrounding natural forests and woodlands. However, during the meetings, the women showed more interest and knowledge about edible species than medicinal species while men were more inclined towards highly economic medicinal species. Only the Siphofaneni communities mentioned threatened species of edible animals, and these were: edible caterpillars, locusts and termites. The implications are that in the other study sites, the species of edible products are either abundant or most likely extinct.

The final programme under the consultations was the review of the SiSwati translated version (SiSwati being the local language) of the new National Forest Policy. Simultaneously, the community participants were requested to share their indigenous knowledge and practical experience with the adjacent natural forests and woodlands. A sum total of 21 issues and recommendations were raised pertaining to local-level sustainable forest management in the four ecological zones of rural Swaziland. Some of the issues raised are addressed by the new Forest Policy. However, in all the study sites, the community participants lacked knowledge of any national policies guiding and governing the use of natural forests and woodlands. This will have to be addressed to align the needs of the communities with the requirements of the Forest Policy.

There was very little variation in the issues between sites and as such most of the issues were similar, and these may be made into national issues surrounding the sustainable use and management of natural forests and woodlands for NTFPs. In addition, the main issues raised in the community consultations are in line with issues addressed by the activities and objectives of the following national policies:

1. National Forest Policy
2. National Tourism Policy
3. National Biodiversity Strategy and Action Plan
4. National Land Policy
5. National Resettlement Policy
6. National Energy Policy
7. National Action Plan for the Convention to Combat Desertification

Collaboration between the institutions responsible for the implementation of the above national programmes is imperative for local-level sustainable forest management to be successful. The results of the study reflected a total lack of functional traditional species management and utilization systems in all the study sites. This change also affected the political institutions. Chiefs were relegated to powerless figures, as people were more enlightened about their rights. However, the new National Forest Policy give hope to resuscitate the Chief's powers in monitoring and regulating harvesting of natural forests and woodlands. Consequently, the hypothesis "There are no existing traditional forest management plans that can complement the National Policies" is accepted.

CONCLUSIONS AND RECOMMENDATIONS

Based on the findings of the study, it can be concluded that there is heavy reliance of local communities on natural forests and woodlands through harvesting of NTFPs. The absence of local level NTFP management systems and plans and the absence of national and international policies and legislation governing the sustainable use of NTFPs have lead to an increase in the number of NTFP species that are currently threatened and endangered due to over-exploitation and unsustainable use. The commercialization of NTFPs has lead to the problem of non-resident extractors (traditional healers and wood cavers, as well as other groups) that has been highlighted to be a threat to forest biodiversity in most communities. The un-prescribed and uncontrolled burning systems and heavy infestation by alien invasive species have lead to a rise in veld fires (that is, wild fires) resulting in the decline in forest quality and vitality (manifested by disappearance of important NTFP species) in adjacent natural forests and woodlands.
The following actions are recommended:

1. Local and national governments, through their tenure reform systems, should formulate laws (legally enforceable rights) that will give resource users and communities the power to exclude and sanction other users, so that resources can be used sustainably;
2. Development of new and innovative up to date feasible forest management strategies is crucial, if the current forest degradation and land conversion is to be checked. Securing community rights to trees and other forest products is a recommended route to the greater participation of communities in natural resources management;
3. Local and national governments should develop strong but clear policies and legislation safeguarding the benefits flow. Rights to forests and forest resources should be linked to management responsibilities. Negotiated agreements should be clear, internalised with local resource use ownership;
4. Local and national governments and development partners should develop and implement policies and legislation to provide secure access and benefits to the people whose livelihoods are dependent on or supplemented by NTFPs; and
5. Local and national governments and concerned development partners should ensure that stakeholders, particularly collectors, growers and traders, are provided with incentives to sustainably manage NTFPs.

In addition to the aforementioned recommendations, there is a need to adopt and implement the following steps of contemporary approaches to sustainable use and conservation of NTFPs (Crafter et al., 1997; Geldenhuys, 2003):

a. Building NTFP use into management plans;
b. Defining the role of governments and NGO's in community participation;
c. Addressing policy, legal and institutional issues (policy and legislation, property rights and land tenure systems);
d. Promoting education and training at all levels of society.

ACKNOWLEDGEMENTS

The authors would like to thank the following individuals and organizations: Dr Isla Grundy, Professor G van Wyk, the University of Stellenbosch, the University of Swaziland, the Fort Cox College of Agriculture and Forestry, Tibiyo TakaNgwane, all Communities involved in the research, Government Foresters in the Ministry of Agriculture and Cooperatives in Swaziland, subject matter specialists interviewed in the various Government Departments and Sections in Swaziland.

REFERENCES

Abubakr S, Kilgroe MA, Chapman K (1997). New Approaches to Forest Sustainability. Environmental Conference and Exhibit/477. Minneapolis Convention Centre, Minneapolis, Atlanta, USA. MN. 5-7, May.

Appasamy PP (1993). Role of Non-Timber forest products in a subsistence economy. The case of a joint forestry project in India. Econ. Bot., 47 (3): 258-267.
Campbell BM, Lucket M, Swenes I (1997).Local-level valuation of Savanna Resources: A case strudy from Zimbabwe. Econ. Bot., 51(1): 59-97.
Crafter SA, Awimbo J, Broekhoven AJ (1997). Non-timber forest products: Value, use and management issues in Africa, including examples from Latin America. Proceedings of a workshop on Non-timber Forest Products. Naro Moru, Kenya, 8-13 May, 167p.
DANCED (Danish Corporation for International Development) (1999). Forest Resource Assessment. Ministry if Agriculture and Co-operatives, Mbabane, Swaziland, 42p.
Dlamini CS (1997). Germination pre-treatment in the seeds of Sclerocarya birrea (A. Rich) Hochst.sub-species caffra (Sond). Kokwaro (ANACARDIACEAE) - The Marula. Unpublished BSc. Hons. Mini Project, University of Stellenbosch, Stellenbosch. South Africa.
Dlamini CS (1998). Provenance and family variation in germination, early seedling growth, fruit composition and seed mass in Sclerocarya birrea (A. Rich) Hochst. sub-species caffra (Sond). Kokwaro (ANACARDIACEAE) - The Marula. M.Sc. Thesis. University of Stellenbosch, Stellenbosch, 99p.
Dlamini CS (2007). Towards the Improvement of Policy and Strategy Development for the Sustainable Management of Non-timber Forest Products: Swaziland: A Case Study. Ph.D. Dissertation.University of Stellenbosch.RSA, 249p.
Dlamini CS (2010a). Towards the improvement of policy and strategy development for the sustainable management of non-timber forest products: A synthesis of recent studies from Swaziland. In: Tevera, D.S. and Matondo, J.I (eds). Socio-economic Development and the Environment in Swaziland. Department of Geography, Environmental Science and Planning, University of Swaziland.
Dlamini CS (2010b). Provenance and family variation in germination and early seedling growth in Sclerocarya birrea sub-species caffra. J. Hort. For., 2 (9): 229-235. www.academicjournals.org/JHF/abstracts/abstracts2010/Nov/Dlamini.htm
Dlamini CS, Geldenhuys CJ (2009). The Socio-economic Status of the Non-timber Forest Product Sub-sector in Swaziland. Southern For., 71(4):311-318.
Dlamini CS, Geldenhuys CJ (2011a). (in press). The potential impacts of national, regional and international policies and legislation on the sustainable forest management for NTFPs: A Swaziland Case Study. J. Agric. Sci. Technol., pp. 579-585
Dlamini CS, Geldenhuys CJ (2011b). An assessment of non-timber forest products in the four ecological zones of Swaziland as a basis for policy and strategy development. J. Agric. Sci. Technol. (In press).
Dovie BKD (2003b). Rural Economy and livelihoods from the non-timber forest products trade. Compromising sustainability in southern Africa? Int. J. Sustain. Dev. World Ecol., (10): 247-262.
Dovie BKD, Shackleton CM, Witkowski TF (2001). Valuing non-timber forest products-indicator for interplay between poverty, livelihoods and the environment. Presented at an open meeting of the Global Environment Change Research Community, Rio de Janeiro, pp. 6-8.
FAO (2001). Non-Wood Forest Products in Africa: A Regional and National Overview. FAO Forestry Working Paper FOPW/01/1. FAO, Rome, 303p.
Geldenhuys CJ (2002). Concepts and approach towards development of sustainable resource use of non-timber forest products: the example of bark harvesting for traditional medicine. In: Seydack, A.H.W., Vorster, T., Vermeulen, W.J. and Van der merve, I.J. (eds). Multiple use management of natural forests and savanna woodlands: Policy refinements and scientific progress. Proceedings of Natural Forests and Savanna Woodlands Symposium III 6-9 May 2002, Berg-en-Dal, Kruger National Park, pp. 91-104.
Geldenhuys CJ (2003). Bark Harvesting for Traditional Medicine: from illegal resource degradation to participatory management. Scand. J. For. Res., 19(4): 103-115.
Geldenhuys CJ (2004). Meeting the demand of Ocotea bullata bark. Implications for the conservation of high value and medicinal tree species. In: Lawes, M.J., Eeley, H.A.C., Shackleton, C.M. and Geach, B.G.S (Eds). Indigenous Forests and Woodlands in South Africa: Policy, People and Practice. University of KwaZulu-Natal Press. Scottsville, South Africa, pp. 517-550.
Glass GV, Peckham PD, Sanders JR (1972). Consequences of Failure to meet assumptions underlying the fixed effects analyses of variance and covariance. Rev. Educ. Res., 42(3): 237-288.
Godoy RA, Bawa KS (1993). The economic value and sustainable harvest of plants and animals from the tropical forest. Assumptions Hypotheses and Methods. Econ. Bot., 47(3): 215-219.
Godoy RA, Lubowski G, Anil M (1993). A method for the economic valuation of non-timber tropical forest products. Workshop: Economic valuation and sustainable management of non-timber tropical forest products. Econ. Bot., 47(3): 220-223.
Godoy RA, Wilkie D, Overman A, Cubas G, Demmer J, McSweety K, Brokaw N (2000). Valuation of consumption and sale of forest goods from a Central American rain forest. Nature, 406: 62-63.
GOS (1999). National Environmental Policy. Ministry of Tourism, Environment and Communications, Government of Swaziland, Mbabane.
GOS (2001). National Biodiversity Strategy and Action Plan. Swaziland Environment Authority, Ministry of Tourism, Environment and Tourism, Government of Swaziland, Mbabane.
GOS (2002). National Forest Policy. Ministry of Agriculture and Cooperatives, Government of Swaziland, Mbabane.
Grundy, I,. and Breton, Gus-Le. (1998). The SAFIRE MITI Programme - A New Approach to Natural Resource Management in Communal Areas of Zimbabwe. ODI Rural Development Forestry Network, Network Paper, p. 22e.
Hall P, Bawa, K.S. (1993). Methods to assess the impacts of extraction of non-timber tropical forest products. Econ. Bot., 47(3): 234-247.
Hassan RM, Mbuli P, Dlamini CS (2002). Natural resource accounts for the state and economic contribution of forests and woodlands resources in Swaziland. CEEPA Discussion Paper No.4, CEEPA, University of Pretoria, Pretoria. 66p.
Hedge R, Suryaprakash L, Achoth L, Bawa KS (1996). Extraction of Non-timber Forest Products in the Forests of Biligiri Rangan Hills, India. Contribution to Rural Income. Econ. Bot., 50(3): 243-251.
Hess P, Forster H, Gwaitta-Magumba D (1990). National Forest Inventory of Swaziland results and interpretation. Swazi German Forest Inventory and Planning Project. Report No.5. Mbabane. pp.5
Lasschuit P (1994). Rural household energy strategies in Swaziland. University of Amsterdam, Interfaculty Department of Environmental Science, Amsterdam, 64p.
Lasschuit PE (1995). Urban Household Energy Strategies in Swaziland. University of Amsterdam, IVAM Environmental Research, Amsterdam, 81p.
Nibbering JW, Samyn JM (2002). Integrated participatory forest management in a densely populated mountain region, NWFP, Pakistan. Proceedings of the Research Course "The Formulation of Integrated Management Plans (IMPs) for Mountain Forests" Bardonecchia, Italy-30 June-5 July 2002.
Prasad, R. (1999). Joint forest management in India and the impact of state control over non-wood forest products. Unasylva, 198(50): 58-62.
Qureshi MH, Kumar S (1998). Contributions of common lands to household economics in Hiryana, India. Environ. Conserv., 25(4): 342–353.
SAS (1999). SAS/STAT User's Guide, Version 8, 1st printing, Volume 2. SAS Institute Inc, SAS Campus Drive, Cary, North Carolina, 2: 27513.
Shackleton CM (1996). Potentail Stimulation of Local Rural Economies by Harvesting Secondary Products: A case Study of the Central Transvaal Lowveld, South Africa. Ambio 25(1): 33-38.
Shackleton CM, Shackleton SE (1997). The use and potential for commercialization of veld products in the Bushbuckridge Area. DANCED-Community Forestry Project in the Bushbuckridge Area, Department of Water Affairs and Forestry, 104p.
Shackleton CM, Shackleton SE (2000). Direct Use Values of Secondary Resources Harvested from Communal Savannas in the Bushbuckridge Lowveld, South Africa. J. Trop. For. Prod., 6(1):28-47
Shackleton CM, Shackleton SE (2002). Household wealth status and natural resource in the Kat River valley, Eastern Cape. Unpublished

report, Rhodes University, Grahamstown.
Shackleton CM, Shackleton SE (2004). The importance of NTFPs in rural livelihood security and as safety net: a review of evidence from South Africa. South Afr. J. Sci., (100): 658-664.
Shackleton CM, Shackleton SE (2006). Household wealth status and natural resource in the Kat River valley, South Africa. Ecol. Econ., 57(306): 317- 307
Shackleton CM, Shackleton SE, Ntshudu M, Ntzebeza J (2002). The Role and Value of Savanna Non-Timber Forest Products to Rural Households in the Kat River Valley, South Africa. J. Trop. For. Prod., 8(1): 45-65.
Shapiro SS, Wilk MB (1965) An Analysis of Variance Test for Normality (complete samples). Biometrika, (52): 591-611.
Sreedharan CK (2002). An Integrated Approach to Conservation of Forests with Development of Villages in Tamil Nadu. A paper presented at the IUFRO Science/Policy Interface Task Force regional meeting held in Chennai, India at the M. S. Swaminathan Research Foundation, 16-19 July.
Wild CP (2004). Commercial products from the wild [on line]. www.cpwild.co.za., March 25th.
Wily LA (2002). Participatory forest management in Africa. An overview of progress and issues. Draft report.

Assessment of surface water receiving sewage effluents for gardening irrigation purposes: A case study of Ona River in Ibadan Southwest Local Government area of Oyo State, Nigeria

Adetayo, Adewale Olusola

Institute of Agricultural Research and Training, Obafemi Awolowo University, P. M. B. 5029, Ibadan, Oyo State, Nigeria.
E-mail: wale.agromet@yahoo.com.

The analysis of chemical quality of Ona River in Ibadan Southwest Local Government (IBSWLG) area of Oyo State was undertaken in this study to assess its suitability for irrigation purposes by considering two major seasons in the ecological zone. In order to achieve the objectives, twenty sampling points were selected along the stream at about 20 m interval. The water samples collected were analyzed in the laboratory using methods of Ademoroti (1996). The result showed that despite the large content of sewage effluents characterized by some of the areas, the quality of the surface water remains within the safe limit for irrigation purposes; thus, Ona River water can be developed for all-year round crop production in the study area. However, the need to continuously monitor and assess surface water quality for irrigation is stressed.

Key words: Sewage, effluents, water quality, sampling points, quality assessment.

INTRODUCTION

The utilization of flood plain for irrigation in the urban areas drained by perennial rivers and stream is on the increase in Nigeria. This is due to the population and consequent increase in chemical for vegetable and some other animal crops. Furthermore, the erratic pattern of occurrence of rainfall as a result of global warming has made it absolutely necessary to augment water supply from rainfall with other available sources of water supply in order to meet the crop water requirement especially during the storage of supply.

Conceptually, water quality refers to the characteristics of a water supply that will influence its suitability for a specific use, that is, how the quality will meet the needs of the user. Quality is defined by certain physical, chemical, and biological characteristics. Specifically, we have different quality needs and one water supply is considered, if it produces better result or cause fever problems, then an alternative water supply is considered. The suitability of water for irrigation is determined not only by soil, climate, and crop but also by the skill and knowledge of water user (Beecroft, 1982). The soil problems most commonly encountered and used as a basis to evaluate water quality are those related to salinity (salt in water to an extent yield is affected), water infiltration rate (caused by relatively high sodium or low calcium), specific ion toxicity, for example, sodium chloride or boron and excessive nutrient. It is observed that urban drainage surface tend to flow all year round partly because they are fed by water from home and industries.

Surface water in cities may be advantageous for irrigation since it is largely sewage effluents that may contain valuable plant nutrients. Martins and Bello (1997) warned that wart-water that drains into surface water is likely to increase the quantities of solutes thereby voicing the concentration of certain ions that may ultimately lower the quality of water for irrigation.

Since variation is hydrologic parameter, particularly rainfall and evaporation affects the quality of stream runoff; it is necessary to examine the seasonal variation of chemical constituents of surface water receiving sewage effluent in order to ascertain their quality and suitability for irrigation purposes throughout the year. Therefore, attempts are made in this study to analyze the seasonal variation of surface water receiving sewage

Table 1. Results of some physicochemical parameters of Ona surface water receiving sewage effluents collected during the wet season.

pH	EC (dS m^{-1})	TDS (mg l^{-1})	Ca^{2+} (me l^{-1})	Mg^{2+} (me l^{-1})	Na^{2+} (me l^{-1})	CO_3^- (mel^{-1})	HCO_3^- (me l^{-1})	SO_4^{2-} (mel^{-1})	Cl (me l^{-1})	NO_3^- (mg l^{-1})	K^+ (mg l^{-1})
6.2	0.90	500	2.21	0.24	9.42	0.01	0.6	12.4	20.4	6.4	0.9
6.4	1.20	470	2.36	0.14	8.64	0.01	1.2	15.2	11.4	6.2	0.4
5.8	0.40	550	1.02	0.94	7.56	0.01	1.2	9.8	9.2	6.4	0.5
6.4	0.30	850	0.92	1.20	6.06	0.03	2.4	10.2	11.4	5.2	0.6
6.2	0.01	700	1.02	1.23	10.2	0.04	2.6	10.4	11.2	4.8	1.0
5.4	0.05	725	0.86	1.49	9.46	0.40	2.9	11.2	9.6	2.4	1.0
6.2	0.10	800	2.42	1.64	8.45	0.90	3.2	12.4	10.2	3.2	1.0
6.0	0.96	820	2.34	2.01	10.24	0.02	4.0	11.4	11.0	3.6	0.9
7.0	0.17	810	3.01	2.92	10.23	0.92	5.2	12.0	10.1	4.8	0.8
6.2	1.00	815	2.96	2.94	11.14	0.84	6.2	10.4	9.6	6.2	0.4
5.8	1.01	805	2.98	0.96	12.22	0.62	4.8	12.4	8.2	6.6	0.9
5.2	1.02	810	2.90	1.24	10.24	0.48	9.2	10.8	9.6	7.2	0.6
5.8	1.02	820	2.64	1.36	11.04	0.52	6.6	11.4	10.2	2.5	0.7
6.2	0.96	800	3.21	1.49	11.12	0.64	1.6	12.0	6.4	3.4	0.8
6.4	0.99	790	2.91	1.64	10.04	0.92	2.4	11.2	8.6	4.4	0.6
5.8	0.94	800	3.14	2.01	9.96	0.86	5.2	12.4	8.9	4.3	0.6
7.4	1.02	810	2.40	2.02	10.02	0.92	4.8	11.4	9.2	5.2	0.9
6.2	1.04	850	1.96	0.14	9.22	0.96	5.2	9.6	10.5	6.2	1.0
6.0	1.06	450	0.94	0.92	10.92	0.02	6.4	10.2	11.2	6.0	0.1
5.2	0.02	700	1.94	0.64	9.04	0.04	7.2	8.1	12.4	7.1	0.9
5.4	0.04	780	2.08	1.04	9.62	0.05	6.8	7.1	6.4	2.1	1.0
6.6	0.08	500	1.46	1.24	10.02	0.44	7.8	10.2	6.6	2.2	1.1
6.2	1.02	550	1.52	1.26	11.24	0.12	6.8	11.2	7.4	6.4	1.2
6.4	1.01	570	2.56	1.34	10.02	0.32	6.0	6.4	8.4	4.8	0.9
5.8	0.94	580	3.04	1.04	11.02	0.45	7.4	5.2	9.2	5.2	0.6
5.4	0.64	550	3.52	0.96	12.02	0.64	6.2	8.1	6.2	6.2	0.6
5.4	0.75	580	3.43	0.92	9.64	0.92	4.5	9.2	11.4	2.3	0.9
6.0	0.66	800	3.49	0.94	10.02	0.56	6.8	10.2	10.2	3.4	0.2
7.2	0.78	860	3.42	1.46	11.42	0.64	7.2	10.5	12.4	5.5	0.3
7.1	0.92	860	3.24	1.05	12.45	0.44	6.8	10.6	13.1	6.6	0.9
6.2	1.02	920	1.45	1.52	10.04	0.52	7.2	11.2	8.6	4.8	0.6
6.2	1.46	850	1.94	1.62	11.04	0.62	6.8	6.4	9.4	5.2	0.8
5.2	1.52	870	0.96	1.72	12.45	0.72	6.5	8.2	10.2	4.8	0.6
5.4	0.66	625	2.24	2.02	10.64	0.84	7.5	9.2	11.1	5.2	0.9
5.5	0.92	600	2.36	2.24	10.24	0.90	1.3	9.6	12.1	4.8	1.1
5.8	0.92	610	2.92	2.00	10.24	0.09	2.5	10.2	6.4	4.7	1.0
6.1	1.02	620	3.01	0.12	9.24	0.92	3.0	9.8	8.4	5.1	0.9
5.7	0.92	710	3.20	0.92	9.92	0.90	4.0	10.2	9.6	5.2	0.4
7.1	0.42	780	4.10	1.06	10.02	0.62	5.2	11.4	9.9	6.2	0.6
6.7	0.64	750	2.90	1.10	12.12	0.64	6.4	12.2	10.0	6.6	0.9

effluents vis-à-vis comparing it with Food and Agriculture Organization (FAO) standard for irrigation purposes.

MATERIALS AND METHODS

The study was carried out on Ona river in Ibadan Southwest Local Government Area (7°25"N, 3°25"E) Oyo State, Nigeria. The city has a tropical wet and dry climate with distinct dry period of about 130 days; mean annual rainfall and temperature are about 1205 mm and 28°C respectively while estimated potential evapotranspiration is 1100 mm (Martins and Bello, 1997). The city is underlain by crystalline pre-Cambrian basement complex of igneous and metamorphic origin (Fagoyinbo, 1966).

Twenty sampling points were located along the stream at about 20 m interval. Twenty samples were collected during each season (wet and dry) making a total of forty samples. The samples for wet season were taken at the peak of raining season (July 2010) while the sample for dry season was taken at the peak of the dry season (January 2011).

The water samples collected were analyzed in the laboratory using the method by Ademoroti (1996). The parameters considered for analysis includes electrical conductivity, total dissolved solids, calcium, magnesium, sodium potassium carbonate, bicarbonate, chloride sulphate, nitrate, phosphate, boron acid/ basicity and sodium absorption ratio. The results obtained from the analysis were compared with the Food and Agriculture Organization (FAO) standard for irrigation purpose to ascertain its suitability for irrigation purposes.

RESULTS AND DISCUSSION

Tables 1 and 2 showed the result of the investigation of

Table 2. Results of some physicochemical parameters of Ona surface water receiving sewage effluents collected during the dry season.

pH	EC (dS m^{-1})	TDS (mg l^{-1})	Ca^{2+} (me l^{-1})	Mg^{2+} (me l^{-1})	Na^{2+} (me l^{-1})	CO_3^- (me l^{-1})	HCO_3^- (mel^{-1})	SO_4^{2-} (mel^{l-1})	Cl (me l^{-1})	NO_3^- (mgl^{-1})	K^+ (mg l^{-1})
6.2	1.02	650	9.2	1.82	12.4	0.54	4.2	16.2	11.8	8.9	0.9
5.8	1.47	700	7.6	2.62	18.2	0.66	4.0	15.8	12.8	9.6	0.8
5.5	1.48	780	6.8	1.94	9.8	0.84	3.8	10.5	16.4	9.2	0.9
5.6	1.34	840	6.4	3.14	15.2	1.02	3.8	10.7	10.8	9.6	0.9
5.6	0.96	800	7.2	1.52	6.4	0.84	4.0	11.7	10.6	8.4	1.2
5.6	1.62	780	6.8	1.62	10.8	0.94	4.0	12.2	12.4	7.2	1.4
5.8	1.06	758	9.0	1.82	11.2	0.42	3.8	13.3	12.8	8.6	1.8
6.0	1.09	720	8.4	1.08	9.8	0.48	3.6	12.4	11.2	9.2	1.6
6.1	1.08	680	8.6	1.04	10.2	0.66	4.2	12.2	12.2	9.6	1.6
6.0	1.05	700	9.2	2.54	11.8	0.48	3.8	16.1	12.8	9.2	1.4
6.0	1.62	720	8.8	3.02	15.8	0.52	2.2	9.2	12.5	9.3	1.6
6.2	1.64	740	8.4	1.66	18.2	0.71	3.8	10.2	12.8	9.4	0.9
6.2	1.48	840	8.2	1.49	21.2	0.68	3.2	10.1	11.4	12.8	0.9
6.2	1.53	650	6.8	1.50	20.8	0.94	3.4	10.8	11.4	12.2	1.2
6.1	1.55	600	7.2	1.58	21.2	0.92	3.8	11.4	10.8	14.2	1.4
5.9	1.50	624	6.4	1.62	18.2	0.66	3.2	11.4	12.0	12.4	1.8
5.8	1.62	550	4.8	2.64	19.1	0.34	3.8	11.8	13.0	12.8	1.1
5.7	1.72	450	7.4	2.58	8.4	0.36	3.2	16.4	9.8	8.8	1.4
5.5	1.84	450	8.2	2.46	12.6	0.42	3.0	12.2	9.8	8.6	1.5
6.2	1.62	300	9.2	2.41	12.8	0.47	3.2	12.4	9.0	9.2	1.6
6.2	1.81	350	9.0	2.41	13.2	0.45	3.8	13.5	9.6	10.1	1.8
6.2	1.06	320	10.1	2.52	14.8	1.00	4.2	13.6	9.8	10.2	1.5
6.3	1.24	300	11.2	2.48	14.6	0.66	1.9	14.8	10.2	12.1	1.2
6.2	1.36	340	9.2	2.62	15.2	0.71	1.6	16.4	10.8	13.2	1.3
6.5	1.06	320	6.2	2.48	9.6	0.61	2.4	12.4	9.8	14.1	1.4
6.0	1.08	310	6.8	2.64	10.2	0.81	2.8	8.8	12.8	13.1	1.5
5.8	1.22	320	6.0	2.52	10.8	0.76	5.2	13.8	12.8	14.2	1.6
5.8	1.46	300	5.8	3.04	11.6	0.94	6.4	14.2	12.9	12.4	1.8
5.7	1.52	450	9.2	2.98	12.8	0.86	7.2	18.4	10.6	10.8	1.9
5.8	1.50	510	10.1	2.90	8.9	0.94	8.6	19.2	10.2	12.2	1.0
5.8	1.48	520	10.0	1.46	10.2	0.48	9.6	10.8	9.8	12.8	0.9
5.8	1.50	330	10.1	2.92	9.8	0.94	3.9	9.2	10.2	12.4	1.0
6.2	1.52	340	9.3	2.94	10.4	0.48	3.8	9.4	10.6	14.2	1.1
6.0	1.62	350	9.4	2.01	11.8	0.39	3.9	12.4	9.8	13.8	1.2
6.1	1.48	320	9.8	2.62	12.2	0.45	4.2	12.1	9.8	12.4	0.9
6.0	1.40	300	9.6	2.84	13.8	0.47	4.1	15.2	10.2	10.9	0.9
6.1	1.52	350	9.2	2.94	14.2	0.94	4.2	15.4	14.2	12.2	1.2
6.0	1.64	310	9.2	3.06	15.1	0.98	4.8	16.4	11.6	9.8	1.1
6.5	1.72	350	9.2	3.10	16.2	0.64	4.9	18.1	9.2	8.2	1.2
6.2	1.70	450	10.8	2.98	10.8	0.72	5.0	9.2	10.8	11.4	1.3

the quality assessment of Ona River water samples collected during the rainy season (July 2010) and the dry season (January 2011) respectively. The difference in the concentration may be due to the difference in the quantity (volume) of water during the two seasons. Result obtained shows that the concentrations of the physio-chemical parameters were higher during the dry season than during the wet season. The range of concentration of electrical conductivity, total dissolved solids, calcium, magnesium, sodium, carbonate, bicarbonate, chloride, sulphate, nitrate, potassium and Ph are 0.01 to 1.52 ds/m, 450 to 920 mg/l, 0.86 to 3.43 me/l, 0.12 to 2.94 me/l, 6.06 to 20.4 me/l, 2.1 to 7.2 me/l, 0.09 to 1.2 mg/l and 5.2 to 7.4 mg/l respectively during the dry season and were 0.96 to 1.84 ds/m, 300 to 84 mg/l, 5.8 to 11.2 me/l, 1.08 to 3.14 me/l, 8.4 to 21.2 me/l, 0.34 to 1.02 me/l, 1.6 to 9.6 me/l, 8.8 to 19.2 me/l, 9.0 to 16.4 me/l, 7.2 to 14.2 mg/l, 0.90 to 1.92 mg/l and 5.5 to 6.5 mg/l during the wet season respectively.

Result showed that during both rainy and dry season, the concentration of soluble cations and anions vis-à-vis the interacting effect on salinity and toxicity hazard of water are low and within acceptable levels for irrigation of crops.

Conclusion

It can be concluded that Ona River during both the rainy and the dry season are safe for irrigation for crops. Ona River can be developed for an all-year round irrigation for crop production in the study area.

However, the need to continuously monitor and assess surface water quality for irrigation purposes is hereby recommended.

REFERENCES

Ademoroti CMA (1996). Standards methods for water and effluent analysis 3rd edition Foludex press ltd. Ibadan.

Beecroft JA (1982). Development of water resources in resources in England and Wales. J. Institut. Water Engine., 24:155-177.

Fagoyinbo AO (1996). Evaluation of shallow groundwater reserves in basement complex areas: a case study of southwestern Nigeria. J. Mining Geol., 1: 65-79.

Food and Agricultural Organization (1990). Guidelines for irrigation water standards (Appendix 1).

Martins AO, Bello NJ (1997). Assessment of surface water quality for gardening irrigation in Abeokuta, Nigeria. Nig. J. Sci., 31:151-158.

APPENDIX

Appendix 1. Guidelines for irrigation water standards

Water parameter	Usual range in irrigation water
Electrical conductivity (ds/m)	0 - 3
Total dissolved solids (mg/l)	0 - 2000
Acid /basicity (pH)	6.0 - 8.5
Calcium (Ca^{+}) (me/l)	0 - 20
Magnessium (Mg^{2+}) (me/l)	0 - 5
Sodium (Na^{+}) (me/l)	0 - 40
Boron (B) (mg/l)	0 - 2
Carbonate (CO_3) (me/l)	0 - 1
Biccarbonate (HCO_3) (me/l)	0 - 10
Chloride (Cl) (me/l)	0 - 30
Sulphate (SO_4^{2-}) (me/l)	0 - 20
Nitrate (NO^{3-}) (mg/l)	0 - 10
Ammonium (NH^{4-}) (mg/l)	0 - 5

Source: Food and Agriculture Organization (FAO) 1990.

Performance of 40 poinsettia cultivars grown under two different temperatures

Bruce L. Dunn[1]*, Carla Goad[2] and Stephen Stanphill[1]

[1]Department of Horticulture and Landscape Architecture, Oklahoma State University, 358 Ag Hall, Stillwater, OK 74078, USA.
[2]Department of Statistics, Oklahoma State University 301 MSCS Building, Stillwater, OK 74078, USA.

This study evaluated forty poinsettia cultivars to determine which cultivars perform well under cool growing temperatures. Poinsettias were grown in separate greenhouses at temperatures of 20°C during the day and 17°C at night for cold production, or 25°C during the day and 22°C at night for warm production. Number of lateral shoots leading to inflorescences, number of bracts, canopy diameter, and stem diameter was determined. A significant (P= 0.03) cultivars by greenhouse temperature interaction existed for all parameters measured. Cultivars were then separated into classes based on greatest values for each trait. Poinsettia cultivars with the largest stem diameter in both warm and cold production systems included '39-02B', '7-07', 'Classic™ White', and 'Enduring™ Red'. Poinsettia cultivars 'HC-18B', 'Winter Blush', 'Cortez™ Burgundy, 'Winter Rose™ Early Red', '7-07', 'Silverstar™ Red', 'Topez'™, 'Classic™ Red', 'Enduring™ White', Novia™ Red', Advent™ Red', 'Silverstar™ Marble', and '1232' had high bract numbers in both warm and cold production. Only one cultivar, '1266', produced a large number of lateral inflorescences in both production systems. Poinsettia cultivars 'Classic™ White', 'Mira™ White', 'Novia™ Red', Early 'Orion™ Red', 'Mars'™ Pink', and 'Enduring™ Red' had larger canopy diameters when grown in cool temperatures than under warmer conditions.

Key words: Greenhouse production, temperature, *Euphorbia pulcherrima*, plant physiology, plant architecture

INTRODUCTION

Poinsettias are greenhouse-grown using a precise schedule under specific horticultural conditions and practices to produce crops of uniform size and quality (Fisher et al., 1996). According to the most recent floricultural crop survey, poinsettias represent an industry valued at over $154 million (USDA 2009). The popularity of this traditional holiday plant has been sustained by releasing new poinsettia cultivars with improved post-harvest life, growth habits, color patterns, and varying maturity dates. Profit margins are small for poinsettia production due to the number of inputs required to produce a quality crop. The two most significant costs associated with poinsettia production are labor and energy (Williams, 2007).

Energy conservation is achieved by increasing facility insulation, starting production later, consolidating production, installing energy curtains, contracting fuel, or switching to alternative fuel sources; however, the strategies may require substantial investments in time, costs, and include certain risks. A simple solution would be to decrease the growing temperatures (Lopez, 2008). It is possible to grow quality poinsettias with less energy (cooler temperatures) using cultivars adapted to such conditions. Syngenta Flowers and Paul Ecke Ranch, both with poinsettia breeding programs, provide cold growing cultivar recommendations for cultivars that perform well under cool growing conditions (Ecke, 2008; Syngenta, 2009).

Benefits of growing cool temperature tolerant poinsettias include, cost savings on energy, less need for plant growth regulators and insecticides, more intense bract color, more compact plants that are more durable in shipping, and earlier developing cultivars may be sold later in the season along with longer developing varieties grown under normal warmer production (Lang, 2009). Lopez (2008) reported that growers in the north and on the east coast could potentially save between 24 and

*Corresponding author. E-mail: bruce.dunn@okstate.edu.

52% on energy costs by growing at 18/14°C (65/57°F) compared to 24/19°C (75/67°F) for daytime and nighttime temperatures, respectively; however, Faust et al. (2009) reported that due to increased production time associated with cooler growing, actual savings is cultivar and market date dependent.

Grower specific information, on plant growth using cooler temperatures throughout production, is not readily available. The present study was conducted to determine the effect of growing 40 different poinsettia cultivars under cold and warm production temperatures. The goal of this research was to determine cultivars that would lend themselves to a reduced energy production program, that would reduce costs associated with production and ultimately increase profitability.

MATERIALS AND METHODS

Twenty-three trial poinsettia cultivars were received from Paul Ecke Ranch (Encinitas, CA, U.S.A.) on 8 August 2009. The 23 cultivars from Ecke Ranch consisted of eight reds ('Prestige™ Red', 'Advent™ Red', 'Enduring™ Red', 'Prestige™ Early Red', 'Classic™ Red', 'Freedom™ Early Red', '1232'); six whites ('Polar Bear', 'Classic™ White', 'Enduring™ White' and 'Freedom™ Early White', '1266', '7-07'); three pinks ('Classic™ Pink' and 'Polly's Pink'™, 'HC-18b'); one marble ('Winter Blush'); six specialty cultivars ('Winter Rose™ Early Red', 'Freedom™ Peppermint', 'Orange Spice'™ and 'Ice Punch', '146-06', '39-02b'), as well as six unnamed selections labeled: 'HC-18b', '146-06', '1232', '39-02b', '1266'' and '7-07'. Seventeen different trial poinsettia cultivars were received from Syngenta Flowers Inc. (Boulder, CO, U.S.A) on 17 August, 2009. The poinsettias from Syngenta included: eight reds ('Silverstar™ Red', 'Orion™ Red', 'Early Orion™ Red', 'Mira™ Red', 'Cortez™ Red', 'Sonora™ Red', 'Mars™ Red' and 'Novia™ Red'); three white cultivars ('Sonora™ White', 'Mira™ White' and 'Mars™ White'); one pink ('Mars™ Pink'); two marble ('Silverstar™ Marble' and 'Mars™ Marble'); and three specialty cultivars ('Cortez™ Burgundy', 'Topaz'™ and 'Cinnamon Star™'). All cuttings were rooted in wedge-shaped foam plugs prior to shipping, and were placed under intermittent mist (16 s every 32 min) until transplanting on 19 August, 2009.

Cuttings were transplanted into 15.2 cm Elite Azalea Pots (ITML Horticultural Products, Middlefield OH) filled with Metro Mix 702 media (SunGro Horticulture Distribution Inc., Bellevue, WA). Each poinsettia cultivar was then divided into two equal groups. The first group was placed in a greenhouse with a night temperature of 17°C and a daytime temperature of 20°C [average daily temperature (ADT) of 18°C]; whereas, the second group was placed in a greenhouse with a night temperature of 22°C and daytime temperature of 25°C (ADT of 24°C). Poinsettias were grouped by cultivar and spaced 25.4 to 38.1 cm apart. Poinsettias were grown in the Oklahoma State University research greenhouses, which have fiberglass, semi-translucent walls and a clear polycarbonate roof, at Stillwater, Oklahoma.

No supplemental lighting was used during the 18 week study. Plants were fertigated as needed with non-pressure compensated 0.2 cm inside diameter on/off drip tube-weight emitters (Chapin Watermatics Inc. Watertown, NY) until media was saturated. One emitter with a flow rate ranging from 0.3 to 0.4 L per minute was placed in each pot. Plants were fertilized at each irrigation with 100 ppm 13N-2P-13K (Peter's Excel® 13N-2 P_2O_5-13 K_2O, +6% Ca +3% Mg, Plug Special Basic Fertilizer, Scotts-Sierra Horticultural Products Co.; Marysville, OH).

On 16 September, 2009, all poinsettias were pinched to 5 nodes. The pH and EC were recorded weekly. The pH was high (average < 6.50) and EC was low (average 1.4 $dS{\cdot}m^{-1}$). So fertilizer was changed to apply 250 ppm 20N-10P-20K (Jack's Professional® General Purpose 20N-10P_2O_5-20K_2O acidic fertilizer, J.R. Peters Inc., Allentown, PA). On 22 September 2009, cultivars were treated once with 1 L magnesium sulfate, 0.1 L Soluble Trace Element Mix (J.R. Peters Inc., Allentown, PA) and ½ g molybdenum, mixed into 19 L of 250 ppm N 20N-20P-20K (Peter's Excel® General Use 20N-20P_2O_5-20K_2O fertilizer, J.R. Peters Inc., Allentown, PA). Two days later, all plants received a drench of 750 ppm B-9 (Fine Americas Incorporated, Walnut Creek, CA) and 750 ppm Cycocel (OHP industries, Mainland, PA). Insecticide treatments of either TriStar WPS 70 (Cleary Chemicals Company, Dayton, NJ) or a mixture of Enstar II (Wellmark International, Schaumburg, IL) and Mavrik (Wellmark International, Schaumburg, IL) were applied every 14 days starting 7 September 2009. Poinsettias were treated with a mixture of 2 g/L calcium chloride mixed with 30 mL of a Spread-It (Parkway Research Corp., Houston, TX) sticker spreader occurring weekly starting on 9 November, 2009.

Data collection consisted of the number of lateral shoots producing inflorescences, diameter (mm) of the tallest shoot at the median position of the main stem, the total number of bracts (that either started or had changed color) on the tallest shoot, and canopy diameter (cm) determined as the average of two measurements taken at the widest and narrowest diameter on 7 December, 2009.

Each poinsettia species was analyzed separately using a completely randomized design with six plants per cultivar as experimental units. All data were analyzed using PROC MIXED or PROC GLIMMIX of SAS (version 9.2; SAS Institute, Cary, NC). Fixed effects in the analyses were cultivar and growing temperature. The MIXED procedure was used for the continuous variables (and a normal distribution was appropriate), and the GLIMMIX procedure was used to model the discrete random variable responses using a Poisson distribution. Dunnett's multiple comparison method was used to determine a best treatment combination group, maximizing the response variables since a higher number was considered more favorable.

RESULTS AND DISCUSSION

A significant (P< 0.03) interaction occurred between cultivars and greenhouse temperatures for all measured traits. 'HC-18B' had the greatest number of bracts in both warm and cold production (Table 1), though bract size was nearly a quarter to a third of the size of a traditional bract. Bracts are modified leaves that are the showy part of the plant, and vary in number, size, and color among cultivars (Ecke et al., 2004). Bract size is heavily influenced by growing temperature, especially during bract expansion (Tsujita and Craig, 1980; Albrecht and Ladd, 1984). Poinsettia cultivars '1232', '7-07', 'Advent™ Red', 'Classic™ Red', 'Enduring™ White', 'Novia™ Red', 'Silverstar™ Marble', 'SilverStar™ Red', 'Topaz™', 'Winter Blush', and 'Winter Rose™ Early Red' had a large number of bracts in both warm and cool production systems (Table 1). Within this same group, ten cultivars grown under cool production temperatures, had more bracts than the same cultivar grown under warmer temperature (Table 1). Sink and Carlson (1970) reported an increase in the number of bracts of poinsettia cultivar 'Stoplight' when grown at 16 or 18°C than at 21°C. Bract color was delayed by as many as nine days for '146-06' under cool production, but the other cultivars displayed first color within two days in either warm or cool production (data not shown).

Table 1. Poinsettia cultivars grown under warm and cold production temperatures with the highest number of bracts on the tallest shoot.

Variety	Company	Greenhouse*	Bract number**
HC-18B***	Ecke	Warm	40.7
HC-18B***	Ecke	Cool	37.0
Winter Blush****	Ecke	Cool	23.7
Cortez™ Burgundy	Syngenta	Warm	21.7
Classic™ White	Ecke	Cool	21.7
Cortez™ Burgundy	Syngenta	Cool	21.7
Early Orion™ Red	Syngenta	Cool	21.3
Mira™ White	Syngenta	Warm	21.0
Winter Rose™ Early Red	Ecke	Warm	20.8
7-07	Ecke	Cool	20.3
7-07	Ecke	Warm	20.3
Mars™ Pink	Syngenta	Cool	20.3
Silverstar™ Red	Syngenta	Warm	20.3
Topez™	Syngenta	Warm	20.2
Classic™ Red	Ecke	Warm	19.8
Enduring™ White	Ecke	Cool	19.8
Mars™ Marble	Syngenta	Warm	19.8
Sonora™ Red	Syngenta	Cool	19.7
Freedom™ Early White	Ecke	Cool	19.2
Novia™ Red	Syngenta	Warm	19.2
Advent™ Red	Ecke	Warm	19.0
Enduring™ Red	Ecke	Cool	19.0
Silverstar™ Marble	Syngenta	Warm	19.0
Silverstar™ Red	Syngenta	Cool	19.0
Winter Blush	Ecke	Warm	18.8
Winter Rose™ Early Red	Ecke	Cool	18.8
Mars™ Red	Syngenta	Warm	18.8
Enduring™ White	Ecke	Warm	18.7
Freedom™ Peppermint	Ecke	Warm	18.7
1232	Ecke	Cool	18.3
Classic™ Pink	Ecke	Warm	18.3
146-06	Ecke	Cool	18.0
1232	Ecke	Warm	18.0
Advent™ Red	Ecke	Cool	17.8
Classic™ Red	Ecke	Cool	17.8
Prestige™ Early Red	Ecke	Warm	17.8
Cortez™ Red	Syngenta	Cool	17.8
Polar Bear	Ecke	Cool	17.5
Topez™	Syngenta	Cool	17.3
Sonora™ White	Syngenta	Warm	17.2
Novia™ Red	Syngenta	Cool	17.0
Mira™ Red	Syngenta	Warm	17.0
Prestige™ Red	Ecke	Warm	16.8
Silverstar™ Marble	Syngenta	Cool	16.8
Ice Punch	Ecke	Cool	16.4

*Plants grown at either 65°F (18-19°C) during the day and 60°F (15-16°C) at night (cool); or plants grown at 75°F (23-24°C) during the day and 70°F (21-22°C) at night (warm).** Mean of six plants.*** In top class and was significantly greater (a =0.05) than all other cultivars.****Identifies the mean to which all cultivar with fewer bracts are compared using Dunnett's multiple comparison with this next "best" cultivar (α = 0.05). Means listed below this in this table are not significantly different from this cultivar.

Table 2. Poinsettia cultivars grown under warm and cold production temperatures with the highest number of lateral shoots producing inflorescences.

Variety	Company	Greenhouse*	Lateral shoots**
HC-18B***	Ecke	Warm	9.3
Cortez™ Burgundy	Syngenta	Cool	8.5
1266	Ecke	Warm	8.0
Classic™ Pink	Ecke	Cool	7.8
1266	Ecke	Cool	7.7
Polar Bear	Ecke	Warm	7.7
Sonora™ White	Syngenta	Cool	7.7
Cortez™ Red	Syngenta	Warm	7.5
Mars™ Pink	Syngenta	Warm	7.5

* Plants grown at either 65°F (18-19°C) during the day and 60°F (15-16°C) at night (cool); or plants grown at 75°F (23-24°C) during the day and 70°F (21-22°C) at night (warm).** Mean of six plants. *** Identifies the mean to which all varieties with smaller means are compared using Dunnett's multiple comparison with this "best" cultivar ($\alpha = 0.05$). Means listed in this table are not significantly different from this variety

Table 3. Poinsettia cultivars grown under warm and cold production temperatures with the largest stem diameter taken from the tallest shoot.

Variety	Company	Greenhouse*	Mean (mm)**
39-02B***	Ecke	Warm	6.9
7-07***	Ecke	Warm	6.9
39-02B	Ecke	Cool	6.5
7-07	Ecke	Cool	6.4
Enduring™ White	Ecke	Warm	6.4
Winter Rose™ Early Red	Ecke	Cool	6.4
Enduring™ Red	Ecke	Warm	6.3
Advent™ Red	Ecke	Warm	6.2
Classic™ White	Ecke	Cool	6.0
Enduring™ Red	Ecke	Cool	6.0
Classic™ White	Ecke	Warm	5.9
Silverstar™ Marble	Syngenta	Cool	5.8
Prestige™ Red	Ecke	Warm	5.7
Prestige™ Early Red	Ecke	Warm	5.6
Freedom™ Peppermint	Ecke	Warm	5.5

* Plants grown at either 65°F (18-19 °C) during the day and 60°F (15-16 °C) at night (cool); or plants grown at 75°F (23-24°C) during the day and 70°F (21-22 °C) at night (warm).** Average of six plants.*** Identifies the mean to which all varieties with smaller means are compared using Dunnett's multiple comparison with this "best" cultivar ($\alpha = 0.05$). Means listed in this table are not significantly different from this cultivar.

Within the group for the most lateral shoots producing inflorescences, eight cultivars had mean inflorescence numbers ranging from 7.5 ('Cortez™ Red' and 'Mars™ Pink') to 9.3 ('HC-18B') (Table 2). Poinsettia cultivar '1266' produced several inflorescences under both production systems, whereas, three cultivars ('Classic™ Pink', 'Cortez™ Burgundy', and 'Sonora™ White') in this highest group produced more inflorescences during cool production than in warm production. Faust and Heins (1996) report that poinsettia cultivars differ in their capacity to branch after pinching, and growing temperature affects axillary bud development. Thus cool temperatures favored lateral shoot and inflorescence development in these three cultivars.

Stem diameter varies among cultivars (Goreta et al., 2008). In this study, stem diameter ranged from 2.7 mm ('Polar Bear') to 8.3 mm ('Advent™ Red') for warm production, and 3.5 mm ('HC-18B', '1266', and 'Freedom™ Peppermint') to 7.4 mm ('Enduring™ Red') for the cool production system. Poinsettia cultivars '39-02B', '7-07', 'Classic™ White', and 'Enduring™ Red' had the largest stem diameter in warm and cold production systems (Table 3). Two cultivars ('Silverstar™ Marble' and 'Winter Rose™ Early Red') had larger stem diameters in cold production than warm production (Table 3). Lang (2009) reported that poinsettias grown under cooler temperature resulted in thicker more compact stems, allowing plants to avoid stem breakage associated with handling and shipping.

No cultivar had the largest canopy diameter in both

Table 4. Poinsettia cultivars grown under warm and cold production temperatures with the largest canopy diameter.

Variety	Company	Greenhouse*	Mean (cm)**
Classic™ White***	Ecke	Cool	42.1
Classic™ Red	Ecke	Warm	41.4
Enduring™ White	Ecke	Warm	40.3
Mars™ Red	Syngenta	Warm	39.8
Mira™ White	Syngenta	Cool	39.6
Novia™ Red	Syngenta	Cool	39.3
Polly's Pink	Ecke	Warm	38.6
Orion™ Red	Syngenta	Warm	38.2
Early Orion™ Red	Syngenta	Cool	37.9
Mars™ Pink	Syngenta	Cool	37.8
Cortez™ Burgundy	Syngenta	Warm	37.8
Mars™ Marble	Syngenta	Warm	37.7
Enduring™ Red	Ecke	Cool	37.5
Prestige™ Early Red	Ecke	Warm	37.5

* Plants grown at either 65°F (18-19°C) during the day and 60°F (15-16°C) at night (cool); or plants grown at 75°F (23-24°C) during the day and 70°F (21-22°C) at night (warm).** Average of six plants with measurements taken at the widest and shortest diameter. *** Identifies the mean to which all varieties with smaller means are compared using Dunnett's multiple comparison with this "best" or "top" variety ($\alpha = 0.05$). Means listed in this table are not significantly different from this variety.

warm and cold production. Six cultivars ('Classic™ White', 'Enduring™ Red', 'Mars™ Pink', 'Mira™ White', 'Novia™ Red', and 'Early Orion™ Red') grown under cold production had the largest canopy diameter class (Table 4). These cultivars show promise for cool production since poinsettias grown under cooler conditions often have smaller growth habits (Gislerod and Litlere, 1976; Runkle and Faust, 2008). Lang (2009) also reported 'Early Orion™ Red' had sufficient height and bract development when finished at 17°C or 18°C. Although the majority of plants grown under cool temperatures were smaller, they can serve a market for a medium-sized plant desired for travel or table space limitations.

Temperature clearly affects poinsettia growth and morphology; however, this study indicates that certain Syngenta Flowers and Paul Ecke Ranch poinsettia cultivars are less affected than other cultivars for measured traits when grown continuously cooler temperatures (Tables 1, 2, 3, 4). No cultivar, grown under cooler conditions, were placed in the top class for all measured traits; however several cultivars outperformed similar plants grown under warmer conditions for one trait or another (Tables 1, 2, 3 and 4). There are a number of environmental and production factors that influenced these results. Growing poinsettia cultivars under cooler average daily temperatures did not improve growth or quality enough to compensate for reduced fuel usage; however, growers who market them as environmentally friendly due to reduced fuel and chemical usage may get a premium price for the product.

Cooler temperatures delay flowering, so cultivars with long response times are recommended (Lang, 2009). Faust and Kehoe (2008) and Faust et al. (2008) suggest choosing cultivars that naturally flower in November due to delayed development, have medium to high growth rates to compensate for slower growth, and select cultivars with large bracts to overcome any effects on bract size. Nutrient and water uptake as well as shoot and root growth are reduced by the cooler temperatures, therefore, plant growth regulator application will likely need to be reduced.

ACKNOWLEDGEMENTS

We would like to thank Syngenta Flowers and Paul Ecke Ranch for the donation of rooted cuttings used in this trial, as well as, Sun Gro Horticulture for donation of media.

REFERENCES

Albrecht ML, Ladd DL (1984). Comparison of 5 poinsettia cultivars grown under different temperature regimes. Hort. Sci., 19: 438-439.

Ecke (2008). Energy efficient poinsettias. http://www.ecke.com/html/tibs/energy_efficient_poinsettia.htm.

Ecke III, P, Faust JE, Higgins A, Williams J (2004). The Ecke Poinsettia Manual. Ball Publishing. Batavia, IL., p. 1-287.

Faust J, Lewis KP, Rapaka VK, Knauer R (2008). Exploring the lower limits of cold poinsettia production. GPN Magazine. http://www.gpnmag.com/Exploring-the-Lower-Limits-of-Cold-Poinsettia-Production-article8963.

Faust J, Heins RD (1996). Axillary bud development of poinsettia 'Eckesoint Lilo' and 'Eckesoint Red Sails' (*Euphorbia pulcherrima* Willd.) is inhibited by high temperatures. J. Am. Soc. Hort. Sci., 121: 920-926.

Faust J, Kehoe R (2008). Battling rising fuel costs with cold poinsettia production: turn a bigger profit by turning the thermostat down. Everything under the Sun. November. http://www.flowerandplant.org/files/66-CF&PA%20Newsletter%2011-08%20email.pdf.

Faust J, Odula O, Lewis K, Turoop L (2009). New Concepts for fuel-efficient poinsettia production. GPN July, p. 30-36.

Fisher PR, Heins RD, Lieth JH (1996). Quantifying the relationship between phases of stem elongation and flower initiation in poinsettia. J. Am. Soc. Hort. Sci., 121: 686-693.

Gislerod, HR, Litlere B (1976). Temperature effects on poinsettia under long day conditions. Acta Hort., 64: 205-209.

Goreta, S, Batelja K, Perica S (2008). Growth of poinsettia as affected by cultivar, thinning, and pot size. Hort. Technol., 18: 122-129.

Lang H (2009). Cold growing poinsettias. Big Grower, September, pp. 18-20.

Lopez, RG (2008). Cold and sustainable poinsettia production. Greenhouse Grower, July, p. 116.

Runkle, E, Faust J (2008). Energy-efficient poinsettia production. GPN Magazine. http://www.gpnmag.com/Energy-Efficient-Poinsettia-Production-article9526.

Sink K, Carlson W (1970). The effect of temperature on sub-terminal bract coloration of poinsettia 'Stoplight'. Hort. Sci., 5: 15-16.

Syngenta (2009). Energy efficient poinsettia production. http://www.syngentaflowersinc.com/pdf/cultural/FINALPoinsettiaColdGrowing.pdf.

Tsujita MJ, Craig WE (1980). Reduced night temperature effects on poinsettias. J. Hort. Sci., 55: 45-47.

USDA [NASS] (2009). Floriculture crops - 2008 summary. Sp Cr 6-1. http://usda.mannlib.cornell.edu/usda/current/FlorCrop/FlorCrop-04-23-2009.pdf.

Williams J (2007). Energy efficient poinsettia crops. Greenhouse Canada. September. http://www.greenhousecanada.com/content/view/845/38/.

Simulation of water uptake and redistribution in growing media during ebb-and-flow irrigation

R. Anlauf*, P. Rehrmann and H. Schacht

Osnabrueck University of Applied Sciences, P. O. Box 1940, 49009 Osnabrueck, Germany.

To optimize ebb-and-flow irrigation on concrete floors, the relative importance of substrate type, flooding depth and flooding time on water uptake of growing media in containers is important information for the grower. Describing water uptake and distribution in the container with a dynamic simulation model may overcome the disadvantages of the static parameters such as container capacity and air capacity. Water uptake and redistribution was investigated for two different growing media, a coarse white peat and a fine seedling substrate, two flooding depths (1 and 4 cm) and three flooding durations (5, 10 and 15 min). The results were used to evaluate the use of the simulation model HYDRUS1D to describe water uptake and redistribution. The hydraulic functions water retention curve and hydraulic conductivity needed for the simulation model were determined in the laboratory. The results show that substrate properties and flooding height are the main parameters determining water uptake during ebb-and-flow irrigation while flooding time has a minor effect only. Therefore, the slope of concrete floors should be at a minimum to ensure an even flooding depth for all containers. The longer flooding times and the longer drainage durations have a very small effect on water uptake. The simulation model HYDRUS1D is able to describe water uptake and redistribution in containers filled with the two growing media sufficiently well only if the hysteresis of the water retention curve is taken into account.

Key words: Horticultural substrate, hysteresis, water retention curve, HYDRUS1D, simulation model.

INTRODUCTION

Ebb-and-Flow irrigation (also called flood and drain irrigation) is often used in horticultural practice to irrigate plant containers on concrete floors. Often, this irrigation method is more economic compared to flooded tables.

The depth of water during irrigation is usually 1 to 2 cm but can be as much as 5 cm (Aendekerk, 1997; Raviv and Lieth, 2008). It is very important that the irrigation water is distributed quickly and evenly to realize a rather uniform water level for all containers on the floor. The water is drained after about 10 to 20 min and collected in a storage tank (Raviv and Lieth, 2008).

It is known from practical experience that uneven water distribution in containers at different positions on a concrete floor happens frequently. One of the main reasons is the necessary slope of the concrete floor (0.2 to 0.4%) to realize quick drainage of water (Raviv and Lieth, 2008). Containers on a lower position will receive a higher irrigation depth and a longer irrigation time compared to containers at higher positions. This may result in either drought stress for the plants in some positions or too much moisture at others which may cause aeration and disease problems.

Other factors influencing water uptake and redistribution in plant containers are the shape of the container bottom, the type of plant and the physical properties of the growing media. The variability of the container bottom and that of the plant species on a concrete floor will be very low. Therefore, the influence of the physical properties of the growing media and that of different irrigation depths and irrigation durations will be investigated in this study.

The physical properties of growing media describing

*Corresponding author. E-Mail: r.anlauf@hs-osnabrueck.de.

the quality of the substrate for different horticultural uses are usually measured at standard conditions, such as the container capacity and the air capacity, both measured at a matric potential of -10 hPa. These methods are standardized (DIN EN 13041, 2010) and are widely accepted. However, horticultural practice shows that in some situations producers have big problems related to low oxygen supply even with growing media with a high air and container capacity, while in other cases plants with good growth and quality are produced with substrates of minor quality. This is not surprising because the standard parameters describe a static condition whereas in horticultural practice we have a dynamic system where the actual water and air content fluctuates considerably with time mainly due to irrigation practices. Therefore, the use of methods to describe the dynamic behavior of water and air in growing media, such as simulation models, seem to be a promising tool to describe the actual situation in a container more realistically (Fonteno, 1993; Heinen and de Willigen, 1995; Palla et al., 2008).

The use of simulation models to calculate water uptake and redistribution in plant containers depending on irrigation depth, time and type of material could be a cost effective method. Computer models are frequently and effectively used to simulate water movement in mineral soils. Model applications for growing media are much less frequent and the quality of the simulation is usually less good compared to mineral soils. Possible reasons are shrinking and swelling processes and the hysteretic behavior of the hydraulic functions. Both processes are of much higher significance in growing media compared to mineral soils.

The objectives of this study were (1) to measure the water content and water uptake during ebb-and-flow irrigation for two different growing media depending on flooding depth and flooding duration, (2) to deduct proposals to optimize the shape of concrete floors used for ebb-and-flow irrigation, and (3) to test the application of the frequently used HYDRUS1D simulation model (Simunek et al., 2008) to describe water content and water uptake and redistribution in the two materials.

THEORETICAL BACKGROUND

Water retention curve

The physical properties of growing media are described with different specific values. However, the use of these values is not always consistently used in literature. In this study, we used the following definitions (De Boodt and Verdonck, 1972; Raviv and Lieth, 2008): total porosity (TP) (combined volume of the liquid and gaseous phase of the medium), air capacity (AC) (volumetric percentage of the medium filled with air at a matric potential of 10 hPa, that is, the difference of the water content at 10 hPa and the total porosity), container capacity (CC) (volumetric water content at 10 hPa), easily available water (EAW) (difference of the volumetric water content at 10 and 50 hPa) and the water buffering capacity (WBC) (difference of the volumetric water content at 50 and 100 hPa).

Basis for modeling the behavior of water in porous media is the media's pore structure, which is reflected in the water retention curve and the unsaturated hydraulic conductivity function. Both functions are necessary to simulate changes in the content and fluxes of water.

A commonly used parametric model to relate volumetric water content to the matric potential was proposed by van Genuchten (1980) and has been used in several studies on growing media (Fonteno, 1989; Jones and Or, 1998; Raviv et al., 2001; Caron and Nkongolo, 2004). This formulation is also implemented in the HYDRUS1D model:

$$\theta_\psi = \theta_r + \frac{\theta_s - \theta_r}{\left(1 + |\alpha \cdot \psi|^n\right)^m}$$

where θ_ψ is the water content at matric potential ψ, θ_r is a minimum residual water content, which is either fitted (Simunek et al., 2008) or used as the water content at 300 hPa (Raviv and Lieth, 2008), θ_s is the saturation water content (that is, the total porosity), α, n and m are parameters without a physical meaning describing the shape of the function where m is usually fixed as m=1-1/n (Simunek et al., 2008).

The parametric formulation of van Genuchten (1980) for the water retention curve can be used in combination with an equation of Mualem (1976) to describe the unsaturated hydraulic conductivity function (Simunek et al., 2008). This formulation is also implemented in the HYDRUS1D model:

$$K_\psi = K_s \cdot S_e^{\,l} \cdot \left[1 - \left(1 - S_e^{1/m}\right)^m\right]^2$$

where K_ψ is the hydraulic conductivity at matric potential ψ, K_s is the saturated hydraulic conductivity, S_e is the effective water content $((\theta-\theta_r)/(\theta_s-\theta_r))$, l is a parameter describing the pore structure of the soil, usually set to 0.5, and m is fixed as m=1-1/n (Simunek et al., 2008; Raviv and Lieth, 2008).

Hysteretic properties of the water retention curve

After drying usually a poor rewetting of growing media can be observed. Thus, the water content in the substrate by rewetting at a given matric potential will not reach its value measured during drying. This is due to air

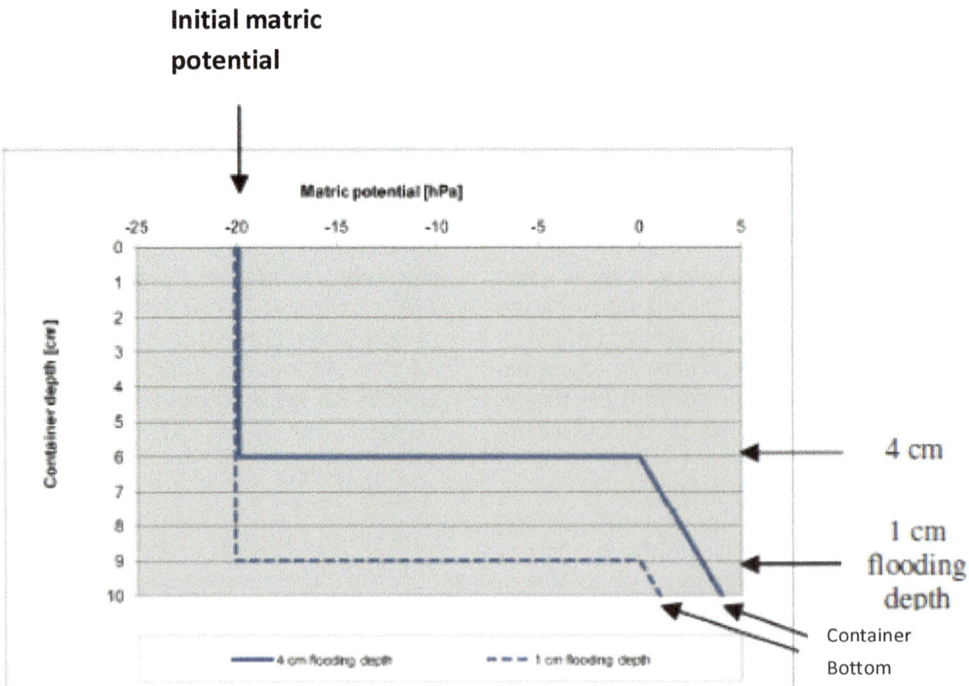

Figure 1. Initial conditions for the simulation (example for the seedling substrate).

inclusions or wetting problems of the organic material and is called hysteresis of the water retention curve (Raviv and Lieth, 2008). Hysteresis becomes especially important in irrigation systems where the water is applied from the bottom of the container, such as ebb-and-flow irrigation (Wever et al., 1997).

Applications of unsaturated flow models often assume unique, non-hysteretic functions for $\theta(\psi_m)$ and $K(\psi_m)$ to characterize the hydraulic properties of a material. Such a simplification may be acceptable for flow simulations in mineral soils; growing media usually require a more realistic description involving hysteresis in the soil hydraulic properties (Heinen and Raats, 1999; Naasz et al., 2005; Raviv and Lieth, 2008)

The procedure for modeling hysteresis in the retention function in the HYDRUS model requires that both the main drying and main wetting curves are known. θ_r, θ_s and n are assumed constant for the drying and wetting curves (Simunek et al., 2008). Thus, α is the only parameter to change when describing the main wetting and drying curves.

Simulation model, initial and boundary conditions

HYDRUS-1D is a software package for simulating water, heat and solute movement in one-dimensional variably saturated media (Simunek et al., 2008). The HYDRUS-1D program numerically solves the Richards' equation for variably-saturated water flow and convection-dispersion type equations for heat and solute transport. The program may be used to analyze water and solute movement in unsaturated, partially saturated, or fully saturated porous media.

The initial water content (or the matric potential) for all depths must be specified at the start of the simulation. For the flooding cycle, a matric potential corresponding to the measured initial water content was specified for the depths above the water surface (-20 hPa or 32 %vol and -5.000 hPa or 18.7 % vol for the seedling substrate and white peat, respectively). At the water surface during the flooding cycle (4 and 1 cm above the container bottom), a matric potential of 0 hPa was specified and at the container bottom a matric potential of +4 and +1 hPa respectively. Between the container bottom and the water surface, a linear interpolation of the matric potential was carried out (Figure 1).

For the drainage cycle, the simulated water content at the end of the flooding cycle was used as initial condition. For both, flooding and drainage cycle, a constant zero flux was used as upper boundary condition that is the assumption that there is no evaporation from the surface of the growing media:

$$q = 0 \qquad \text{at } h = 0 \text{ cm}$$

where q is the water flux and h is the height of the container counted positively downward. The lower boundary condition during the flooding cycle is a constant matric potential of zero at the water surface and of +4 hPa or +1 hPa at the bottom of the container for a flooding depth of 4 and 1 cm, respectively:

$\psi_m = +4$ hPa at h= 10 cm (for 4 cm flooding depth)

$\psi_m = +1$ hPa at h= 10 cm (for 1 cm flooding depth)

During the drainage cycle, the lower boundary condition changes to seepage face boundary condition: as long as the lower end of the container is saturated, the pressure head becomes zero and water may leave the container. If the matric potential becomes negative (unsaturated conditions), no more water may leave the container and the flux becomes zero. Thus, the boundary condition changes from a constant zero potential to a constant zero flux (Simunek et al., 2008).

$\psi_m = 0$ hPa at h= 10 cm (if $\theta \geq \theta_s$)

$q = 0$ at h= 10 cm (if $\psi_m < 0$)

Model quality evaluation

To evaluate the quality of the simulations with HYDRUS-1D different quality measures were applied: for a quick overview of the modeling quality, graphs measured against the simulated values were drawn together with the linear regression, the correlation coefficient and the 1:1 line. Without any model error, the measured and simulated values are identical and all points should lie on the 1:1 line. The points of good quality simulations should lie close to the 1:1 line, the slope of the linear regression should be close to one and the correlation coefficient should be close to one.

Numerical measures of agreement between the measured and simulated values were used as follows: a simple method to quantify the average difference between the measured and simulated values is the bias (Wallach, 2006):

$$bias = \frac{1}{N}\sum_{i=1}^{N}(X_i - P_i)$$

where N is the number of observations, X_i are the measured values and P_i the simulated (predicted) values. There should be no bias that is no over- or under-prediction of the values on an average. However, a bias close to zero is not sufficient to quantify model quality, because this could be also a result of a good prediction, or large over- and under-prediction may simply cancel each other.

A measure which avoids compensation between over- and under-prediction is the mean absolute error (MAE) (Wallach, 2006):

$$MAE = \frac{1}{N}\sum_{i=1}^{N}\left|(X_i - P_i)\right|$$

The MAE should be close to zero. Both bias and MAE have the same units as the measured and simulated data. A widely used measure of agreement between measured and simulated values is the root mean squared error (Wallach, 2006):

$$RMSE = \sqrt{\frac{1}{N}\sum_{i=1}^{N}(X_i - P_i)^2}$$

The root mean square error (RMSE) also has the same units as the measured and simulated values. However, large differences are weighed much higher than small differences between measured and simulated values. A variant of the RMSE is the relative roots mean squared error (RRMSE), which is the RMSE divided by the average of the observed values (Wallach, 2006):

$$RRMSE = \frac{\sqrt{\frac{1}{N}\sum_{i=1}^{N}(X_i - P_i)^2}}{X_{av}}$$

where X_{av} is the average of the measured X_i values. It is a meaningful measure to compare simulation quality of data with highly different averages and it is independent of the units used (for example, water content in $cm^3\,cm^{-3}$ or in %vol). To compare completely different data or different models, a widely used measure is the modeling efficiency (EF) (Wallach, 2006):

$$EF = 1 - \frac{\sum_{i=1}^{n}(X_i - P_i)^2}{\sum_{i=1}^{n}(X_i - X_{av})^2}$$

EF calculates the advantage of model results compared to using one average value. If the model gives perfect results, the predicted values P_i will be equal to the measured values X_i and, thus, EF = 1. If the average of the measured values is used as a predictor for every case, EF = 0. A model which is a worse predictor than the average may result in EF < 0. A model with acceptable quality should have EF > 0.5.

MATERIALS AND METHODS

Experimental setup

The experimental container (Figure 2) is made of plastic with an inner diameter of 10 cm and is similar to a commercial 10 cm container. It is made up of six elements that can be put together

Figure 2. Experimental container with 10 cm inner diameter and 15 cm height.

Figure 3. Flooding tub with experimental containers.

similar to those used by Kritz and Khaled (1995). The total height is 15 cm, so that the shrinking of the substrate during irrigation and drying can be considered. The topmost ring is 5 cm high; the following five rings have a height of 2 cm each. The bottom of the container is replaced by a piece of gauze, which is stretched around the lower ring. The volume of the experimental container is 1.178 L. Three flooding times, two flooding depths and 3 drainage durations were used. This results in 36 treatments:

Flooding time: 1 = 5 min, 2 = 10 min, 3 = 15 min
Irrigation depth: 1 = 1 cm, 2 = 4 cm
Drainage duration: 1 = 10 min, 2 = 30 min, 3 = 90 min

The flooding depths and durations are selected according to usual practice. Each treatment is carried out in four replications. To replace for the greenhouse floor, a specially made flooding tub is used (Figure 3).

Growing media used in the study

Two types of substrate typically used in horticultural practice were chosen to carry out this investigation: Lithuanian white peat and a commercially available seedling substrate. The seedling substrate contains white peat, black peat and finely sieved cocos fiber. It is a substrate which is used specially for salinity sensitive ornamental seedlings in trays.

The properties of growing media also depend on the substrate charge and on storage conditions and duration. Therefore, the material was mixed from different original bags and stored airtight until the measurements.

The chemical properties of the substrates used in this study (EC, pH) were determined after DIN EN 13037 (2009) and DIN EN 13038 (2009). Organic matter was determined as ignition loss (DIN EN 13039, 2009): air dried and ground substrate samples are burnt at 550°C for 24 h in a muffle furnace. The residue is con-sidered as mineral material and the ignition loss as organic matter. Both are expressed as weight percent and related to dry matter content.

The particle size distribution of each substrate was determined using three 100 g oven dry samples. Each sample was placed on a series of 5 sieves (ranging from 4 to 0.5 mm) and shaken for 5 min at 160 shakes per min. Portions of substrate samples remaining on each screen were weighed and expressed as the percentage of total sample weight. The mean weight diameter (MWD) was calculated as;

Table 1. Some chemical properties of the studied materials and its mixes.

	pH	EC (mS/m)	DB (g/cm³)	DP (g/cm³)	OM (g/g)
White peat	3.9	10	0.130	1.57	0.969
Seedling substrate	5.5	25	0.139	1.63	0.886

pH (1:10); EC (1:10); DB, bulk density; DP, particle density; OM, organic matter.

Table 2. Particle size distribution of the two materials (mass fraction; g/g).

Material	> 40 (mm)	40-20 (mm)	20-10 (mm)	10-3.15 (mm)	3.15-2 (mm)	2-1 (mm)	1-0.063 (mm)	< 0.063 (mm)	MWD* (mm)
Seedling	0.0	0.0	0.5	3.8	7.5	18.9	67.3	2.0	1.15
Peat	0.0	3.3	16.4	25.0	8.9	11.0	33.5	2.0	5.65

*MWD, Mean weight diameter.

$$MWD = \sum_{i=1}^{n} x_i f_i$$

where x_i is the mass retained on the sieve divided by the total medium mass, f_i is the average particle size, and n is the number of classes (Kemper and Rosenau, 1986). Bulk density was determined after VDLUFA (1991). Particle density (DP) was estimated from the gravimetric organic matter content and the gravimetric mineral matter content (DIN EN 13041, 2010). The saturated hydraulic conductivity was determined with an Eijkelkamp constant head permeameter (DIN 19683, 1998).

Water retention drying curves were determined using an Eijkelkamp standard sand box apparatus (Gabriels and Verdonck, 1991; DIN EN 13041, 2010) to measure the water content at pF 0.7, 1.0, 1.5 and 2.0. The TP was calculated as TP = (1 – DB/DP), where DB is bulk density (g cm^{-3}) and DP is the particle density (g cm^{-3}). The water retention curve was parameterized after van Genuchten (1980) from total porosity and the volumetric water content at the measured pF values.

As growing media are known to show intensive hysteretic behavior, water retention wetting curves were determined with the above mentioned experimental container made up of six elements. The rings were filled with growing media with the same initial water content as in the experiments (-20 hPa or 32%vol and -5.000 hPa or 18.7%vol for the seedling substrate and white peat respectively) and the same bulk density, and a flooding depth of 1 cm was installed. Under the assumption that the water content at equilibrium is analogous to the water retention curve (Raviv and Lieth, 2008), the water content of the media in the rings 2, 4, 6, and 8 cm above the water surface will correspond to pF log(2) [pF 0.3], pF log(4) [pF 0.6], pF log(6) [pF 0.78], and pF log (8) [pF 0.9]. The water retention wetting curve was parameterized by adjusting the parameter α (Simunek et al., 2008). The other van-Genuchten parameters were the same as in the drying curve.

RESULTS

Properties of the growing media

Table 1 shows some basic properties of the studied materials. According to the particle size distribution of the studied materials (Table 2) the seedling substrate had a lower percentage of macro particles (> 2 mm; 11.8%), while the white peat had 53.6%. The mean weighted diameter was 1.15 mm for the seedling substrate and 5.65 for the white peat. Consequently, one can expect a higher percentage of small pores, more easily available water and lower air capacity of the seedling substrate compared to the white peat. This should also be reflected in the water retention curve.

Water retention curves

Major differences were observed in the water holding parameters of the two substrates (Table 3). The total porosity and the water buffering capacity were nearly equal. CC (0.88 and 0.71 $cm^3 cm^{-3}$) at 10 hPa and EAW (0.44 and 0.26 $cm^3 cm^{-3}$) was significantly higher in the seedling substrate compared to the peat, while the AC (0.03 and 0.21 $cm^3 cm^{-3}$) was significantly lower for the seedling substrate compared to the white peat. This agrees well with the particle size distribution which showed a high percentage of small particles in the seedling substrate compared to the peat. The measured saturated hydraulic conductivity was slightly lower in the seedling substrate compared to the peat which also agrees

Table 3. Physical properties of the two materials.

Material	Total porosity (TP) $cm^3\ cm^{-3}$	Container capacity (CC) $cm^3\ cm^{-3}$	Air capacity (AC) $cm^3\ cm^{-3}$	Easily avail. water (EAW) $cm^3\ cm^{-3}$	Water buffering capacity (WBC) $cm^3\ cm^{-3}$	Sat. hydraulic conductivity (Ks) $cm\ s^{-1}$
Seedling substr.	0.91	0.88	0.03	0.44	0.05	0.097
White peat	0.92	0.71	0.21	0.26	0.06	0.121

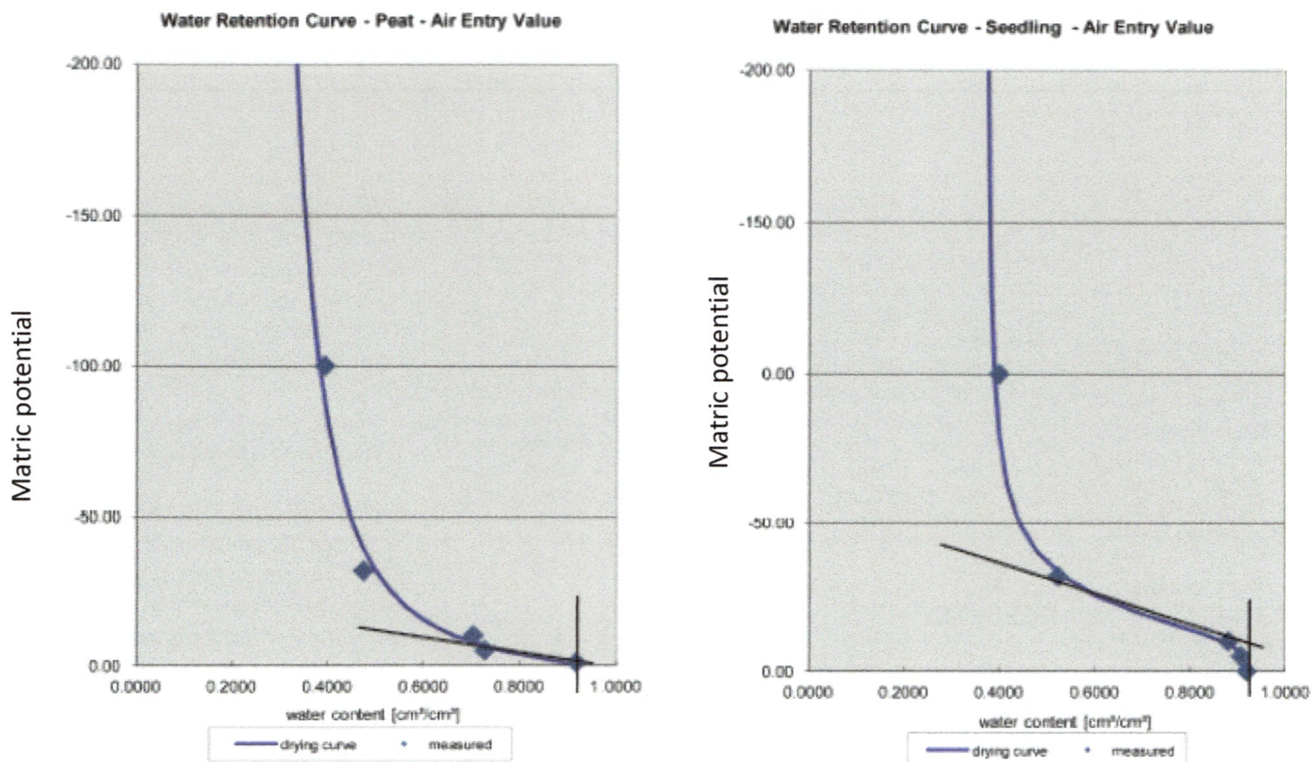

Figure 4. Drying water retention curves for the peat and the seedling substrate and the graphical determination of the air entry value.

with the smaller particle size.

Figure 4 shows the drying water retention curves for both media and the graphical determination of the air entry value (Konyai et al., 2009). The van-Genuchten parameters θ_r, α_w and n were fitted to the measured data by least square regression whereas θ_s was fixed as the water content at saturation. The van-Genuchten parameters and the air entry values are given in Table 3.

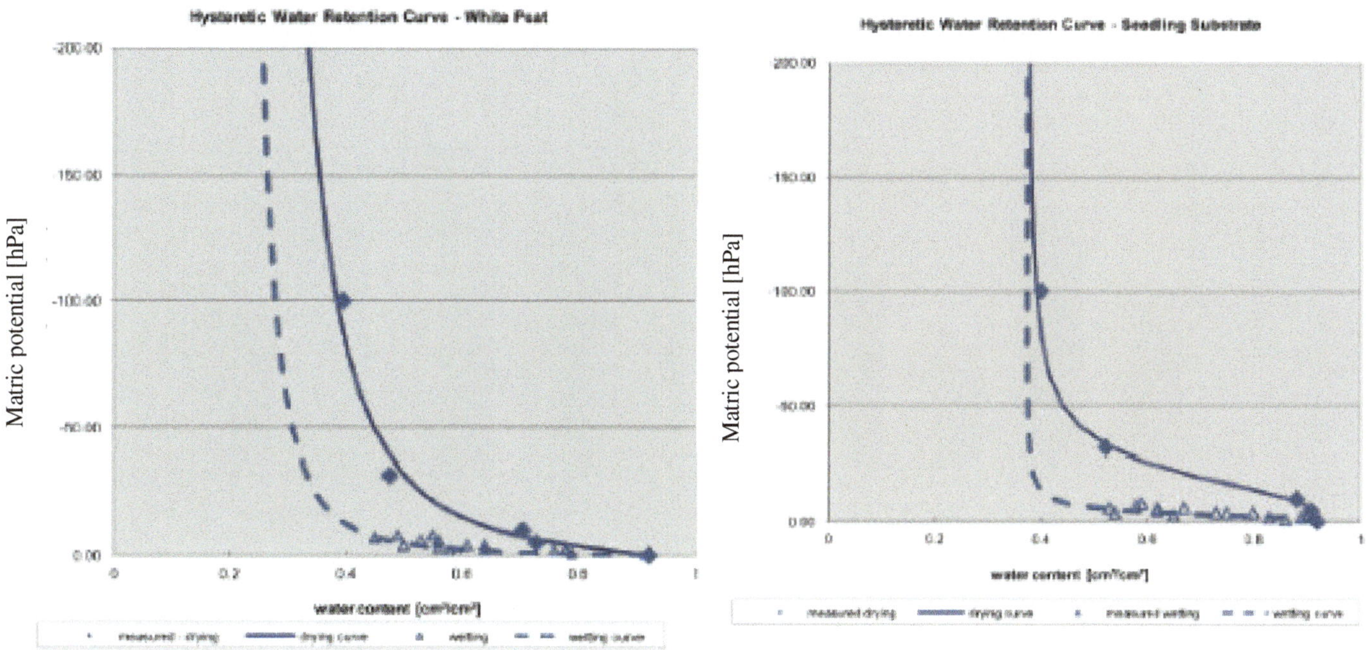

Figure 5. Water retention curves (drying and wetting curves) for the white peat and the seedling substrate.

Table 4. van-Genuchten parameters for the water retention drying and wetting curves for the peat and the seedling substrate.

Parameter	White Peat	Seedling substrate
θ_s [cm³ cm⁻³]	0.920	0.910
θ_r [cm³ cm⁻³]	0.187	0.373
α_d	0.232	0.055
α_w	1.600	0.320
α_d/α_w	6.90	5.82
n	1.411	3.022
Air entry value [hPa]	-1	-8

The air entry value characterizes the matric potential at which the largest pores drain during drying and air starts entering the material. The air entry value of the peat (-1 hPa) corresponds to a maximum pore size of approximate 3 mm (Hillel, 1998). The air entry value of the much finer seedling substrate (-8 hPa) corresponds to a maximum pore size of approximate 0.375 mm.

The residual water content θ_r of the van-Genuchten parameterization (0.37 and 0.19 $cm^3 cm^{-3}$ for the seedling substrate and the peat respectively) also characterizes a much higher water content at low matric potentials and, thus, lower air content.

The wetting curves show a distinct hysteresis for both substrates (Figure 5). To describe hysteresis, the α values were fitted to the measured data by least square regression (Table 4). The parameter α_w describing the shape of the wetting curve was about 6 times larger than the parameter α_d describing the drying curve. The measured water content at high matric potentials (-5 to -10 hPa), where hysteresis shows its main effect, was about 20% higher for both growing media for the drying curve compared to the wetting curve. This corresponds well with investigations of hysteresis in peat substrates of other authors (Wever et al., 1997; Aendekerk, 1997).

Water uptake and redistribution

Figure 6a shows the measured water uptake for the two substrates, two flooding depths and the three flooding times. All factors have a statistically highly significant effect ($p < 0.01$) on the water uptake. The largest difference can be observed for the different flooding depths. The average water uptake is 198 and 132 cm³ for a flooding depth of 1 cm for the seedling substrate and the white peat, respectively, while the average water

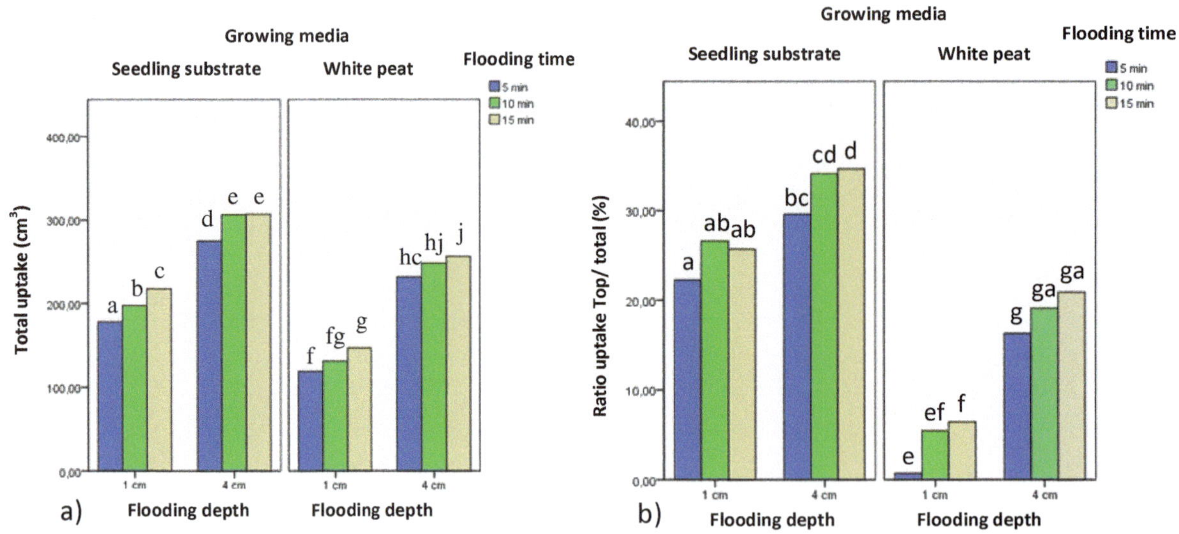

Figure 6. a) Total Water Uptake (cm³ per container) for flooding depth 1 and 4 cm and flooding time 5, 10 and 15 min for the two growing media; b) Ratio Water Uptake of the top half of the container to the total water uptake for flooding depth 1 and 4 cm and flooding time 5, 10 and 15 min for the two growing media. Different letters denote statistically significant differences.

uptake is 296 and 246 cm³ for a flooding depth of 4 cm. The influence of the substrate is high as well (average water uptake is 246 cm³ for the seedling substrate and 189 cm³ for the white peat respectively) while the effect of the different flooding times is much less pronounced (average water uptake is 201, 221 and 232 cm³ for 5, 10 and 15 min flooding time respectively). The differences of the water uptake after 10 and 15 min are statistically not significant (with the exception of seedling substrate, 1 cm flooding depth) indicating that the water uptake is a relatively fast process taking place mainly in the first 10 min.

The water uptake of the white peat at 4 cm flooding depth (246 cm³) is between that of the seedling substrate at 1 and 4 cm flooding depth (198 and 296 cm³ respectively). A similar water uptake as the white peat at 4 cm flooding depth can be expected at a depth of somewhere between 1 and 4 cm for the seedling substrate.

Figure 6b shows the ratio of the water uptake of the topmost 5 cm of growing media in relation to the total water uptake (10 cm height). 22 to 35% of the total water uptake goes to the upper half of the container filled with the seedling substrate whereas, this ratio is only 1 and 21% for the white peat. The effect of the different influence factors substrate, flooding depth and flooding time is similar to that for the total water uptake: all factors are statistically highly significant ($p < 0.01$); substrate and flooding depths show the highest effect whereas the influence of flooding time is less pronounced. There is no significant difference in the percentage of the water uptake of the topmost 5 cm in relation to the total water uptake after 10 and 15 min of flooding. This agrees well with the results of the total water uptake (water uptake occurs mainly in the first 10 minutes).

After the end of an irrigation cycle the redistribution of water through capillary rise from the bottom to the top of the container is expected. There is a slight but not significant tendency that the total water uptake will increase with time after the end of the irrigation (Figure 7a).

The ratio of the water uptake of the upper part of the container to the total water uptake is shown in Figure 7b. The increase of this ratio is negligible in the time period from 10 to 30 min, whereas there is a, however not significant, tendency, that this ratio will increase between 30 and 90 min after irrigation ends.

Simulation results

In a first step the water uptake and redistribution was simulated without taking hysteresis into account. Figure 8a and b show as an example the results of the simulation for 15 min flooding duration, flooding depth of 1 cm and a drainage time of 90 min. Obviously, the simulated water content is much higher than the measured one. In the bottommost layer, the simulated water content exceeds the measured by 10 to 20%vol. In the topmost layer, the simulated water content is about 40 to 50%vol higher than the simulated one. Thus, the model highly over-estimates the amount of water entering the substrate by capillary forces. This fact is also reflected in the drying and wetting water retention curves which differ significantly (Figure 5). To describe water content and water uptake correctly with the model, the hysteresis of the water retention curve must be taken into account.

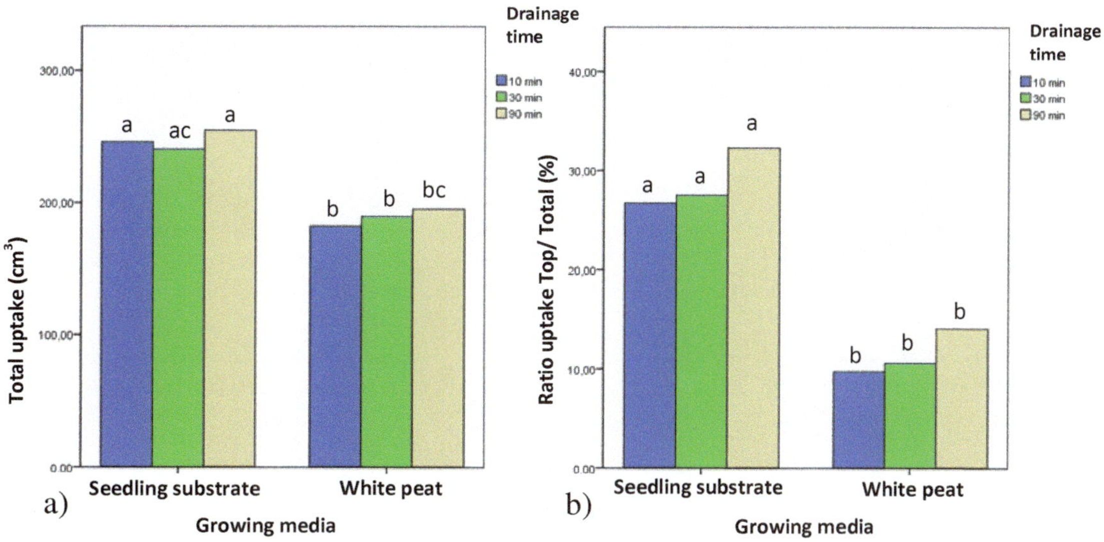

Figure 7. a) Total Water Uptake (cm³ per container) at 10, 30 and 90 min after the end of the irrigation cycle for the two growing media; b) ratio Water Uptake of the top half of the container to the total water uptake at 10, 30 and 90 min after the end of the irrigation cycle for the two growing media. Different letters denote statistically significant differences.

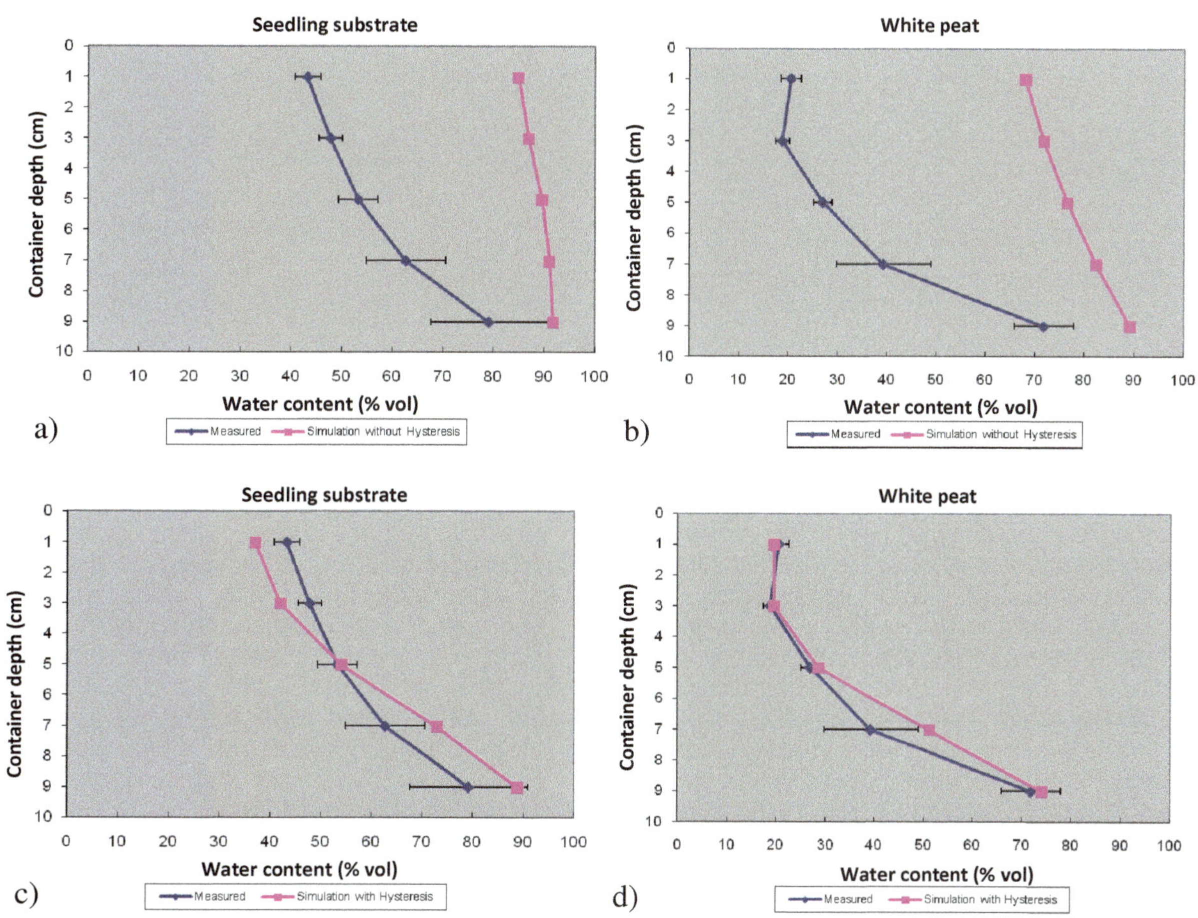

Figure 8. Water content (%vol) after 15 min of flooding, flooding depth 1 cm and after 90 min of drainage simulated without hysteresis (a, b) and with hysteresis (c, d) for the seedling substrate and the white peat.

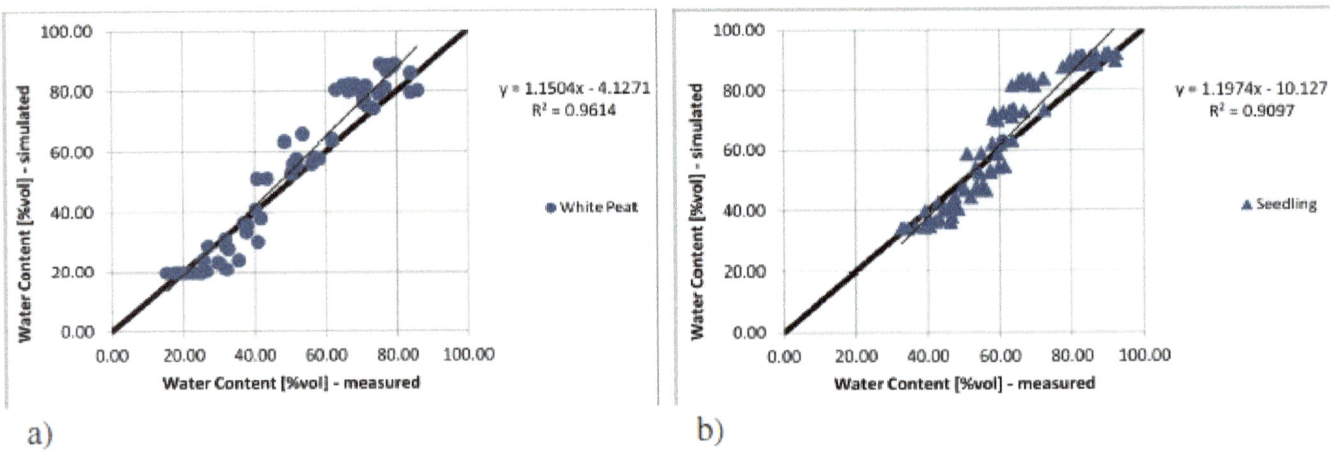

Figure 9. Measured against simulated water content (%vol) for the white peat (a) and the seedling substrate (b).

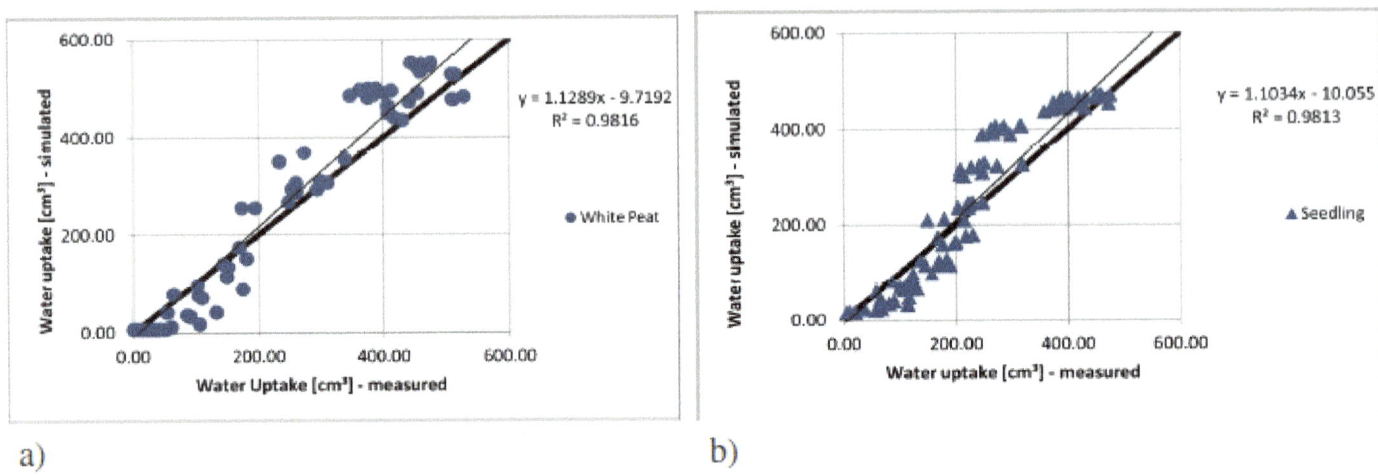

Figure 10. Measured against simulated water uptake (cm³) for the white peat (a) and the seedling substrate (b).

Considering hysteresis, the simulation yields a completely different picture (Figure 8c and d). Due to hysteresis, the water uptake into the initially relatively dry substrate during the flooding period is much smaller. The measured and simulated values agree much better. Therefore, the simulations for all cases measured were carried out using the hysteretic water retention curves.

The quality of the simulation was evaluated using the above mentioned quality measures. Figure 9 shows the measured against simulated water content for both growing media together with the 1:1 line. The correlation coefficient (0.96 and 0.91 for the white peat and seedling substrate respectively) is high. The slope of the regression line (1.15 and 1.20) however indicates that there is a tendency that the model under-estimates low values and over-estimates high values.

Similar but slightly better results can be seen for the water uptake (Figure 10), where the correlation coefficient is 0.98 and the slope is 1.13 and 1.10 for the white peat and seedling substrate, respectively.

Table 5 shows numerical quality measures for the simulation of the water content in the containers. For growing media and for flooding depths, there is a negative bias indicating that the simulated values are on an average 2.31 and 2.40 %vol higher for the seedling substrate and the white peat. This bias is less pronounced for the 1 cm flooding depth simulation compared to the 4 cm flooding depth. MAE and RMSE give similar results in that respect that the simulation quality for the 1 cm flooding depth is better than that for the 4 cm flooding depth, and that there are no big differences in the simulation quality between the two growing media investigated. The RRMSE ranges from about 10 to 17%. The modeling EF ranges from 0.70 to 0.97 indicating good to reasonable simulation quality.

Numerical quality measures for the simulation of the water uptake are given in Table 6. The maximum total water uptake for the different measurements in the

Table 5. Quality measures for the simulation of the water content.

measure	Seedling substrate			White peat		
	1 cm flooding depth	4 cm flooding depth	All	1 cm flooding depth	4 cm flooding depth	All
Bias (%vol)	-0.32	-4.30	-2.31	-0.36	-4.44	-2.40
MAE (%vol)	2.66	7.15	4.90	5.72	6.63	6.18
RMSE (%vol)	3.56	8.96	6.81	6.67	8.36	7.56
RRMSE (%)	10.01	17.91	13.63	11.66	11.99	10.84
EF	0.973	0.844	0.905	0.827	0.698	0.766

MAE, Mean absolute error; RMSE, root mean squared error; RRMSE, relative root mean squared error; EF, modeling efficiency.

Table 6. Quality measures for the simulation of the water uptake.

Measure	Seedling substrate			White peat		
	1 cm flooding depth	4 cm flooding depth	All	1 cm flooding depth	4 cm flooding depth	All
Bias (cm^3)	-12.65	-168.67	-90.66	-13.96	-174.24	-94.10
MAE (cm^3)	25.27	168.67	96.97	38.96	174.24	106.60
RMSE (cm^3)	35.88	187.64	135.09	53.61	182.60	134.57
RRMSE (%)	16.24	45.79	32.97	16.26	36.98	27.25
EF	0.994	0.948	0.958	0.994	0.966	0.975

MAE, Mean absolute error; RMSE, root mean squared error; RRMSE, relative root mean squared error; EF, modeling efficiency.

experimental containers is approximately 550 cm^3. The negative bias ranges from -12.7 to -174 cm^3 indicating that the simulated water uptake on an average over-estimates the measured water uptake. Similar to the simulation of the water content, the bias for the water uptake is less pronounced for the 1 cm flooding depth simulation compared to the 4 cm flooding depth. MAE and RMSE give similar results as for the simulated water content: the simulation quality for 1 cm flooding depth is better than for 4 cm flooding depth, and there are no big differences in the simulation quality between the two growing media. The RRMSE for the water uptake ranges from about 16 to 46%. The modeling EF is very high ranging from 0.95 to 0.99.

DISCUSSION

Water uptake and redistribution

The high water uptake of the seedling substrate shows its good ability for a capillary upward movement of irrigation water, but also the possible disadvantage of resulting low air contents. The white peat shows a distinctively different behavior due to its different physical properties. The results also show that a lower ability for an upward water movement in case of peat may be compensated by a higher irrigation depth.

After some time of cultivation, plants in containers usually have roots throughout the whole container. Right after planting, however, the availability of water in the top half of the container is important to secure the water supply of the plants. There is a tendency that the total water uptake will increase with time after the end of the irrigation. Obviously, after the flooding phase stopped and water drained from the concrete floor, there can be no more water uptake with time. A possible reason could be the fact, that right after water is removed at the end of the irrigation cycle some water is stored by capillary forces between the lower end of the container itself and the supporting surface. By removing the container to sample and determine the water content of the different layers, this amount of water will be left on the supporting surface. With longer time until sampling the containers, part of this water may enter the substrate by the higher capillary forces of the substrate. This may result in slightly, but statistically negligible increasing water contents after longer drainage times.

There is a tendency for a slight upward movement of water by capillary forces inside of the substrate in the container. However, this seems to be of very little importance leaving the substrate properties and the flooding height as the main parameters determining water uptake during ebb-and-flow irrigation.

Because the type of growing media used by horticultural producers is often determined by the type of cultivation and the production workflow, the results indicate that the main emphasis to optimize ebb-and-flow irrigation systems should be given to the flooding depth, while flooding time is of minor importance, at least for the growing media investigated in this study.

In practice, concrete floors have a slope of 0.2 to 0.4% to realize quick drainage of water. The results show however that flooding time is not important for water uptake but, apart from the physical properties of the media themselves, flooding depth is the main factor responsible for water uptake. A practical implication from the physical point of view for water uptake is that the slope of concrete floors should be at a minimum to ensure an even flooding depth for all containers. This however will result in longer flooding times, but they will have a very small effect on water uptake, and also longer drainage durations, which also have a negligible effect on water uptake. Possible negative effects of longer flooding times on higher susceptibility to plant diseases could be minimized by either the shape of the container bottom or by rills in the concrete floor similar to those used on flooded benches.

Simulation

The model highly over-estimates the amount of water entering the substrate by capillary forces. This fact is also reflected in the drying and wetting water retention curves which differ significantly. To describe water content and water uptake correctly with the model, the hysteresis of the water retention curve must be taken into account

During the flooding period, the model mainly used the wetting water retention curve, whereas during the drainage and redistribution phase, both curves and their connections were used in the lower parts of the container, the water content mainly decreased during drainage and redistribution resulting in the use of the drying curve whereas in the upper part of the container, some capillary rise of water takes place resulting in the use of the wetting curve.

Conclusions

Water uptake and redistribution during ebb-and-flow irrigation

The results of this study show that substrate properties and the flooding height are the main parameters determining water uptake during ebb-and-flow irrigation while flooding time has a small effect only. There is no significant change of the total water content in the time period after 10 min of drainage indicating that drainage is a fast process even in the seedling substrate with more fine pores compared to the white peat. This has several implications for optimizing ebb-and-flow irrigation: The physical properties of the growing media are very important factors to influence water content in the substrate. This is not surprising and is usually considered by horticultural producers who select an appropriate substrate suitable for the respective plant needs. Because flooding depth is the main factor responsible for water uptake and flooding time is not important, the slope of concrete floors should be at a minimum to ensure an even flooding depth for all containers.

Simulation with HYDRUS1D

The simulation results show that the simulation model HYDRUS1D is able to basically describe water uptake and redistribution in containers filled with the two growing media investigated. However, as these substrates show high hysteresis of the water retention curve, it is absolutely necessary that the model takes this hysteresis into account. The regression lines of the simulated against the measured values indicate that on average the higher water contents are slightly over-estimated whereas the lower water contents are under-estimated. A possible reason is that the hysteresis of the water retention curve is not described completely correct because this hysteresis mainly governs the water uptake and distribution in the container. The description of hysteresis is simplified in two aspects: 1) the wetting curve is described only by changing the parameter α of the van-Genuchten parameterization while the other parameters are kept constant. This is done because the HYDRUS1D version used has only this option to consider hysteresis while other authors suggest that hysteresis of growing media needs a more complex formulation (Naasz et al., 2005); 2) the determination of the wetting curve was done in a very simple way to measure the water content in different layers after water uptake took place and relating these values to the theoretical matric potential at equilibrium. This method should be evaluated and optimized in future investigations. However, the use of HYDRUS1D seems to be a promising method to overcome the pure static description of physical properties of growing media, such as available water capacity and air capacity, towards a dynamic description of the water movement in containers with the ultimate goal to optimize the production of horticultural plants produced in growing media.

ACKNOWLEDGEMENTS

We wish to thank our former students Kirsten Hoppe, Jasmin Koritke, Yvonne Rondot and Emma Wagner for conducting part of the analyses in the framework of a students' project.

REFERENCES

Aendekerk TGL (1997). Decomposition of peat substrates in relation to physical properties and growth of Chamaecyparis. Acta Hortic., 450: 191-198. http://www.actahort.org/books/450/450_22.htm.

Caron J, Nkongolo NV (2004). Assessing gas diffusion coefficients in growing media from in situ water flow and storage measurements. Vadose Zone J., 3: 300-311.
De Boodt M, Verdonck O (1972). The physical properties of the substrates in horticulture. Acta Hort., 26: 37-44. http://www.actahort.org/books/26/26_5.htm.
DIN 19683 (1998). Methods of Soil Investigations for agricultural engineering – Physical laboratory tests – Part 9: Determination of the saturated hydraulic water conductivity in the cylindrical core-cutter (in German). Beuth, Berlin, 9p.
DIN EN 13037 (2009). Soil Improvers and growing media – Determination of pH. German Version prEN 13037, Beuth, Berlin,
DIN EN 13038 (2009). Soil Improvers and growing media – Determination of electric conductivity. German Version prEN 13038, Beuth, Berlin.
DIN EN 13039 (2009). Soil Improvers and growing media – Determination of organic content and ash. German Version prEN 13039, Beuth, Berlin.
DIN EN 13041 (2010). Soil Improvers and growing media – Determination of physical properties – Dry bulk density, air volume, water volume, shrinkage value and total pore space. German Version prEN 13041, Beuth, Berlin.
Fonteno WC (1989). An approach to modeling air and water status of horticultural substrates. Acta Hortic., 238: 67-74. http://www.actahort.org/books/238/238_7.htm.
Fonteno WC (1993). Problems and considerations in determining physical properties of horticultural substrates. Acta Hortic., 342: 197-204. http://www.actahort.org/books/342/342_22.htm.
Gabriels R, Verdonck O (1991). Physical and chemical characterization of plant substrates: towards a European standardization. Acta Hort. 294: 249-259. http://www.actahort.org/books/294/294_27.htm.
Heinen M, de Willigen P (1995). Dynamics of water and nutrients in closed, recirculating sand bed systems: Modeling water flow, nutrient transport, root water and nutrient uptake. Acta Hortic., 401:501-508. http://www.actahort.org/books/401/401_60.htm.
Heinen M, Raats PAC (1999). Hydraulic properties of root zone substrates used in greenhouse horticulture. *In* M.Th. Van Genuchten et al. (ed.) Proc. Int. Workshop on the characterization and measurement of the hydraulic properties of unsaturated porous media. Univ. of California, Riverside, pp. 467–476.
Hillel D (1998). Environmental Soil Physics. Academic Press, San Diego, 771p.
Jones SB, Or D (1998). Design of porous media for optimal gas and liquid fluxes to plant roots. Soil Sci. Soc. Am. J., 62: 563-573. doi:10.2136/sssaj1998.03615995006200030002x
Kemper WD, Rosenau RC (1986). Aggregate stability and size distribution. In: Klute, A. (ed), Methods of soil analysis, part 1. 2nd ed Agron. Monogr. No 9. ADSA and SSSA, Madison WI, pp. 425-441.
Konyai S, Sriboonlue V, Trelo-Ges V (2009). The Effect of Air Entry Values on Hysteresis of Water Retention Curve in Saline Soil. Am. J. Environ. Sci., 5(3): 341-345. http://www.scipub.org/fulltext/ajes/ajes53341-345.pdf.
Kritz G, Khaled T (1995). Method for measuring air space and moisture content at different pot depths. Acta Hort., 401: 107-114. http://www.actahort.org/books/401/401_12.htm.
Mualem Y (1976). A new model for predicting the hydraulic conductivity of unsaturated porous media. Water Resour. Res., 12: 513–522.
Naasz R, Michel JC, Charpentier S (2005). Measuring Hysteretic Hydraulic Properties of Peat and Pine Bark using a Transient Method. Soil Sci. Soc. Am. J., 69: 13–22. doi:10.2136/sssaj2005.0013.
Palla A, Lanza LG, La Barbera P (2008). A green roof experimental site in the Mediterranean climate. Proc. 11th Conf. on Urban Drainage, Edinburgh, UK, 31. Aug.-5. Sept., 10p.
Raviv M, Lieth JH (2008). Soilless Culture: Theory and Practice. Elsevier, 587p.
Raviv M, Lieth JH, Burger DW, Wallach R (2001). Optimization of transpiration and potential growth rates of 'Kardinal' rose with respect to root zone physical properties. J. Am. Soc. Hortic. Sci., 124: 205-209.
Simunek J, Sejna M, Saito H, Sakai M, van Genuchten MT (2008). The HYDRUS-1D Software Package for Simulating the Movement of Water, Heat, and Multiple Solutes in Variably Saturated Media, Version 4.0, HYDRUS Software Series 3, Department of Environmental Sciences, University of California Riverside, Riverside, California, USA., 240p.
van Genuchten MTh (1980). A closed form equation for predicting the hydraulic conductivity of unsaturated soils. Soil Sci. Soc. Am. J., 49: 12-19. doi:10.2136/sssaj1980.03615995004400050002x
VDLUFA (1991). Bestimmung der Rohdichte (Volumengewicht) von gaertnerieschen Erden und Substraten ohne sperrige Komponenten. Methodenbuch Band I: Die Untersuchung von Boeden. Darmstadt.
Wallach D (2006). Evaluating Crop models. In: Wallach, D., Makowski, D. and Jones, J.W. (Ed.): Working with dynamic crop models, Elsevier, pp. 11-53.
Wever G, van Leeuwen AA, van der Meer MC (1997). Saturation rate and hysteresis of substrates. Acta Hortic., 450: 287-295. http://www.actahort.org/books/450/450_34.htm.

Effect of cultivar on minor components in Tunisia olive fruits cultivated in microclimate

Ouni Youssef[1]*, Flamini Guido[2], Douja Daoud[1] and Zarrouk Mokhtar[1]

[1]Laboratoire Caractérisation et Qualité de l'huile d'Olive, Centre de Biotechnologie de Borj-Cedria, B. P. 901, 2050 Hammam-Lif, Tunisia.
[2]Dipartimento di Chimica Bioorganicae Biofarmacia, via Bonanno 33, 56126 Pisa, Italy.

This paper evaluates the usefulness of three chemical parameters (composition of volatiles compounds, total phenols and fatty acids) as a tool to discriminate the olive oils obtained from three varieties (Oueslati, Chemlali and Chetoui). These varieties are included among the cultivars permitted by the disciplinary for the production of the "Kairouan olive oil", a Tunisian protected designation of origin (PDO) product. The olives were collected during the year crop 2009/2010 from the same orchard (calcareous soil), in order to eliminate geographical and climatic influences. Analysis of the effect of cultivar on the different analytical values, revealed statistically significant differences in some parameters, mainly in free fatty acid and phenol contents and oxidative stability. Furthermore, most of the quality indices and fatty acid composition showed significant variations among olive varieties. Oueslati variety had the highest values of oleic acid, whereas Chetoui was noteworthy for its high content of phenolic compounds. The major volatile component was the C6 aldehyde fraction whose content varied greatly between the different varieties: the (E)-2-hexenal content ranged from 20.9% in the oil obtained from the Oueslati variety to 7.7% in the case of Chemlali one; the amount of hexanal ranged from 10.2% in Oueslati to 3.7% in Chetoui. These results suggest that the genetic factor (cultivar) influences the volatiles formation.

Key words: Cultivar, volatiles, quality indices, phenols.

INTRODUCTION

In Tunisia, the second VOO exporter and producer after the European Union (Haddada et al., 2007), the two most diffuse cultivars are Chemlali and Chetoui. Chetoui, the second main variety cultivated in Tunisia, is widespread in the North of the Country, occurring in plains as well as in mountain Regions, being highly adaptable to various pedo-climatic conditions. In addition, this oil has a very balanced fatty acid profile and an important amount of phenols, which make it quite astringent. However, this latter characteristic is not accepted by most consumers, especially by kids. Chemlali variety, which is mainly cultivated in the Central and Southern areas of the Country, contributes to 80% of the National olive oil production. It is a productive variety, well adapted to severe environmental conditions. In addition, other minor varieties such as Oueslati, are cultivated in Tunisia. The oil of Oueslati variety is characterised by a good content of total phenols, tocopherols and a good resistance to oxidation.

The extra virgin olive oil is the principal source of fat in the Mediterranean diet with important nutraceutic effects due to its abundance of oleic acid, a monounsaturated fatty acid controlling the cholesterol level and an adequate content of linoleic and linolenic acids, the major essential fatty acids that lower the risk of coronary heart diseases and cancers (Galli and Visioli, 1999). Virgin olive oil is the only vegetal fat that can be eaten raw (also called "olive juice"), without refining operations. This allows the preserving of its natural composition, including the minor, non-saponificable compounds, making up to 1 – 2% of total content, e.g. hydrocarbons, phenols, alcohols, sterols, pigments, tocopherols and vitamins. These compounds are crucial both for the oil oxidative stability (improving the shelf life) and for its unique flavor.

*Corresponding author. E-mail: ouni_youssef@yahoo.fr.

Aroma and phenols are the only parameters that consumers can appraise directly, while other quality features (e.g. chemical composition) are not always reported on the bottle.

Two recent works reviewed the several factors influencing aromatic quality of virgin olive oil, that is, biogenesis and composition of volatiles, relationships with sensory notes, possible influence of agronomic and processing factors and oil oxidation (Kalua et al., 2007; Angerosa et al., 2004). All these findings show that volatiles content, mainly C6- and C5-skeleton compounds from the lipoxygenate pathway, are strongly influenced by the genetic origin (cultivar) for the enzymatic expression and by horticultural and processing parameters for the enzyme activity. The unique flavor of virgin olive oil is mainly attributed to the volatiles that develop during and after oil extraction from the fruit. These compounds become less important during oil storage due to oxidetion. These changes on volatile composition, together with genetic, horticultural and processing influences, explain quality differences in olive oils (Harwood and Aparicio, 2000).

In a previous study (Tura et al., 2007), the antioxidant profile was found to be firstly influenced by the cultivar and then by the site of cultivation. The oxidative stability was correlated to phenols, tocopherols and saturated on unsaturated fatty acids. In another study several olive accessions from a local cultivars collection in the Garda lake area (Northern Italy) were successfully classified by oil sensory profile (Tura et al., 2002). Moreover, these sensorial attributes results correlated to some volatile and phenolic compounds. A similar cultivar assessment was scored by a principal components analysis and eight aromatic compounds were found to be the more significant in varietal characterization, e.g. (Z)-3-hexen-1-ol, hexan-1-ol, pentan-3-one and (E)-2-hexen-1-ol (Pedo et al., 2002).

A study on the volatile profiles of Australian virgin olive oils has shown that the cultivar is the single-most important factor in determining aroma oil quality (Tura et al., 2004). Other works confirmed the cultivar strong effect on the aroma quality. Dhifi et al. (2005) found that the different volatile composition in four Tunisian oils was affected by the cultivar, also showing a close relation with the enzymatic profile that is genetically determined. Luna et al. (2006) characterized many virgin olive oils from several countries by their volatile compounds and sensory attributes. Berlioz et al. (2006) analyzed the volatile and flavor compositions of several French oils from Protected Designation of Origin (PDO) districts and standard commercial olive oils, developing a chemometric method able to discriminate the oils. Baccouri et al. (2007) demonstrated that the volatile profiles of oleaster oils (olive cultivars selected from wild olives, Olea europea var. oleaster) were different from standard European and Tunisian virgin oils from Olea europea var. sativa. Other peculiar differences in the composition of volatiles in Tunisian and French PDO oils were found by Haddada et al. (2007) who demonstrated that the building up of metabolites in oils from different cultivars was related to genetic origin. Vichi et al. (2003) found signifycant differences in volatile composition in oils of different cultivars and geographical origin in Northern Italy.

Some authors compared different analytical techniques to assess the volatile compounds and/or sensory attributes in olive oils. Procida et al. (2005) arranged a chemometric approach to correlate volatile molecules with oil sensory defects. Also, Morales et al. (2005) studied the correlations between volatile profiles and defects by olfactometry techniques. Cavalli et al. (2003) and Kanavouras et al. (2005) compared several methods for volatiles extraction from oils, that is, static and dynamic headspace, solid phase micro-extraction (SPME), sorptive extraction and thermal desorption. Garcia-Gonzalez et al. (2004) tested the electronic nose coupled with SPME to distinguish different olive oils. Contini and Esti (2006) checked the effectiveness of HS-SPME for the volatiles analysis of virgin olive oils at different dilutions and Jimenez et al. (2006) applied this method to carry out a quality control of virgin olive oils from fruit picked either from the tree or from the ground.

In Tunisia, the presence of typical varieties, the peculiar microclimate conditions and precise olive orchard management, lead to the production of very valuable olive oils with a distinctive taste. The aim of this study is the global characterization of some olive varieties, cultivated in microclimate calcareous soil in the centre of Tunisia, by analysis of their antioxidants content, oxidetive stability, antiradical activity, volatile compounds and chemometric analysis of their virgin olive oils. To avoid the influence of other factors in the characterization, olive trees were cultivated under the same pedoclimatic conditions, olive fruits were picked at the same maturity stage, and their oils were extracted with the same processing system.

MATERIALS AND METHODS

Samples

The olives were hand-picked, at the same stage of maturity, from 3 varieties cultivated in the same condition in the Centre of Tunisia and immediately processed with the same laboratory mill to obtain the olive oil samples. Only drupes not damaged, fresh and healthy were selected. The olives were washed and deleafed, crushed with a hammer crusher, and the paste mixed at 2°C for 30 min, centrifuged without addition of warm water and then transferred into dark glass bottles and stored in the dark at 4°C until analysis.

Quality parameters

Determination of the regulated physicochemical quality parameters (free acidity, peroxide value and UV absorption characteristics, K270 and K232), was carried out according to the analytical methods described by Regulation EEC/ 2568/91 and EEC/ 1429/92 of European Union Commission (EUC, 1991, 1992).

Free acidity, given as percentage of oleic acid, was determined by titration of a solution of oil dissolved in ethanol/ether (1:1) with

0.1 M potassium hydroxide in ethanol.

Peroxide value, expressed as milli-equivalents of active oxygen per kilogram of oil (meq/kg), was determined as follows: a mixture of oil and chloroform–acetic acid was left to react with a solution of potassium iodide in the darkness; the free iodine was then titrated with a sodium thiosulfate solution.

K270 and K232 extinction coefficients were calculated from absorption at 270 and 232 nm, respectively, with a UV spectrophotometer (Hewlett–Packard, HP 8452 A), using pure cyclohexane as a blank.

Fatty acids composition

The determination of fatty acids was performed as described in the EEC regulation (European Union Commission Regulation 2568/91). Methyl-esters were prepared from olive oil after room temperature saponification by vigorous shaking of a solution of oil in hexane (0.2 g in 3 ml) with 0.4 ml of 2 N methanolic potassium hydroxide and analysed by gas chromatography (GC) with a Hewlett–Packard (HP 4890 D, Hewlett-Packard Company, Wilmington, DE) chromatograph equipped with a capillary column (Supelcowax 10: 30 × 0.53 × 0.25 µm), a split-splitless injector and a FID detector.

Pigment content

Chlorophyll and carotenoids were determined colorimetrically operating as described by Minguez-Mosquera et al. (1991). The maximum absorption at 670 nm is related to the chlorophyll fraction and that at 470 nm is related to carotenoid fraction. The values of the coefficients of specific extinction applied were E0 = 613 for pheophytin as a major component in the chlorophyll fraction and E0 = 2,000 for lutein as a major component in the carotenoids fraction. Thus the pigment contents were calculated as follows:

Chlorophyll (mg/kg) = (A670 × 106) / (613 × 100 × d)

Carotenoid (mg/kg) = (A470 × 106) / (2,000 × 100 × d)

Where A is the absorbance and d is the spectrophotometer cell thickness (1 cm).

Total phenols

Total phenols and o-diphenols were quantified colorimetrically (Ranalli et al., 1999). Phenolic compounds were isolated by triple extraction of a solution of oil (10 g) in hexane (25 ml) with a methanol-water mixture (60:40, v/v). The Folin-Ciocalteau reagent (Merck Schuchardt OHG, Hohenbrunn, Germany) was added to a suitable aliquot of the combined extracts and the absorption of the solution at 725 nm was measured. Values are given as milligrams of caffeic acid per kilogram of oil (Gutfinger, 1981). Ortho-diphenols were also measured colorimetrically at 370 nm after adding 5% (w/v) sodium molybdate in 50% ethanol to the extract (Gutfinger, 1981). Results are given as milligrams of caffeic acid per kilogram of oil.

Volatile compound analyses

Sampling

Solid phase micro extraction was used as a technique for headspace sampling of virgin olive oils. Supelco SPME devices coated with polydimethylsiloxane (PDMS, 100 µm) were used to sample the headspace of 2 ml of virgin olive oil inserted into a 5 ml glass vial and allowed to equilibrate for 30 min. After the equilibration time, the fiber was exposed to the headspace for 50 min at room temperature. Once sampling was finished, the fiber was withdrawn into the needle and transferred to the injection port of the GC and GC-MS system (Campeol et al., 2001).

Identification

GC analyses were accomplished with an HP-5890 series II instrument equipped with a DB-5 capillary column (30 m × 0.25 mm, 0.25 µm film thickness), working with the following temperature programme: 60°C for 10 min, ramp of 5°C/min to 220°C; injector and detector temperatures, 250°C; carrier gas, helium (2 ml/min); detector FID; splitless injection. The identification of the components was performed by comparison of their retention times with those of pure authentic samples and by means of their linear retention indices (LRI) relative to the series of n-hydrocarbons. The relative proportions of the constituents were obtained by FID peak area normalization.

GC-EIMS analyses were performed with a Varian CP 3800 gas-chromatograph equipped with a DB-5 Capillary column (30 m × 0.25 mm; coating thickness = 0.25 µm) and a Varian Saturn 2000 ion trap mass detector. Analytical conditions were as follows: injector and transfer line temperature at 250 and 240°C respectively; oven temperature was programmed from 60 to 240°C at 3°C/min; carrier gas, helium at 1 ml/min; splitless injection. Identification of the constituents was based on comparison of the retention times with those of authentic samples, comparing their linear retention indices relative to the series of n-hydrocarbons and on computer matching against commercial (NIST 98 and ADAMS) and home made library mass spectra built from pure substances and components of known oils and MS literature data (Massada, 1976; Jennings and Shibamoto, 1980; Davies, 1990; Adams, 1995). Moreover, the molecular weights of all the identification substances were confirmed by GC-CIMS, using MeOH as CI ionizing gas (Flamini et al., 2003).

Oil stability

Oxidative stability was evaluated by the Rancimat method (Gutierrez, 1989). Stability was expressed as the oxidation induction time (hours), measured with the Rancimat 743 apparatus (Metrohm Co., Basel, Switzerland), using an oil sample of 3.5 g warmed to 100°C and air flow of 10 l/h.

RESULTS

Determination of oil quality parameters

The values of the quality parameters were comprised within the ranges established for "extra virgin olive oil" (EVOO) category for all the oils. At the sensory analysis, acidity is considered an important quality index, which has been exclusively used as a traditional criterion for classifying olive oil. The percentage of free acidity (Table 1), did not exceed the upper limit of 0.8% established by the IOOC norm and corresponding to the EVOO class. The lowest value was found for Chetoui with 0.25% while for the oil obtained from Oueslati it was higher (0.83%). These low values of acidity were due the use of fresh and healthy olives, harvested at the optimal ripening stage,

Table 1. Quality parameters of the monovarietal olive oils from plants cultivated in microclimate.

Parameters	Minor varieties	Major varieties		EVOO
	Oueslati	Chetoui	Chemlali	
FA (%C18:1)	0.83 ± 0.03	0.25 ± 0.10	0.3 ± 0.02	0.8
Pv (meq O2 kg^{-1})	6.2 ± 0.73	2.92 ± 0.36	5.83 ± 1.03	20
K_{270}	0.2 ± 0.01	0.18 ± 0 .03	0.09 ± 0.01	0.22
K_{232}	2.09 ± 0.06	1.69 ± 0.16	1.77 ± 0.04	2.5
Chlorophyll (mg/kg)	5.15 ± 0.29	4.51 ± 0.39	1.83 ± 0.29	-
Carotenoids (mg/kg)	2.5 ± 0.06	2.15 ± 0.30	0.72 ± 0.06	-
Phenols (mg/kg)	510.42 ± 0.40	828.27 ± 13.2	190.09 ± 0.4	-
o-diphenols (mg/kg)	185.62 ± 0.20	282.82 ± 40.95	91.13 ± 0.20	-
Stability (h)	103.62 ± 0.02	73.25 ± 5.81	41.12 ± 0.52	-

Table 2. Fatty acids composition (%) of the monovarietal olive oils from plants cultivated in moicroclimate.

Composition	Minor varieties	Major varieties		EVOO
	Oueslati	Chetoui	Chemlali	
C16:0	12.08 ± 0.20	13.55 ± 0.16	18.59 ± 0.26	7.5 – 20.0
C16:1	0.62 ± 0.35	0.31 ± 0.00	2.20 ± 0.24	0.3 -3.50
C18:0	3.55 ± 0.03	2.28 ± 0.17	2.71 ± 0.87	0.5 – 5.0
C18:1	70.51 ± 0.90	68.85 ± 0.43	55.87 ± 0.56	55.0 – 83
C18:2	12.58 ± 0.23	14.01 ± 0.34	18.05 ± 0.52	3.5 - 21.0
C18:3	0.63 ± 0.03	0.55 ± 0.02	0.97 ± 0.09	≤1.0
C20:0	0.32 ± 0.11	0.40 ± 0.05	0.51 ± 0.05	≤ 0.6
C18:1/C18:2	5.60 ± 0.170	4.91 ± 0.14	3.09 ± 0.170	-
SFA	15.95 ± 0.32	16.23 ± 0.12	21.81	-
MUFA	71.13 ± 0.62	69.16 ± 0.21	58.07 ± 0.62	-
PUFA	13.21 ± 0.79	15.56 ± 0.18	19.02 ± 0.79	-
MUFA/ PUFA	5.51	4.38	3.05	-

C16:0: palmitic; C16:1 palmitoleic; C18:0: stearic, C18:1: oleic; C18:2: linoleic, C18:3; linolenic, C20:0: arachidic; SFA: Saturated fatty acid, MUFA: monounsaturated fatty acid; PUFA: polyunsaturated fatty acid; EVOO: Extra virgin olive oil.

followed by immediate extraction without olive storage. Some authors reported that at the later stage of olive ripening, the oil obtained presents a higher percentage of free acidity because of the increase of enzymatic activity, especially by lipolytic enzymes, favored by olive tissue damages (Salvador et al., 2001). The peroxide value and UV characteristics are the two main parameters that represent the assessment of the progress state of the different stages that lead to oil rancidity. As indicated in Table 3, all the oil samples presented lower peroxide values, ranging from 2.92 to 6.2 meqO2/kg, belonging to the EVOO category. UV values were generally markedly lower than the limit established for the EVOO (Table 1).

Fatty acid composition

The fatty acid composition of the oil may differ depending on the variety. As shown in Table 2, numerous fatty acids were detected in the EVOOs of the examined varieties. Palmitic, oleic and linoleic acids were the major ones, while palmitoleic, stearic, linolenic and arachidic acids were present in smaller amounts. The levels of palmitic acid ranged from 12.08 ± 0.20% for Oueslati olive oil to 18.59 ± 0.26% for the Chemlali one. These values are potentially high, generally with an average of 13.55% for Chetoui olive oil. Oleic acid, the major MUFA, was high especially in Oueslati and Chetoui oils (70.51 ± 0.9 and 68.85 ± 0.43%, respectively), while it was lower for Chemlali (55.87 ± 0.44%). Concerning linoleic acid, which is much more susceptible to oxidation than MUFAs, its highest percentage was observed in Chemlali oil (18.05 ± 0.52%), whereas its lowest one is in Oueslati (12.58 ± 0.23%). For the other fatty acids, palmitoleic (C16:1), stearic (C18:0), linolenic (C18:3) and arachidic (C20:0), although their percentages changed from one olive oil to the other, they were quite small. All these levels are close to that of the IOOC norm (2001). These findings are in

Table 3. Composition of volatile obtained from plants cultivated in moicroclimate.

Constituent	l.r.i.	Minor varieties	Major varieties	
		Oueslati	Chemlali	Chetoui
hexanal	804	10.2	4.8	3.7
(E)-2-hexenal	855	20.9	7.7	–
(E)-3-hexenol	857	–	–	26.1
(E)-2-hexenol	862	–	–	–
1-hexanol	873	1.2	63.7	19.9
1-heptanol	967	1.2	–	0.5
1-octen-3-ol	980	–	–	–
3-octanone	982	–	–	–
6-methyl-5-hepten-2-one	987	0.7	1.4	0.7
decane	1000	–	–	–
octanal	1003	–	–	–
(E,Z)-2,4-heptadienal	1006	6.3	4.7	–
(Z)-3-hexenyl acetate	1009	5.9	–	13.0
1-hexyl acetate	1011	3.8	–	3.7
(E,E)-2,4-heptadienal	1015	–	–	–
limonene	1032	0.3	0.2	0.9
(Z)-β-ocimene	1041	tr	0.4	–
(E)-β-ocimene	1052	1.2	2.6	2.4
α-terpinene	1062	0.8	0.5	0.3
1-octanol	1070	–	–	–
(Z)-6-nonenal	1097	0.8	2.1	–
undecane	1100	0.5	tr	0.3
nonanal	1103	1.0	0.9	0.6
(E)-2-nonenal	1162	0.6	0.8	–
dodecane	1200	0.4	0.2	–
decanal	1206	1.9	3.0	14.2
(E)-2-decenal	1266	1.8	0.6	–
(E,Z)-2,4-decadienal	1295	0.1	–	–
tridecane	1300	–	0.1	–
(E,E)-2,4-decadienal	1318	–	–	–
cyclosativene	1370	1.9	–	–
longicyclene	1373	1.3	–	–
α-copaene	1376	17.1	0.2	tr
1-tetradecene	1393	0.3	–	0.1
tetradecane	1400	2.7	1.5	1.8
dodecanal	1409	0.3	–	–
(Z)-α-farnesene	1444	–	–	0.4
(E)-geranylacetone	1457	–	–	–
pentadecane	1500	–	–	–
trans-g-guaiene	1502	3.7	–	–
(E,E)- α -farnesene	1508	6.8	1.9	7.4
hexadecane	1600	1.9	0.7	0.8
Total identified		95.8	98.0	96.8

LRI: linear retention indices; tr: < 0.1%. Data values expressed in mg/kg.

good agreement with those of other authors working on other Tunisian olive oil varieties, like Sayali, Gerboui and Zalmati (Abaza et al., 2002). In comparison to the EVOO of the main Tunisian varieties (Chemlali and Chetoui), the Oueslati variety produced oils with a higher level of oleic acid (around 70%) and lower linoleic and palmitic acids, (both about 12%).

Furthermore, some differences in the fatty acid percenttages were present with respect to Cornicabra olive oil, the main cultivar grown in Castilla La-Mancha (Spain). Its oil contains a high percentage of oleic acid, with an average value of 80.4% and a low percentage of palmitic and linoleic acids, with mean values of 9.22 and 4.46%, respectively (Aranda et al., 2004). The percentages of saturated fatty acids (SFA), MUFA and PUFA of the studied olive oils were also calculated and it was observed that Chemlali was rich in total SFA (21.81 ± 0.23%) essentially because of its higher content in palmitic acid, which represents the main acid of the SFA fraction. Regarding the total MUFA, Oueslati olive oil contained the highest percentage (71.13 ± 0.62%) because of its high oleic acid content. These values are in good agreement with to those reported by some authors for other olive oil varieties (Ranalli et al., 2002; Aranda et al., 2004).

Pigment contents

The olive oil color is directly related to their chlorophyll and carotenoid content and it has been proposed as a characterizing factor and as a quality index related to the oil extraction method and to the olive variety (Minguez-Mosquera et al., 1991). In the studied oils, chlorophyll pigments were found at concentrations comprised between 1.83 and 5.15 mg/kg (Table 1). Oueslati oils contained the highest amount, while Chemlali samples the lowest one. Carotenoid pigments contents varied between 0.72 and 2.50 mg/kg according to cultivars (Table 1). The highest amounts were observed in the oils obtained from Oueslati and Chetoui varieties, while the lowest ones were noted in Chemlali oils. Therefore, the different cultivars were characterized by significant differences in their pigments contents. The chlorophyll content of EVOO from Oueslati was 5.15 mg/kg, while EVOO from Cornicabra variety presented higher amounts than the Tunisian varieties; in fact chlorophyll and carotenoids have been found to range from 2 to 27 mg/kg and from 2 to 14 mg/kg, respectively (Salvador et al., 2001).

Total phenols and *o*-diphenols

The phenolic compounds contained in EVOOs are one of the bases of the nutritional importance of this oil (Beltran et al., 2005). In Table 1 the changes in total phenols and *o*-diphenols are reported to the variety. In these oils, the phenol content varied between 190.09 to 828.27 mg/kg. Chemlali varieties contained the lowest amount of total phenols in comparison with Chetoui and Oueslati ones. Chetoui olive oil contained the highest value, followed by the Oueslati variety (510.42 mg/kg). Thus, these samples were relatively rich in total phenols (> 500 mg/kg). The bitterness of olive oils is related to their phenols content; however, this characteristic is not appreciated by the majority of the consumers, despite their beneficial effect. In the case of *o*-diphenols, the highest content of was observed in Chetoui oil (282.82 mg/kg), followed by Oueslati (185.62 mg/kg), while Chemlali had the lowest amount (91.13 mg/kg). These observations suggest the existence of a genotype effect, as the extraction system, the olive ripeness and pedoclimatic and agronomic conditions were all the same for all the varieties.

Oxidative stability

Rancid is a common sensory characteristic of oils and fats that have undergone a process of autoxidation caused by a prolonged contact with the air. The defects are due to inadequate fruit preservation before olive oil processing while the last one is produced during olive oil storage (Morales et al., 2005). The oxidative stability index (OSI) measured by the Rancimat apparatus is a very useful parameter for the prediction of the shelf life of the sample. Under the accelerated conditions, Oueslatii oil showed a quite high resistance to oxidation (103.62 h), followed by Chetoui oil (73.25 h). The oil obtained from Chemlali varieties showed intermediate values, (41.12 h) (Table 1).

Aroma profiles

Volatile compounds are low molecular weight compounds (less than 300 Da), which readily vaporise at room temperature. Some volatile compounds reach the olfactory epithelium, dissolve into the mucus and may bond with olfactory receptors to give an odour sensation (Angerosa, 2002). The aroma of olive oil is attributed to aldehydes, alcohols, esters, hydrocarbons, ketones, furans and probably, other as yet unidentified volatile compounds. The major volatile compounds reported in virgin olive oils are the C6 and the C5 volatile compounds. Hexanal, *(E)*-2-hexenal, hexan-1-ol and 3-methylbutan-1-ol are found in most virgin olive oils in Tunisia (Issaoui et al., 2009).

For the Tunisian monovarietal olive oils involved in this study, *(E)*-2-hexenal was the major volatile in Oueslati olive oils (20.9%). Our results are in agreement with those of Haddada et al. (2007). Chemlali contained a considerably lower amount (7.7%) of this compound that was not found at all in Chetoui olive oil (Table 3). Other C6 compounds were identified in the headspace of the, samples, such as hexanal. However, Oueslati contained higher amounts of this chemical (10.2%) with respect to the major varieties cultivated in Tunisia, such as Chetoui and Chemlali (3.7 and 4.8% respectively). This compound is usually associated to oxidation in vegetable oils. However, it was also detected in the initial virgin olive oil flavor and produced from linoleic acid through the LOX pathway (Morales et al., 1997). In the studies of Aparicio et al. (1996) and Morales et al. (1995, 1996), authors demonstrated that hexanal is an important flavor compound of virgin olive oil and contributes to a sweet perception. Furthermore, it has been postulated that hexanal content in virgin olive oil is positively correlated with the overall acceptability of potential and habitual consumers of virgin olive oil (McEwan, 1994). The third C6 compound detected in the oils was hexanol (1.2 - 63.7%). The highest percentage was observed in Chemlali oil (63.7%), followed by Chetoui (19.9%), while Oueslati contained the lowest amount (1.2%), (Table 3). The C6 aldehydes hexanal and *(E)*-2-hexenal, as well as hexanol, contribute to the typical green sensory perception of Tunisian virgin olive oils. Produced via the LOX pathway from polyunsaturated fatty acids (linolenic acids and linoleic acids), hexanal and *(E)*-2-hexenal accumulate in virgin olive oils during physical extraction procedures (Olias et al., 1993). The latter is derived from *(Z)*-3-hexenal, which undergoes isomerization to a more stable compound that can then be further reduced to *(E)*-2-hexen-1-ol (Iraqi et al., 2005). Hexyl acetate and *(E)*-3-hexenyl acetate are present in the aroma of volatiles of all Tunisian olive oil samples, but at different levels according to the respective varieties. These esters are synthesized by alcohol acyltransferase within the LOX pathway. Moreover, the low level of esters in the Chemlali and Oueslati cultivars also indicates a lower content of alcohol acyltransferase in the olive oils compared with Chetoui and its fruits. Other minor volatile compounds were observed in some of the virgin olive oils samples.

Among them, nonanal is obtained via autoxi-dation reactions (Morales et al., 1997) that inevitably start after virgin olive oil has been extracted. For the varieties involved in this study, mono- and sesquiterpene hydrocarbons were detected as the main aroma chemical class of the oils. The role of these components in the definition of the flavor is not clear (Baccouri et al., 2007). In fact, in the literature, only very few papers (Bentivenga et al., 2001; Flamini et al., 2003; Vichi et al., 2003; Baccouri et al., 2007) report the presence of these compounds, which could play a very important role in the fragrance of this valuable food (Baccouri et al., 2007). The presence of 6-methyl-5-hepten-2-one is attributed to the activity of some microorganisms (*Pseudomonas* sp.) present on all the fruits before extraction. Also, the levels of the alkanes from C9 to C17 found in the samples did not allow their differentiation, in spite of their absence in some of the samples.

The volatile fraction of the Oueslati oil was the richest in aroma compounds in comparison with the other major varieties of Tunisia (29 compounds in its aroma profiles) in which *(E)*-2-hexanal was the dominant one (20.9%) followed by α copaene (17.1%), hexanal (10.2%), *(E,E)*-α-farnesene (6.8%), *(E,Z)*–2-4-heptadienal (6.3%) and *(Z)*-3-hexenyl acetate (5.9%). The second aroma compounds-richest oil was that of the Chemlali variety, in which 1-hexanol (63.7%) was the main volatile, followed by *(E)*-2-hexenal (7.7%) and hexanal (4.8%). In the Chetoui olive oil headspace, 18 different compounds were identified, accounting for 96.8% of total volatiles. It contained *(E)*-3-hexanol, 1-hexanol, decanal, *(E)*-2-hexenal, *(Z)*-3-hexenyl acetate and *(E,E)*-α-farnesene among its major components (26.1, 19.9, 14.2, 13.0, 7.4 and 19.0%, respectively) (Table 3). It is also important to note the increasing level of 1-hexanol (detected at considerable lower levels in Oueslati), which emphasizes the perception of green pleasant attributes. As reported by Angerosa and Basti (2003), these results should be considered in a very positive way because the activity of alcohol acyltransferase seems to be little compromised, since esters contribute to the smooth green notes that are generated from the LOX pathway (Grosch, 1993). In fact, Angerosa and Basti (2003) studying the strict dependence on the enzymatic store for the production of the volatile compounds suggested that, when more cultivars are milled, some interactions and/or synergisms could occur among the enzymes involved in the LOX cascade. It is likely to attribute these interactions and synergisms to their changed activity in the olive paste, probably ascribable to the new relative concentration of each enzyme involved in the LOX pathway fixed by the genetic stores of the two cultivars.

Conclusion

The chemical characterization of EVOO from some major varieties and one minor varieties cultivated in an arid Region in the center of Tunisia, showed that, with the exception of the Chemlali variety, which presented a high level of linoleic acid, all the other varieties produced EVOO with good chemical characteristics with regard to their fatty acid composition, which complies with commercial standards, as well as for their appreciable amounts of phenolic and volatiles compounds. Furthermore, the Oueslati and Chetoui varieties presented an improved oil fatty acid composition as compared to Chemlali, the main olive variety in Tunisia. These two cultivars could be recommended to Tunisian olive growers for a large-scale plantation in the future.

ACKNOWLEDGEMENT

This work had been done as apart of a National Research Project. We thank the Ministry of Higher Education, Scientific Research and Technology for financially supporting this program. This work was supported by our Ministry of High Education, Scientific Research and Technology. Part of this work was carried out at the Dipartimento di Chimica Bioorganicae Biofarmacia, Universita' di Pisa, Italy.

REFERENCES

Abaza L, Msallem M, Daoud D, Zarrouk M (2002). Caractérisation des huiles de sept variétés d'olivier Tunisiennes. OCL Ol. Corps gras Lipides, 9(2–3): 174–179.

Adams RP (1995). Identification of essential oil components by gas chromatography/mass spectroscopy. IL: *Allured, Carol Stream.*

Angerosa F (2002). Influence of volatile compounds on virgin olive oil quality evaluated by analytical approaches and sensor panels. J. Lipid Sci. Technol., 104: 639–660.

Angerosa F, Basti C (2003). The volatile composition of samples from the blend of monovarietal olive oils and from the processing of mixtures of olive fruits. Eur. J. Lipid Sci. Technol. 105: 327–332.

Angerosa F, Servili M Selvaggini R, Taticchi A, Esposto S, Montedoro G F (2004). Review. Volatile compounds in virgin olive oil: occurrence and their relationship with quality. J. Chromatogr. A, 1054: 17–31.

Aparicio R, Morales MT, Alonso MV (1996). Relationship between volatile compounds and sensory attributes of olive oil by the sensory wheel. J. Am. Oil Chem. Soc., *73*: 1253–1264.

Aranda F, Gomez-Alonso S, Rivera DA, Salvador MD, Fregapane G (2004). Triglyceride, total and 2-position fatty acid composition of Cornicabra virgin olive oil: Comparison with other Spanish cultivars. Food Chem., 86: 485–492.

Baccouri B, Temime SB, Campeol E, Cioni PL, Daoud D, Zarrouk M (2007). Application of solid-phase microextraction to the analysis of volatile compounds in virgin olive oils from five new cultivars. Food Chem., 102: 850–856.

Beltran G, Aguillera MP, Delrio C, Sanchez S, Marinez L (2005). Influence of fruit ripening process on the natural antioxidant content of Hojiblanca virgin olive oil. Food Chem., 89: 207– 215.

Bentivenga G, D'auri M, Deluca E, De bona A, Mauriello G (2001). The use of SPME-GC-MS in the analysis of flavor of virgin olive oil. Riv. Ital. Sostanze Grasse, 78: 157–162.

Berlioz B, Cordella C, Cavalli JF, Lizzani-Cuvelier L, Loiseau AM, Fernandez X (2006). Comparison of the amounts of volatile compounds in French Protected Designation of Origin virgin olive oils. J. Agric. Food Chem., 54: 10092–10101.

Campeol E, Flamini G, Chericoni, S, Catalano, S, Cremonini R (2001). Volatile compounds from three cultivars of Olea europea from Italy. J. Agric. Food Chem., 49: 5409–5411

Cavalli JF, Fernandez X, Lizzani-Cuvelier L, Loiseau AM (2003). Comparison of static headspace, headspace solid phase microextraction, headspace sorptive extraction, and direct thermal desorption techniques on chemical composition of French olive oils. J. Agric. Food Chem., 51: 7709–7716.

Contini M, Esti M (2006). Effect of the matrix volatile composition in the headspace solid-phase microextraction analysis of extra virgin olive oil. Food Chem. 94,143–150.

Davies NW (1990). Gas chromatographic retention indexes of monoterpenes and sesquiterpenes on methyl silicone and carbowax 20M phases. J. Chromatogr., 503: 1–24.

Dhifi W, Angerosa F, Serraiocco A, Oumar I, Hamrouni I, Marzouk B (2005). Virgin olive oil aroma: characterization of some Tunisian cultivars. Food Chem., 93: 697–701.

Flamini G, Cioni PL, Morelli I (2003). Volatiles from leaves, fruits, and virgin oil from *Olea europaea* Cv. Olivastra Seggianese from Italy. J. Agric. Food Chem., 51: 1382–1386.

Galli C, Visioli F (1999). Antioxidant and other activities of phenolics in olives/olive oil, typical components of the mediterranean diet. Lipids, 34(supplement): S23– S26.

Garcia-Gonzalez DL, Barie N, Rapp M, Aparicio R (2004). Analysis of virgin olive oil volatiles by a novel electronic nose based on a miniaturized SAW sensor array coupled with SPME enhanced headspace enrichment. J. Agric. Food Chem., 52: 7475–7479.

Grosch W (1993). Determination of potent odorants in foods by aroma extract dilution analysis. Trends food. Sci. Technol., 4: 68–73.

Gutfinger T (1981). Polyphenols in olive oils. J. Am. Oil Chem. Society, 58(11): 966-968.

Gutierrez F (1989). Determinación de la estabilidad oxidativa de aceites de oliva vírgenes: Comparación entre del método A.O.M. yel método Rancimat. Grasas Aceites, 40: 1–5.

Haddada FM, Manai H, Daoud D, Fernandez X, Lizzanicuvelier L, Zarouk M (2007). Profiles of volatiles compounds from some monovarietal Tunisian virgin olive oils. Comparison with French PDO. Food Chem., 103: 467–476.

Harwood J, Aparicio R (2000). Handbook of Olive Oil: Analysis and Properties. Aspen Publishers, Inc., Gaithersburg, Maryland.

Iraqi R, Vermeulen C, Benzekri A, Bouseta A, Collin S (2005). Screening for key odorants in Moroccan olives by gas chromatography-Olfactometry/Aroma extract dilution analysis. J. Agric. Food Chem. *53*: 1179–1184.

Jimenez A, Aguilera MP, Beltran G, Uceda M, (2006). Application of solid-phase microextraction to virgin olive oil quality control. J. Chromatogr. A, 1121: 140–144.

Kalua CM, Allen MS, Bedgood Jr, DR, Bishop AG, Prenzler PD, Robards K (2007). Olive oil volatile compounds, flavour development and quality. A critical review. Food Chem., 100: 273–286.

Kanavouras A, Kiritsakis A , Hernandez RJ (2005). Comparative study on volatile analysis of extra virgin olive oil by dynamic headspace and solid phase microextraction. Food Chem., 90: 69–79.

Luna G, Morales MT, Aparicio R (2006). Characterisation of 39 varietal virgin olive oils by their volatile compositions. Food Chem., 98: 243–252.

Issaoui M, Ben Hassine K, Flamini G, Brahmi F, Chehab H, Ouni Y, Beligh M, Mokhtar Z, Hammami M (2009). Discrimination of some Tunisian olive oil varieties according to their oxidative stability, volatiles compounds and chemometric. J. Food Lipids, 16: 164–186.

Massada Y (1976). Analysis of essential oils by gas chromatography and mass spectrometry. New York: Wiley.

Mcewan JA (1994). Consumer attitudes and olive oil acceptance: The potential consumer. Grasas Aceites, 45: 9–15.

Minguez-Mosquera MI, Rejano L, Gandul B, Sanchez AH, Garrido J (1991). Color pigment correlation in virgin olive oil. J. Am. Oil Chem. Soc., 68: 332–336.

Morales MT, Alonso MV, Rios JJ, Aparicio R (1995). Virgin olive oil aroma: Relationship between volatile compounds and sensory attributes by chemometrics. J. Agric. Food Chem., 43: 2925–2931.

Morales MT, Calvente JJ, Aparicio R (1996). Influence of olive ripeness on the concentration of green aroma compounds in virgin olive oil. Flavour Frag. J., (11): 171–178.

Morales MT, Luna G, Aparicio R (2005). Comparative study of virgin olive oil sensory defects. Food Chem., 91: 293–301.

Morales MT, Rios JJ, Aparicio R (1997). Changes in the volatile composition of virgin olive oil during oxidation: Flavors and off-flavors. J. Agric. Food Chem., 45: 2666–2673.

Pedo` S, Tura D, Failla O, Bassi D (2002). Quality aspect of mono-varietal olive oils in Northern Italy (Garda lake). In: Proceeding of "Convegno Internazionale di olivicoltura" International Conference on Olive Growing. Spoleto (PG), Italy, 22–23 April, (in Italian), pp. 503-8.

Procida G, Giomo A, Cichelli, Conte LS (2005). Study of volatile compounds of defective virgin olive oils and sensory evaluation: a chemometric approach. J. Sci. Food Agric., 85: 2175–2183.

Ranalli A, Pollastri L, Contento S, Di Loreto G, Jannucci E, Lucera L, Russi F (2002). Acylglycerol and fatty acid components of pulp, seed, and whole olive fruit oils. Their use to characterize fruit variety by chemometrics. J. Agric. Food Chem., 50: 3775–3779.

Salvador MD, Aranda F, Gomez-Aonso S, Fregapane G (2001). Cornicabra virgin olive oil: A study of five crop seasons. Composition, quality and oxidative stability. Food Chem., 74: 267–274.

Tura D, Failla O, Bassi D, Serraiocco A (2002). Sensory and chemical analyses of monovarietal olive oils from lake Garda (Northern Italy). Acta Hortic., 586: 595– 598

Tura, D, Failla O, Pedo` S, Gigliotti C, Bassi D, Serraiocco A (2004). Effect of seasonal weather variability on olive oil composition in Northern Italy. In: Proceedings of the fifth International Symposium on Olive Growing, Turkey, 27 September–2 October, Acta Hortic.

Tura D, Gigliotti C, Pedo` S, Failla O, Bassi D, Serraiocco A (2007). Influence of cultivar and site of cultivation on levels of lipophilic and hydrophilic antioxidants in virgin olive oils (Olea europea L.) and correlations with oxidative stability. Sci. Hortic., 112: 108–119.

Vichi S, Pizzale L, Conte LS, Buxaderas S, Lopez-Tamames E (2003). Solid-phase microextraction in the analysis of virgin olive oil volatile fraction: characterization of virgin olive oils from two distinct geo-graphical areas of Northern Italy. J. Agric. Food Chem., 51: 6572–77.

Methods of braking seed dormancy on germination and early seedling growth of African locust bean (*Parkia biglobosa*) (JACQ.) Benth.

A. I. Okunlola*, R. A. Adebayo and A. D. Orimogunje

Department of Crop Soil and Pest Management, Federal University of Technology, P. M. B. 704, Akure Ondo State, Nigeria.

An experiment was conducted under a net house in the department of Crop, Soil and Pest Management of the Federal University of Technology Akure. This was to investigate the most effective pre-sowing treatments to break seed dormancy, to stimulate seed germination and to identify morphological traits for the assessment of *Parkia biglobosa* seedling vigour. Matured seeds of *P. biglobosa* were collected from farmers in Akoko in Southwest Nigeria, dried at room temperature and tested for viability by floatation. The seeds were then subjected to pre-treatment that includes: soaking in 100% sulphuric acid for one minute, mechanical scarification with sandpaper, soaking in boiling water for two minutes, soaking in cold water (at room temperature) for 24 h. At 12 weeks, the seedlings growth parameters were assessed: seedling height, leaf production, and diameter and branch production. Result obtained shows that all the treatments except soaking in boiling water improved seed germination and reduced Mean Germination Time (MGT). The mechanical scarification appeared to be the most effective method of pre-treatment having the highest germination percentage 91.7%. Plant height appeared to be the strongest morphological trait to identify vigorous seedlings of *P. biglobosa.* Higher value of plant heights, number of leaves, branches and stem girth were recorded on the seedlings treated with scarification while hot water treated seeds performed low. This study revealed that seeds mechanically scarified improved seed germination and seedling growth. It is therefore recommended that mechanical scarification of the seeds of *P. biglobosa* may be effective for breaking dormancy and improving the seedling vigour which can enhance the domestication and cultivation of these valuable seeds in the environment.

Key words: *Parkia biglobosa*, seed pre-treatment, mean germination time, seedling vigour index, scarification, morphological traits.

INTRODUCTION

The “locust bean”, *Parkia biglobosa* (Jacq.) Benth, ccurs in a diversity of agro-ecological zones, ranging from tropical forests with high rainfall to arid zones, from the lower Sudan Savanna Southwards to the derived Savanna and the lowland forest where mean annual rainfall may be less than 400 mm (Gbadamosi, 2005). The tree has the capacity to withstand drought conditions because of its deep tap root system and an ability to restrict transpiration. The “locust bean” tree is an important food tree and plays a very vital role in the rural economics of West African countries; virtually every part of the species is of value as food or fodder (Gbadamosi, 2002). The leaves are rich in nitrogen and used as manure. The yellow pulp surrounding the seed is edible in many forms and the seeds are made into condiments (iru), used extensively as flavouring and additives to soups and stews. The seed contain 54% fat and 30% protein in addition to vitamins and minerals such as calcium, potassium and phosphorus (Aliero, 2004). The branches are sometimes lopped for firewood, the pods and roots are used as sponges and as strings for musical instruments. The tree also serves as wind break and provides shade.

The tree is however, propagated mostly by seeds which

*Corresponding author. E-mail: aiokunlola@yahoo.co.uk.

Table 1. Germination percentage of the treated seeds.

Treatment	% mean at 7 DAS	% mean at 14 DAS	% mean at 21 DAS	% mean at 28 DAS	% mean at 35 DAS
Concentrated H_2SO_4 (A)	41.7[b]	91.7[a]	100[a]	100[a]	100[a]
Mechanical scarification (B)	83.3[a]	91.7[a]	91.7[a]	91.7[a]	91.7[a]
Hot water (C)	0[c]	8.3[b]	25[b]	25[b]	25[b]
Coldwater (D)	0[c]	50[ab]	75[a]	83.3[a]	100[a]
Control (E)	0[c]	50[ab]	91.7[a]	91.7[a]	91.7[a]

Mean values with the same letters are not significantly different from each other.

has commercial value and highly sought by people for food. Hall et al. (1997) discovered that *P. biglobosa* possesses an exogenous dormancy in which the hard seed coat prevents its germination. Due to poor seed germination there is rapid depletion of the natural population of the tree. The numerous importance of the tree has however, called for its domestication and conservation and this can be achieved when there is an alternative means of breaking the dormancy characteristics posed by the seed of *P. biglobosa* the study aimed to assess the most effective pre-sowing treatment methods and its effect on the seedling growth.

MATERIALS AND METHODS

This experiment was conducted during the wet season (June) 2008 at the screen house of the Crop, Soil and Pest Management Department, Federal University of Technology, Akure, Ondo-State, Nigeria. The topsoil used was obtained from the Crop Type Museum in the department. The materials used for the experiment were polypots filled with the top soil in the screen house. Matured seeds of *P. biglobosa* were collected from the natural forest of Oka-Akoko, Akoko South West Local Government area of Ondo State Nigeria. The seeds were washed, sun dried and stored in trays for weeks in a cool dry place. The stored seeds of *P. biglobosa* were subjected to different treatments, which includes:

1. Acid treatment: The seeds were randomly selected and soaked in 100% concentrated H_2S0_4 for one minute and immediately air dried before sowing according to Aleiro (2004).
2. The mechanical treatment: The selected seeds were rubbed against the rough surface of the sand paper until slight exposure of the cotyledon of the seed.
3. Hot water treatment: The seeds treated were soaked in 100°C hot water for about two minutes and air-dried before sowing according to Aleiro (2004).
4. Cold water treatment: The seeds were soaked overnight for 24 h in cold water at room temperature and air-dried before sowing.
5. The seeds were sown without any treatment (as control)

The experiment was laid out in a completely randomized design and replicated three times. The seeds were watered at 3-5 days interval and weeding was done manually by hand pulling.

Data collection

Data were collected on the following parameters:

Germination rate (%): This was evaluated as the proportion of germinated seed expressed as germination percentage of the total number of seeds sown (ISTA, 1985) at 7, 14, 21, 28 and 35 day after sowing (DAS) that is,

(No of germinated seed / No of planted seed) x 100

Seedling assessment was carried out at weekly intervals beginning from the 24 DAS. Parameters assessed include: seedling height (using meter rule), numbers of leaves, numbers of branches and stem girth (using veneer caliper). The numbers of branches and numbers of leaves were counted manually. The data collected in percentage were transformed using the arc sin formula and then analyzed using the SPSS data package and the Duncan Multiple Range Test (DMRT) was used to separate the means found to differ significantly. Graphs were drawn using excel package.

RESULTS

The different treatments showed substantial variation in germination percentage at seven days after sowing (DAS) (Table 1). Seeds mechanically scarified with sandpaper had germination of 83.3% followed by seeds treated with concentrated acid with 41.7%, while seeds treated with hot water, cold water and the control had no germination at 7 days after sowing (DAS) (Table 1). By the 14th day after sowing, seeds treated with 100% concentrated H_2SO_4 and those mechanically scarified with sandpaper had 91.7% germination while seeds treated with cold water and the control had 50% germination and those treated with hot water treatment had 8.3% germination (Table 1). At 21 (DAS) seeds treated with concentration H_2SO_4 had 100% germination while those scarified with sandpaper and the control both had 91.7%. Seed treated with cold water had 75% germination and those treated with hot water had 25% germination. At 28 days after sowing (DAS) seeds treated with concentrated H_2SO_4 had 100% germination, those mechanically scarified and the control had 91.7%, while

Table 2. Mean values for seedling characteristics at (21 DAS).

Treatments	Plant height (cm)	Number of leaves/plant	Stem girth (cm)	Number of branched/plt
A	15.26[a]	9.25[ab]	0.1[a]	2[a]
B	14.28[a]	10.3[a]	0.1[a]	2[ab]
C	4.36[a]	4.2[d]	0.5[a]	1[b]
D	8.12[c]	7.1[c]	0.1[a]	1[b]
E	11.41[b]	8.2[bc]	0.1[a]	1[b]

Mean values with the same letters are not significantly different from each other.

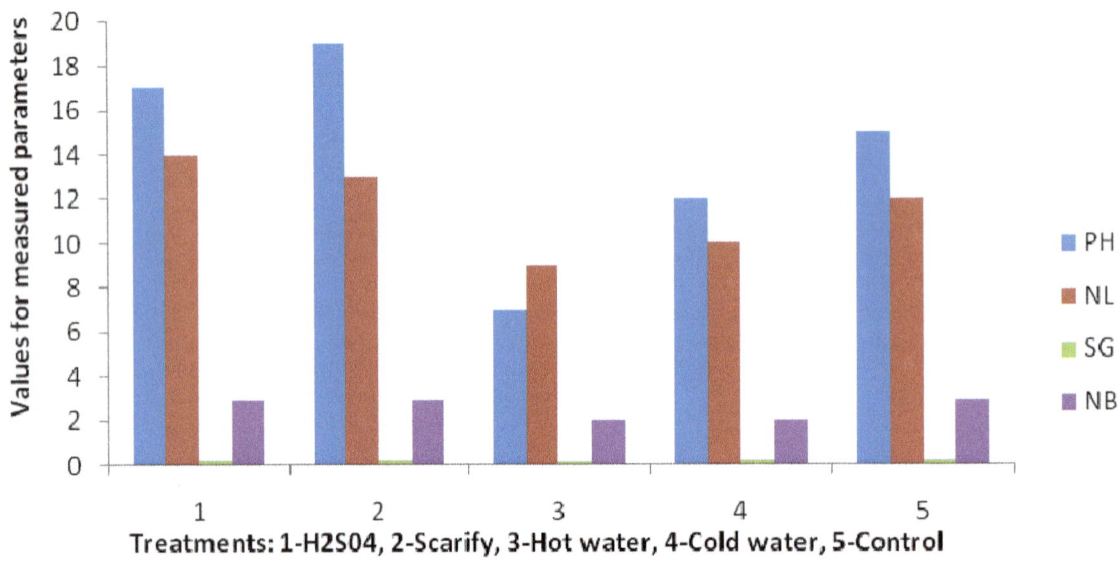

Figure 1.Seedlings growth parameters at 30 days after planting.

the cold water treated seeds had 83.3% and seeds treated with hot water had 25% germination. At 35 DAS, there was no change in seed germination over that recorded on the 28th day after sowing for the seeds subjected to the different pre-treatments (Table 1).

Seedling characteristics

The mean heights of seedlings from each treatment shown in Table 2 were significantly different from each other except for treatment A, B and C which had no significant difference. The mean values for the number of leaves in treatment A and B were not significantly different. Likewise the number of leaves for seedlings under treatment A and E both differed significantly from that of the seedlings from treatment C. There were no significant differences between the number of leaves in treatment D and E in (Table 2). The stem girth of all seedlings in (Table 2) showed no significant difference among all the treatments. At two weeks after sowing, the plant height of seedlings from all the treatments varied amongst each other in with seedlings from scarified seeds attained highest height (Figure 1). The numbers of leaves were more on the seedlings from H_2SO_4 treated seeds. The stem-girth of seedlings from treatments was the same but the number of branches varied from one treatment to the others with seedlings from seed treated with H_2SO_4, scarification and control higher than those treated with hot water and cold water respectively.

In Figure 2, plant height of seedlings from seeds treated with H_2SO_4 and scarification recorded same height which was better than that treated with cold water. Lowest numbers of leaves was recorded on seedlings from hot water treated seeds. The stem-girth of seedlings in all treatment remains relatively the same at the end of 3 weeks. However, the numbers of branches for the seedlings from control and hot water treated seeds was similar but higher than the number obtained from cold water treated seeds. From Figure 3, plant height obtained from scarified seeds was better than all other treatments. Shortest plant height was recorded on seedlings treated with hot water. The numbers of leaves was also highest in the seedlings from seeds scarified while cold water

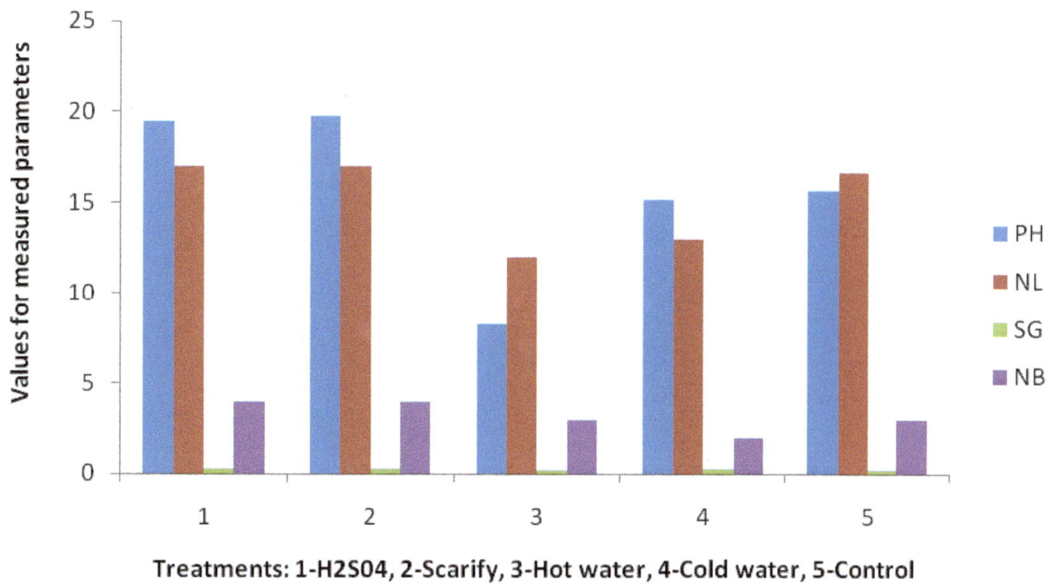

Figure 2.Seedlings growth parameters at 37 days after planting.

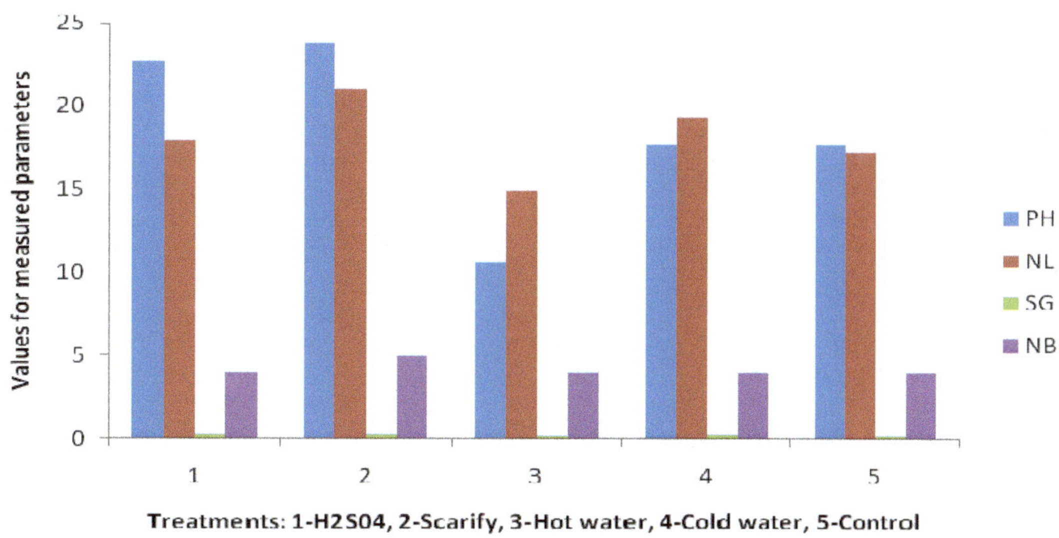

Figure 3. Seedlings growth parameters at 44 days after planting.

treated seeds also performed better than that treated with H_2SO_4. The stem-girth for all the seedlings does not differ from each other up to the end of 4th week. At 55 days after sowing, it was observed that highest plant heights were recorded on seedlings germinated from scarified, H_2SO4, and cold water treated seeds respectively while the lowest height was recorded on seeds treated with hot water (Figure 4). The numbers of leaves recorded also followed the same trend as obtained for plant height. Numbers of branches obtained from seedlings from all treatments were the same except for that obtained from scarified seeds which were higher than the others. The stem-girth of seedlings from seeds treated by scarification remains the biggest at the end of the 5th week while others showed similar result.

DISCUSSION

The differences observed in the germination percentage of seeds subjected to the different treatments (Table 1) implies a significant impact of the various pre-treatments on breaking the seed dormancy. The mechanical scarification by sandpaper had the highest speed of

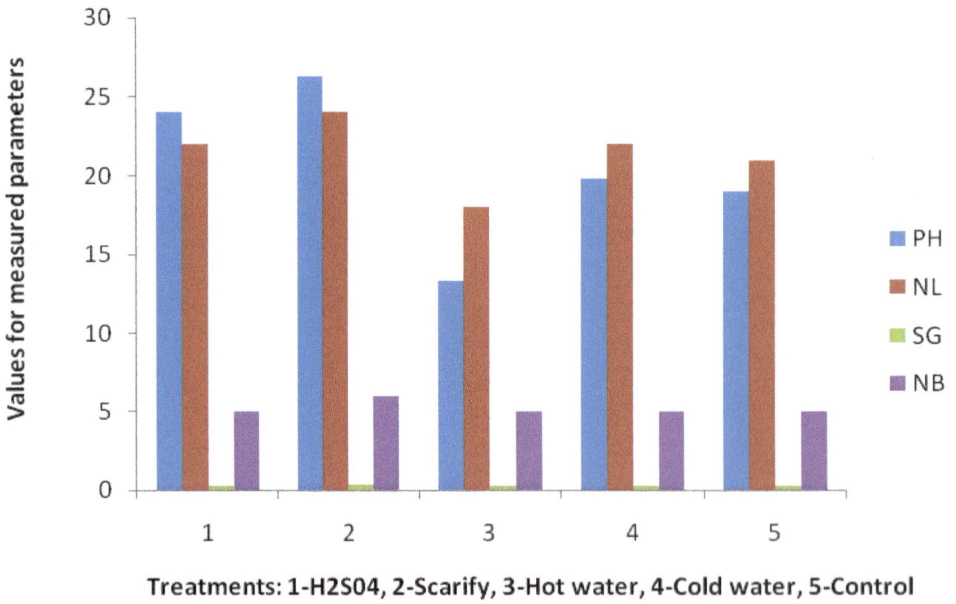

Figure 4. Seedlings growth parameters at 51 days after planting.

germination, which agrees with the report of Tomlinson et al. (2000) that seed dormancy resulting from an impermeable seed coat may be overcome by peeling off the coat. Germination must have occurred as a result of the partial exposure of the cotyledons of the seeds which permits the process of hydrolysis whereby hormones such as auxins and ethylene which could increase nucleic acid metabolism and protein synthesis are released (Uwaegbute, 1996). The seeds treated with concentrated H_2SO_4 also induced germination progressively which must have occurred as a result of its effect on the seed coat. It can be inferred from the result that dormancy of the seeds of *P. biglobosa* were probably associated with the seeds coat, since the treatment that induce germination were those that can effect disruption of the seed coat. Immersion of seed in highest concentrated sulphuric acid disrupts the seed coat. Aleiro (2004) concluded that 98% concentrated sulphuric acid gave the highest percentage of germination for *P. biglobosa* and within the shortest period as compared to 90, 70 and 50%, which indicates that the more rapidly the seed coat is ruptured the faster the rate of germination.

The seeds of *P. biglobosa* treated with hot water showed the least performance in germination percentage as a result of its effect on the seed coat that must have ruptured or damaged the seeds embryo. Sudden dip of dry seed in boiling water may lead to the rupture of the seed coat allowing water to permeate the tissues causing physiological changes and subsequent germination of the embryo (Agboola and Etejere, 1991; Agboola ad Adedire, 1998; Sabongari, 2001) but can as well be detrimental when in excess leading to the death of the seeds. The seeds of *P. biglobosa* treated with cold water for 24 h increased in germination percentage with time and at 35 DAS had attained 100% germination. The effect of this treatment is similar with the placing of seeds in a beaker containing tap water for 24 h (Aleiro, 2004). The seeds under control experiment gave similarly high germination but seeds took a longer period to germinate. The variations observed among the seedling growth characteristics such as plant height, stem-girth, numbers of leaves and numbers of branches can be attributed to the effect of the treatments the seeds were subjected to. Seedlings with the highest plant height, stem-girth, number of leaves and number of branches may be as a result of the early germination by the seedlings induced by the method of dormancy breakage. Significant relationships were observed between seed morphological characteristics and seedling vigour in *Terminalia invorensi* A. Chev (Oni, 1991). These parameters may be very useful in the promotion of rapid production of vigorous seedlings for nursery establishment or species for plantation establishment. Based on the result of this experiment in Table 2, Figures 1, 2, 3 and 4 seedlings from seeds that were mechanically scarified with sandpaper had performed best in comparison with time of germination and highest mean values for seedlings growth characteristics. The seedlings from seed treated with concentrated H_2SO_4 and cold water treatment differed significantly in terms of time of germination, plant height and number of branches (Figure 4). Furthermore, seeds treated with hot water showed the least performance both in germination percentage and seedling growth characteristics compared to the seedling under control experiment. Aliero (2004) reported that seed germination of *P. biglobosa* decreases when seeds were soaked in hot water for more than 4 s, suggesting that embryo may get destroyed on contact with boiling water for a prolonged period. However, it can be deduced from the experiment that the mechanically scarified seeds using the sandpaper had the best performance while the least performed treatment is the hot water treatment.

CONCLUSION AND RECOMMENDATION

The result from the study showed that seeds mechanically scarified improved seed germination and seedling growth. It is therefore recommended that mechanical scarification of the seeds of *P. biglobosa* may be effective for breaking dormancy and improving the seedling vigour which can enhance the domestication and cultivation of these valuable seeds in the environment.

REFERENCES

Agboola DA, Etejere EO (1991). Studies on seed dormancy of selected economic tropical forest species. Nig. J. Bot., 4: 115-125.

Aliero BL (2004). Effects of sulphuric acid, mechanical scarification and wet heat treatment on germination of seeds of African locust bean tree *Parkia biglobosa*. Afr. J. Biotechnol., 3(3): 179–181.

Gbadamosi AE (2002). Domestication of *Enantia chlorantha* (oliv) – A medicinal plant Ph.D Thesis University of Ibadan, p. 192.

Gbadamosi AE 2005. Genetic variation in *Enantia chlorantha* (oliv). A medicinal plant. J. Food Agric. Environ., (JFAE) 3(1): 153-156.

Gbadamosi AE, Faboye OO, Oni O (2005). Seed morphological traits, germination and early seedling growth of *Parkia biglobosa* (Jacq.) R. Br.Ex.G.don from four provenances in Nigeria. Nig. J. For., 35(2): 129–137.

Hall TTJ, Ikeneboneh MJ, Kok R, Ingram JM, Antai SP, Ibrahim MH (1997). Distribution, Conservation and use of *Parkia biglobosa* in Western Africa: J. Sci. Food Agric., 42: 253-262.

Obizoba IC (1998). Fermentation of African Locust Bean, Text on Nutri-quality of plant fruits. Post Harvest Research Unit, Department of Biochemistry, UNIBEN, Nigeria, pp. 160–198.

Oni O (1991). Effects of seed size on seedlings vigour in Idigbo (*Terminalia ivorensis*). J. Trop. For. Sci., 4(3): 215-224.

Tomlinson H, Ng'andwe M, Nikiema A (2000). Field and *in vitro* methods of propagation of the African locust bean tree (*Parkia biglobosa* (Jacq)(Benth.). J. Hortic. Sci. Biotechnol., 75: 42-49.

Uwaegbute AC (1996). African locust bean in: Food from legumes and Oil seeds, pp. 21-23.

Commons becoming non-commons in the efforts for reconciliation between conservation and livelihoods: A case study of northern Pakistan

S. R. Khan[1], S. A. Rahman[2]* and T. Sunderland[2]

[1]Natural Resources Institute, Clayton H. Riddell Faculty of Environment, Earth, and Resources, University of Manitoba, 70 Dysart Rd., Winnipeg, MB, Canada R3T 2N2.
[2]Center for International Forestry Research (CIFOR), Bogor Barat 16680, Indonesia.

The paper examines the impact of converting previous open-access common land into protected areas on the community resource management system in Shimshal, northern Pakistan. We explored three main questions in this regard. First, what happens to community resource management under strict government control of the resource? Secondly, what has happened to the centuries-old practice of the traditional yak herding system of the Shimshal community as a result of the introduction of new policies leading to the creation of protected areas? Thirdly, can the community retain resource control of the protected area (state property), and how can we conceptualize it under the property regime? The research draws the following three conclusions: (1) a clear divide exists between the local government and the community on resource management and resource use, (2) the Shimshal community relies significantly on yaks for their livelihood, and the imposition of new rules and regulations in protected areas has significantly altered the Shimshal community's traditional yak herding practices, and (3) community control over resources in protected areas would be a new experiment under the state property regime. The study concludes that it is possible to bridge the gap between what have been regarded as irreconcilable principles of protected areas and livelihoods. This arrangement needs to be contextualised by giving specific importance to the community and their management practices. The state would achieve its conservation goals by monitoring the local community's activities and ensuring that their livelihoods and conservation efforts do not negatively affect each other.

Key words: Commons, property rights, yak herding, protected areas, livelihoods.

INTRODUCTION

The current issues pertaining to the conversion of commons into protected areas (a state-controlled resource) emerged with the notion that free access to common resources leads to degradation and areas of environmental value should be protected from human activity. As Hardin (1988) concludes, "freedom in the commons brings ruin to all." This view advocates the allocation of full authority to an external agency to regulate the commons. In other words, a state-controlled property regime can reduce overexploitation of common property resources (Hardin, 1988). This perspective could be supported by Hardin's argument, which McCay and Acheson (1987) have succinctly summarized as: "all resources owned in common are, or eventually will be, overexploited." In this context, protected areas represent a clear example of conversion from common property regime to state property. Many governments in developing countries have initiated the process conserving resources in this fashion. However, there has been a growing body of literature on common property resources to show that users are able to restrict access and establish rules among themselves for the sustainable use of natural resources (Berkes, 1986, 1989; McCay and Acheson, 1987; Berkes et al., 1989; Ostrom, 1990;

*Corresponding author. E-mail: sumonsociology@yahoo.com.

Bromley et al., 1992; Cronkleton et al., 2008; Ojha et al., 2008). A wide range of common resource systems have been developed and maintained by local communities in many different societies in the past. Examples can be seen in various regions of India and Mexico, where communities are managing their common resources through sharing forest resources and utilizing different land classification systems, respectively (Nayak and Berkes, 2008). In the context of northern Pakistan, common resources have remained under community control since the time of the Mirs (rulers), where communities established the traditional management system and customary laws to regulate their rights and access to the resources in an effective manner at the local level (Bilal et al., 2003).

However, conflicts have emerged between communities and the state as a result of disagreement over resource-use rights, access, and control of natural resources. In examining the foundations of these conflicts, it is important to examine the issues from two different angles. Firstly, according to the view of those who claim that communities' access and user rights would disrupt the areas of biological importance, state control of natural resources is the only effective way to conserve biodiversity (Locke and Dearden, 2005). The second perspective is a more people-centred approach, "based on the premises that local populations have a greater interest in the sustainable use of resources than the state; and that they are more able to effectively manage those resources through local or 'traditional' forms of access" (Brosius et al., 1998). This has led to debates about whether state or community control over resources would be more effective in resource management. The existing 'blueprint' approaches and strategies to natural resources management have increasingly proved to be unsuitable (Agrawal and Gibson, 1999). Recent literature suggests the need for a shift in the approach to natural resources management so that it might focus increasingly on people-centred policies, bottom-up planning processes, and decentralized governance (Chambers, 1994; Agrawal and Ostrom, 2001; Cronkleton et al., 2008; Ojha et al., 2008). The need for such a shift became obvious as the ill effects of the centralized control of resources became apparent. It led to the alienation of resource users and incited a growing demand for a change in approach towards greater community participation and empowerment. Another management option, co-management, where responsibilities are shared between communities and the state, has been mooted as an alternative means to sustainably manage resources in many countries. However, this management arrangement has been criticised for the gap between design and reality, particularly in relation to participation and control (Agarwal, 2001; Lele 2000).

This study examines the impacts of protected area creation on the local management of the pasture resources of the Shimshal community in northern Pakistan. Our research has explores three main questions:

(1) What happens to community resource management under strict government control of the resources?
(2) What has happened to the centuries-old practice of the traditional yak herding system of the Shimshal community as a result of new policies on protected areas?
(3) Can the community retain control of the protected area (state property)?

We analyse the impact of PA's (protected areas) on the community resource management system, and describe how community control over resources in protected areas would be reflected under the state property regime.

STUDY AREA AND METHODS

The study was conducted in Shimshal, upper Hunza (Gojal), in the Gilgit district of northern Pakistan. It is approximately 62 km from Pasu, located in the Central Karakuram Mountains, in northern Pakistan. This site is characterized by a dry alpine habitat comprised of juniper, shrubby vegetation, community plantations on lower lands, and permanent snowfields in the higher altitudes. Shimshal is comprised of 112 households, with a total population of approximately 1500 individuals living in five adjoining small villages (hamlets).

These villages are permanent settlements in the Khunjerab National Park. Seasonal settlements of the Shimshal community are located in the vast alpine pastures of Pamir, Ghujerab and Lupgar. This is due, in part, to the community's exclusive control of 2700 km^2 of high altitude pasture land (Butz, 1996). The community owns agricultural land; they complement their irrigated agriculture with extensive herding of sheep, goats, cattle and yaks. Their enormous knowledge in raising livestock is well recognized. Livestock plays a major role in their livelihoods (Ali and Butz, 2003), and their herding system represents a high altitude pastures.

The Shimshal community is well organized and is well-known for the people's hardiness. This is the only community in northern Pakistan that has remained in isolation without access to roads until recently. In 2003, after twenty years of hard labour, the community was able to construct 62 km of a single-lane road, with some support from both the government and the Aga Khan Foundation.

Field research was conducted in all five villages in Shimshal. Data were collected by the Participatory Rural Appraisal approach and Focus Group Discussions. Two village level (two annual) and four household level (four quarterly) surveys were carried out to obtain household level information. For village and household surveys, a random sampling method with a sample size of ≥ 50% was selected. A total of 70 households were surveyed at the village level. Semi-structured interviews were conducted with household heads, and focus group discussions were carried out with community groups. The participatory resource mapping tool was used to identify pasture cycles, and this helped to analyze the linkages, patterns, and inter-relationships between land use and their livelihood systems.

CONTEXT OF PROTECTED AREAS (PA'S) AND CONFLICTS TO COMMONS

The establishment of protected areas has been seen as a viable solution for protecting natural ecosystems by excluding humans from those areas (Hutton et al., 2004). Bennett and Lopoukhline (1998) stated that the concept of protected areas has been acknowledged as the basic method for biodiversity conservation since it makes use of regulatory frameworks to prevent disturbance activities. Most ecological zones in Pakistan are not adequately represented within the protected area system (Virk et al., 2003). In addition, many of these areas do not have an effective management system and they employ management approaches that are confined only to protective measures (IUCN, 2003). In Pakistan, PA's have so far followed a top-down approach, which has resulted in conflicts with the communities over access to and use of natural resources (Mock, 1997). According to Mock and O'Neil (1996), the Government of Pakistan recognises only three categories of protected areas: national parks, wildlife sanctuaries, and game reserves. Up to now, the country has established 225 Protected Areas, comprising 14 National Parks, 99 Wildlife Sanctuaries, 96 Game Reserves, and 16 unclassified areas (private, proposed or recommended). The number of PA's continues to grow each year. These conservation measures, however, have not always been based on detailed biological information or criteria and many of these PAs are too small and fragmented to guarantee species survival, especially of wide-ranging fauna such as the snow leopard (endangered species) and Himalayan lynx. Boundaries have not been worked out according to ecological considerations, and the main focus has remained on game species (MACP, 1999; IUCN 2006).

Mock (1997) describes protected areas in Pakistan as islands in a sea of different land uses and of strongly altered ecosystems; such islands may have difficulty maintaining species diversity and may not incorporate functional ecosystems. Studies have shown that PA's have not succeeded in meeting their management objectives with specific reference to Mountain Protected Areas, and the reasons include indigenous peoples' rights to resource extraction for their livelihoods; the improper design, legal establishment, boundary demarcation, resource assessment and setting of objectives for protected areas; and the often negative relationship between forest officials and the local people. These findings reflect flaws in the concept of protected areas which fail to recognise and respect the rights and values of local and indigenous peoples. Over centuries, these communities have developed a way of life which is broadly in balance with the setting in which they live (Mock, 1997). In most cases, the protection of community livelihood systems and cultures and the conservation of the resources upon which they are based have not been seen as special features during the establishment of protected areas. The unique features of the communities' livelihoods are compatible with conversation objectives but do not necessarily adhere to a Western notion of conservation (Mock, 1997).

Generally, conflicts arise as a result of separating natural resource protection from human use and of not adequately addressing livelihood issues (Hough, 1994; Hutton et al., 2004). Many of these conflicts are related to resource access and the rights of the communities, which see their livelihood security being undermined. Many communities have been deprived of access to common resources because the PA's have been based on external principles of protection, while ignoring traditional knowledge and practice founded on learning developed over hundreds of years of experience and adaptation. These communities have traditionally followed laws and regulations set up and enforced by themselves to protect their resources. However, the laws and regulations of the protected areas overthrew the traditional management systems and heightened conflicts over access, use or control of the resources (Davey, 1998).

The Khunjerab National Park (KNP), in northern Pakistan, is a notable example of state control over common resources, in an area where the local community has followed sustained herding practices for centuries (Butz, 1996; Knudsen, 1999). The state views that protected land is an entity that belong to the State (that is, the authoritative bureaucratic-politcal entity) rather than the local community. This has led to conflicts between the community and the state, and possibly will alter the important pastoral systems that help maintain the park landscape (Knudsen, 1999). Ali (2003) points out that "the original park plans would fragment this local management system and upset pastoral migrations which were fully compatible with the aims of conservation defined under the IUCN's guidelines 1979 for conservation in mountain environments".

Wegge (1988) confirms that the Shimshalies' herd management system seemed to be a typical 'traditional practice' and there was no evidence of over-grazing. But the state perceived the local community and their traditional practices of herding as a threat to the National Park, which resulted in alienating the community from their hereditary common resources. Knusen (1999) stresses that it is dangerous to write the community off before their potential role has been established. The concerns here are that the existing policies have put an end to the pastoral system and that the park landscape might not be upheld (Knudsen, 1999). There is a possibility that the conflict will bring detrimental results both to the state and to the community if the conflict is not resolved. For the state, the wildlife it aims to conserve will be lost as a result of aggravated actions, and for the community, the loss of spiritual and cultural affiliation to their heritage seems obvious.

GOVERNING THE COMMONS IN NORTHERN PAKISTAN

In northern Pakistan, historically, common resources were divided among tribes and lineage groups by the Mirs (rulers), where the communities established traditional management systems and customary laws to regulate their rights and access to the resources in a manner that was effective in terms of management at the local level. Primarily, two property systems exist throughout the region:

(i) Tribal (Shinaki)
(ii) Principality (Rajgiri) systems (Bilal et al., 2003).

Under the Tribal system, a council of elders dealt with the issues of excludability, whereas under the Principality system, the representatives of the Mirs decided the excludability and subtractability. However, under both systems, local communities were part of the decision-making process, and there was an arrangement to include the collective decisions of the communities regarding the extent of exclusion and subtraction (Bilal et al., 2003). The representative (Wazir) of the Mir used to convene a gathering of village elders to discuss issues of relevance. The community would vote on each decision, and if greater than a majority was expressed, the decision was considered final. If the Wazir gave a final decision that was against the decision taken by the community, then it would have the opportunity to approach the Mir by visiting him and bringing him a small gift, generally, livestock – a yak or bull - and animal products such as Poshoro (5 to 10 kg butter) or Philam (woollen cloth). Then the Mir could decide in favour of the community; he sometimes based his decision on an outsider's advice or opinion.

However, the situation changed after the 1970's, with the abolition of the Mir's regime. The principalities were declared state property, while in areas such as Darel and Tangir within the Diamer district, where there was a tribal system, the tribal councils negotiated with the government to retain their commons (Bilal et al., 2003). The state passed the Forest Act 1975 and, under this act, the state then had entitlement to all the resources except the forests in a few parts of the Northern Areas, such as in Darel of Diamer district, where the communities own the forest. However, with the end of the Mir's regime and the establishment of state control, many communities in northern Pakistan continued to follow their traditional user rights to most of the common resources, specifically, on pastures. Many of the pastures remain under the exclusive control of particular communities, where the community decides who has the right of use and who is excluded. One of the best examples is the Shimshal community in northern Pakistan, which has exclusive control over 7200 km^2 of high altitude pasture land that includes Pamir, Ghujerab, Lupgar and many other pastures (Butz, 1996). The community has been utilizing these pasture resources for many centuries through their unique yak herding system. However, the establishment of Khunjerab National Park by the state has led to conflicts related to the Shimshal community's exclusive rights and their herding practices.

SHIMSHAL YAK HERDING SYSTEM

Shimshal yak herding predominantly follows a traditional pattern dictated by the climate and seasons, the topography of the land, and social and cultural influences. Their traditional system relies on accumulated centuries of experience, including knowledge of the pastures' productivity, availability of water during the summer and winter, accessibility, vulnerability to predators, and religio-cultural influences embedded in their self-identity as 'Shimshali', a community that is highly devoted to maintaining their culture and hereditary resource. The community affiliation with livestock, especially yak, and with the pasture reflects their devotion to their legacy and their livelihood security.

For most of the community, livestock plays a major role in their household economy, their consumption of dairy and non-dairy products, and their meat requirements. Animals are also sold to obtain cash income to cover several months of their household expenses.

The Shimshali yak herding practice, the use of different pastures in different seasons and periods, the number of livestock herded in particular pastures, and the periodic movement of the animals to higher or lower elevations, are akin to and represent a 'modern' rotational grazing system. This herding system, which Butz (1996), referred to as 'interwoven transhumant cycles', reflects generations of decisions villagers have made about their herding practice. Confidence in their best pasturing system without employing any scientific knowledge comes from the decision making process they employ at three levels, as described by Butz (1996): first, at the household level, followed by the community level, and the third, at the pasture-cycle level. Households need to herd their livestock to ensure their livelihood security. They are also motivated to maintain their relationship with hereditary

Table 1. Key mechanisms of community management of common resources in Shimshal.

Mechanism	Purpose
Village level decisions: all village heads and household heads have a role in decision making process. Village elder (lumberdar) makes the decision.	Involve all village heads, household heads in decision making and provide an equal opportunity to every household head. Have every individual abide by collective decisions and the regulations set by the community.
Household level decisions: all members of the household have a role in household level decisions.	Involve all household members in decision making and provide an equal opportunity to every member. Ensure their commitment, availability and contribution in labour.
Pasture-cycle level decisions	Informed decision based on climate conditions, accessibility, availability of fodder and safety of livestock.
Pasture management: A series of pasturing on specific pastures, pasturing cycles, time of pasturing, determining duration and livestock numbers.	Attain maximum benefit from the resources. Maintain pasture quality by ensuring the continuous supply of fodder for the livestock in different seasons; facilitate seed regeneration, and new growth of vegetation. Retain traditions and heritage as proud symbols.

Source: Focus group discussion with the community elder group.

pastoral land, which Butz (1996) referred to as instrumental and symbolic values. To put this into practice, households have to make a conscious decision. We observed that the household members get together to plan how many livestock are required to send to pastures. The decisions are based on several factors: availability of number of persons (labour), affordability of cost in terms of cash or kind (material), and the number of milking animals available.

At the community level, various actors are involved in the decision making process. These include but are not limited to the appropriateness of the pasture for the specific number of livestock, mapping of pastoral movements based on their years of experience, religious festivals and other ceremonies. The community level decisions ensure an equal opportunity is provided to all households in the shared resource system. It is important to note that the community's main priority is to conserve their resources, and this is reflected in their resource-use activities. At the pasture-cycle level, herdsmen have to make conscious decisions based on their experience of weather conditions, such as access during winter; climatic conditions always dictate their decisions. The herdsmen have to make sure that the livestock herd is safe from predators, that enough fodder is available to feed on, that the water requirement is fulfilled, and, especially, that attention is given for the timely departure to other pastures as well as for the village to attend the special ceremony called 'Kutch'. This event reflects the community's gratitude to God for His blessings and for their safe arrival with their animals and accumulated wealth.

The mechanism of the herding practice and the management of the resources are described in Table 1.

SUMMER PASTURING SYSTEM

Summer pasturing represents the combination of ecological knowledge about the climate and vegetation with the decision making process. Knowledge of the pasture, vegetation, and caring capacity plays a key role in pasturing. In summer pasturing, the major role that women play in livestock herding is well recognized. After village level decisions are made about who would be going with the livestock, the herders have to leave the village by the first week of May to travel to Shujerab, the nearest pasture. This arrangement is made based on two reasons: firstly, the availability of new vegetation in pastures for the livestock is considered and, secondly, the village has already completed the cultivation of agricultural crops and the fields have been closed for grazing.

The next move, to Shwert pasture, is determined by climatic conditions and the availability of new vegetation in the next pastures. There is also a time limit (period) set for each pasture, with a time cushion of three to four days. The pasture cycle continues, with several stays made at Sher Lakhsh, Furzin-i-Dasht, Gorjerav, Sher Bulak, Ghrsar and Sher-a-lik, until the final destination, Pamir, is reached, where they spend two and half months. By September 10th, they have to leave Pamir. Any delay or early return would put the livestock at risk. Delays in the return would make them face cold temperatures or snow at high altitudes, and an early return would put the livestock at risk because they would have to cross water routes several times, and this would be made difficult by peak water flows. Their arrival to the village after spending over six moths in pastures is celebrated with an event called 'Kutch'. This is a special occasion to thank God for their safe return with the accumulated wealth. The celebration continues for a week, with friends and families invited to their homes. Similar summer pasturing is done by other herders in other main pastures, such as the Gujerab, Lupgar, and Yazghel.

Most of the dairy products such as butter, qurut (a milk product), cheese and many other products are produced during summer pasturing. These products are consumed in their main diet, and are also sold for cash or bartered.

WINTER PASTURING SYSTEM

Winter pasturing is done predominantly with yaks. Normally, herders stay with their livestock to protect it from being attacked by wolves or snow leopards, especially at the time of calving, and to prevent the herd from straying. Given the fact that the Shimshal community has limited land available in the village to feed their livestock, especially yaks, their dependency on pasture resources is obvious. However, yak herding in Shimshal has another purpose related to the proper utilization of pastures. Although it is very tough to graze their yaks in the winter, the community has been practicing winter pasturing for centuries and in this fashion they maintain the pasture resource. It is their perception, based on experience, that if certain pastures remain untouched after grazing their livestock in the summer, then those pastures will have low productivity in the following year. Yaks are grazed during winter in those pastures, and their pattern represents a well established grazing system of high altitude pastures.

Common to both summer and winter pasturing is the caring of their ancestral resource, which they affiliate with the founder of Shimshal. According to one of the popular histories, Mamu Singh, a Burusho (brushski-speaker) from Baltit (Central Hunza), found this region four centuries ago. Later his son, Sher, discovered all the other territories, including Pamir, which the lineage of the Shimshal—Gazikator, Bakhtikator and Baqikator—claim as their ancestral land (SNT, 2007). However, the community practices have been disrupted as they have been forced to abandon the use of certain pastures upon which their livelihoods depend.

COMMON RESOURCES AS LIVELIHOOD SECURITY, SPIRITUAL AND CULTURAL WELL-BEING

The commons have been seen as a resource that will eventually be degraded by free access, as Garret Hardin (1988) hypothesized. Many scholars argued against and criticised this view, which led to more thorough examinations of the commons. An entire array of information has been collected and knowledge has been gained through various studies related to the scale, complexity, institutions, self-organization and many other matters associated with the commons. However, the focus has never been on linking the commons to livelihood security and to spiritual and cultural integrity. If we look at common property with a view to "poverty" rather than from a "resource" perspective, there are thousands of people that survive on common resources, mostly in hilly areas. We may be able to capture a part of the answers to the above questions by understanding the livelihood systems of communities and recognizing how their management systems demonstrate the complexity of commons management. By looking at the example of the Shimshal community and their traditional systems, the commons play a pivotal role in maintaining the community's culture and the spiritual values that enrich their spiritual life and cultural well-being. These are deeply embedded in their self-identity as 'Shimshali', a community that is highly devoted to maintaining their culture and hereditary resources. These communities work collectively to secure their livelihoods.

For many communities, the commons are the only resources to which the poor may get an equal share in resource use, and this gives them a sense of security. As Kirkby (2000) affirms, poor communities are mainly dependent on the physical environment, and especially on common property resources, because they cannot afford access to any other resources. The poorest therefore suffer the most from the effects of resource degradation. Fenny at al. (1998) expresses their concern over the effects of restricting poor people's access to natural resources. According to the authors:

"Poor people tend to be most dependent upon the environment and the direct use of natural resources for their livelihood opportunities, and therefore are the most severely affected when the environment is degraded or their access to natural resources is limited" (Fenny et al., 2002).

There is ample field-based evidence that the communities have effectively managed their common resources. Perspectives on the commons have changed from the initial perception of Hardin (1968) that the commons are everyone's resource, which will eventually be degraded, to the perspective that the commons traditionally are a well managed and maintained resource that replenishes itself as long as human activity acts to maintain the landscape.

THE NEW HORIZON: LINKING CONSERVATION TO LIVELIHOODS

There is a potential to bridge the gap between what have been regarded as irreconcilable principles in the management of protected areas and the pursuit of livelihoods. Recently, a growing literature has emerged

Table 2. Community managed state property.

Practices	Under state control	Under new arrangement
Livestock grazing	Very limited or no access	Supervised grazing
Collection of medicinal plants	Very limited access or no access	Supervised collection
Trophy hunting	Completely banned	Limited, based on trophy size and violability of the population
Eco-tourism, trophy hunting	Allowed free entry There is no provision for trophy hunting under the PA category of National Park	Allowed with payment of a nominal fee, fund will be used for the development of the community and resources. Revenue from the trophy hunting will be used in building schools at the community level and community health centres.
Institutional level management	Run by a single authoritative person	Committee members comprised of elders and the well-educated
Benefits	State gets all the benefits	Sharing of benefit: community will get the benefit from protecting it.
Regular monitoring	With very limited funds available to the government there is no monitoring	Community will provide regular monitoring though selected individuals and with partner organizations (WWF, IUCN) to conduct wildlife census and determine caring capacity.
Protection of culture	Not a mandate	Retaining culture would be the main priority.

Source: semi-structured interviews, group discussions, official documents/records.

that aims to link conservation and livelihoods through partnerships between the government, community groups and non-government organizations. Some examples of such partnerships can be widely seen in northern Kenya, India, Nepal, Bangladesh and many other countries. However, there are very few examples in Pakistan in which the community has participated in state-controlled resources through private organizations (MACP, 1999). A new emerging and potentially promising community livelihoods-centred approach, which appears to offer a way forward to reduce state-community conflict, would be a positive step in managing resources in areas with high resource-dependent communities. Such an approach would view indigenous cultures and resource-use practices as mutually compatible and key parts of a system. The community-driven conservation regime, with a focus on their livelihoods diversification and advancement, would bring a sense of responsibility, respect and ownership of resources. The goal would be to provide an option to the community to continue their resource-use practices in harmony with the natural resource base and also encourage them to gain skills and knowledge for the best use of the resource. A comparison between the state-controlled and the new management arrangements is outlined in Table 2.

In the case of Shimshal, the community is already involved in eco-tourism activities. This can be further explored by utilizing their local knowledge on high altitude alpine pastures, including the traditional yak herding system, and also by understanding their socio-cultural systems.

It would be appropriate for the state to see the community as stewards that conserve the resources. The state could play a role as an advisor to achieve conservation goals by ensuring the livelihood security of the local community. This would be made possible by enabling government policy directives that emphasize community involvement and benefit-sharing in the resource management. The challenge would be related to how such an arrangement would work under the property regime.

STATE PROPERTY UNDER COMMUNITY CONTROL (NEW MANAGEMENT ARRANGEMENT)

The literature is clear about how property can be held as distinct resources within four different categories: open access, private property, state property, and communal property (Berkes and Farvar, 1989; Feeny et al., 1998). However, there is a gap in terms of possible combinations of categories of property, which leads us to understand that property rights only make sense if they are held as exclusive rights. This kind of understanding of property is especially critical for analysing property rights in the context of protected areas (state property) under community management and control.

Similarly, co-management arrangements cannot be

Table 3. Bundle of rights under the new category of property regimes.

Bundle of rights	Under state control	New category (community controlled state property)
Access	x	xx
Withdrawal	x	xx
Exclusion	-	xxx
Management	-	xxx

'x' = Limited access, 'xx'= Moderate access, 'xxx'= More access, '-' = No access to resources.

clearly defined either as state or common property based on the given definitions. Several scholars have defined the nature of these regimes (Berkes, 1989; Feeny et. al., 1998; McCay and Acheson, 1987; Bromley, 1992; Ostrom, 1990) through developing the following explanations. Open Access is the absence of well-defined property rights, where access to the resource is open to everyone. With Private Property, individual rights to exclude others and regulate the use of the resource are known as private property. State Property rights are vested in the government to decide access to and levels of exploitation of the resource. Common Property is held by an identifiable community of interdependent users who exclude outsiders while regulating use by members of the local community.

State property under community control would be a new experiment under property regimes. This arrangement would allow the community to manage the resource and exclude others from using it (grazing of pastures or extraction of the resource). The state has the authority to allow a specific number of livestock per season based on the carrying capacity of the pastures or the specific number of visitors per year to the park.

Any income generated through the park would be invested in community development schemes, the development of pastures, the procurement or creation of educational material and the pursuit of related research. The bundle of rights, access, withdrawal, exclusion and management under the new category would give ownership to the community under the guardianship of the state. Table 3 outlines the bundle of rights that would exist under the new management arrangement.

What makes this arrangement different from co-management is that government decision making is made in consideration of the community's rights, allowing the community to negate government officials' decisions. This is particularly important where the communities are not aware of their rights. In the case of the above arrangement, the community enjoys the power of making decisions related to their resource and the state acts as an advisor.

CONCLUSION

Given the diverse cultures and languages of northern Pakistan, it would be difficult to generalize our conclusions to the whole region, but in a specific context like the Shimshal valley, or other similar communities, our conclusion is grounded in realities and presents some viable alternative strategies for conservation. Our observations support that pastures are an important natural capital for the Shimshal community, but inappropriate management strategies have increased their livelihood insecurity.

The concept of protected areas in the context of the mountains in northern Pakistan needs to be re-examined to avoid bringing about severe consequences, since it often neglects the role that humans have long played in shaping and maintaining the landscape. Based on the experiences and the consequences of protected areas, we need to rethink our current understanding of conservation in consideration of the context. The current system of protected areas needs to be adjusted to reconcile conservation and livelihoods. There is a possibility of bridging the gap between what have been regarded as irreconcilable principles of protected areas and livelihoods. This will require a change in the management system from state control to community control by allowing local communities to serve as stewards of the resource.

Community control of protected areas would represent a new experiment in northern Pakistan. However, this arrangement needs to be contextualised by giving specific importance to the community and their management practices. The state would achieve its conservation goals by monitoring the local communities' activities to ensure that livelihood and conservation pursuits are not negatively affecting each other. Such an arrangement would not only strengthen community livelihoods but also give a boost to the management of the resource itself. This would reduce conflicts between the state and the community and simultaneously lead to effective conservation of local natural resources.

ACKNOWLEDGMENTS

We would like to thank the Shimshal Community for their participation in the work. The field research was supported by the Aga Khan Foundation (AKF), Geneva.

REFERENCES

Agarwal B (2001). Participatory exclusions, community forestry, and gender: An analysis for south Asia and conceptual framework. World Dev., 29 (10): 1623-1648.

Agrawal A (2002). Common resources and institutional sustainability, in: Ostrom E, Dietz T, Dolsak N, Stern PC, Stonich S, Weber EU (eds.), The drama of the commons. National Academy Press, Washington DC.

Agrawal A, Gibson C (1999). Enchantment and disenchantment: the role of community in natural resources management. World Dev., 27(4): 269-649.

Agrawal A, Ostrom E (2001). Collective action, property rights and decentralization in resource use in India and Nepal. Polit. Soc., 29(4): 485-514.

Ali I, Butz D (2003). The Shimshal governance model - a community conserved area, a sense of cultural identity, a way of life. Policy Matters, 12:111-120.

Bennett G, Lopoukhine N (1998). The cultural heritage: a Canadian approach to cultural resource management. UNESCO.

Berkes F (1989). Common property resources: ecology and community-based sustainable development. Belhaven Press, London.

Berkes F (1986). Local-level management and the commons problem: a comparative study of Turkish coastal fisheries. Mar. Policy, 10: 215-229.

Berkes F, Farvar TM (1989). Common property resources: ecology and community-based sustainable development. Belhavan Press, London.

Berkes F, Feeny D, McCay BJ, Acheson JM (1989). The benefits of commons. Nature, 340: 91-93.

Bilal A, Haque H, Moore P (2003). Customary laws. IUCN-The World Conservation Union, Pakistan.

Bromley DW, Feeny D, McKean M, Peters P, Gilles J, Oakerson R, Runge CF, Thomson J (eds.) (1992). Making the commons work: theory, practice, and policy. ICS Press, Oakland, CA.

Bromley DW (1992). The commons, property and common property regimes, in Bromley DW (ed.), Making the commons works theory, practice and policy. Institute of Contemporary Studies, San Francisco.

Brosius JP, Tsing AL, Zerner C (1998). Representing communities: histories and politics of community-based natural resource management. Soc. Nat. Resour., 11: 157-168.

Butz D (1996). Sustaining indigenous communities: symbolic and instrumental dimensions of pastoral resource use in Shimshal, Pakistan. The Canadian Geogr., 40(1): 36-53.

Chambers, R., (1994). 'Paradigm Shifts and the Practice of Participatory Research and Development', Working paper 2, Institute of Development Studies, University of Sussex.

Creswell JW (1994). Research design: Qualitative and quantitative approaches. Sage Publications, Thousand Oaks, CA,

Cronkleton P, Taylor P, Barry D, Stone-Jovicich S, Schmink M (2008). Environmental governance and the emergence of forest-based social movements. Occasional Paper 49, CIFOR, Bogor, Indonesia.

Davey A (1998). National system planning for protected areas. IUCN-International, Switzerland.

Ellis F (2000). Rural livelihoods and diversity in developing countries. Oxford University Press, London.

FAO (2000). Informal working group on participatory approaches and methods to support sustainable livelihoods and food security (iwg-pa) of the food and agriculture organization of the United Nations (FAO). Accessed from www.fao.org/participation/default.html on August 2009.

Feeny D, Berkes F, McCay BJ, Acheson JM (1998). The tragedy of the commons: twenty-two years later. Hum. Ecol., 18: 1-19.

Hardin JL (1988). Obstacles to effective management of conflicts between national parks and surrounding human communities in developing countries. Environ. Conserv., 15(2):129-136.

Hough JL (1994). Institutional constraints to the integration of conservation and development: a case study of Madagascar. Soc. Nat. Resour., 7:119-124.

Hutton JM, Adams WM, Murombedzi JC (2005). Back to the barriers? changing narratives in biodiversity conservation. Forum Dev. Stud., 2: 341-370.

IUCN (2003). Strategy for sustainable development, northern areas. IUCN, Pakistan.

IUCN (2006). Database of protected areas, Pakistan. IUCN, Pakistan.

Kirkby J (2000). Environmental security, livelihoods, and entitlements: what is the relationship between land degradation and poverty. ETC Netherlands.

Knudsen A (1999). Conservation and controversy in the Karakorum: Khunjerab national park Pakistan. J. Pol. Ecol., 56: 1-29.

Lele S (2000). Godsend, sleight of hand, or just muddling through: Joint water and forest management in India. ODI Natural Resource Perspectives, no. 53. Oversees Development Institute, UK.

Locke H, Dearden P (2005). Rethinking protected area categories and the new paradigm. Environ. Conserv., 32: 1-10.

MACP (1999). Mountain area conservancy project document. IUCN, Pakistan.

McCay BJ, Acheson JM (1987). The question of the commons: the culture and ecology of communal resource. University of Arizona Press, Tueson.

Mock J (1997). Mountain protected areas in northern Pakistan: the case of the national parks. Proceedings of the Third International Hindukush Cultural Conference, Karachi, Pakistan.

Mock J, O'Neil K (1996). Survey on eco-tourism potential in the biodiversity project area. IUCN, Pakistan.

Morgan D (1997). Focus groups as qualitative research. Sage, London,

Morgan D (1998). Focus group as qualitative research methods. Sage Publications, New Delhi.

Nayak P, Berkes F (2008). Politics of co-optation: community forest management versus joint forest management in Orissa, India. Environ. Manage., 90: 70-88.

Nelson J (2004). Conservation and communities in central Africa: The need to secure indigenous rights and biodiversity. accessed from http://www.forestpeoples.org/documents/africa/cefdhac_5_ip_rights_ biodiversity_ may04_eng.shtml#_Conservation's_failure_in_Central_A on March 2009.

Ojha HR, Timsina NP, Chhetri R, Paudel K (2008). Communities, forests and governance: policy and institutional innovations from Nepal. Adroit Publishers, New Delhi.

Ostrom E (1990). Governing the commons: The evolution of institutions for collective action. Cambridge University Press, London.

Ostrom E (1987). Institutional arrangements for resolving the commons dilemma: some contending approaches, in McCay BJ, Acheson JM (eds.), The question of the commons. University of Arizona Press, USA.

SNT (2007). Shimshal history. accessed from http://www.snt.org.pk/aboutus.html on October 2008.

Virk A, Sheikh K, Marwat A (2003). Northern areas strategy for sustainable development. IUCN Pakistan.

Wegge P (1988). Assessment of Khunjerab national park and environs, Pakistan. IUCN, Gland.

Use of aeroponics technique for potato (*Solanum tuberosum*) minitubers production in Kenya

M. W. Mbiyu[1]*, J. Muthoni[1]*, J. Kabira[1], G. Elmar[2], C. Muchira[1], P. Pwaipwai[1], J. Ngaruiya[1], S. Otieno[1] and J. Onditi[1]

[1]Kenya Agricultural Research Institute, National Potato Research Centre -Tigoni P. O. Box 338-00217, Limuru Kenya.
[2]International Potato Centre (CIP), sub-Saharan Africa Regional Office P. O. Box 25171-00603 Nairobi, Kenya.

In Kenya, potato productivity is severely constrained by limited high quality seed tubers; this is precipitated by inefficiencies in various stages in the seed production system. Production of certified seed starts with meristem tip cuttings in the tissue culture. The resultant plantlets are grown in pots in the greenhouse and/or screen-house for production of minitubers (generation 0). In Kenya, production of minitubers through aeroponics systems has been introduced. This paper discusses the production of potato minitubers using aeroponics. Important considerations that should be addressed before setting-up an aeroponics based potato minituber production system are also discussed. Research areas to optimize production and improve field performance of mini-tubers produced through aeroponics are also presented.

Key words: Aeroponics, clonal multiplication, hydroponics, minituber production, rapid multiplication techniques, tissue culture.

INTRODUCTION

Aeroponics is a plant culture technique in which mechanically supported plant roots are either continuously or periodically misted with nutrient solution (Barak et al., 1996). The international union of soil-less culture defines aeroponics "as a system where roots are continuously or discontinuously in an environment saturated with fine drops (a mist or aerosol) of nutrient solution" (Nugali et al., 2005). The basic principle of aeroponics is to grow plants in a closed or semi-closed environment by spraying the plant's roots with a nutrient-rich solution. Ideally, the environment should be kept free from pests and diseases so that the plants may grow healthier and quicker than plants grown in a soil medium. However, because most aeroponic environments are not properly closed, pests and disease may still be a threat. As aeroponics is conducted in air combined with micro-droplets of water, almost any plant can grow to maturity because of an abundant supply of oxygen, water and nutrients.

Perennial plants have been maintained for as long as 13 months in the aeroponics system (Peterson and Krueger, 1988).

In addition to commercial crop production, aeroponics system has been used for studies on root system especially root micro-organisms (Hung and Sylvia, 1988; Khan and Sinclair, 1992; Sylvia and Jarstfer, 1992; Wagner and Wilkinson, 1992), root response to drought (Mavoungou et al., 1982; Hubick et al., 1986; Robertson et al., 1990a, b), effects of oxygen concentrations on root growth (Shtrausberg and Rakitina, 1970; Soffer and Burger, 1988) and plant cultivar differences in root growth (Truong and Beunaid, 1978). Aeroponics offers researchers a non-invasive means to examine plant roots during development. It also allows researchers a large number and a wide range of experimental parameters to use in their work (Stoner, 1983). The ability to precisely

*Corresponding author. E-mail: jayney480@yahoo.com.

control the root zone moisture levels and the amount of water delivered makes aeroponics ideally suited for the study of water stress. Hubick et al. (1986) evaluated aeroponics as a means of producing consistent, minimally water-stressed plants for use in drought or flood physiology experiments. Use of aeroponic systems to evaluate plant response to nutrient regimes has been infrequent, nutrient uptake rates in aeroponics systems is unknown (Weathers and Zobel, 1992). Barak et al. (1996) suggested that the nutrient uptake rates could be calculated by measuring the concentrations and volumes of influx and efflux solutions in cranberry plants.

This paper discusses the production of potato minitubers using aeroponics in Kenya. Important considerations that should be addressed before setting-up an aeroponics-based potato minituber production system are also discussed. Research areas to optimize production and improve field performance of mini-tubers produced through aeroponics are also presented.

HISTORY OF AEROPONICS USE

Techniques of growing plants without soil were first developed in the 1920s by botanists who used primitive aeroponics to study plant root structure; aeroponics has long been used as a research tool in root physiology (Barker, 1922). In the early 1940s, the technology was largely used as a research tool rather than an economically feasible method of crop production. Carter (1942) was the first researcher to study air culture growing and described a method of growing plants in water vapor to facilitate examination of roots. Fifteen years after the study of Carter (1942) and Went (1957) named the air-growing process in spray culture as "aeroponics".

Aeroponics has been used successfully in production of several horticultural and ornamental crops (Biddinger et al., 1998). Aeroponic system has been applied successfully in Korea for potato seed tuber production (Kang et al., 1996; Kim et al., 1999). Ritter et al. (2001) demonstrated that minituber production using aeroponics under temperate conditions substantially improved yields. Farran and Mingo (2006) reported a minituber yield of 800 tubers/m^2 at a plant density of 60 plants/m^2 over a five month period with weekly harvests. This translates into a multiplication rate of 1:13. They also found the field performance of aeroponically produced tubers to be similar to minitubers produced from the pots. At the International Potato Centre (CIP) in Peru, yields of over 100 tuberlets /plant were obtained (Otazu, 2010). Lommen and Struik (1992) found that the number and timing of the harvests were the key factors in optimizing minituber production.

Aeroponics technology is being tested in several African countries for the production of potato minitubers(Lung'aho et al., 2010). In Kenya, pilot units have been set up at KARI-Tigoni, Genetics Technology International Limited, Kisima Farm, Suera Farm and Agricultural Development Farm, Molo.

OVERVIEW OF POTATO SEED PRODUCTION IN KENYA

Shortage of good quality seed potato is one of the most important factors limiting potato production in Kenya (MoA/GTZ, 2009). The annual national certified seed potato requirement was estimated to be 300,000 tons (MoA/GTZ, 2009). Basic potato seeds are produced by Kenya Agricultural Research Institute (KARI) Tigoni whose physical capacity is limited and Kisima Farm (a private company) (Riungu, 2011). Consequently, Agricultural development cooperation (ADC) can only produce 1% of the national certified seed requirement (Ayieko and Tschirley, 2006). Due to limited supply, the certified potato seeds are highly priced and the cost of seeds account for 42% of the total production costs (Kaguongo et al., 2008). As a result, farmers depend on seed from informal sources which include farm-saved (self supply), local markets, and neighbours (Muthoni et al., 2010). Self-supply is the major source of potato seed tubers for most farmers (Kaguongo et al., 2008). This informal system leads to the use of poor quality seeds and often accelerates the spread of seed-borne diseases (viruses, bacteria wilt, and nematodes) (Ng'ang'a et al., 2003).

Production of basic and pre-basic seed tubers in Kenya

Production of disease-free potato seed tubers starts with tissue culture using meristem tip culture (KARI, 2007). The *in-vitro* plantlets produced are then multiplied 3 to 4 times in the laboratory using nodal cuttings. Six to seven weeks after the last multiplication step, the plantlets are transferred into the sand trays where excess media is carefully removed from the roots using tap water and they are then transferred into seedling trays containing sand substrate. The water and nutrient solution is added in the ratio of 1:1 and watering is done each day until they are transplanted into the aeroponic boxes or pots for production of pre-basic seeds (generation 0) (Muthoni et al., 2011). The generation 0 seed tubers obtained are then multiplied in the field for three generations to produce basic seeds. The three generations (1, 2 and 3) are only meant to increase the amount of seeds. The basic seeds are then supplied to Agricultural Development Corporation (ADC) to produce certified seeds; the certified seeds are then sold to farmers for production of ware potatoes (KARI, 2007). During pre-basic seed production, the germplasm is indexed regularly to ascertain the absence of viruses. According to Muthoni et al. (2010), efficiency in pre basic seed potato production is important for certified seed potato production. At KARI-Tigoni however, most pre-basic seeds are produced using the pot method. This technique requires soil sterilization leading to high fuel costs. In addition, the technique has a low multiplication rate (6 to 8 tubers /plant) unlike aeroponics (50 to 100 tubers per plant (Otazu, 2010; Muthoni et al., 2011). Because of this, an aeroponic unit was set up at Kenya Agricultural Research Institute-Tigoni in 2008 by the international potato centre (CIP) to enhance prebasic (minitiuber) production.

AEROPONICS SYSTEM FOR POTATO MINITUBER PRODUCTION AT KARI-TIGONI

The greenhouse/screen-house facility

An aeroponics system for seed tuber production is housed in a tall screenhouse; the conventional screen house is too low for incorporation of the aeroponics system (Otazu, 2010). When using aeroponics, it is desired to maximize the vertical dimensions of the screen house. This is because both the root system and the foliage usually grow longer than in the pots (Otazu, 2010). In the aeroponics unit at KARI, Tigoni, potatoes have grown up-to 110 to 140 cm tall. Shorter greenhouses also tend to get hotter. The screen-house is roofed with asbestos sheets to keep out dust and fungal spores. The roof is then covered with shading net to lower the temperature in the greenhouse. The roof and doors are properly sealed to keep off insects. The sides are be made of fine net to keep off insects.

Growth boxes

The growth boxes are placed inside the screen house. The boxes to be used depend on the capacity of the screen house and the amount of minitubers required. In KARI-Tigoni, boxes of 60 × 45 × 85 cm depth are used. The growth boxes (chambers) have a framework of steel bars; the lid, sides and bottom are made of Styrofoam. The lids are removable. The lids have 2 cm diameter holes at a spacing of 10 × 10 cm into which short perpex pipes (½ an inch in diameter) were inserted. The plants are fitted into these pipes. The sides are also movable to allow frequent harvesting and monitoring of the root system. At the floor, the chambers are fitted with drainage pipes; these are inclined to allow flow of the nutrients back to a reservoir tank. The tank is placed underground to maintain low temperatures and to allow free flow of the nutrient solution.

Temperature regulation

Temperature in the aeroponic unit should be regulated according to Otazu (2010); the roof needs to be covered by a shade net to regulate the temperature inside the screen house. Day temperature should not be higher than 30°C and lower than 4°C. The best temperature for tuberization is 10 to 15°C during the night and around 20°C during the day. At KARI-Tigoni, the temperatures are regulated using a shade net and maintained at 18 to 23°C during the day and 14 to 15°C at night.

Power

Power is needed to operate the fertigation system. In KARI-Tigoni, electricity is the main source of power. A fuel-driven generator is used when there is no electricity.

Crop management

After fixing the plants in the perpex pipes, a small piece of thin sponge is wrapped around the crown of the plant to seal off the hole completely and to hold the plant in an upright position. The sponge is then covered with a piece of black polythene paper fastened with masking tape. This is to ensure that the boxes are dark inside. For tuber formation, total darkness is required; otherwise, the stolons will develop into foliage. Six weeks later, the lower leaves of the plant are cut and the plant pulled slightly down inside the box to promote more stolon development. When tending the crop, sanitation procedures observed include washing hands with disinfectant, dipping shoes into a footbath and sterilizing all equipment used using common disinfectant as this is to avoid spreading diseases.

Nutrient management

Fog nozzles (36 L/h) are fitted on the underside of the lid of the growth chamber. The aeroponic system is calibrated to mist for 5 min at 15 min intervals. The draining solution flows back into the reservoir tank by gravity. Farran and Mingo (2006) used systems that mist for 10 s at 20 min intervals. The nutrient solution has an EC which does not exceed 2.0 mS/cm. In potato minituber production at KARI-Tigoni, nutrient solution of EC values between 2.0 and 2.5 dSm^{-1} has been used. The pH of the nutrient solution does not exceed 7.3; the optimum is 6.5 to 6.8. The nutrient solution is changed in every four weeks to replenish the nutrients and maintain the correct pH. According to Farran and Mingo (2006), the following nutrient solution gives satisfactory minituber yield: KNO_3 (0.4 me/l); Ca $(NO_3)_2$ (3.1 me/l); NH_4NO_3 (4.4 me/l) and $MgSO_4$ (1.5 me/l), at a pH of 5.7. The following nutrient solutions have been used with satisfactory results in the Kenyan potato program (Table 1).

Quality control

Leaf samples were collected from the plants for virus

Table 1. Composition of 500 L nutrient solution for seed potato production (Farran et al., 2006).

Nutrient	Quantity
KNO_3	252 g
$Ca(NO_3)_2$	118 g
KH_2PO_4	68 g
$MgSO_4$	246
Fe(EDTA) Fe 6%	9 g
Micro (fetrilon)	12
pH	5.7

*Fetrilon combi is a commercial foliar micronutrient powder that has the following formulation: 9% MgO, 3% S, 4% Fe, 4% Mn, 1.5% Cu, 1.5% Zn, 0.5% B, and 0.1% Mo.

indexing using ELISA (enzyme linked Immuno-sorbent assay) tests. At least two ELISA tests were conducted; the first test was conducted one month after transplanting and the second test two to three months after transplanting.

Harvesting and handling of minitubers

Two weeks after transplanting, the early maturing potato varieties began to produce tuberlets. Harvesting of minitubers commenced when they attain at least 8 g. This corresponds to the size of 0.5 cm in diameter. It is better to harvest in the morning when it is cool (Otazu, 2010). During the first harvest, all minitubers larger than 1.5 cm in diameter are removed. This is done in order to enhance more tuber initiation. Subsequent harvests are done every 10 to 14 days. Harvesting continues until plants are about six months old. Harvested minitubers are left to cure for 2 weeks before cold storage. This is because they have imbibed a lot of water during development and they can rot if they are taken to cold store immediately.

FACTORS AFFECTING PRODUCTION OF MINITUBERS IN AEROPONIC SYSTEM

1. Temperature: it should be 18 to 20°C during the day and 14 to 15°C at night (Otazu, 2010).
2. Nutrition: Calcium is required for stolon tip development and to promote tuber initiation (Balamani et al., 1986). When N supply is in excess, there is an excessive vegetative growth and delayed tuberization (Kang et al., 1996).
3. Mechanical stress: Lugt et al. (1964) reported that aeroponics cultures perform better when roots grow in air without mechanical obstruction.
4. Light: Absolute darkness is necessary for tuber formation, with minimum light penetration; the stolons develop into new stems.
5. Lowering of plant: Hand pulling down the plants after each harvest is necessary to increase the formation of new stolons with tubers (Lommen, 1995).
6. Harvesting: Repeated harvesting increases yield especially tuber number. This may be due to the removal of the dominant large tubers, which allows initiation of new tubers (Lommen, 1995).

DISCUSSION

Aeroponics technique is a rapid multiplication technology (RMT) able to produce large numbers of minitubers in one generation, thus, allowing bulking of large number of potato seeds. This eliminates the need for field generations 1, 2 and 3 thereby reducing costs and saving time.

Aeroponics system offer several advantages over other methods. These include:

(i) Aeroponics optimizes root aeration resulting in more yields than classical hydroponics (Soffer and Burger, 1988).
(ii) Aeroponics uses little water that is, 1/10th to 1/30th of the water used in field production of the same amount of potatoes.
(iii) There is good nutrient recirculation, control of nutrients and pH.
(iii) No sterilization of growing media is required thus, minimizing the costs. There are no soil-borne diseases, weeds and nematodes in aeroponic systems.
(iv) An aeroponics system as a whole, allows for uniform water availability to plants.
(iv) Extended growing season that is, crops can be grown all the year around and hence more yields.
(v) Intensive production in a small space that is, more plants per unit area and hence, more yields.
(vi) No loss of fertility and so no crop rotation is needed in field grown crops.
(vii) Closer plant spacing is possible and moveable plant channels allow greater production from equal areas for some crops.
(viii) Use of biological controls including beneficial insects and safe methods of insect control are possible and practical in a controlled environment system.
(ix) According to Muthoni et al. (2010) and CIP (2010), the method is up to ten times more effective than with the conventional techniques.
(x) The minitubers can be harvested at any size seed the user wants, from 5 to 30 g. Spraying fertilizers directly on to the roots makes it possible for the growth phase to continue for more than 180 days without interruption, which does not happen with conventional techniques.
(xi) Harvesting of minitubers in aeroponic systems is convenient, clean, and permits a greater size control through sequential harvesting. It also increases yield especially tuber number; this may be due to the removal of the dominant large tubers, which allows initiation of

new tubers as well as the development of existing ones (Lommen, 1995; Ritter et al., 2001). Aeroponics also has got its share of disadvantages. These include:

(a) Minimum light penetration of the stolons develop into new stems. Maintaining total darkness is a problem because the sides are regularly opened to monitor the root growth and harvest minitubers; therefore absolute darkness is necessary for tuber formation.
(b) It is labour and cost intensive for example, the staking of plants, manual harvesting and pulling down of the plants after each harvest lowering of plant. Hand pulling down the plants after each harvest is necessary to increase the formation of new stolons with tubers. This calls for a lot of work.
(c) Start-up costs of aeroponics are also high. It costs about 24,000 dollars to set up a simple system able to produce 80,000 minitubers in a year.
(d) Aeroponic systems are dependent on low volume of nutrient solution that is available to the root system and any losses of power to pumps can produce irreversible damages including complete loss of plantlets.
(e) There are variations in nutrient requirements by various varieties; hence toxicities and deficiencies are likely to occur.
(f) Lack of water and chemical buffering capacities in the system when power cuts occur must be compensated by security systems (alarms, pumps), high infrastructure costs, high technology and a specialized organization of growers.
(g) Functional system including rooting due to tubers being touching the box floor leading to blockage of the drainage system.
(h) For large-scale minituber production in aeroponic system, an economic evaluation is necessary for soil-less culture and particularly aeroponics, to validate the technology as low cost for disease-free minituber production.
(i) Temperature: Ideal temperature should be 18 to 22°C during the day and 14 to 18°C at night. Maintaining this temperature is a challenge because the screen house does not have automatic control.

CONCLUSION AND RECOMMENDATIONS

To optimize aeroponics system, various issues need to be addressed:

(a) Harvesting intervals should be longer to increase the number of larger seed minitubers. Plants from small minitubers are sensitive to heat and drought stress due to the small size of mother tubers. They are also less vigorous and less competitive against weeds that plants from larger tubers (Struik and Lommen, 1990).
(b) An economic evaluation is also necessary to ascertain the viability of aeroponics systems in the developing countries for minituber production.
(c) Studies on appropriate nutrient solutions, plant densities, number of harvests and harvest intervals, as well as all interactions between production factors need to be investigated.
(d) Effects of lowering the plants by hand pulling down of plants after each harvest (as a way of increasing the formation of new stolons and tubers) on overall yield should be investigated.
(e) Best production time needs to be determined in each area depending on weather and planting season since day time with temperatures higher than 30°C are too warm for aeroponic plants and these temperatures are not desirable for tuberization to occur.
(f) Pest and disease control need to be optimized as they are different from those encountered in field potato production.

ACKNOWLEDGEMENT

The authors acknowledge International Potato Centre (CIP), Sub-Saharan Africa Regional Office for constructing the aeroponic unit at KARI-tigoni.

REFERENCES

Ayieko M, Tschirley D (2006). Enhancing access and utilization of improved seed for food security in Kenya. Working paper No. 27 Tegemeo Institute of agricultural policy and development, Egerton University, Kenya.
Balamani V, Veluthambi K, Poovaiah BW (1986). Effect of calcium on tuberization in potato. Plant physiol. 80:856-858
Barak P, Smith JD, Krueger R, Peterson A (1996). Measurement of short-term nutrient uptake rates in cranberry by aeroponics. Plant Cell Environ. 19:237-242.
Barker BTP (1922). Studies on root development. Long Ashton Res. Station Ann. Rep. 1921:9-57.
Biddinger EJ, Liu CM, Joly RJ, Raghothama KG (1998). Physiological and molecular responses of aeroponically grown tomato plants to phosphorous deficiency. J. Am Soc. Hortic. Sci. 123:330-333.
Carter WA (1942). A method of growing plants in water vapor to facilitate examination of roots. Phytopathology 732:623-625.
Centro Internacional de la Papa International Potato Centre (CIP) (2010). Press Available at Releases.cip-webmaster@cgiar.org. Accessed on March 1st 2012. CIP, Lima, Peru.
Farran I, Mingo-castel AM (2006). Potato minituber production using aeroponics: Effects of density and harvest intervals. Am. J. Potato Res. 83(1):47-53.
Hubick KT, Drakeford DR, Reid DM (1986). A comparison of two techniques for growing minimally water-stressed plants. Can. J. Bot. 60:219-223.
Hung LLL, Sylvia DM (1988). Production of vesicular-arbuscular mycorrhizal fungus inoculum in aeroponic culture. Appl. Environ. Microbiol. 54:353-357.
Kaguongo WP, Gildemache P, Demo P, Wagoire W, Kinyae P, Andrade J, Forbes G, Fuglie K, Thiele G (2008). Farmer practices and adoption of improved potato varieties in Kenya and Uganda. Social Sciences Working Paper 5. CIP, Lima, Peru.
Kang JG, Kim SY, Om YH, Kim JK (1996). Growth and tuberization of potato (*Solanum tuberosum L.*) cultivars in aeroponic, deep flow technique and nutrient film technique culture films. J. Korean Soc. Hort. Sci. 37:24-27.
Kenya Agricultural Research Institute (2007). National Potato Research

Center, Tigoni, annual report. Tigoni, Kenya: National Potato Research Center.

Khan M, Sinclair JB (1992). Pathogenicity of sclerotia- and nonsclerotia-lorming isolates of *Colletotrichum truncation* on soybean plants and roots. Phytopathology 82:314-319.

Kim HS, Lee EM. Lee MA, Woo IS, Moon CS, Lee YB, Kim SY (1999). Production of high quality potato plantlets by autotrophic culture for aeroponic systems. J. Korean Soc. Hort. Sci. 123:330-333.

Lommen WJM, Struik PC (1992). Potato Production of potato minitubers by repeated harvesting: Plant productivity and initiation, growth and reobsorption of tubers. Netherlands J. Agric. Sci. 40:342-359.

Lommen WJM (1995). Basic studies on the production and performance of potato minitubers. PhD Thesis, Wageningen University, Wageningen, The Netherlands.

Lugt C, Bodlaender KBA, Goodijik (1964). Observation on the induction of second growth in potato tubers. Eur. Potota J. 7:21:9-227

Lung'aho C, Nyongesa M, Mbiyu MW, Ng'ang'a NM, Kipkoech DN, Pwaipwai P, Karinga J (2010). Potato (*solanum tuberosum*) minituber production using aeroponics: another arrow in the quiver?. In: Poceedings of the 12th biennial conference of the Kenya Agricultural Research Institute.

Mavoungou ZC, Beunard P, Tr-uong B (1982). Test de resistance a la transpiration de quatr-e varietes de soja en culture aeroponique. E'Agt (*momie Tropicate*, Paris) 37:288-294.

Ministry of Agriculture and Deustsche Gessellschaft fur Technische Zusammenarbeit. (2009) National Potato taskforce report. Final report. MoA/GTZ, Nairobi,Kenya

Muthoni J, Mbiyu MW, Nyamongo DO (2010). A review of potato seed systems. J. Agric. Food Inf. 11:157-167.

Muthoni J, Miriam M, Jackson NK (2011). Up-scaling production of certified potato seed tubers in Kenya: Potential of aeroponics technology. J. Hortic. For. 3(8):238-243.

Ng'ang'a NM, Kinyae P, Walingo A, Wakahiu MW (2003). Establishing effective seed potato prices for small scale farmers in Kenya. Unpublished report. KARI. Nairobi, Kenya.

Nugali Yadde MM, De Silva HDM, Perera, R, Ariyaratna D, Sangakkara UR (2005). An aeroponic system for the production of pre-basic seed potato. Ann. Sri Lanka Department Agric. 7:199-288.

Otazu V (2010). Manual on quality seed potato production using aeroponics. International potato Cenre (CIP).Lima, Peru. 2010; 44 retrived January 24th 2012, from http://cipotato.org/resources/publications/manual/manual-on-quality-seed-potato production-using-aeroponics

Peterson LA, Krueger AR (1988). An intermittent aeroponics systems. Crop Sci. 28:712-713.

Ritter E, Angulo B, Riga P, Herran C, Relloso J, San JM (2001). Comparison of hydroponics and aeroponics cultivation systems for the production of potato minitubers. Potato Res. 44:127-135.

Riungu C (2011).The small holder cash crop of the future. Horticultural News, East Afr. Fresh Prod. J., 19:15-38

Robertson JM, Hubick KT, Yeung EC, Reid DM (1990a). Developmental responses to drought and abscisic acid in sunllower roots. I. Root growth, apical anatomy, and osmotic adjustment. J. Exp. Bot. 41:325-337.

Robertson JM, Yeung EC, Reid DM, Hubick KT (1990b). Developmental responses to drought and abscisic acid in sunflower roots. Mitotic activity. J. Exp. Bot. 41:339-350.

Shtrausberg DV, Rakitina EG (1970). On the aeration and gas regime of roots in aeroponics and water culture. Agrokhitniia p. 101-110.

Soffer H, Burger DW (1988). Effects of dissolved oxygen concentration in aeroponics on the formation and growth of adventitious roots. J. Am. Soc. Hortic. Sci. 113:218-221.

Stoner RJ (1983). Aeroponics Versus Bed and Hydroponic Propagation. Florists Rev. 1(173):4477.

Sylvia DM, Jarstter AG (1992). Sheared-root inocula of vesicular-arbuscular mycorrhizal *fung]*. Appl. Environ. Microbiol. 58:229-232.

Struik PC, Lommen WJM (1990). Production, Storage and use of micro- and minitubers. Abstracts of the 11th Triennial Conf. Eur. Ass. Potato Res. Edinburg 11:122-123.

Truong B, Beunard P (1978). Etude de la croissanee racinaire de six varietes de riz pluvial en culture aeroponique. Premiers resultats. E'Agronomie Tropicale (Paris) 33:231-236.

Wagner RE, Wilkinson HT (1992). An aeroponics system for investigating disease development on soybean taproots infeeted with *Phytophthora. sojae.* Plant Dis. 76:610-614.

Weathers PJ, Zobel RW (1992). Aeroponics for the culture of organisms, tissues and cells. Biotechnol. Adv. 10:93-115.

Went FW (1957). The experiment control of plant growth. New York: Ronald Press.

Investigation of biofertilizers influence on quantity and quality characteristics in *Nigella sativa* L.

Sayed Alireza Valadabadi[1] and Hossein Aliabadi Farahani[2]

[1]Associate Professor, Islamic Azad University, Shahr-e-Qods Branch, Tehran, Iran.
[2]Young Researchers Club, Islamic Azad University, Shahr-e-Qods Branch, Tehran, Iran.

To reduce farmers' dependency on mineral fertilizers, to increase water use efficiency, and to increase households' incomes, participatory on-farm research trials to test the beneficial impact of biological fertilizers application on black cumin (*Nigella sativa* L.) some yield characters were investigated. Our objectives in this study were the interactive effects of biofertilizer application on quality and quantity yields at Iran in 2010. In this respect, the experimental unit had been designed by the achieved treatments in factorial on the basis of completely randomized block design with three replicates. Certain factors including four levels of animal manure (0, 10, 20 and 30 ton/ha respectively) and two levels of azotobacter (non-application and application) were studied. The final statistical analysis indicated that in the 20 ton/ha animal manure and azotobacter application, yield components were significantly higher. Those such as: biological yield, seed yield, essential oil yield, stem dry weight, capsule dry weight, leaf dry weight and plant height were higher in unites by 20 ton/ha animal manure and azotobacter application. The finding may give applicable advice to farmers for management and concern on fertilizer strategy and carefully estimate chemical fertilizer supply by biofertilizer application.

Key words: Azotobacter, animal manure, essential oil yield, *Nigella sativa* L.

INTRODUCTION

Nigella sativa is an annual flowering plant, native to southwest Asia. It grows to 20 to 30 cm (7.9 to 12 inch) tall, with finely divided, linear (but not thread-like) leaves. The flowers are delicate, and usually coloured pale blue and white, with 5 to 10 petals. The fruit is a large and inflated capsule composed of 3 to 7 united follicles, each containing numerous seeds. The seed is used as a spice. Origian Black cumin is rarely available so *N. sativa* is widely used instead, in India *Carum carvi* is the substitute. {cumins are from Apiaceae or Umbelliferae (both names are allowed by the ICBN) family but *N. sativa* is from Ranunculaceae family} Black cumin (not *N. sativa*) seeds come as paired or separate carpels, and are 3 to 4 mm long. They have a striped pattern of nine ridges and oil canals, and are fragrant (Ayurveda says"Kaala jaaji sugandhaa cha" =Black cumin seed is fragrant itself), blackish in colour, boat-shaped, tapering at each extremity, with tiny stalks attached; has been used for medicinal purposes for centuries, both as a herb and pressed into oil, in Asia, Middle East, and Africa. It has been traditionally used for a variety of conditions and treatments related to respiratory health, stomach and intestinal health, kidney and liver function, circulatory and immune system support, as analgesic, anti-inflammatory, antiallergic, antioxidants, anticancer, antiviral and for general well-being. *N. sativa* oil (not Black cumin seed oil) contains nigellone, which protects guinea pigs from histamine-induced bronchial spasms (Oxford Uni, 2000). Azotobacter is a genus of usually motile, oval or spherical bacteria that form thick-walled cysts, and may produce large quantities of capsular slime. Azotobacter is an aerobic, free-living soil microbe which fixes nitrogen from the atmosphere. Beyond Azotobacter's use as a model organism it has biotechnological applications. Examples are its use for alginate production and for nitrogen production in batch fermentations. Polyhydroxybutyrate is produced under certain conditions. Azotobacter is Gram-

*Corresponding author. E-mail: farahani_1362@yahoo.com.

Table 1. The results of soil analysis.

Soil texture	Sand (%)	Silt (%)	Clay (%)	K (mg/kg)	P (mg/kg)	N (mg/kg)	Na (Ds/m)	EC (1: 2.5)	pH	Depth of sampling(cm)
Ca.L	35	30	35	142.2	5.2	38.7	0.05	0.18	7.9	0-30

negative bacteria (Dixon and Kahn, 2004). Biofertilizers are products containing living cells of different types of microorganisms (Vessey, 2003; Chen, 2006) that have an ability to convert nutritionally important elements from unavailable to available form through biological processes (Vessey, 2003) and are known to help with expansion of the root system and better seed germination. Biofertilizers differ from chemical and organic fertilizers in that they do not directly supply any nutrients to crops and are cultures of special bacteria and fungi. Some microorganisms have positive effects on plant growth promotion, including the plant growth promoting rhizobacteria (PGPR) such as *Azospirillum*, *Azotobacter*, *Pseudomonas fluorescens*, and several gram positive *Bacillus* spp. (Chen, 2006). The diazotrophic rhizobiocoenosis is an important biological process that plays a major role in satisfying the nutritional requirements of the commercial medicinal plants (Deka et al., 1992). The strong and rapidly stimulating effect of fungal elicitor on plant secondary metabolism in main crops has attracted considerable attention and research efforts (Zhao et al., 2005). *Azotobacter* and *Azospirillum* are free-living N_2-fixing bacteria that in the rhizospheric zone have the ability to synthesize and secret some biologically active substances that enhance root growth. They also increase germination and vigour in young plants, leading to improved crop stands (Chen, 2006). Various *Pseudomonas* species have shown to be effective in controlling pathogenic fungi and stimulating plant growth by a variety of mechanisms, including production of siderophores, synthesis of antibiotics, production of phytohormones, enhancement of phosphate uptake by the plant, nitrogen fixation, and synthesis of enzymes that regulate plant ethylene levels (Abdul-Jaleel et al., 2007). Good soil fertility management ensures adequate nutrient availability to plants and increases yields. High above-ground biomass yield is obviously accompanied by an active root system, which releases an array of organic compounds into the rhizosphere (Mandal et al., 2007). It is well known that a considerable number of bacterial and fungal species possess a functional relationship and constitute a holistic system with plants. They are able to exert beneficial effects on plant growth (Vessey, 2003) and also enhance plant resistance to adverse environmental stresses, such as water and nutrient deficiency and heavy metal contamination (Wu et al., 2005). An experiment was conducted to evaluating the effects of *Azotobacter* inoculant on the yield of wheat (cv. Kanchan). The treatments were T_0 (control), T_1, T_2, T_3, T_4 and T_5. Except 1000 grain weight, all the yield components of wheat viz. plant height, filled spikelets spike^{-1}, spike length and the number of grain spike^{-1} were influenced significantly by the treatments. The highest grain yield of 780 mg plant^{-1}, that is, 84% increase over the control (425 mg plant^{-1}) was obtained due to the treatment of T_5 which did not differ significantly from the yield obtained (687, 732 mg and 715 mg plant^{-1}) with the application of T_1, T_3 and T_4, respectively. There was 18% increase in grain yields due to using *Azotobacter* inoculant only over the control, which was not statistically significant. The straw yields showed a similar pattern. *Azotobacter* inoculation also influenced the root growth significantly. Total N uptake in grain, straw and root increased significantly due to different treatments. The highest N uptake (23.17 mg plant^{-1}) was recorded with the treatment T_5 and the lowest with the T_0 (control), (11.03 mg plant^{-1}). The total N uptake was increased by 89, 36, 101, 88 and 109% over the control due to T_1, T_2, T_3, T_4 and T_5, respectively. *Azotobacter* either alone or in combination with urea N had some beneficial effects on the yield of wheat, which amounted to saving about 20% of urea N (Kader et al., 2002). The experiment of Sudhakar et al. (2000) was conducted in Berhampore, India, from 1993 to 1996 to evaluate *Azotobacter* as a biofertilizer under graded levels of inorganic nitrogen fertilizer to reduce the cost of mulberry leaf production. *Azotobacter* was applied as biofertilizer (farmyard manure plus inoculum with more than 10^8 cells/g carrier). This application of biofertilizer with 150 kg N/ha per year as inorganic nitrogen gave as much leaf as the recommended dose (300 kg N/ha per year) of inorganic nitrogen fertilizer alone. Further reduction in inorganic nitrogen fertilizer, however, reduced leaf yield and quality. Application of *Azotobacter* with 50% of the recommended inorganic nitrogen had no effect on rearing and cocoon characters of silkworm. Therefore, the objective of this study was to evaluate the biofertilizers influence on quantity and quality characteristics of black cumin (*N.sativa* L.).

MATERIALS AND METHODS

This study was conducted on experimental field of Islamic Azad University, Shahr-e-Qods Branch at Iran (27°38' N, 40°21' E; 1417 m above sea level) during 2010, with clay loam soil (Table 1), mean annual temperature (31°C) and rainfall in the study area is distributed with an annual mean of 215 mm. The experimental unit had designed by achieved treatments in factorial on the basis completely randomized block design with three replicates. Certain factors including four levels of animal manure (0, 10, 20 and 30

Table 2. Analysis of variance.

Sources of variation	df	Mean square					
		Biological yield (kg/ha)	Seed Yield (kg/ha)	Capsule dry weight(kg/ha)	Essential oil yield(kg/ha)	Leaf dry weight(kg/ha)	Plant height(cm)
Replication	2	0.007	15.931 **	189143.313	0.005	5.414 **	0.084 **
Azotobacter	1	1.423 **	10.087 **	1520201.243**	0.071 **	41.633 **	0.041 **
Animal manure	3	1.515 **	11.797 **	1089514.354**	0.058 **	67.903 **	0.022 **
Azotobacter× animal manure	3	0.027	0.008	211820.317 *	0.056	0.267	0.004
Error	18	0.009	0.032	86199.201	0.003	1.541	0.002
CV(%)		6.92	4.54	8.22	2.77	9.64	8.08

* and ** : Significant at 5% and 1% levels respectively

ton/ha respectively) and two levels of azotobacter (non-application and application) were studied. The soil consisted of 25% clay, 30% silt and 45% sand (Table 1) and further the field was prepared in a 15 m^2 area (5 × 3 m).

At the end of growth stage we collected 10 plants from each plot randomly for determination of plant characteristics and selected 100 g seed from each plot for determination of essential oil percentage by Clevenger. Finally, essential oil yield was determined by the following formula (Farahani et al., 2008).

Essential oil yield = Essential oil percentage × Root yield

Data were subjected to analysis of variance (ANOVA) using Statistical Analysis System [SAS, 1988] and followed by Duncan's multiple range tests. Terms were considered significant at $P < 0.05$.

RESULTS

Final results of plants values showed that azotobacter significantly affected biological yield, seed yield, essential oil yield, stem dry weight, capsule dry weight, leaf dry weight and plant height in $P \leq 0.01$ (Table 2) which indicated the highest biological yield (9206 kg/ha), seed yield (1815 kg/ha), stem dry weight (3728 kg/ha), capsule dry weight (305.2 kg/ha), leaf dry weight (3476 kg/ha), essential oil yield (17.5 kg/ha) and plant height (84 cm) were obtained by azotobacter application (Table 3). Also, animal manure significantly affected biological yield, seed yield, essential oil yield, stem dry weight, capsule dry weight, leaf dry weight and plant height in $P \leq 0.01$ (Table 2). The highest biological yield (10080 kg/ha), seed yield (2279 kg/ha), stem dry weight (4231 kg/ha), capsule dry weight (363.5 kg/ha), essential oil yield (20.1 kg/ha), leaf dry weight (1714 kg/ha) and plant height (82 cm) were achieved by application of 30 ton/ha animal manure (Table 3). Interaction of the azotobacter and animal manure had significant effect on capsule dry weight in $P<0.05$ (Table 2) and highest biological yield (9643 kg/ha), seed yield (2047 kg/ha), stem dry weight (3979 kg/ha), capsule dry weight (334.3 kg/ha), essential oil yield (18.8 kg/ha), leaf dry weight (1595 kg/ha) and plant height (103 cm) were obtained under application of azotobacter and 20 ton/ha animal mature (Table 4).

DISCUSSION

In this study, increases in agronomic criteria were observed following inoculation with azotobacter. This may be due to better utilization of nutrients in the soil through inoculation of efficient microorganisms. A positive effect of azotobacter on yield and yield components has been reported in the literature (Migahed et al., 2004). In addition, higher dry matter production by the inoculated plant might be because of the augmented uptake of N which in turn was a consequence of the root proliferation. Also, the increased growth parameters in hyssop might be due to the production of growth hormones by the bacteria (Ratti et al., 2001). The results showed that application of azotobacter and 20 ton/ha animal mature increased yield and yield components of black cumin, because **N** is a primary constituent of proteins, is extremely susceptible to loss when considering that average recovery rates fall in the range of 20 to 50% for dry matter production systems in plants. N generally because deficiency of potassium increased carbohydrate storage and reduced proteins, alteration in amino acid balance and consequently change in the quality of proteins and are a main element in chlorophyll production. Toxic concentrations of nitrogen fertilizers cause characteristic symptoms of nitrite or nitrate toxicity in plants, particularly in the leaves. Although pre plant fertilizer applications decrease the potential for nutrient deficiencies in early stages of growth, presence of residual soil NO_3–N (plant-available mineral N from the previous season) may pose a risk to the environment. The water of soil be salt by inordinate N application and increase its potential. Finally, the plant use high energy for absorb of salt water that be causes dry matter reduces in this condition. For example the experiment of Manjunatha et al. (2002) was conducted to study the effect of bio-fertilizers on growth, yield and essential oil content in Patchouli (*Pogostemon cablin* Pellet.) at the Medicinal and Aromatic Crops Section, Department of Horticulture, University of Agricultural Sciences, Bangalore during 1999 to 2000. The cultivar 'Johore' was used for the experiment. The fifteen treatments included three levels

Table 3. Means comparison.

Treatments		Biological yield(kg/ha)	Seed yield (kg/ha)	Stem dry weight (kg/ha)	Capsule dry weight (kg/ha)	Essential oil yield(kg/ha)	Leaf dry weight (kg/ha)	Plant height(cm)
Azotobacter	Application	7433 b	1352 b	3225 b	247 b	14.8 b	1238 b	92 a
	Non- application	10080 a	2279 a	4231 a	363.5 a	20.1 a	1714 a	58 b
Animal manure	Non- application	7803 b	1433 b	3491 ab	260.5 b	16.5 b	1207 b	82 a
	10 ton/ha	7473 b	1303 b	3330 b	246.2 b	16.6 b	1181 b	78 a
	20 ton/ha	9206 a	3815 a	3728 a	305.2 a	17.5 a	1476 a	84 a
	30 ton/ha	5555 c	673 c	2608 c	166 c	12.6 d	1787 c	63 b

Means within the same column and rows and factors, followed by the same letter are not significantly difference (P<0.05)

Table 4. Means comparison of interaction.

Survey instance qualification		Biological yield(kg/ha)	Seed yield (kg/ha)	Stem dry weight(kg/ha)	Capsule dry weight (kg/ha)	Essential oil yield(kg/ha)	Leaf dry weight(kg/ha)	Plant height(cm)
Application of azotobacter	Non -application	7618 d	1392 cd	3358 bcd	253.7 c	15.6 c	1222 c	100 a
	10 ton/ha	7453 d	1327 d	3277 cd	264.6 c	15.7 c	1209 c	95 a
	20 ton/ha	8319 c	1583 ab	3476 abcd	276.1 c	16.1 c	1357 b	103 a
	30 ton/ha	7325 d	1298 d	3293 d	242.4 c	15.7 d	1107 c	82 b
		8941 b	1856 ab	3861 ab	312 ab	18.3 b	1460 b	63 de
Non – application of azotobacter	Non -application	8776 bc	1791 abc	3780 abcd	304.8 ab	18.2 b	1447 b	59 def
	10 ton/ha	9643 a	2047 a	3997 a	334.3 a	18.8 a	1595 a	66 cd
	20 ton/ha	7718 d	1476 d	3419 d	264.7 c	16.3 c	1250 c	49 f
	30 ton/ha	7618 d	1392 cd	3358 bcd	253.7 c	15.6 c	1222 c	100 a

Means within the same column and rows and factors, followed by the same letter are not significantly difference (P<0.05).

each of nitrogen (N) and phosphate (P_2O_5) [520, 75 and 100% of recommended dose] with Potash (K_2O) at constant level of 50 kg/ha along with different biofertilizers (Azotobacter, Azospirillum, Phosphorus Solubilising Bacteria (PSB) and Vesicular Arbuscular Mycorrhizal Fungi (VAM) in combination. The results revealed that the treatments differed significantly; among the treatments, 75% NP + 100 K + Azotobacter + Azospirillum + VAM recorded significantly superior values for plant height (80.14 cm), number of leaves (357.75), number of branches (22.04), plant spread (76.12 cm), leaf area (4075.66 cm^2), yield of fresh herbage [10.733 t/ha against 7.277 t/ha in control] and essential oil yield[71.74 l/ha as compared 47.5 l/ha in control] which was followed by 50% NP + 100 K + Azotobacter + PSB + VAM and 50% P + NK + PSB + VAM. On the whole, the treatment with 75% NP + 100 K + Azotobacter + Azospirillum + VAM emerged as one of the best treatments and effected the saving of fertilizers to the extent of 25%. Also, an investigation was carried out under Madurai conditions of Tamil Nadu, India to study the influence of nitrogen, application of nitrogen 93.75 kg ha^{-1} and phosphorus 93.75 kg ha^{-1} along with *Azospirillum* gave the highest plant height, number of laterals, fresh and dry weight of shoot, dry matter production, fresh herbage yield and essential oil yield. The maximum fresh and dry weight of root was obtained by application of nitrogen 93.75 kg ha^{-1} and phosphorus 93.75 kg ha^{-1} along with VAM. Application of inorganic nutrients and biofertilizers had no significant effect on essential oil constituents of davana. Therefore, dry matter reduced under application of chemical fertilizer application because injured roots and was reduced the absorption. Our results were similar to the findings of Stark and Tindall (1992); Wuest and Cassman (1992) and Knowles et al (1994).

Conclusion

In general it appears that, as expected, application of azotobacter improved yield and other plant criteria. Therefore, it appears that application of azotobacter could be promising in production of black cumin by reduction of chemical fertilizer application. Our finding may give applicable advice to farmers for management and concern on fertilizer strategy and carefully estimate chemical fertilizer supply by azotobacter application.

REFERENCES

Abdul Jl C, Manivannan P, Sankar B, Kishorekumar A, Gopi R, Somasundaram R, Panneerselvam R (2007). *Pseudomonas fluorescens* enhances biomass yield and ajmalicine production in *Catharanthus roseus* under water deficit stress. Colloids Surf. B: Biointerf. 60:7–11.

Aliabadi FH (2006). Investigation of arbuscular mycorrhizal fungi (AMF), different levels of phosphorus and drought stress effects on quantity and quality characteristics of coriander (*Coriandrum sativum* L.). MSc Thesis, Department of Agriculture, Islamic Azad University of Takestan branch, Iran. p. 231.

Chen J (2006). The combined use of chemical and organic fertilizers and/or biofetilizer for crop growth and soil fertility. International Workshop on Sustained Management of the Soil-Rhizosphere System for Efficient Crop Production and Fertilizer Use.16 – 20 October, Thailand.

Deka BC, Bora GC, Shadeque A (1992). Effect of *Azospirillum* on growth and yield of chilli (*Capsicum annuum* L.) cultivar Pusa Jawala, Haryana. J. Hort. Sci. 38:41–46.

Dixon R, Kahn D (2004). Genetic regulation of biological nitrogen fixation. Nat Rev Microbiol. 2(8): 621-31.

Domestication of plants in the Old World (3 ed.). Oxford University Press. 2000. p. 206.

Kader MA, Mian MH, Hoque MS (2002). Effects of Azotobacter Inoculant on the Yield and Nitrogen Uptake by Wheat. J. Biol Sci., 2(4): 259-261.

Knowles TC, Hipp BW, Graff PS, Marshall DS (1994). Timing and rate of topdress nitrogen for rainfed winter wheat. J. Prod. Agric. 7:216–220.

Kumar TS, Swaminathan V, Kumar S (2009). Influence of nitrogen, phosphorus and biofertilizers on growth, yield and essential oil constituents in ratoon crop of davana (*Artemisia pallens* Wall.). Electronic Journal of Environmental, Agric. Food Chem., 8(2): 86-95.

Mandal A, Patra AK, Singh D, Swarup A, Ebhin Masto R (2007). Effect of long-term application of manure and fertilizer on biological and biochemical activities in soil during crop development stages. Biores. Technol. 98: 3585–3592.

ManjunathaR, Farooqi AA, Vasundhara M, Srinivasappa KN (2002). Effect of biofertilizers on growth, yield and essential oil content in patchouli (*Pogostemon cablin* Pellet.). Indian Perfumer. 46(2): 97-104.

Migahed HA, Ahmed AE, Abd El-Ghany BF (2004). Effect of different bacteial strains as biofertilizer agents on growth, production and oil of *Apium graveolense* under Calcareous soil. J. Agric. Sci. 12: 511-525.

Ratti N, Kumar S, Verma HN, Gautams SP (2001). Improvement in bioavailability of tricalcium phosphate to *Cymbopogon martini* var. motia by rhizobacteria, AMF and Azospirillum inoculation. Microbiol. Res. 156: 145-149.

Stark JC, Tindall TA (1992). Timing split applications of nitrogen for irrigated hard red spring wheat. J. Prod. Agric. 5:221–226.

Vessey JK (2003). Plant growth promoting rhizobacteria as biofertilizers. Plant Soil. 255: 571–586.

Wuest SB, Cassman KG (1992). Fertilizer-nitrogen use efficiency of irrigated wheat: I. Uptake efficiency of preplant versus late-season application. Agron. J. 84:682–688.

Wu SC, Caob ZH, Lib ZG, Cheung KC, Wong MH (2005). Effects of biofertilizer containing N-fixer, P and K solubilizers and AM fungi on maize growth: a greenhouse trial. Geoderma. 125:155–166.

Zhao J, Lawrence T, Davis C, Verpoorte R (2005). Elicitor signal transduction leading to production of plant secondary metabolites. Biotechnol. Adv. 23:283–333

Access to and utilisation of forest resources: Evidence from common property forest management in Swaziland

Getrude Sithole

AMADI-Midlands State University 1st floor Embassy House, Gwamile Street/Msakato Street P. O. Box 6146 Mbabane Kingdom of Swaziland.

A major challenge facing Swaziland is how to maximise the use of forest resources while still maintaining their sustainability. Forest resources are being depleted and degraded due to poor access controls, inequities in land-tenure and user rights. The research was conducted using the Institutional Analysis and Development (IAD) framework as the underlying scheme. The main objective of this study was to investigate access to and utilisation of forest resources in the selected chiefdoms in Mahlangatsha Inkundla. The survey methodology whose target population was from the two chiefdoms were household heads was adopted in this study (N = 185). Data were collected using questionnaires. Stratified random sampling technique which combined both stratification and randomization were used. Findings revealed that, access levels of households to forest resources were different. Whilst access to community forest resources was restricted, the natural forest resources were open to extraction by anyone. Institutional and community rules that are used to regulate access to forest resources lack effectiveness. Consequently, there are problems of deforestation, degradation, illegal harvesting, and lack of involvement by community members to manage forest resources. The study recommends that, local people should be educated about rules that regulate access to forest resources and common rules be set at the local level.

Key words: Utilisation of forest resources, Accces of forest resources, forest management, sustainability.

INTRODUCTION

The contribution of forest resources to the livelihood strategies of poor people has long been appreciated as significant. Most rural poor people rely directly and indirectly on forests for their livelihoods. How to ensure that poor people have rights and opportunities to access forest resources, as well as responsibilities for the sustainable management of forest resources, has become a central question in debates (Shimizu and Trudel, 2006; Tevera and Mukora, 2007). People in most developing countries often depend on extracting resources from nearby forests for their livelihoods, whether for consumption or fuelwood, or as a source of income. Community or social forestry relates to forestry activities by rural people in the local environment. It involves community and individual participation in the planning, execution and management of a variety of social and economic forestry elements (Tevera and Mukora, 2007) It is people's needs which, in theory, determine tree planting and harvesting priorities. People are seen as partners in the planning and management of

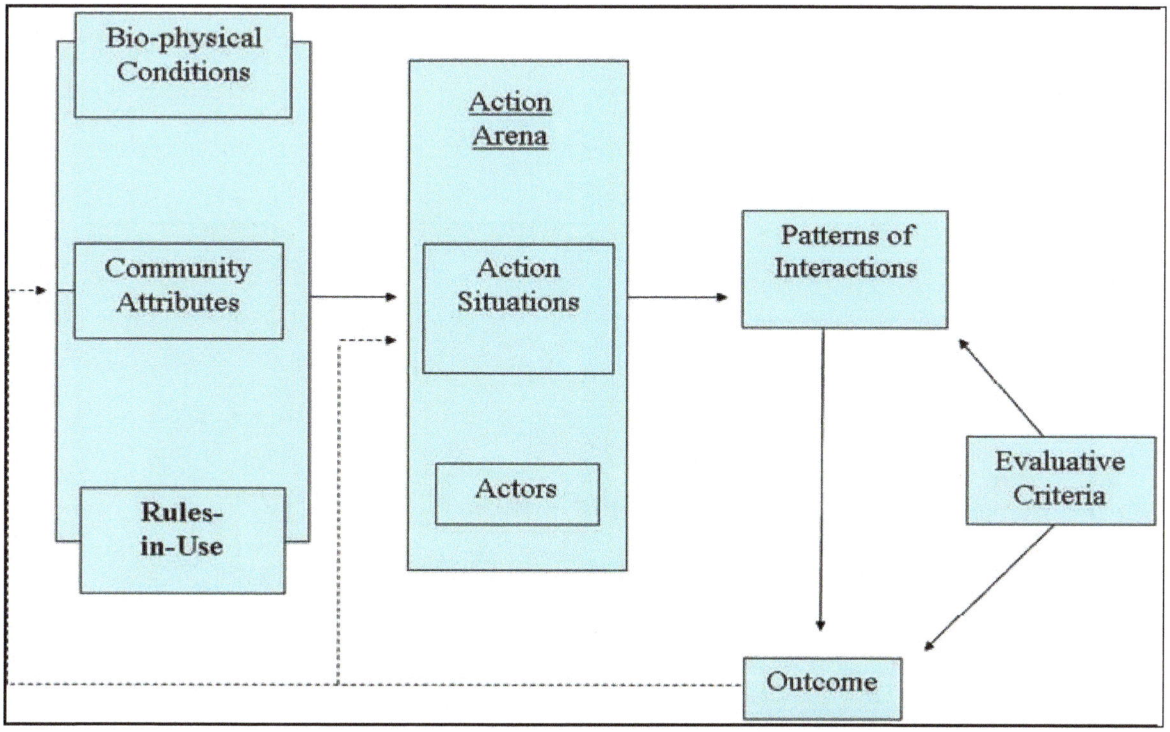

Figure 1. The focal level of analysis of the IAD framework. Source: Ostrom et al. (2007).

forest resources, rather than as resource exploiters to be controlled or excluded Forestry can play a significant role for the well being of the people living in and around the forest areas, and conversely, these people can play a major part in making the forests around them more productive. Managing forests primarily with a view to protecting, developing and utilising them is sustainable forest management.

To this end, the requirements for fuelwood, fodder and construction timber required by the people for their consumptive and productive purposes have been regarded as important benefits to communities (Fisher, 2004). According to a National Forest Policy Green Paper (2000) in Swaziland, forests are even more important for the immediate beneficiaries, the people that are employed in the forest industry, or make a living from trade in forestry products. All rural people directly benefit from the forest as they depend on a range of forest products which are derived from their immediate environment. Thus, the most important aspect of their role is to act as custodians and ensure sustainable management of the forest resources.

Conceptual framework

The Institutional Analysis and Development (IAD) framework was used for this study to explore access to and utilisation of forest resources in selected chiefdoms in Mahlangatsha Inkundla (Figure 1). A key feature is its structure, which extends from the operational level, where decisions directly affect resource access and use, to the collective-choice level, where the rules that govern resource access and use are designed, to the constitutional level, where decisions affect the rules that govern how decisions are taken at the collective-choice level. The IAD framework is one of the most distinguished and tested frameworks in the field of natural resource management (Gibson et al., 2005; Rudd, 2004; Carlsson, 2000). It has notably been used as a basis for developing a theory of common-pool resource management and has been supported in this field by a strong record of empirical research and theoretical development (Ostrom et al., 2007). The IAD framework places emphasis on institutions, rules and actors that create patterns of interactions.

Study area

The study area covers Sibovu, and Mpolonjeni chiefdoms in Mahlangatsha Inkhundla, which is located in the Manzini District of Swaziland. According to the Government of Swaziland in 2007, this region of Swaziland is the most heavily settled and the increase in population is having a negative impact on the natural vegetation. Swaziland is divided into four regions. These regions are; Hhohho; Manzini, Lubombo, and Shiselweni. The regions are shown in Figure 2. The Mahlangatsha Inkundla is situated in the southern part of Manziniregion.

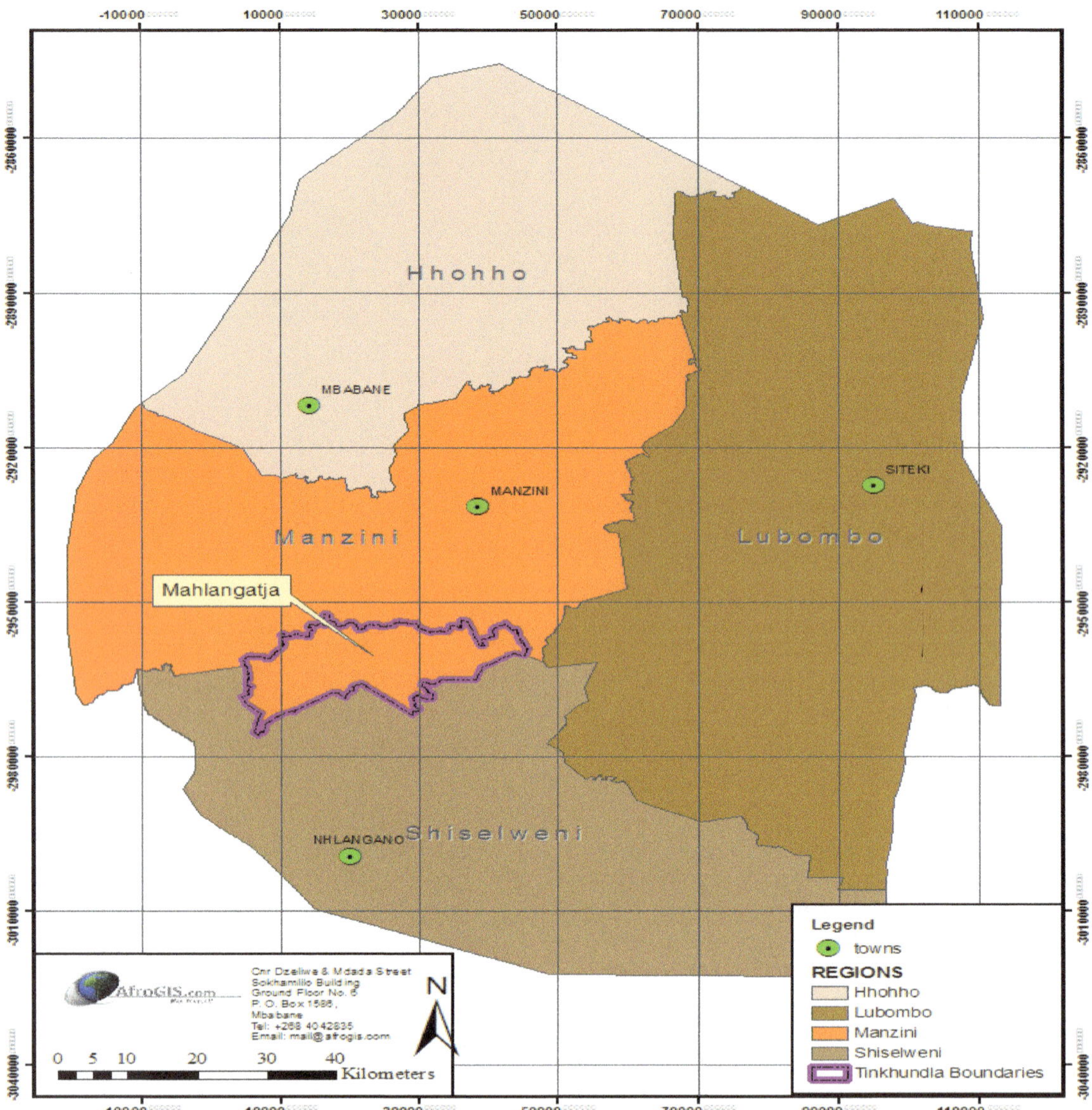

Figure 2. Map of Swaziland showing the location of Mahlangatsha Inkundla in Swaziland. Source: AfroGIS Swaziland (2008).

Figures 3 and 4 show the location of Mpolonjeni and Sibovu in Mahlangatsha Inkundla, respectively. The study area lies between longitude 31° 00' to 31° 15' E and latitude 26° 45' to 27° 00' S (Figure 2). Sibovu and Mpolonjeni are located between Motane and Tungolubi rivers. The selected chiefdoms are in Mahlangatsha Inkundla and they have natural trees and community woodlots. Mahlangatsha *Inkhundla* is in the Middleveld of Swaziland and is located in an area designated as Swazi Nation Land (SNL).

Population

The Mahlangatsha Inkundla has 11 chiefdoms. The study focuses on Sibovu, and Mpolonjeni chiefdoms. The chiefdoms were selected for the study because there are both community woodlots and indigenous forests. The study area has a total population of 3320 people (Government of Swaziland, 2007).

Soils and vegetation

Sibovu has soils which are generally acidic. Eucalyptus trees are generally grown in Sibovu chiefdom and constitute 15 ha (Government of Swaziland, 2007). According to the Government of Swaziland in 2007, there are also wattle trees which constitute 80 ha. Common forms of plants include small woodlots and fruit trees.

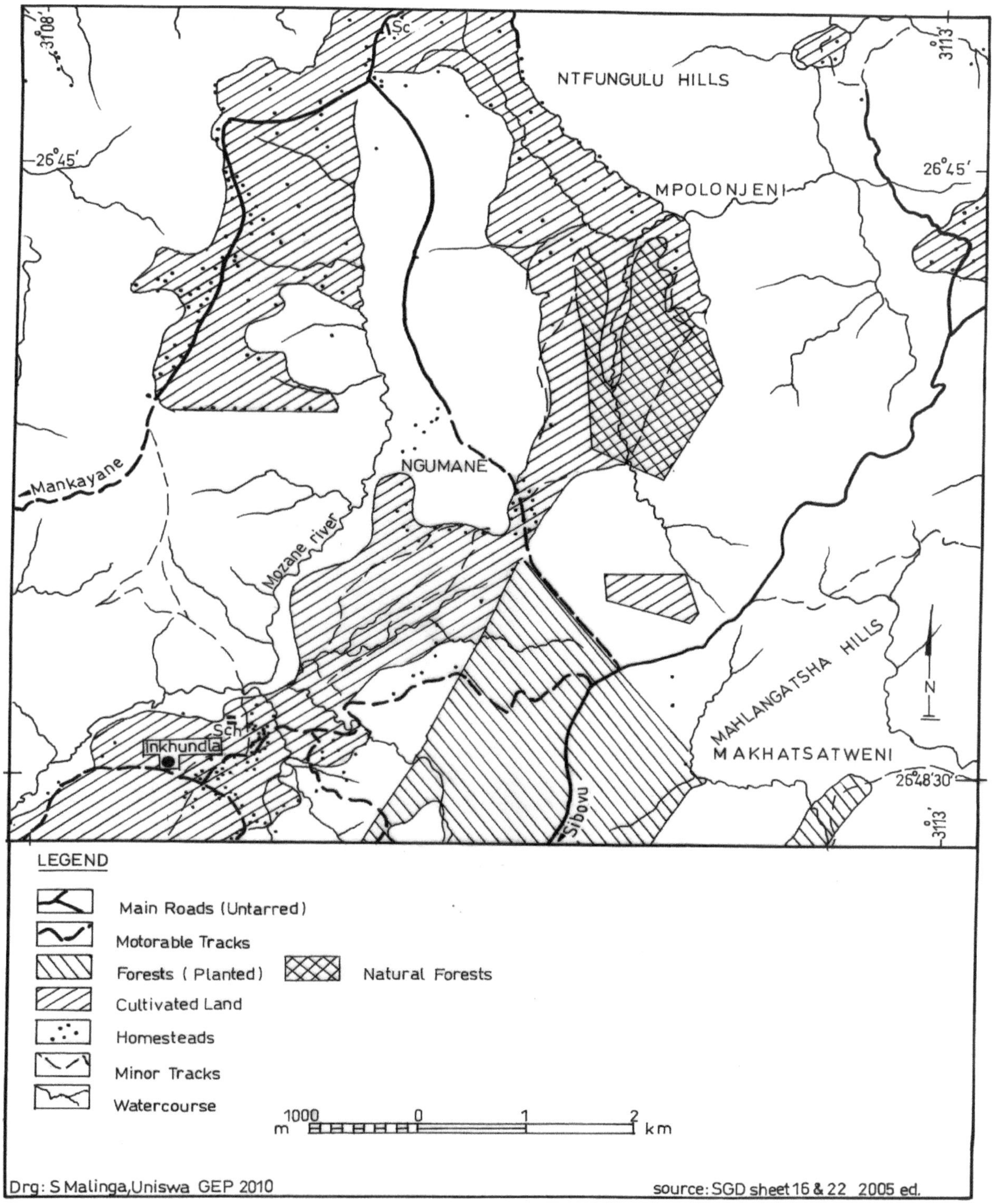

Figure 3. Map showing location of Mpolonjeni (Mahlangatsha Inkundla) Source: SGD Sheet 16 and 22 2005 edition.

Most commonly planted woodlots consist of two introduced wattle species *umtfolo and umtfolowesi lungu (Acacia mearnsii* and *Acacia Decurrens*, respectively*)* which may also be referred to as the Black Wattle. The natural trees which are found in Mpolonjeni chiefdom include *Dalbegia, Sygium Cummunii, Kaya Nyassica* and others. According to Government of Swaziland in 2007, Mpolonjeni has 120 ha of eucalyptus, 50 ha wattle and 30 ha natural forest. Exotic plantations were introduced in the study area to curb land degradation and to ensure that local communities get products for use. Sibovu and Mpolonjeni chiefdoms practise afforestation using

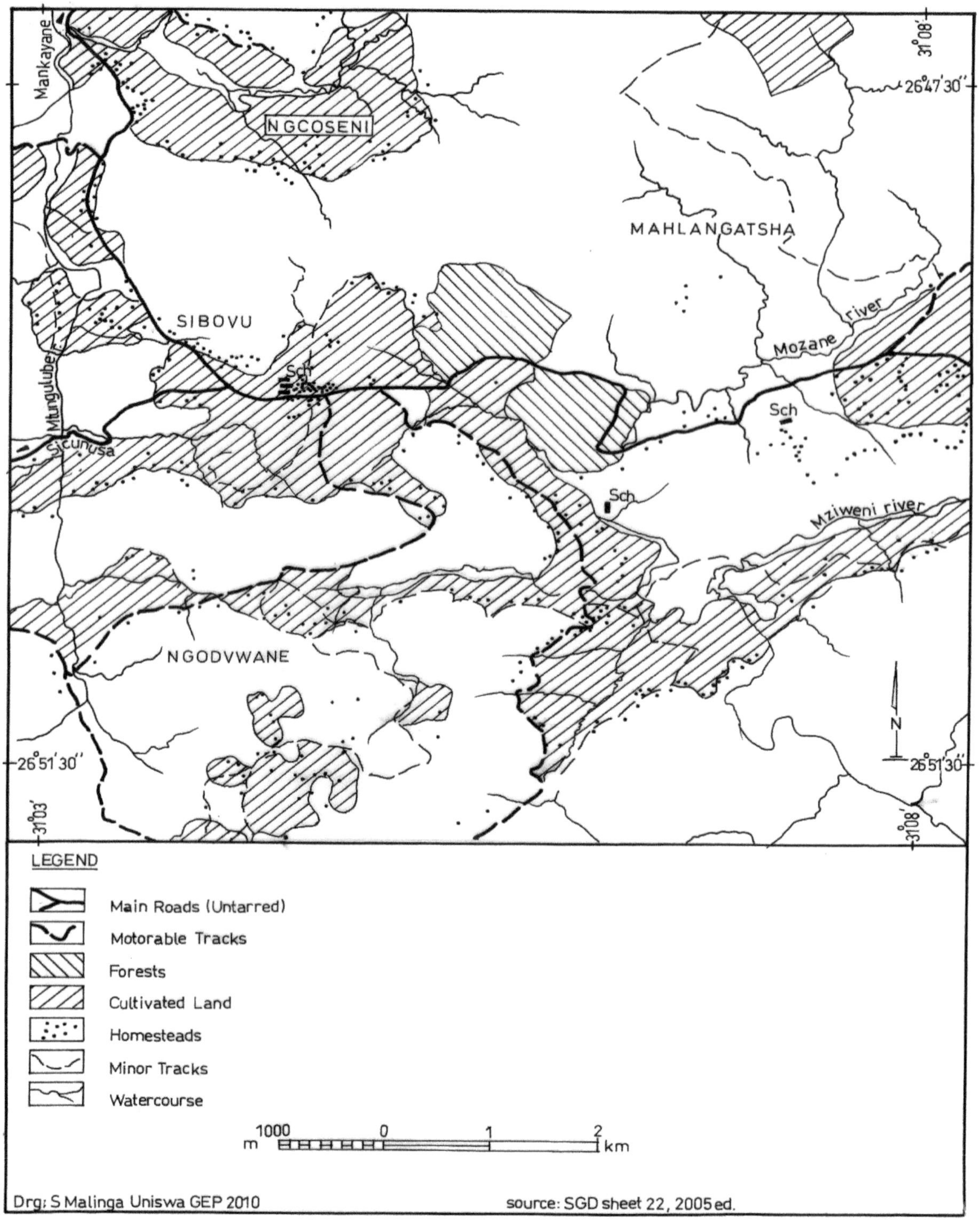

Figure 4. Map showing location of Sibovu. Source: SGD Sheet 22 2005 edition.

seedlings provided by the Mahlangatsha Rural Development Area (RDA) programme which helps in community forestry. Centralized government nurseries have been established by the forestry department for raising seedlings, primarily for the community and individual woodlots on the SNL.

The woodlots are expected to produce wood fuel and poles as well as contributing to soil conservation. Seedlings are issued free of charge by the Swaziland Government and technical expertise were provided by the extension workers on the establishment and management of tree crops. Community woodlots are

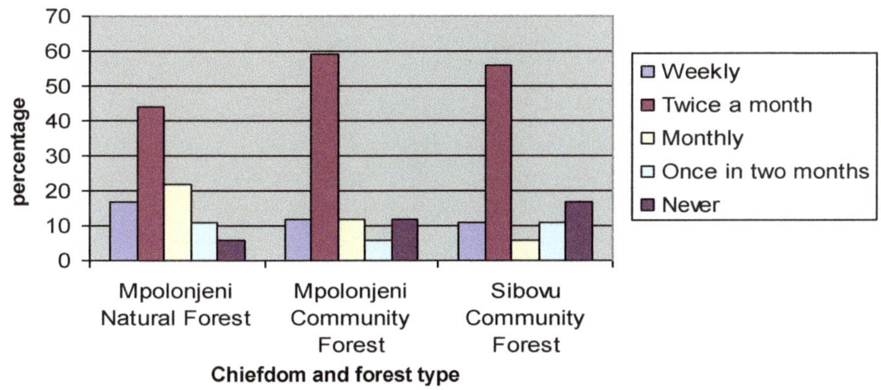

Figure 5. Frequency of getting to the forest to harvesting forest resources Source: Fieldwork 2009/2010.

established under the supervision of local chiefs, through participatory efforts. The chief nominates a special village committee to select sites for woodlots and mobilize people for the establishment, protection and tending of the woodlots.

RESEARCH METHODS

Research design

A quantitative approach was adopted for this study. A household survey incorporating 185 respondents was carried out within the study area. The sampling frame was obtained from the RDA office at Sibovu to get the number of households in the study area. Some names that did not appear on the RDA's office were obtained from the chief runners (tindvuna *and* libandla). A total of 617 households were identified where Sibovu has 425 households and Mpolonjeni has 192 households. In order to make sure that there was meaningful representation from each chiefdom, the researcher selected 30% from each chiefdom (Sibovu-125; Mpolonjeni-60) to make a total of 185 households for the purpose of the study. The researcher got assistance from the RDA's office and members of the tindvuna and libandla to locate the households and selected in the sample.

To enable meaningful representation within the chiefdoms the researcher selected 30% sample size from each chiefdom. Therefore, vast majority (125 respondents) were drawn from Sibovu and 60 respondents were drawn from Mpolonjeni. A probabilistic type of sampling namely stratified random sampling, which combined both stratification and randomisation was employed in selecting the study's respondents. This was to enhance the extent of representativeness of the sample. The primary data collecting instrument used was a questionnaire which was administered to heads of households.

Sampling

The sampling method that was used in this study is stratified random sampling. Stratified random sampling was done by chiefdom. This sampling method was selected for this study to ensure that, a minimum number of households were chosen from each chiefdom. Stratified random sampling was used so that, there will be assurance of enough cases from each group to make meaningful subgroup inferences. Within each of the chiefdom, simple random sampling was carried out to ensure that each household was chosen by chance.

Data collection

The researcher conducted a survey research. A questionnaire was prepared for each household in the different chiefdoms and responses were recorded from the household head. The questionnaire used in the study had open-ended questions and close-ended questions (Appendix A).

After coding the findings and ensuring that the data was accurate, a computer program, Statistical Package for Social Sciences (SPSS) was used for statistical analysis. This enabled the production of tables, graphs and means. Analytical tables were also produced to present the data that could not be presented graphically.

Research findings

Very low proportions of respondents in the study area indicated that, they did not go and harvest forest resources from the natural and community forests (Figure 5). The general pattern which is emerging from the interviewed households is that, due to poverty, most households depend on forest products for their day to day use.

Access to community forest resources in the chiefdoms studied is open to the local communities and to the outside communities though there are some controls done by some traditional leaders to access the forest resources. The results of this study showed that, more than 80% of the community members seek permission to access forest resources from Sibovu and Mpolonjeni community forests whilst the same proportion of respondents (80%) also indicated that, community members need to seek permission to access products from the natural forest in Mpolonjeni. However, permission was sought to help in controlling access to forest resources. As Figure 6 portrays, outsiders need permission to access forest resources then either pay a fee or freely access the forests, whilst community, members either pay a fee or enter freely.. The percentage of respondents who said outsiders seek permission to access forest resources from the community forest is highest in Mpolonjeni (71%) whilst Sibovu constitutes 56%.

Results reveal that, few household heads in Mpolonjeni (18%) and Sibovu (11%) indicated that, it was free to get access into the community forests. It should be noted, however that outsiders pay a nominal fee of E10 and E15 per pole in Mpolonjeni and Sibovu,

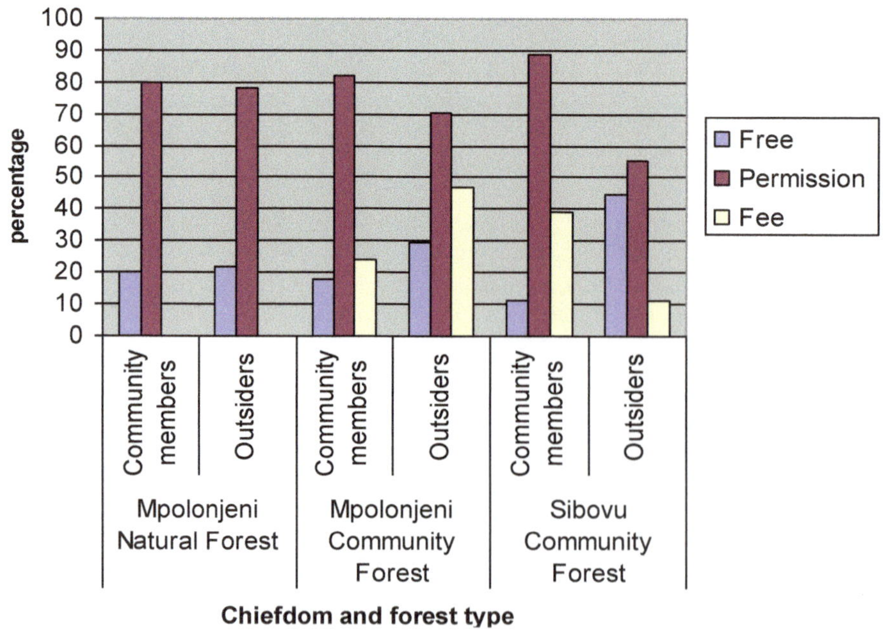

Figure 6. How to get access to the community forests. Source: Fieldwork 2009/2010.

Table 1. Forest products harvested and used.

Resource obtained from the forest	Mpolonjeni community forest (%)	Sibovu community forest (%)	Mpolonjeni natural forest (%)	Source	Uses
Firewood	88	94	88	Trunk	Cooking
Poles	59	83	71	Trunk	Construction, fencing
Seeds	-	-	47	Branches	Necklace making
Medicine	-	-	59	Bark, leaf, fruit	Curing diarrhoea
Fruits	-	-	76	Branches, roots	Food
Leaves	-	-	41	Branches	Medicinal, basket making
Roots	-	-	76	Roots	Medicinal
Honey	35	33	21	Trunk and branches	Food

Source: Fieldwork 2009/2010.

respectively, to access community forest products whilst some local community members do not pay any nominal fee since they are custodians of the community forests. Some of the respondents argued that, although the idea of paying to harvest from the forest sounded plausible theoretically, it was not very practical because some households are too poor to pay the fees, hence end up accessing the forests free of charge.

With respect to forest products harvested and used high proportions of respondents in Sibovu (94%) and Mpolonjeni (88%) said they harvest firewood from the community forest whilst the proportion is also very high in Mpolonjeni natural forest (88%) (Table 1). The proportion of those who harvest poles is highest in Sibovu community forest (83%) whilst in Mpolonjeni, 71% harvest the poles from the natural forest. It was noted that, community members and outsiders need poles from the natural forest in Mpolonjeni for constructing carts which are commonly used as a mode of transport. It is important to emphasize that, most of the products are collected for self-consumption and community rules require the community members to make a formal request before they can cut trees.

Most respondents (88%) indicated that, if forest products are harvested illegally from Mpolonjeni community forest the culprits are arrested. The percentage of respondents in this category is also high in Sibovu community forest (50%). To the contrary in Mpolonjeni natural forest, only 25% of the respondents said if harvesting is done illegally from natural forest the culprit will be arrested. A high percentage of respondents in Sibovu (50%) indicated that, illegal harvesting of forest resources led to payment of a fine and in Mpolonjeni chiefdom, a very small percentage (12%) indicated that, there was payment of a fine if there was illegal harvesting of products from the natural forest. In the chiefdoms studied, culprits can both be arrested and made to pay fines (Figure 7).

It is worth mentioning that 40% of the household heads in Mpolonjeni said nothing was done if forest resources were harvested illegally from the natural forest. However, the study

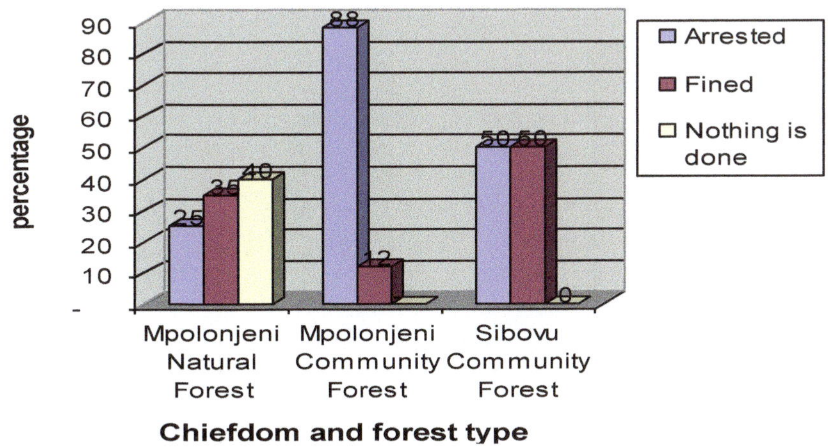

Figure 7. Punitive measures for illegal harvesting of forest resources. Source: Fieldwork 2009/2010.

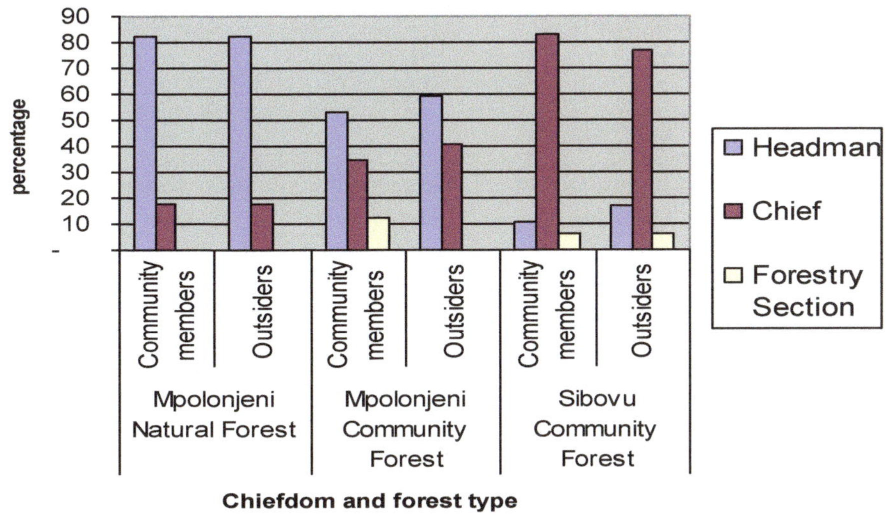

Figure 8. Granting of permission to access forest products. Source: Fieldwork 2009/2010.

established that, the practice of illegal harvesting of forest resources was a result of lack of enforcement of existing rules by traditional leaders in the natural forests in Mpolonjeni.

In the chiefdoms studied, the results reveal that, the local traditional leaders control access to forest resources. A high percentage of respondents in Sibovu said permission for outsiders (78%) and community members (83%) to access forest products from the community forest is obtained from the chief. 53% of the respondents in Mpolonjeni pointed out that, access to the community forest by community members is obtained from the headman and 59% of the respondents said outsiders get permission from the headman. Very few household heads said permission to harvest from Sibovu community forest (6%) and Mpolonjeni community forest (12%) is given by the state through the forestry section. More than 80% of the respondents indicated that, permission to access forest resources from Mpolonjeni natural forest by outsiders and community members is given by the headman (Figure 8).

The results of this study show that, there were conflicts between the management of Government and local institutions in Sibovu community forest (Figure 9). Higher percentage of respondents indicated that, there were no conflicts in Mpolonjeni community forest (54%) and Mpolonjeni natural forest (57%). It emerged from the study that conflicts in Sibovu community forest occurred because of marginalisation which was caused by the inner council of chiefs (Bandlancane). The inner council of chiefs allowed some community members, who were closer to them, to go and harvest forest resources but lacked the expertise on proper ways of harvesting and tended to over harvest which was opposed to the forestry section's rules of harvesting. The study also noted that, conflicts were prevalent in Sibovu between the forestry section and few individual community members who grazed their livestock in the community forest thereby degrading the forests. In addition, in Mpolonjeni it was indicated that, there was a marked communication gap and distrust between the state and the traditional leaders.

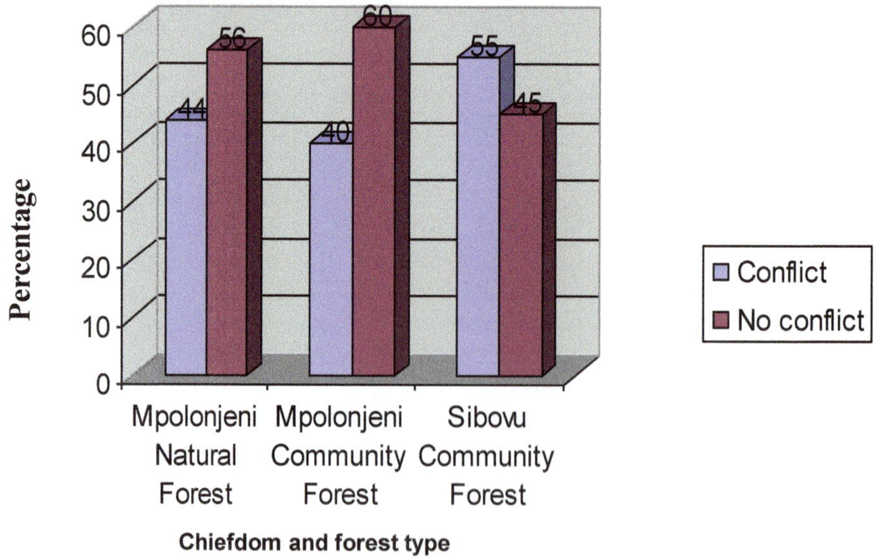

Figure 9. Conflicts between the management of organisations and local institutions. Source: Fieldwork 2009/2010.

Table 2. Who is responsible for collection of forest resources?

Parameter	Mpolonjeni natural forest		Mpolonjeni community forest		Sibovu community forest	
Forest resource collected	Male (%)	Female (%)	Male (%)	Female (%)	Male (%)	Female (%)
Firewood	43	58	60	5	60	4
Poles	77	21	48	2	48	5
Honey	53	58	32	-	32	2
Medicine	45	52	-	-	-	-
Fruits	53	57	-	-	-	-
Others	44	58	-	-	-	-

Source: Fieldwork 2009/2010.

The results show that, different members of the households are responsible for collecting different forest resources. Analysis of who is responsible for firewood collection (Table 2) showed that, the percentage of females who collect forest products is highest from Sibovu community forest (60%) and Mpolonjeni community forest (60%) whilst the proportions of respondents is pronounced in Mpolonjeni natural forest (58%). It is important to note that, the proportion of males who collect poles is highest in Mpolonjeni natural forest (77%), followed by those who collect from Sibovu community forest (72%) and Mpolonjeni community forest (72%). The results portray that, the collection of honey from Sibovu community forest (60%) and Mpolonjeni community forest (60%) is more pronounced to males whilst in Mpolonjeni natural forest it is more pronounced to females (58%). Many respondents indicated that, women collect fruits in the natural forest and the proportion is equally high for males (53%). Analysis of collection of forest resources in Mpolonjeni natural forest reveal that, majority of females collect medicine (52%), fruits (57%), and other products (58%).

Analysis of the findings based on the responses also showed that, there are other forest products that are collected such as seeds for necklace-making and leaves for basketry. A further important dimension of the survey is that, among the identified activities the collection of firewood for own consumption is the main forestry activity carried out by female members of the household in Sibovu, while male members carry out most of the income-generating activities. On the other hand, it seems that among the activities carried out by men some of them can also be carried out by women in that, they are not high energy-consuming, such as honey harvesting, although they may be time-consuming.

RESULT AND DISCUSSION

Access levels of households to forest resources

The study has revealed that, community forests and the natural forest in the study area are located on SNL thus, are governed through traditional system. This finding is consistent with that of Gamedze and Jaeger (2006) where all land and resources under SNL are administered and managed by the Chiefs who allocate it to subjects on behalf of the King. However, unlike previous studies, although it emerged that the community forests are owned by the community whilst natural forests are owned by the state, it is important to note that, few people in Mpolonjeni respect the boundaries of the natural forest whilst boundaries of community forests are well respected by a majority of people. Non respect of natural forest boundaries has led to high cases of illegal harvesting of forest resources.

Community members in the study area as well as outsiders can get into the forest and have access to forest resources from the community and natural forests. From the results it has emerged that, there is access to forest resources in the forests for both community members and outsiders though permission has to be sought first from the traditional leaders. There is restricted access to community forests for both community members and outsiders. As a way of helping to control access into the community forests a nominal fee of E15 and E10 for a pole is paid to the traditional leaders in Sibovu and Mpolonjeni chiefdoms, respectively. This payment is made by outsiders whilst community members do not pay any fee since they are the custodians to the forest resources. On the contrary, community members and outsiders do not pay any fee to harvest forest resources from Mpolonjeni natural forest. Consequently, those who lacked cash to buy the forest resources resorted to stealing the products from community forests. It should be noted that, forest resources in Mpolonjeni natural forest are open to extraction to anyone from the community and outsiders thereby creating the situation of the forest resources becoming open access goods.

Generally, most households in the study area depend on accessing forest resources for their day to day use. There are some variations in distance travelled and time taken to go and harvest forest resources but results from the study show that, people still make an effort to access the products because they depend on them for a living. Community forests have controlled access to community members and outsiders because there are mechanisms in place which allow the communities to exclude outsiders from its forest resources. Forest resources in the natural forest are accessible to any community member and outsiders thereby leading to the forest resource being open to extraction to anyone. The forest resources that are open to extraction include firewood poles and fruits. It is also worth mentioning that, the rules that regulate access to forests are not well enforced by traditional leaders and the state.

Forest resources utilised

The results show that, multiple uses of some indigenous trees which were initially few have led to reduction of the tree species. Consequently, people are now taking more time and travelling longer distances in search of the tree species so that they can get the required forest resources. The majority of people harvest firewood for cooking from the forests whilst poles are harvested for construction and fencing. Residents use several forest products from the forests for their own consumption, and do not get much out of the forest which can be sold outside. However, analysis of the results showed that, amongst the forest resources that are harvested, some of the products are sold locally, with a significant proportion being sold to urban people whilst small proportions are sold to people from outside the country.

Similarly, Yadav et al. (2003) state that, in Nepal forest, people rely on forests and trees for fodder and bedding materials, for timber and poles for houses and agricultural implements such as ploughs and for fuel wood, which is the most important, and often the only source of energy for cooking and heating for most rural households. In the study area firewood, poles and honey are harvested from the community forests whilst the natural forest provides firewood, poles, seeds, medicine, fruits, leaves, roots, and honey. This heavy reliance on the natural forest for poles arises from the need to construct carts which are used to transport poles and other products. However, it is important to note that, this has contributed to deforestation in the chiefdom.

Institutional and community rules that regulate access to forest resources

The results indicate that, there are rules in use which traditional leaders and the forestry section apply in order to regulate, access, and protect the forest resources. If forest resources are illegally harvested from the community forest, the culprit is arrested and taken to umphakatsi where forest resources are confiscated or a fine is paid.

Illegal harvesting of forest resources is high in both forest management systems. This study has found that, nothing is done when forest resources are illegally harvested from the natural forest and the frequency of illegal harvesting of forest resources is highest in this forest management system. This is due largely to lack of effective rules that protect the forests as well as lack of enforcement of the rules that regulate access to the forest resources. Surprisingly, the rules governing access and protect forest resources exist, but differences exist in knowledge levels of the rules by people and enforcement levels by the local leadership. The state through the forestry section is there to conserve and manage forest resources in the community forest. Based on this the study, it was concluded that, there is lack of involvement by any organisation in managing the natural forest whilst both community and natural forests are not managed by any outside organisations. This is contrary to the observation by Ostrom (2001) that in some countries, community- based forest management institutions have received increasing attention from governments, donors and Non Governmental Organisations in the past. These organisations see themselves as stakeholders in community managed forests and want to understand how community-based institutions work and how they can be supported, reoriented or recreated to advance particular environment and development goals.

Studies by Odera (2004) note that, virtually all Sub Saharan African countries are experiencing difficulties in managing their forests sustainably in the face of rising challenges and pressures. This has been made worse by increased patronage and state-people conflicts on who owns controls and manages the forest. This situation whereby conflicts exist is evident in Sibovuchiefdom between the forestry section officials and the traditional leaders. Additionally, conflicts also exist between the government and the community members on lack of ownership of the natural forest in Mpolonjeni by the local members. It seems insecurity of land and tree tenure has led to general degradation of the forests in the study area. Consequently, the same characteristic has led to degradation of the forests as evidenced by the unsustainable ways of harvesting forest resources and lack of involvement in management of the forests by community members.

Gender roles in collection of forest resources

Different members of the households were found to be responsible for collecting different forest resources. Collection of firewood, medicine, fruits and other forest resources is a responsibility of women. Males on the other hand are responsible for collection of poles and honey with the exception of Mpolonjeni where the majority of women are responsible for collecting honey from the natural forest. Generally, the collection of firewood for own consumption is the main forestry activity carried out by female members of the household in Sibovu, while male members carry out most of the income-generating activities. On the other hand, among the activities carried out by men some of them can also be carried out by women such as honey harvesting.

Conclusion

The results obtained from the study have helped to cast some light in the understanding of access to and utilisation of forest resources in Swaziland. The approach used was comparative analysis, whereby two chiefdoms were examined. The findings of the study show that forest resources from both community and natural forests can be accessed by community members and outsiders. They also show that, there is restricted access to forest resources from the community forests. Outsiders access forest resources by paying a nominal fee of E10 in Mpolonjeni per pole and E15 per pole in Sibovu whilst community members do not pay any fee. This has implications for those outsiders who do not possess the cash but are in need of the forest resources since they resort to illegal harvesting of the products. At the same time, forest resources from the natural forest are an open access resource to community members and outsiders. This uncontrolled open access system is leading to excessive use and it is difficult to prevent any user from continuing to subtract units from the natural forest. There is no payment of a nominal fee by either community members or outsiders to access forest resources from the natural forest. It can thus be concluded that, natural forests in Mpolonjeni are open access resources and are unsustainable since they are accessible to anyone and are being degraded because traditional leaders are doing little to protect them.

Extraction of forest resources is mainly for own consumption by the local communities and not much out of the forest resources are sold. Some tree species specifically Maesa lanceolata (Umbhongozi) which were initially few and have multiple uses such as provision of poles, firewood, medicine, and fruits have become scarce in the natural forest. The study has revealed that, uncontrolled extraction of firewood for cooking and poles for construction have led to deforestation.

The findings show that, traditional leaders control the access to forest resources in the studied chiefdoms. The natural woodlots are owned by the government while the community woodlots are owned by the community. There are some conflicts between state and traditional leaders in Sibovu. The state and community members in Mpolonjeni natural forest have conflicts on ownership of the forest resources and this has led to lack of involvement in the management of forest resources by community members as is reflected by the high cases of fire outbreaks, and lack of enforcement of the rules that protect the forests.

There are government and community rules that protect forest resources and regulate access and utilisation. Most of the government rules that protect natural forests are not known by community members whilst community rules are fairly known but findings reveal that, they are not effective in protecting the forest resources. What can be inferred from this is that, there is lack of enforcement of rules by traditional leaders and government which could be the leading reasons for high frequency of illegal harvesting of forest resources from the natural forest. What is striking is that, community and government rules that protect community forests are well known and are effective in regulating access and utilisation. The study also noted that, there are no outside organisations that deal with conservation and management of forest resources in the study area.

RECOMMENDATIONS

Based on the results it was clear that, people were not involved in the management of the natural forest since they made accusations on lack of ownership of the natural forests. The study, therefore, recommends that, there should be improvement of relationships between local community members and the forestry section authorities. The study also recommends that, the relationship between the local institutions and forestry section officials should be improved. Local people are not aware of the rules that protect the natural forests for example, people should harvest stipulated quantities. The study thus recommends that, traditional leaders and the forestry section officials educate people about the rules that regulate access to forest resources and protect the natural forest.

Furthermore, to ensure that the natural forest is protected and well managed, the study recommends that, local community members are provided with economic incentives by the government so that there is maximum involvement in sustainable management of the forest resources. It is crucial that, commonly understood rules are set at the local level with local people generally agreeing upon what rules they should follow and there is need for successful enforcement of the rules by traditional leaders and the government through the forestry section. Agrawal and Goyal (2001) note that, without this agreement, there is less incentive to comply with rules. They further state that, moderately-sized communities who agree on a general set of rules regarding forest use can better afford to share monitoring duties and thus enjoy better forest resources.

REFERENCES

AfroGIS (2008). Biophysical environmental analysis and land use mapping of Mahlangatja. Swaziland Forest Report. Mbabane.

Agrawal A, Goyal S (2001). Group size and collective action. Comp. Polit. Stud. 34(1):63-93.

Carlsson L (2000). Policy networks as collective action. Policy Stud. J. 28(3):502-520.

Fisher M (2004). Household welfare and forest dependence in Southern Malawi. Environ. Dev. Econ. 9(2):135-154.

Gamedze S, Jæger M (2006). Swaziland Forest Policy and Legislation Project, Retrieved August, 10, 2009, http://dwafapp4.dwaf.gov.sz.

Gibson CC, Andersson K, Ostrom E, Shivakumar S (2005). The samaritan's dilemma. Polit. Econ. Dev. Aid 13(2):105-124.

Government of Swaziland (2007). MahlangatshaInkundlaProfile, Ministry of Agriculture, Mbabane.

National Forest Policy Green Paper (2000). Environmental Consulting Services, Mbabane.

Odera J (2004). Lessons learnt on sustainable forest management in Africa. African Forest Research Network (AFORNET). African Academy of Sciences (AAS).

Ostrom E (2001). Institutions, collective action and effective forest management: Learning from studies in Nepal, In E. Webb & G.Shivakoti (eds.), Environment and Society pp. 193-223. New Delhi, Sage: India.

Ostrom E, Janssen MA, Anderies JM (2007). Going beyond panaceas.Proceedings of the NationalAcademy of Science in the United States of America. 104(39).

Rudd MA (2004). An institutional framework for designing and monitoring ecosystem-based fisheries management policy Experiments. Ecol. Econ. 48(1):109-124.

Shimizu T, Trudel M (2006). Methodology and case studies on linkages between poverty and forestry: Afghanistan, Iran, Kyrgyzstan and Turkey. FAO LSP WP 35. Access to Natural Resources Sub-Programme.

Tevera DS, Mukora CM (2007). Common Property Resources.In DS. Tevera CM. Mukora (eds.), Environmental Policy Planning and Management in Southern Africa. Dept of Geography and Environmental Science: University of Zimbabwe. pp. 50-57.

Yadav NP, Dev OP, Springate-Baginski O, Soussan J (2003). Forest Management and Utilization under Community Forestry. For. Livelihood 3(1):37-50.

APPENDIX A

Questionnaire for Household survey

Demographic and socio economic data

1.0. Household data.
1.1. Name of the household head …………………
1.2. Area of household ………………………

1.3. Gender:
1.3.1. Male……………….
1.3.2. Female…………………

1.4. Marital Status
1.4.1. Single …………….
1.4.2. Married …………..
1.4.3. Divorced …………
1.4.4. Widowed …………

1.5. Age of the household head
1.5.1. < 20 years ……….
1.5.2. 20-39 years……….
1.5.3. 40-59 years………
1.5.4. >60 years ……….

1.6. Household composition.

Age (years)	Male	Female
<18		
18-55		
>55		

1.7. Household main economic activities.

Activity	Yes/No	Who is involved
Farming		
Employment		
Fishing		
Small business		
Others(specify)……		

1.8 Education level
1.8.1. No formal education ………………………..
1.8.2. Adult education ……………………………
1.8.3. Primary education …………………………
1.8.4. Secondary …………………………………
1.8.5. Tertiary……………………………………..
1.8.6 Other (specify)……………………………..

2.0 FOREST RESOURCE ACCESS

A. Community forests

2.1. Who owns the community forests?
2.1.2. Central Government……………………
2.1.2. Community (Name)…………………………
2.1.3. Others (Specify) ……………………………………

2.2. Do you know the boundaries of the forests?
2.2.1. Yes …………………
2.2.2. No. …………

2.3. If yes, are they respected?
2.3.1 Yes …………………..
2.3.2. No. ………….

2.4. What happens to a person who enters into the community forests and harvest forest products illegally?
2.4.1. Arrested…………………………………………….
2.4.2. Fine (Amount)………………………………………..
2.4.3. Others (specify)……………………………………..

2.5 How frequently do people enter and harvest community forests illegally?
2.5.1. Everyday……………………………………………
2.5.2. Once every week……………………………………
2.5.3. Once per month……………………………………..
2.5.4. Once in two months…………………………………
2.5.5. Never………………………………………………..

2.6 What measures are taken to protect the forest? (Tick the appropriate box/es)
2.6.1. Use of community police
2.6.2. Use of firebreaks
2.6.3. Fencing
2.6.4. Having specific times of harvesting
2.6.5. Limiting the quantities of resources harvested
2.6.6. Other (Specify)…………………………………..

2.7. What are the main names of trees found in community forests?

Name:

2.8. Do you get into the forest to collect forest products?
2.8.1. Yes---------------
2.8.2. No.---------------

2.9. If yes, how do you get access to the community forest?
2.9.1. Free--------------------------
2.9.2. Permission-------------------
2.9.3. Fee (specify)-----------------

2.10. Who gives you permission to access the forest products?
2.10.1. Headman…………………
2.10.2. Chief……………………..
2.10.3. State (specify)…………..

2.10.4. Other (specify).............

2.11. Can people from outside the area access forest products from the community forest
2.11.1. Yes----------------
2.11.2. No.----------------

2.12. If yes, how do they get access to the community forest?
2.12.1. Free---------------------------
2.12.2. Permission--------------------
2.12.3. Fee (specify)------------------

2.13. Who gives them permission to access the forest products?
2.13.1. Headman.....................
2.13.2. Chief..........................
2.13.3. State (specify)...............
2.13.4. Other (specify)..............

2.14. What is the approximate distance to the forest?
2.14.1.<2km -------------------------
2.14.2. 2-5km-------------------------
2.14. 3. >5km -------------------------

2.15. What is the approximate time taken to get to the forest?
2.15.1 <15 minutes ....
2.15.2. 16-30minutes................
2.15.3. 31- 60minutes...............
2.15.4. more than 1hour.............

2.16. Who is responsible for forest product collection?

Product	Collected by:		
Gender	Male	Female	All
Firewood			
Poles			
Medicine			
Honey			
Fruits			
Others (specify)			

B. NATURAL FORESTS

2.17. Who owns the natural forests?
2.17.1 Central Government..............................
2.17.2. Community (Name)..............................
2.17.3..Others (Specify) ....

2.18. Do you know the boundaries of the forests?
2.18.1. Yes ....
2.18.2. No. ....

2.19. If yes are they respected?
2.19.1 Yes ....
2.19.2. No. ....

2.20. What happens to a person who enters into the natural forests and harvest forest products illegally?
2.20.1. Arrested...
2.20.2. Fine (Amount).....................................
2.20.3. Others (specify)...................................

2.21. How frequently do people enter and harvest natural forests illegally?
2.21.1. Everyday.................................
2.21.2. Once every week.........................
2.21.3. Once per month..........................
2.21.4. Once in two months......................
2.21.4. Never....................................

2.22. What measures are taken to protect the forest? (Tick the appropriate box(es)
2.22.1. Use of community police
2.22.2. Use of firebreaks
2.22.3. Fencing
2.22.4. Having specific times of harvesting
2.22.5. Limiting the quantities of resources harvested
2.22.6. Other (Specify).............................

2.23. What are the main names of trees found in natural forests?

Name:

2.24. Do you get into the forest to collect forest products?
2.24.1. Yes----------------
2.24.2. No----------------

2.25. If yes, how do you get access to the natural forest?
2.25.1. Free---------------------------
2.25.2. Permission--------------------
2.25.3. Fee (specify)------------------

2.26. Who gives you permission to access the forest products?
2.26.1. Headman....................
2.26.2. Chief........................
2.26.3. State (specify)..............
2.26.4. Other (specify)................

2.27. Can people from outside the area access forest products from the natural forest?
2.27.1. Yes----------------
2.27.2. No----------------

2.28. If yes, how do they get access to the natural forest?
2.28.1. Free---------------------------
2.28.2. Permission--------------------
2.28.3. Fee (specify)------------------

2.29. Who gives them permission to access the forest products?

2.29.1. Headman...................
2.29.2. Chief.......................
2.29.3. State (specify)..............
2.29.4. Other (specify)................

2.30. What is the approximate distance to the forest?
2.30.1. <2 km -------------------------
2.30.2. 2-5 km -------------------------
2.30.3. >5 km -------------------------

2.31. What is the approximate time taken to get to the forest?
2.31.1. <15 min...........................
2.31.2. 16-30 min..........................
2.31.3. 31- 60 min.........................
2.31.4. More than 1hour.......................

2.32. Who is responsible for forest product collection?

Product/Collected by	Male	Female	All
Firewood			
Poles			
Medicine			
Honey			
Fruits			
Others (specify)			

3.0. ROLE OF INSTITUTIONS IN FOREST MANAGEMENT

A. Community forests

3.1 Can you list organizations dealing with forest conservation and management in the area

Name of organization

3.2. Are there any outside organizations involved in conserving forest resources?
3.2.1. Yes............................
3.2.1. No....................................

3.3. Are there any conflicts between the management of these organizations and those which are under the local institutions?
3.3.1. Conflict...
3.3.2. No conflict...

3.4. List all the rules that you know regarding protection of the community forests

Government rules	Who enforces them	Are they effective?
Community rules		

3.5. Do you know any cases of fire outbreak in the community forests?
3.5.1.Yes ....
3.5.2. No. ....

3.6. If yes what are the reasons of fire in the forests?
....

3.7. What is the role of the community when there is fire?
....
....

3.8. Are people being involved in the management of the forests?
3.8.1 Yes..............................
3.8.2. No..................................

3.9. If yes, how?
....
....

3.10. If people are involved in forest management does that increase the level of access to forest resources?

B. Natural forests

3.11. Can you list organizations dealing with forest conservation and management in the area.

Name of organization

3.12. Are there any outside organizations involved in conserving forest resources?
3.12.1. Yes.............................
3.12.1. No....................................
3.13. Are there any conflicts between the management of these organizations and those which are under the

local institutions?
3.13.1. Conflict……………………………………………
3.13.2. No conflict…………………………………………
3.14. List all the rules that you know regarding protection of the natural forests.

Government rules	Who enforces them	Are they effective?
Community rules		

3.15. Do you know any cases of fire outbreak in the natural forests?
3.15.1. Yes ………………………
3.15.2. No. …………………………

3.16. If yes what are the reasons of fire in the forests?
……………………………………………………………

3.17. What is the role of the natural when there is fire?
………………………………………………………………………
………………………………………………………………………

3.18. Are people being involved in the management of the forests?
3.18.1 Yes……………………………
3.18.2. No……………………………..........
3.19. If yes how?
……………………………………………………………
………………………………………………………………………

3.20. If people are involved in forest management does that increase the level of access to forest resources?...

4.0 FOREST RESOURCE UTILIZATION

A. Community forests

4.1. How often do you go to the community forest?
4.1.1. Weekly………..........................
4.1.2. Twice a month……………....
4.1.3. Monthly………........................
4.14. Once in 2 months………….....
4.1.5. Never………….........................

4.2. What are the forest products you harvest and use and where do you get them? Product:

Type	Source	Uses

4.3. Who decides on when to harvest resources from the community forests?
4.3.1. Chief ……..
4.3.2. Headman……….
4.3.3. State (specify)…………………
4.2.4. Other (specify)………..

4.4. Who decides on how to harvest resources from the community forests?
4.4.1. Chief …......
4.4.2. Headman……….
4.4.3. State (specify)…………………
4.4.4. Other (specify)………..

4.5. Are forests resources from community forests sold?
4.5.1. Yes ……………….
4.5.2. No…………………
4.5.3. No idea.......................

4.6. If yes, what resources are sold?
……………………………………………………………

4.7. To whom are the forest resources sold?
4.7.1. Local people……………………..4.8.2. People from urban areas…………..
4.6.2. Other (specify)……………………

4.8. How are the cash benefits used?…………………………………………………………
……………………………………………………………

4.9. What is the main fuel energy you use?
4.9.1. Firewood---------------------
4.9.2. Charcoal----------------------
4.9.3. Kerosene---------------------
4.9.4. Others (Specify) -------------

B. Natural forests

4.10. How often do you go to the natural forest?
4.10.1. Weekly………..
4.10.2. Twice a month………..
4.10.3. Monthly………..
4.10.4. Once in 2 months…….
4.10.5. Never…………..

4.11. What are the forest products you harvest and use and where do you get them?
Product :

Type	Source	Uses

4.12. Who decides on when to harvest resources from the natural forests?
4.12.1. Chief ....
4.12.2. Headman...........
4.12.3. State (specify)...........
4.12.4. Other (specify)............

4.13. Who decides on how to harvest resources from the natural forests?
4.13.1. Chief ....
4.13.2. Headman...........
4.13.3. State (specify)............
4.13.4. Other (specify)............

4.14. Are forests resources from natural forests sold?
4.14.1. Yes ....
4.14.2. No..............
4.14.3. No idea.......................

4.15. If yes, what resources are sold...........................

4.16. To whom are the forest resources sold?
4.16.1. Local people.........
4.16.2. People from urban areas...................
4.16.2. Other (specify)...................................

4.17. How are the cash benefits used?........................

4.18. What is the main fuel energy you use?
4.18.1. Firewood----------------------
4.18.2. Charcoal-----------------------
4.18.3. Kerosene----------------------
4.18.4. Others (Specify) -------------

Initiatives of tropical agroforestry to sustainable agriculture: A case study of Capasia Village, Northern Bangladesh

S. A. Rahman[1]*, F. D. Paras[2], S. R. Khan[3], A. Imtiaj[4], K. M. Farhana[5], M. M. Toy[6], M. B. Akhand[7] and T. Sunderland[1]

[1]Center for International Forestry Research (CIFOR), Bogor Barat 16680, Indonesia.
[2]Department of Social Forestry and Forest Governance, College of Forestry and Natural Resources, University of the Philippines, Los Banos, Philippines.
[3]Natural Resources Institute, University of Manitoba, Canada.
[4]Department of Botany, University of Rajshahi, Rajshahi, Bangladesh.
[5]Department of Sociology, University of Padova, Italy.
[6]Shanto-Mariam University of Creative Technology, Dhaka, Bangladesh.
[7]Intervida (Bangladesh Country office), Dhaka, Bangladesh.

A relatively large percentage of the population in Bangladesh lives under the poverty line and is affected by the country's degrading natural resources. Agroforestry has been seen as one of the few options to lift people out of poverty. Research into the costs and benefits of agroforestry was undertaken in Capasia Village in Northern Bangladesh. Initial results indicate that agroforestry may not only be an optimal livelihood solution for poor farmers, biodiversity conservation and environmental sustainability but agroforestry systems also provide good economic rates of return. Thus the farmers who engage in agroforestry are benefited in different ways.

Key words: Agroforestry, poverty, farmers, livelihoods, conservation.

INTRODUCTION

At present in Bangladesh, many environmental problems stem from poverty- often contributing to a downward spiral in which poverty exacerbates environmental degradation and environmental degradation exacerbates poverty (ADB, 2004, 2001). In poor rural areas, for example, there are close links between the lack of family planning, concomitant population growth and extensive deforestation, as local farmers fell tropical forests for firewood and new farmland. Forests are depleted by commercial timber exploitation and gradually converted into pastures, plantations and cultivated fields (Rahman et al., 2007). Recycling periods for slash-and-burn cultivation systems declined from about 20 years in 1900 to about three years in the recent decade, eliminating forest fallows. According to the World Development Report (World Bank, 2000), only 5.6% of the total area of Bangladesh is forested and that area is being depleted at the alarming rate of 3.1% per annum (FAO, 2009). Moreover, due to the loss of firewood resources, rural communities turn to alternative fuels such as cow dung and crop residues that previously served as organic manure on agricultural fields. The shorter fallow periods, reduced organic inputs and repeated removal of nutrients with harvested crops have contributed to soil degradation and consequently, lowered crop yields and farmers' returns leading to more poverty (Rahman et al., 2008, 2007).

Agroforestry is a collective name for land use systems and practices where woody perennials are deliberatly integrated with crops and/or animals on the same land management unit. The integration can either be in a spatial mixture or in temporal sequence. There are normally both ecological and economic interactions between the woody and the non-woody components in

*Corresponding author. E-mail: sumonsociology@yahoo.com.

Table 1. Site characteristics.

Characteristics	Description
Property zoning	Agriculture
Existing agroforestry species	Mango, coconut, paddy and wheat
Access	National highway
Topography	Flood plain
Slop	0 – 5%
Elevation	40ft.
Rainfall	Over 1450 per year
Average tempareture	Summer 37.8, Winter 9.2
Dry season	December- March
Soil condition	Loamy
Water resources	Ponds, mainly natural rainfall

agroforestry. For centuries, agroforestry has been practiced around the world and is most commonly associated with tropical and sub-tropical regions. Agroforestry can provide many different environmental and production benefits for a farm and its surrounding area. Nair (1989) defines agroforestry as a sustainable land management system which increases the overall yield of the land, combines the production of crops (including tree crops) and forest plants and /or animal simultaneously or sequentially, on the same unit of land, and applies management practices that are computable with the cultural practices of the local population.

In Bangladesh, agroforestry has been practised in different areas for a relatively long time. However, formal agroforestry policies were officially initiated only in 1977 to 1978 (Alim, 1980). Agricultural cropping systems grown with in association with planted tree species were found to be successful in various parts of Bangladesh and agroforestry is seen as one of the very few livelihood options to lift people out of poverty (Table 1).

MATERIALS AND METHODS

The research methods were applied primarily and they include:

(1) A survey and rapid rural appraisal (RRA) for the basic social circumstances of the villages.
(2) Short structured interviews with a purposive sample of 110 households focused primarily towards experiences and actual and envisaged costs and benefits of agroforestry.
(3) Participatory rural appraisal (PRA) and field-based surveys implemented by the 60 interviewees.

Other data was gathered by way of interviews with key informants (GO, NGO, firms and PO) and market prospecting. Moreover, secondary data gathered from local and overseas resources and Internet. 'Kapasia' village of Rajshahi District in North Bangladesh were selected purposively. The village was selected based on various criteria. Firstly, there is an ongoing agroforestry programme in the area and local people have experience with agroforestry practices. Secondly, the researchers are familiar with the culture, norms and local language of the study area and have undertaken previous research in the area.

RESULTS AND DISCUSSION

Financial analysis of agroforestry

Research data indicates that agroforestry provides substantially more benefits than traditional agricultural systems. The analysis of collected data is to estimate the value of net returns from agroforestry. Actual costs and benefits vary widely depending on local conditions, however, for the purposes of our analysis, all values are based on early 2010 prices.

Major costs of initiating agroforestry projects

Establishment cost: The establishment cost involves land preparation, seedlings, planting etc., for agroforestry activities. Costs for preparation vary greatly depending on the condition of the site. The farmers of the study area purchased seedlings from private or state sources, and range in price is dependent on species and quantity grown. The average establishment cost is not more than Tk. 15000 per acre. Usually land costs are not included in the projected costs of agroforestry projects as it is assumed that landowner already has secure tenure rights (Table 2).

***Pesticides and fertilizer*:** Pesticides and fertilizer are often used to enhance the trees and crops in early years. After three to five years fertilizer is used often as tree species provide nutrients for understory crops. Usually the pesticides and fertilizer costs are Tk. 3000 per acre in the early stage of the project (Table 2).

Table 2. Some costs in agroforestry project.

Operation	Year	Comments	Approximate cost (Tk.)
Establishment coast	0	----	9000 per acre
Pesticides and fertilizer	0-4	----	3000 per acre
Pruning and thinning	2	Form pruning	Only self labor
	4	pruning 0-2 m	
	5	pruning 2-4 m	
	6	pruning 4-6 m	1500 per acre
Labor cost	Annual	10-15 days	

USD 1 = 69.72 Tk as of September 2010.

Table 3. Net cash flow (Tk.) of agroforestry project per acre (depicted in line graph above).

Year	1	2	3	4	5	6	7	8	9	10	11
Cost	19020	5040	5040	5040	4800	4800	4800	4800	4680	4920	4920
Returns	30000	30000	30000	30600	31900	32520	33000	40080	48000	48000	51120
Net	10980	24960	24960	25560	27100	27720	28200	35280	43320	43080	46200

Pruning and thinning: Pruning and thinning is used to improve the form of trees by removing lower branches or multiple stems in order to optimize timber production and quality and quantity of fruits (Table 2).

Labor cost: Farmers hire labor when it is needed, especially during the harvesting period. An average reasonable labor cost is seen Tk. 60 per day for hired labor. At other times of the year farmers rely on their own labor or that of a family member. For the analysis, it is important to calculate the amount of labor required on a yearly basis in order to give meaningful result. The yearly hired labor cost Tk. 1,500, which represents a maximum 10 to 15 days (Table 2).

Returns from agroforestry

The farmers realized relatively early crop yields from agroforestry in Capasia village (the first 1 to 3 years). However, from 4th year they also earned fruits from tree crops. In such systems it is seen that, from 4th year, the yields of tree crops are increasingly high. Research data indicates in the beginning year the farmers earned average Tk. 10,980 per acre from agroforestry. They earned Tk. 24,960; Tk. 24,960, Tk. 25,560, Tk. 27,100 and Tk. 27,720 from the year of 2nd, 3rd, 4th, 5th and 6th respectively (Table 3). The net income is gradually increased from this project (Graph 1).

The mango (*Mangifera indica*) and coconut (*Cocos nucifera*) are the most commonly planted agroforestry tree species. The wood of the mango tree is used in furniture and for construction and the leaves and bark are used medicinally and are also used as an energy source of cooking and for animal feed. The ripe fruit is eaten raw and can be used to make jam, chutney, deserts, and is sold locally as a major source of income. The juice, puree, and ripe and green fruit are often exported to some other countries. The mango is also an important Hindu religious plant, and its leaves are used in religious ceremonies.

The brown nut of coconut is a major human staple and animal food. The husk can be dried and made into copra, which is an important source of income. Juice from the immature green nut is an important drink and source of safe drinking water in dry areas and in times of drought. The sap from the flower stalk is used to make fresh and fermented toddy. The timber is used for many purposes. The trunk, fronds, shells, and husks are a major source of fuel for cooking. The roots are used for making fish traps, and the husk fiber is used to make cord, that is used for lashing house beams and other purpose. The shell is used to make containers, fishhooks, charcoal and other useful items. The leaves and fronds are used for weaving baskets, hats, fans, floor mats, food containers, roofing, as well as for shade and mulching in taro gardens. The midrib leaflets are used in brooms, as a toothpick and for other purposes. Almost all parts of the tree are used medicinally. Coconut is a plant of spiritual and ritual importance in many areas.

The returns earned from agroforestry provide farmers the means to enrich their economic and social status in the village. For example, the additional income they

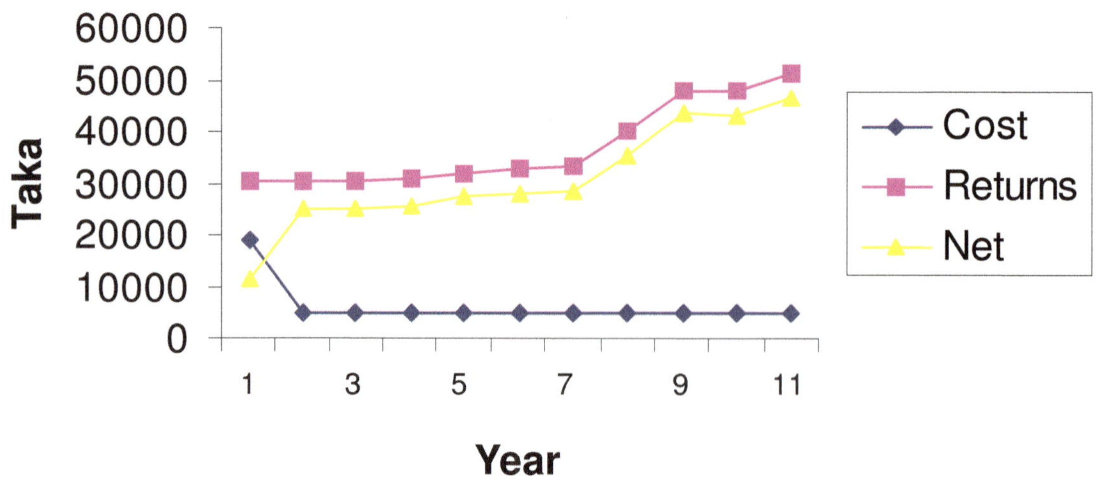

Graph 1. Net cash flow of agroforestry project per acre.

Table 4. Agroforestry systems make more efficient use of natural resources.

Soil nutrients	Trees promote nutrients cycling and efficient use; nutrient status is increased through nitrogen fixation and the uptake of deep soil nutrients.
Sun	Multistoried cropping systems intercept and use sunlight at all level.
Water	Tress can increase water availability in the soil by reducing runoff and evapotranspiration, while increasing water infiltration and soil water-holding capacity.

realize is used for mainly as food, clothes, education and health.

Owners of agroforestry projects in study area are also capture the indirect benefits of forestry such as shade, erosion control, wildlife habitat, aesthetics, land values, etc. However it is difficult to place an exact dollar figure on these additional ecosystem services.

The costs and benefits of agroforestry in the study area vary widely, and depend upon site conditions as well as on many factors such as tree spacing, maintenance techniques, and the rate of growth of trees. To compare an agroforestry project with agriculture, the same analysis should be undertaken for both. Below is the comparable financial viability of two options such as agroforestry and agriculture:

1. Agriculture may not be as profitable, but the returns are annual. Agroforestry has a relatively large cost early in the project, with positive returns occurring many years in the future.
2. Agricultural markets are relatively well known in the short term. Agroforestry markets are more uncertain due to the relatively long time horizon.
3. Most farmers have the skills and confidence for agricultural production. New skills and technologies may have to be learned and understood to successfully carry out a commercial agroforestry product.

Role of agroforestry

Population growth, increasing economic needs, environmental degradation, and a shortage of arable land make the development of sustainable and efficient agricultural systems crucial. Agroforestry practices can improve productivity and reduce inputs, while mitigating some of the environmental damage cause by the past processes of deforestation and the removal of trees from the landscape (Tables 4 to 9).

Improved economic benefits

It is more likely that agroforestry can provide improved economic benefits to landowners. Aside from improving the value of the timber products, there are also a number of strategies that can improve the overall returns from agroforestry. Because of closer landowner involvement, small-scale forestry is uniquely suited to incorporate non-timber yields, which can increase the economic

Table 5. Agroforestry systems provide a more favorable environment for sustained production.

Shade	Filtered shade conserves water, produces evapotranspiration, keeps topsoil cool, and helps maintain beneficial microbial activity
Wind protection	Trees protect crops from wind damage and soil from wind erosion and drying.
Soil conservation	Tree root and mycorrhizal systems reduce nutrient leaching, bind soil, and prevent erosion. Tree leaf litter enhances soil physical, chemical, and biological conditions, which makes soil more resistant to erosion and more able to absorb and hold water.
Nutrient cycling	Through nitrogen- fixing trees and nutrient uptake from deep soil layers, trees promote more closed nutrient cycling and more efficient use of nutrients.
Habitat diversity	Trees provide habitats for birds, insects, and other animals that help maintain pest/ predator balance in the system.

Table 6. Agroforestry systems can be more profitable.

Reduce expenses	Through nutrient cycling and soil and water conservation, tree reduce the need to purchase fertilizer and water. Trees may also reduce the need for pesticides by providing habitat for pest predators.
Diversified products	Mixed cropping systems typically have two or more economic products, reducing dependence on market conditions for a single crop.
Continuous flow of product	Agroforestry systems often combine short term and long term crops, which can lead to a high level of total productivity and year round production.
Greater self reliance	Agroforestry can reduce farmer's dependence on purchased products and resources, as well as reducing vulnerability to changing market conditions.

benefits. The nontimber crops may include short-term crops that are phased out as the forest matures, or permanent production systems of crops, animals, and/or multiple use plants that provide sustained yields for many years.

Earlier returns: An advantage of agroforestry is the possibility of early yields from non timber crops such as herbs, medicinal, fruits, or other crops that are interplanted with the timber trees. By producing economic returns early in the project, non-timber yields can pay back investment costs earlier than timber yields, which can enhance the economics of the participating household as a whole.

More efficient use of space: In non-agroforestry plantings, the space between trees must be mowed or maintained, which is a considerable expense. Using that space for a crop instead may take a similar amount of effort to maintain, but will also provide a return.

Added ecosystem diversity: Generally, the more species included in a system, the lower the risk of devastating problems caused by pests and disease infestations.

Added economic flexibility: Multiple products give the landowner the option to sell products that will give the best overall return in a given year. Product lines can change from year to year based on demand and market conditions.

Short-term non-timber yields (Sequential cropping systems): While forest seedlings are still small, there is considerable open space between them that can be used to grow short-term agricultural crops. By growing such crops during the first 2 to 4 years, the cost of weed control and other maintenance for the forest seedlings can be reduced. Income from these crops can help to off-set part of the establishment costs. This is called "sequential planting", a practice wherein short-term crops are planted with, and eventually replaced by, long term trees. Sequential planting can be a viable system for farmers or forest planters who need a short term return as they invest in long term tree crops (Table 9).

Many combinations of crops and trees have been

Table 7. Agroforestry systems can improve the environment.

Reduced pressure on natural forests	On-farm production of wood and other forest products reduce pressure to exploit natural forests.
Species diversity	Trees and mixed systems provide habitat and support biodiversity (wildlife, micro life, etc.)
Resource conservation	Trees can improve conservation of soil, nutrients, and water in the landscape.
Carbon sequestration	Trees store carbon from the air, helping to reduce carbon dioxide pollution and global climate change.
Decreased pollution	Nutrient cycling of trees may reduce the use of chemical fertilizers, thus reducing chemical and soil run-off.

Table 8. Agroforestry systems are culturally compatible.

Locally based	Agroforestry systems incorporate species and techniques that have been used traditionally in local areas for many generations, in some cases in thousands of years.
Adaptable	Due to timber experience and acceptance over many decades, traditional systems and species provide a strong, locally based framework for future agroforestry development.
Acceotable	By combining production with conservation and land improvement, the agroforestry approach can increase the acceptability and adoption of sustainable practices.

Table 9. Generic sequential cropping system.

Early stage	Initial planting of tree seedlings and crops.
Years 1-5	Trees grow and begin to shade out crops.
Years 6+	Trees close canopy, crops are removed.

combined in this kind of system, with crops including rice, rice corn, beans, root crops, vegetables, fruits, ornamental flowers or foliage, and other cash or subsistence crops. There are several advantages to sequential plantings in agroforestry, including:

1. Greater efficiency in land use (less land area is fallow while tree crops mature).
2. Increased efficiency in labor (because crop maintenance can overlap).
3. Increased total yields.
4. Long term investment made more economically viable through short-term yields.

Sustained non-timber yields: While sequential crops are phased out as the forest matures, there are a number of agroforestry practices that are permanent components of the forestry system. These include shade-loving crops grown under the forest trees, animal integrated with the forest trees, or non-timber products such as fruit and seeds. These supplementary products can provide sustained sources of income for many years over the life of the project. Unlike sequential cropping, many of these systems have fewer forestry trees per acre, and therefore lower timber yields, than a pure stand of forestry trees. However, the earlier, sustained yields from non-timber products make this practice economically attractive to many landowners. In addition, landowners who integrate long-term non-timber yields with their forestry planting have the advantage of a diversified cropping system, which lowers the risk of total loss in the event of environmental or market conditions that may affect a single kind of crop.

Understory crops: Understory crops are grown in the shady environment under the timber trees. Unlike sequential crops, understory crops are installed and managed to produce for many years as the forest matures. Usually the number of timber trees per acre is 25 to 27% less than when timber trees are planted alone, to make space for the understory crops. There are many valuable cash and subsistence crops that thrive in the shady climate under trees. Planting that takes advantage of the understory range from simple systems consisting of one species in the overstory and one in the understory, to complex systems with many layers of trees, shrubs, and herbaceous plants stacked together as appropriate for their needs. Examples of understory crops that can be integrated with forestry include trees crops like guava, along with pepper, ginger etc.

Silvo-pastoral systems: When planned properly, grazing livestock can be integrated with certain tree species. In this scenario, the livestock play two roles. First, the animals are a source of income. The grazing livestock provides biological control of weeds and brush with minimal use of herbicides or mowing. In silvopastural systems, the interactions among timber forage and livestock are managed intensively to simultaniously produce timber commodities, high quality forage and efficient livestock production. For silvopastoral systems, timber species are usually spaced wider than standard to optimize healthy pasture forage underneath. Potential livestock choices include cattle, sheep, goats and chickens.

Silvopasture can provide a relatively constant income from livestock sale. At the same time, the economic risk of the project is reduced because the system yields multiple products. Also, production costs are reduced by distributing management costs between timber and livestock components.

Multipurpose timber species: Some timber trees can provide secondary non-timber products. These can include seeds, fruits, essential oils, medicinal products, resins, flowers, and more. In some cases, harvest of these products can begin after only a few years, and may provide a steady income until the trees are harvested for their timber.

Examples of timber trees that have economically viable secondary products include neem (*Azadirachta indica*), with recognized medicinal and insecticidal properties in the leaf, seeds, and bark; jackfruit (*Artocarpus heterophyllus*), which produces fruit, dyes, and medicinal products; teak (*Tectona grandis*), with many traditional medicinal uses; and mango (*Mangifera indica*), valued for its fruit.

Conclusion

A relatively large percentage of the population in Bangladesh lives under the poverty line and is affected by the countries degrading natural resources. The present research assesses the contribution of tropical agroforestry to livelihoods in a rural area. Agroforestry systems are a land management technique that implies a combination of forest trees with crops. It has now earned a distinct identity as an approach to land use. Results indicate that agroforestry may not only be an optimal solution for poor farmers, species diversity conservation and environmental sustainability but may also have good economic rates of return. The study has led to the conclusion that agroforestry may be considered as one of the major strategies for poverty reduction in Bangladesh, where there is little scope of conserving forest due to the obvious priority for food crop production. Agroforestry not only helps to increase food and fodder but also protects the existing forest where unemployed and poor people rush to earn their livelihood. It is therefore, important that the policy makers, NGOs and Government Organizations should promote sustainable tropical agroforestry in developing environment-friendly and cost-effective technologies for poverty reduction in Bangladesh.

REFERENCES

Alim A (1980). Integrated rural development, protection of forest, use of forest soil potential and connection of rural poor to waste and marginal land. Institute of Forestry, Chittagong University, Bangladesh (Unpublished).

Asian Development Bank (ADB) (2004). Regional study on forest policy and institutional reform. Final Report of Bangladesh Case Study. Metro Manila, Philippines.

Asian Development Bank (ADB) (2001). Chittagong hill tracts region development plan. ADB TA No. 3328. Bangladesh.

Economic and Social Commission for Asia and the Pacific [ESCAP] (2009). 2009 ESCAP population data sheet. Population and Rural and Urban Development Division, Bangkok.

FAO (2009). State of the World's forests 2009. Rome, Italy.

Islam SS, Kibria MG, Chowdhury MH (1998). Financial analysis of agroforestry trial at Ichamati, Chittagong. Bang. J. For. Sci., 2(27): 46-54.

Nair PKR (ed.) (1989). Agroforestry systems in the tropics. Kluwer Academic Publishers and ICRAF, Dordrecht, The Netherlands.

Rahman SA, Rahman MF, Codilan AL, Farhana KM (2007). Analysis of the economic benefits from systematic improvements to shifting cultivation and its evolution towards stable continuous agroforestry in the upland of eastern Bangladesh. Int. For. Rev., 9(1): 536-547.

Rahman SA, Groot W de, Snelder DJ (2008). Exploring the agroforestry adoption gap: Financial and socioeconomics of litchi-based agroforestry by smallholders in Rajshahi (Bangladesh). In: Snelder DJ, Lasco RD (eds.), Smallholder tree growing for rural development and environmental services: Lessons from Asia. Springer, The Netherlands, pp. 227-244.

United Nations Development Programme [UNDP] (2009). Human Development Report 2009. New York.

World Bank (2000). World Development Report 2000. Oxford University Press, Oxford.

Permissions

All chapters in this book were first published in JHF, by Academic Journals; hereby published with permission under the Creative Commons Attribution License or equivalent. Every chapter published in this book has been scrutinized by our experts. Their significance has been extensively debated. The topics covered herein carry significant findings which will fuel the growth of the discipline. They may even be implemented as practical applications or may be referred to as a beginning point for another development.

The contributors of this book come from diverse backgrounds, making this book a truly international effort. This book will bring forth new frontiers with its revolutionizing research information and detailed analysis of the nascent developments around the world.

We would like to thank all the contributing authors for lending their expertise to make the book truly unique. They have played a crucial role in the development of this book. Without their invaluable contributions this book wouldn't have been possible. They have made vital efforts to compile up to date information on the varied aspects of this subject to make this book a valuable addition to the collection of many professionals and students.

This book was conceptualized with the vision of imparting up-to-date information and advanced data in this field. To ensure the same, a matchless editorial board was set up. Every individual on the board went through rigorous rounds of assessment to prove their worth. After which they invested a large part of their time researching and compiling the most relevant data for our readers.

The editorial board has been involved in producing this book since its inception. They have spent rigorous hours researching and exploring the diverse topics which have resulted in the successful publishing of this book. They have passed on their knowledge of decades through this book. To expedite this challenging task, the publisher supported the team at every step. A small team of assistant editors was also appointed to further simplify the editing procedure and attain best results for the readers.

Apart from the editorial board, the designing team has also invested a significant amount of their time in understanding the subject and creating the most relevant covers. They scrutinized every image to scout for the most suitable representation of the subject and create an appropriate cover for the book.

The publishing team has been an ardent support to the editorial, designing and production team. Their endless efforts to recruit the best for this project, has resulted in the accomplishment of this book. They are a veteran in the field of academics and their pool of knowledge is as vast as their experience in printing. Their expertise and guidance has proved useful at every step. Their uncompromising quality standards have made this book an exceptional effort. Their encouragement from time to time has been an inspiration for everyone.

The publisher and the editorial board hope that this book will prove to be a valuable piece of knowledge for researchers, students, practitioners and scholars across the globe.

List of Contributors

T. O. Amusa
Federal College of Wildlife Management, Forestry Research Institute of Nigeria, P. M. B. 268, New Bussa, Niger State, Nigeria

G. M. Legwaila
Department of Crop Science and Production, Botswana College of Agriculture, Private Bag 0027, Gaborone, Botswana

W. Mojeremane
Department of Crop Science and Production, Botswana College of Agriculture, Private Bag 0027, Gaborone, Botswana

M. E. Madisa
Department of Crop Science and Production, Botswana College of Agriculture, Private Bag 0027, Gaborone, Botswana

R. M. Mmolotsi
Department of Crop Science and Production, Botswana College of Agriculture, Private Bag 0027, Gaborone, Botswana

M. Rampart
Department of Crop Science and Production, Botswana College of Agriculture, Private Bag 0027, Gaborone, Botswana

Nsalambi V. Nkongolo
Department of Agriculture and Environmental Science, College of Agricultural and Natural Sciences, Lincoln University, Jefferson City, MO 65102-0029, U.S.A

Mitate Yamada
Crop Production and Postharvest Technology Division, Japan International Research Center for Agricultural Sciences (JIRCAS), Tsukuba Science City, Japan 305-8686

Izumi Nakagawa
Vegetable Cultivation Technology System Laboratory, Department of Farmland Utilization, National Agriculture Research Center, Tsukuba, Ibaraki 305-8666, Japan

Debela Hunde Feyssa
Department of Land Resources and Agricultural Technology, College of Agriculture and Veterinary Sciences, University of Nairobi, Kenya, Home Institute, Jimma University, Ethiopia

Jesse.T. Njoka
Department of Land Resources and Agricultural Technology, College of Agriculture and Veterinary Sciences, University of Nairobi, Kenya

Zemede Asfaw
Department of Biology, College of Natural Sciences, Addis Ababa University, Ethiopia

M. M. Nyangito
Department of Land Resources and Agricultural Technology, College of Agriculture and Veterinary Sciences, University of Nairobi, Kenya

C. N. Egbuchua
Department of Agronomy, Delta State University, Asaba Campus, Asaba, Delta State, Nigeria

E. C. Enujeke
Department of Agronomy, Delta State University, Asaba Campus, Asaba, Delta State, Nigeria

O. O. Famuyide
Forestry Research Institute of Nigeria (FRIN), P. M. B.5054, Jericho hills, Ibadan, Oyo State, Nigeria

O. Adebayo
Forestry Research Institute of Nigeria (FRIN), P. M. B.5054, Jericho hills, Ibadan, Oyo State, Nigeria

K. A. Awe F. Bolaji-Olutunji
Forestry Research Institute of Nigeria (FRIN), P. M. B.5054, Jericho hills, Ibadan, Oyo State, Nigeria

A. Y. Owoeye
Forestry Research Institute of Nigeria (FRIN), P. M. B.5054, Jericho hills, Ibadan, Oyo State, Nigeria

D. O. Awodele
Forestry Research Institute of Nigeria (FRIN), P. M. B.5054, Jericho hills, Ibadan, Oyo State, Nigeria

A. Adeyemo
Forestry Research Institute of Nigeria (FRIN), P. M. B.5054, Jericho hills, Ibadan, Oyo State, Nigeria

D. O. Aworinde
Department of Biological Sciences, Ondo State University of Science and Technology, Okitipupa, Ondo State, Nigeria

S. M. Erinoso
Department of Botany, University of Ibadan, Ibadan, Oyo State, Nigeria

B. O. Ogundairo
Department of Biological Sciences, Federal University of Agriculture, Abeokuta, Ogun State, Nigeria

A. O. Olanloye
Department of Biological Sciences, Federal University of Agriculture, Abeokuta, Ogun State, Nigeria

A. K. Elsafori
College of Forestry and Range Science, Sudan University of Science and Technology, P. O. Box 6146, Post Code 11113, Khartoum, Sudan

A. N. Guma
Faculty of Education, University of Khartoum, Sudan

M. A. El Nour
Faculty of Forestry, University of Khartoum, Sudan

Samuel G. Obae
Department of Plant Science and Landscape Architecture, University of Connecticut, 1376 Storrs Road, Unit 4067, Storrs, CT 06269, USA

Todd P. West
Department of Plant Sciences, North Dakota State University, Dept # 7670, PO Box 6050, Fargo, ND 58108, USA

K. Aculey
Department of Biotechnology, Faculty of Agriculture, University for Development Studies, Tamale, Ghana

A. K. Quainoo
Department of Biotechnology, Faculty of Agriculture, University for Development Studies, Tamale, Ghana

G. Mahanu
Department of Horticulture, Faculty of Agriculture, University for Development Studies, Tamale, Ghana

Jane Muthoni
Kenya Agricultural Research Institute (KARI), National Potato Research Centre, Tigoni, P. O. Box 338 00217, Limuru, Kenya

Miriam Mbiyu
Kenya Agricultural Research Institute (KARI), National Potato Research Centre, Tigoni, P. O. Box 338 00217, Limuru, Kenya

Jackson N. Kabira
Kenya Agricultural Research Institute (KARI), National Potato Research Centre, Tigoni, P. O. Box 338 00217, Limuru, Kenya.

M. Saber
National Research Center, Cairo, Egypt

S. Badr-el-Din
National Research Center, Cairo, Egypt

N. Awad
National Research Center, Cairo, Egypt

Z. Mohammed
Faculty of Science, Cairo University Egypt

M. E. Madisa
Department of Crop Science and Production, Botswana College of Agriculture, Private Bag 0027, Gaborone, Botswana

Y. Assefa
Department of Crop Science and Production, Botswana College of Agriculture, Private Bag 0027, Gaborone, Botswana

Goutam Kumar Dash
Plant Biotechnology Division, Regional Plant Resource Centre, Bhubaneswar- 751 015, Orissa, India

Sunil Kumar Senapati
MITS School of Biotechnology, Plant Biotechnology Division, Infocity, Patia, Bhubaneswar-751 024, India

Gyana Ranjan Rout
Department of Agricultural Biotechnology, College of Agriculture, Orissa University of Agriculture and Technology, Bhubaneswar-751 003, Orissa, India

Rishi Ram Dhakal
Institute of Forestry, Tribhuvan University, Hetauda, Nepal

Gandhiv Kafle
Institute of Forestry, Tribhuvan University, Hetauda, Nepal

Jay N. Yadava
Institute of Forestry, Tribhuvan University, Hetauda, Nepal

J. K. Githiomi
Forest Products Research Centre, Kenya Forestry Research Institute P.O. Box 64636-00620 Nairobi, Kenya

D. N. Mugendi
Department of Environmental Sciences, Kenyatta University, P.O. Box 43844-00100 Nairobi, Kenya

J. B. Kungú
Department of Environmental Sciences, Kenyatta University, P.O. Box 43844-00100 Nairobi, Kenya

A. A. A. Farag
The Central Laboratory for Agricultural Climate, 6 Michail Bakhoum St., Dokki, Giza, Egypt

M. A. A. Abdrabbo
The Central Laboratory for Agricultural Climate, 6 Michail Bakhoum St., Dokki, Giza, Egypt

E. M. Abd-Elmoniem
Soils Department, Faculty of Agriculture, Ain Shams University, Shoubra El-khema, Cairo, Egypt

Shiferaw Abebe
Department of Ecology and Natural Resource Management, Norwegian University of Life Sciences, P. O. Box 5003, N-1432 Ås, Norway
Forestry Agricultural Research Center, Ethiopian Institute of Agricultural Research, P. O. Box 2003, Addis Ababa, Ethiopia Norwegian Institute for Nature Research (NINA), Gaustadallèen 21, No-0349 Oslo, Norway

Kåre Arnstein Lye
Department of Ecology and Natural Resource Management, Norwegian University of Life Sciences, P. O. Box 5003, N-1432 Ås, Norway

Diress Tsegaye
Department of Ecology and Natural Resource Management, Norwegian University of Life Sciences, P. O. Box 5003, N-1432 Ås, Norway
Department of Biology, University of Oslo, P. O. Box 1066 Blindern, N-0316 Oslo, Norway

Y. H. Musa
Department of Agricultural Extension and Management, Taraba State College of Agriculture, Jalingo, Nigeria

J. I Onu
Department of Agricultural Economics and Extension, Federal University of Technology Yola, Adamawa State. Nigeria

I. P. Vosanka
Department of Agricultural Extension and Management, Taraba State College of Agriculture, Jalingo, Nigeria

I. Anonguku
Department of Agricultural Extension and Communication, University of Agriculture, Makurdi, Nigeria

S. A. Enebak
School of Forestry and Wildlife Sciences, Southern Forest Nursery Management Cooperative, Auburn University, AL 36849, USA

T. E. Starkey
School of Forestry and Wildlife Sciences, Southern Forest Nursery Management Cooperative, Auburn University, AL 36849, USA

M. Quicke
School of Forestry and Wildlife Sciences, Southern Forest Nursery Management Cooperative, Auburn University, AL 36849, USA

J. K. Githiomi
Forest Products Research Centre, Kenya Research Institute, P.O. Box 64636-00620 Nairobi, Kenya

D. N. Mugendi
Department of Environmental Sciences, Kenyatta University, P.O. Box 43844-00100 Nairobi, Kenya

J. B. Kungú
Department of Environmental Sciences, Kenyatta University, P.O. Box 43844-00100 Nairobi, Kenya

Cliff S. Dlamini
Department of Forest and Wood Science, Faculty of Agrisciences, University of Stellenbosch, P/B X1, Matieland 7602, South Africa

Coert J. Geldenhuys
Department of Forest and Wood Science, Faculty of Agrisciences, University of Stellenbosch, P/B X1, Matieland 7602, South Africa

Adetayo, Adewale Olusola
Institute of Agricultural Research and Training, Obafemi Awolowo University, P. M. B. 5029, Ibadan, Oyo State, Nigeria

Bruce L. Dunn
Department of Horticulture and Landscape Architecture, Oklahoma State University, 58 Ag Hall, Stillwater, OK 74078, USA

Carla Goad
Department of Statistics, Oklahoma State University 301 MSCS Building, Stillwater, OK 74078, USA

Stephen Stanphill
Department of Horticulture and Landscape Architecture, Oklahoma State University, 58 Ag Hall, Stillwater, OK 74078, USA

R. Anlauf
Osnabrueck University of Applied Sciences, P. O. Box 1940, 49009 Osnabrueck, Germany

P. Rehrmann
Osnabrueck University of Applied Sciences, P. O. Box 1940, 49009 Osnabrueck, Germany

H. Schacht
Osnabrueck University of Applied Sciences, P. O. Box 1940, 49009 Osnabrueck, Germany

Ouni Youssef
Laboratoire Caractérisation et Qualité de l'huile d'Olive, Centre de Biotechnologie de Borj-Cedria, B. P. 901, 2050 Hammam-Lif, Tunisia

Flamini Guido
Dipartimento di Chimica Bioorganicae Biofarmacia, via Bonanno 33, 56126 Pisa, Italy

Douja Daoud
Laboratoire Caractérisation et Qualité de l'huile d'Olive, Centre de Biotechnologie de Borj-Cedria, B. P. 901, 2050 Hammam-Lif, Tunisia

Zarrouk Mokhtar
Laboratoire Caractérisation et Qualité de l'huile d'Olive, Centre de Biotechnologie de Borj-Cedria, B. P. 901, 2050 Hammam-Lif, Tunisia

A. I. Okunlola
Department of Crop Soil and Pest Management, Federal University of Technology, P. M. B. 704, Akure Ondo State, Nigeria

R. A. Adebayo
Department of Crop Soil and Pest Management, Federal University of Technology, P. M. B. 704, Akure Ondo State, Nigeria

A. D. Orimogunje
Department of Crop Soil and Pest Management, Federal University of Technology, P. M. B. 704, Akure Ondo State, Nigeria

S. R. Khan
Natural Resources Institute, Clayton H. Riddell Faculty of Environment, Earth, and Resources, University of Manitoba, 70 Dysart Rd., Winnipeg, MB, Canada R3T 2N2

S. A. Rahman
Center for International Forestry Research (CIFOR), Bogor Barat 16680, Indonesia

T. Sunderland
Center for International Forestry Research (CIFOR), Bogor Barat 16680, Indonesia

M. W. Mbiyu
Kenya Agricultural Research Institute, National Potato Research Centre -Tigoni P. O. Box 338-00217, Limuru Kenya

J. Muthoni
Kenya Agricultural Research Institute, National Potato Research Centre -Tigoni P. O. Box 338-00217, Limuru Kenya

J. Kabira
Kenya Agricultural Research Institute, National Potato Research Centre -Tigoni P. O. Box 338-00217, Limuru Kenya

G. Elmar
International Potato Centre (CIP), sub-Saharan Africa Regional Office P. O. Box 25171-00603 Nairobi, Kenya

C. Muchira
Kenya Agricultural Research Institute, National Potato Research Centre -Tigoni P. O. Box 338-00217, Limuru Kenya

P. Pwaipwa
Kenya Agricultural Research Institute, National Potato Research Centre -Tigoni P. O. Box 338-00217, Limuru Kenya

J. Ngaruiya
Kenya Agricultural Research Institute, National Potato Research Centre -Tigoni P. O. Box 338-00217, Limuru Kenya

S. Otieno
Kenya Agricultural Research Institute, National Potato Research Centre -Tigoni P. O. Box 338-00217, Limuru Kenya.

J. Onditi
Kenya Agricultural Research Institute, National Potato Research Centre -Tigoni P. O. Box 338-00217, Limuru Kenya

Sayed Alireza Valadabadi
Associate Professor, Islamic Azad University, Shahr-e-Qods Branch, Tehran, Iran

Hossein Aliabadi Farahani
Young Researchers Club, Islamic Azad University, Shahr-e-Qods Branch, Tehran, Iran

Getrude Sithole
AMADI-Midlands State University 1st floor Embassy House, Gwamile Street/Msakato Street P. O. Box 6146 Mbabane Kingdom of Swaziland

S. A. Rahman
Center for International Forestry Research (CIFOR), Bogor Barat 16680, Indonesia

F. D. Paras
Department of Social Forestry and Forest Governance, College of Forestry and Natural Resources, University of the Philippines, Los Banos, Philippines

S. R. Khan
Natural Resources Institute, University of Manitoba, Canada

A. Imtiaj
Department of Botany, University of Rajshahi, Rajshahi, Bangladesh

K. M. Farhana
Department of Sociology, University of Padova, Italy

M. M. Toy
Shanto-Mariam University of Creative Technology, Dhaka, Bangladesh

M. B. Akhand
Intervida (Bangladesh Country office), Dhaka, Bangladesh

T. Sunderland
Center for International Forestry Research (CIFOR), Bogor Barat 16680, Indonesia